DATA MODELING AND DATABASE DESIGN

DATA MODELING AND DATABASE DESIGN

Narayan S. Umanath

University of Cincinnati

Richard W. Scamell

University of Houston

THOMSON

COURSE TECHNOLOGY

Australia • Canada • Mexico • Singapore • Spain • United Kingdom • United States

Data Modeling and Database Design
by Narayan S. Umanath and Richard W. Scamell

Acquisitions Editor
Maureen Martin

Senior Product Manager
Alyssa Pratt

Development Editor
Deb Kaufmann

Marketing Manager
Penelope Crosby

Content Project Managers
Marissa Falco, Jill Klaffky

Editorial Assistant
Erin Kennedy

Print Buyer
Julio Esperas

Compositor
GEX Publishing Services

Copy Editor
Harry Johnson

Proofreader
Vicki Zimmer

Indexer
Rich Carlson

To Beloved Bhagwan, the very source of my thoughts, words, and deeds
To my Graduate Teaching Assistants, the very source of my inspiration
To my dear children, Sharda and Kausik, always concerned about their dad overworking
To my dear wife, Lalitha, a pillar of courage I always lean on

Uma

There is a verse that says
Focus on what I'm doing right
and tell me that you appreciate me
so that I learn to feel worthy
and motivated to do more
Led by my family, I have always been surrounded by people (friends, teachers, and students) who
with their kind thoughts, words, and deeds treat me in this way.
This book is dedicated to these people.

Richard

BRIEF CONTENTS

Part IV: Database Implementation Using the Relational Data Model

TABLE OF CONTENTS

Part II: Logical Data Modeling

Part III: Normalization

"Everything should be made as simple as possible—**but no simpler**"
—Albert Einstein

Popular business database books typically provide broad coverage of a wide variety of topics including, for instance, data modeling, database design and implementation, database administration, the client/server database environment, the Internet database environment, distributed databases, object-oriented database development, etc., invariably at the expense of deeper treatment of critical topics such as principles of data modeling and database design. Using current business database books in our courses, we found that in order to properly cover data modeling and database design, we had to augment the texts with significant supplemental material (1) to achieve precision and detail and (2) to impart the depth necessary for the students to gain a robust understanding of data modeling and database design. In addition, we also ended up skipping several chapters as topics to be covered in a different course. We also know other instructors who share this experience. Broad coverage of many database topics in a single book is appropriate for some audiences, but that is not the aim of this book.

The goal of *Data Modeling and Database Design* is to provide core competency in the areas that every Information Systems (IS), Computer Science (CS), and Computer Information Systems (CIS) student and professional should acquire: **data modeling and database design**. It is our experience that this set of topics is the most essential for database professionals, and that, covered in sufficient depth, these topics alone require a full semester of study. It is our intention to address these topics at a level of technical depth achieved in computer science (CS) textbooks, yet make palatable to the business student/IS professional with little sacrifice in precision. We deliberately refrain from the mathematics and algorithmic solutions usually found in CS textbooks, yet we attempt to capture the precision therein via heuristic expressions.

Data Modeling and Database Design provides not just hands-on instruction in current data modeling and database design practices, but gives readers a thorough conceptual background for these practices. We do not subscribe to the idea that a textbook should limit itself to describing what is actually being practiced. Teaching only what is being practiced is bound to lead to knowledge stagnation. Where do practitioners learn what they know? Did they invent the relational data model? Did they invent the ER model? We believe that it is our responsibility to present not only industry "best practices" but also to provide students (future practitioners) with concepts and techniques that are not necessarily used in industry today but can enliven their practice and help it evolve without knowledge stagnation. One of the coauthors of this book has worked in the software development industry for over 15 years with a significant focus on database development. His experience indicates that having a richness of advanced data modeling constructs available enhances the robustness of database design and practitioners readily adopt these techniques in their design practices.

In a nutshell, our goal is to take an IS/CS/CIS student/professional through an intense educational experience starting at conceptual modeling and culminating in a fully implemented database design—**nothing more and nothing less**. This educational journey is briefly articulated in the following paragraphs.

STRUCTURE

We have tried very hard to make the book "fluff-free." It is our hope that every sentence in the book including this preface adds value to a reader's learning (and *footnotes* are no exception to this statement).

The book begins with an introduction to rudimentary concepts of data, metadata, and information, followed by an overview of data management. Pointing out the limitations of file-processing systems, **Chapter 1** introduces database systems as a solution to overcome these limitations. The architecture and components of a database system that makes this possible are discussed. The chapter concludes with the presentation of a framework for the database system design life cycle. Following the introductory chapter on database systems architecture and components, the book contains four parts.

Part I: Conceptual Data Modeling

Part I addresses the topic of conceptual data modeling—i.e., modeling at the highest level of abstraction independent of the limitations of the technology employed to deploy the database system. Four chapters (Chapters 2 – 5) are used in order to provide an extensive discussion of conceptual data modeling. **Chapter 2** lays the groundwork using the *Entity-Relationship (ER) modeling grammar* as the principal means to model a database application domain. **Chapter 3** elaborates on the use of the ER modeling grammar in progressive layers and exemplifies the modeling technique with a comprehensive case called Bearcat, Incorporated. This is followed by a presentation in **Chapter 4** of richer data modeling constructs that overlap with object-oriented modeling constructs. The Bearcat, Incorporated story is further enriched to demonstrate the value of Enhanced ER (EER) modeling constructs. **Chapter 5** provides exclusive coverage of modeling complex relationships that have meaningful real-world significance. At the end of Part I, the reader ought to be able to fully appreciate the value of conceptual data modeling in the database system design life cycle.

Part II: Logical Data Modeling

Part II of the book is dedicated to the discussion of migration of a conceptual data model to its logical counterpart. Since the relational data model architecture forms the basis for the logical data modeling discussed in this textbook, **Chapter 6** focuses on its characteristics. Other logical data modeling architectures prevalent in some legacy systems, the hierarchical data model and the CODASYL data model, appear in Appendix A. An introduction to object-oriented data modeling concepts is presented in Appendix B. The rest of Chapter 6 describes techniques to map a conceptual data model to its logical counterpart. An *Information-Preserving logical data modeling grammar* is introduced and contrasted with existing popular mapping techniques that are information-reducing. A comprehensive set of examples are used to clarify the use and value of the information-preserving grammar.

Part III: Normalization

Part III of the book addresses the critical question of "goodness" of a database design that results from a conceptual and logical data modeling processes. *Normalization* is introduced as the "scientific" way to verify and improve the quality of a logical schema that is available at this stage in the database design. Three chapters are employed to cover the topic of normalization. In **Chapter 7**, we take a look at data redundancy in a relation schema and see how it manifests as a problem. We then trace the problem to its source, viz., undesirable functional dependencies. To that end, we first learn about functional dependencies axiomatically and how inference rules (Armstrong's axioms) can be used to derive candidate keys of a relation schema. In **Chapter 8**, the solution offered by the normalization process to data redundancy problems triggered by undesirable functional dependencies is presented. After discussing First, Second, Third and Boyce-Codd normal forms individually, we examine the side effects of normalization, viz., dependency preservation and lossless-join decomposition and their consequences. Next, we present real-world scenarios of deriving full-fledged relational schemas (sets of relation schemas) given sets of functional dependencies using several examples. The useful topic of denormalization is covered next. Reverse engineering a normalized relational schema to the conceptual tier often forges insightful understanding of the database design and enables a database designer to become a better data modeler. Despite its practical utility, this topic is rarely covered in database textbooks. **Chapter 9** completes the discussion of normalization by examining multi-valued dependency and join-dependency and their impact on a relation schema in terms of Fourth normal form and Project/Join normal form, viz., PJNF (also known as Fifth normal form) respectively.

Part IV: Database Implementation Using the Relational Database Model

Part IV pertains to database implementation using the relational data model. Spread over three chapters, this part of the book covers relational algebra and the ANSI/ISO standard Structured Query Language (SQL). **Chapter 10** focuses on the data definition language (DDL) aspect of SQL. Included in the discussion are the SQL schema evolution statements for adding, altering or dropping table structures, attributes, constraints, and supporting structures. This is followed by the development of SQL/DDL script for a comprehensive case about a college registration system. The chapter also includes the use of INSERT, UPDATE, and DELETE statements in populating a database and performing database maintenance. A discussion of database access control serves to conclude the chapter. Chapters 11 and 12 focus on relational algebra and the use of SQL for data manipulation. The first section of **Chapter 11** concentrates on E.F. Codd's eight original relational algebra operations. The rest of the chapter offers a comprehensive coverage of the SQL SELECT statement to implement these relational algebra operations. **Chapter 12** covers the creation of database assertions, triggers and views, and SQL functions for handling characters, dates and times. Chapter 12 concludes with a comprehensive set of SQL database projects that provide students with real-life scenarios to test and apply the skills and concepts presented in Part IV.

FEATURES OF EACH CHAPTER

Since our objective is a crisp and clear presentation of rather intricate subject matter, each chapter begins with a simple introduction, followed by the treatment of the subject matter, and concludes with a chapter summary, a set of exercises based on the subject matter, and a selected bibliography.

WHAT MAKES THIS BOOK DIFFERENT?

Every book has strengths and weaknesses. If lack of breadth in the coverage of database topics is considered a weakness, we have deliberately chosen to be weak in that dimension. We have not planned this book to be another general book on database systems. We have chosen to limit the scope of this book to data modeling and database design since we firmly believe that this set of topics is the core of database systems and must be learned in depth. Any system designed robustly has the potential to best serve the needs of the users. More importantly, a poor design is a virus that can ruin an enterprise.

In this light, we believe these are the unique strengths of this book:

- It presents conceptual modeling using the entity-relationship modeling grammar including extensive discussion of the enhanced entity-relationship (ER) model.
  We believe that a conceptual model should capture all possible constraints conveyed by the business rules implicit in users' requirement specifications. To that end, we posit that an ER diagram is not an ER model unless accompanied by a comprehensive specification of characteristics of and constraints pertaining to attributes. We accomplish this via a list of semantic integrity constraints (sort of a conceptual data dictionary) that will accompany an ER diagram, a unique feature that we have not seen in other database textbooks. We also seek to demonstrate the systematic development of a multi-layer conceptual data model via a comprehensive illustration at the beginning of each Part. We consider the multi-layer modeling strategy and the heuristics for systematic development as unique features of this book.

- It includes substantial coverage of higher-degree relationships and other complex relationships in the entity-relationship diagram.
  Most business database books seem to provide only a cursory treatment of complex relationships in an ER model. We not only cover relationships beyond binary relationships (e.g., ternary and higher-degree relationships), but also clarify the nuances pertaining to the necessity and efficacy of higher-degree relationships and the various conditions under which even recursive and binary relationships are aggregated in interesting ways to form cluster entity types.

- It discusses the information-preserving issue in data model mapping and introduces a new information-preserving grammar for logical data modeling.
  Many computer scientists have noted that the major difficulty of logical database design, i.e., of transforming an ER schema into a schema in the language of some logical model, is the information preservation issue. Indeed, assuring a complete mapping of all modeling constructs and constraints which are inherent, implicit or explicit in the source schema (e.g., ER/EER model) is problematic since constraints of the source model often cannot be represented directly in terms of structures and constraints of the target model (e.g., Relational schema). In such a case they must be realized through application programs; alternatively, an information-reducing transformation must be accepted (Fahrner and Vossen, 1995). In their research, initially

presented at the Workshop on Information Technologies (WITS) in the ICIS (International Conference on Information Systems) in Brisbane, Australia, Umanath and Chiang (2000) describe a logical modeling grammar that generates an information preserving transformation. We have included this logical modeling grammar as a unique component of this textbook.

- It includes unique features under the topic of normalization rarely covered in business database books:
  - Inference rules for functional dependencies *(Armstrong's axioms) and derivations of candidate keys from a set of functional dependencies*
  - Rich examples to clarify *the basic normal forms (first, second, third, and Boyce-Codd)*
  - Derivation of a complete logical schema *from a large set of functional dependencies considering lossless (non-additive) join properties and dependency preservation*
  - Reverse engineering a logical schema *to an entity-relationship diagram*
  - Advanced coverage of fourth and fifth normal form *(project-join normal form - PJNF) using a variety of examples.*

- It supports in-depth coverage of relational algebra with a significant number of examples of their operationalization in ANSI/ISO SQL.

A NOTE TO THE INSTRUCTOR

The content of this book is designed for a rigorous one-semester course in database design and development and may be used at both undergraduate and graduate levels. Technical emphasis can be tempered by minimizing or eliminating the coverage of some of the following topics from the course syllabus: Enhanced Entity-Relationship (EER) Modeling (Chapter 4) and the related data model mapping topics in Chapter 6 (Section 6.8) on Mapping Enhanced ER Modeling Constructs to a Logical Schema; Modeling Complex Relationships (Chapter 5); and higher normal forms (Chapter 9). The suggested exclusions will not impair the continuity of the subject matter in the rest of the book.

SUPPORTING TECHNOLOGIES

Any business database book can be effective only when supporting technologies are made available for student use. Yet, we don't think that the type of book we are writing should be married to any commercial product. The specific technologies that will render this book highly effective include a drawing tool (such as Microsoft Visio), a software engineering tool (such as ERWIN, ORACLE/Designer, or Visible Analyst), and a relational database management system (RDBMS) product (such as ORACLE, SQL Server, or DB2).

SUPPLEMENTAL MATERIALS

The following supplemental materials are available when this book is used in a classroom setting. All of the teaching tools available with this book are provided to the instructor on a single CD-ROM. Some of these materials may also be found on the Thomson Course Technology Web site at *www.course.com*.

- **Electronic Instructor's Manual**: The Instructor's Manual assists in class preparation by providing suggestions and strategies for teaching the text, chapter outlines, quick quizzes, discussion topics, and key terms.

- **Sample Syllabi and Course Outline:** The sample syllabi and course outlines are provided as a foundation to begin planning and organizing your course.

- **ExamView Test Bank:** ExamView allows instructors to create and administer printed, computer (LAN-based), and Internet exams. The Test Bank includes an array of questions that correspond to the topics covered in this text, enabling students to generate detailed study guides that include page references for further review. The computer-based and Internet testing components allow students to generate detailed study guides that include page references for further review. The computer-based and Internet testing components allow students to take exams at their computers, and also save the instructor time by automatically grading each exam. The Test Bank is also available in Blackboard and WebCT versions posted online at *www.course.com*.

- **PowerPoint Presentations**: Microsoft PowerPoint slides for each chapter are included as a teaching aid for classroom presentation, to make available to students on the network for chapter review, or to be printed for classroom distribution. Instructors can add their own slides for additional topics they introduce to the class.

- **Figure Files**: Figure files from each chapter are provided for the instructor's use in the classroom.

- **Solutions**: A full set of solutions to the end-of-chapter material in the text is available on the Instructor Resources CD-ROM as well as at *www.course.com*.

- **Data Files:** Data files containing scripts to populate the database tables used as examples in Chapters 11 and 12 are provided on the Thomson Course Technology Web site at *www.course.com*, and on the Instructor Resources CD-ROM.

ACKNOWLEDGMENTS

We have never written a textbook before. We have been using books written by our academic colleagues, always supplemented with handouts that we developed ourselves. Over the years we accumulated a lot of supplemental material. In the beginning, we took the positive feedback from the students about the supplemental material rather lightly until we started to see comments like "I don't know why I bought the book; the instructor's handouts were so good and much clearer than the book" in the student evaluation forms. Our impetus to write a textbook thus originated from the consistent positive feedback from our students.

We also realized that, contrary to popular belief, business students are certainly capable of assimilating intricate technical concepts; the trick is to frame the concepts in meaningful business scenarios. The unsolicited testimonials from our alumni about the usefulness of the technical depth offered in our database course in solving real-world design problems reinforced our faith in developing a book focused exclusively on data modeling and database design that was technically rigorous but permeated with business relevance.

Since we both teach database courses regularly, we have had the opportunity to field-test the manuscript of this book over three years at both undergraduate and graduate level information systems courses in the College of Business at the University of Cincinnati and in the C. T. Bauer College of Business at the University of Houston. Hundreds of students—mostly business students—have used earlier drafts of this textbook so far. Interestingly, even the computer science and engineering students taking our courses have expressed their appreciation of the content. This is a long preamble to acknowledge one of the most important and formative elements in the creation of this book: our students.

The students' continued feedback (comments, complaints, suggestions, and criticisms) have significantly contributed to the improvement of the content. In this regard, we must single out Lisa Kerns, a graduate student at the University of Cincinnati. Given her significant industry experience, Lisa offered excellent suggestions to improve the book.

We are grateful to Dr. Iris Junglas at the University of Houston for enthusiastically experimenting with our book in her graduate courses and providing insightful feedback. As we were cycling through revisions of the manuscript, the graduate teaching assistants of Dr. Umanath were a constant source of inspiration. Their meaningful questions and suggestions added significant value to the content of this book. Usha Viriyala, Rishi Khar, and Sugan Narayanan deserve special mention. Dr. Scamell was ably assisted by his graduate assistants Monisha Nag and Sachin Nikam and we are grateful for their service as well.

We would also like to thank the following reviewers whose critiques, comments, and suggestions helped shape every chapter:

Akhilesh Bajaj, *University of Tulsa*

Margaret Porciello, *State University of New York/Farmingdale*

Sandeep Purao, *Pennsylvania State University*

Jaymeen Shah, *Texas State University*

Last, but by no means the least, we gratefully acknowledge the significant contribution of Deb Kaufmann, our development editor—a kind human being, highly professional, always prompt and supportive. Her hands shaped this book for human consumption.

Finally, in the spirit of Einstein's observation, we have endeavored to make this book as simple as possible—***but, no simpler***.

Enjoy!

N. S. Umanath
R. W. Scamell

## CHAPTER 1

Database Systems: Architecture and Components

Data modeling and database design involve elements of both art and engineering. Understanding user requirements and modeling them in the form of an effective logical database design is an artistic process. Transforming the design into a physical database with functionally complete and efficient applications is an engineering process.

To better comprehend what drives the design of databases, it is important first to understand the distinction between data and information. Data consists of raw facts, that is, facts that have not yet been processed to reveal their meaning. Processing these facts provides information on which decisions can be based.

Timely and useful information requires that data be accurate and stored in a manner that is easy to access and process. And, like any basic resource, data must be managed carefully. Data management is a discipline that focuses on the proper acquisition, storage, maintenance, and retrieval of data. Typically, the use of a database enables efficient and effective management of data.

This chapter introduces the rudimentary concepts of data and how information emerges from data when viewed through the lens of metadata. Next, the discussion addresses data management, contrasting file-processing systems with database systems. This is followed by brief examples of desktop, workgroup, and enterprise databases. The chapter then presents a framework for database design that describes the multiple tiers of data modeling and how these tiers function in database design. This framework serves as a roadmap to guide the reader through the remainder of the book.

1.1 Data, Information, and Metadata

Although the terms are often used interchangeably, information is different from data. **Data** can be viewed as raw material consisting of unorganized facts about things, events, activities, and transactions. While data may have implicit meaning, the lack of organization renders it valueless. In other words, **information** is data in context—that is, data that has been organized into a specific context such that it has value to its recipient.

As an example, consider the digits 1713445232. What does this string of digits represent? One response is that they are simply ten meaningless digits. Another might be the number 32 (obtained by summing the 10 digits). A third might be that they represent a person's phone number with the first three digits constituting the area code and the remaining seven digits the local phone number. On the other hand, if the first digit is used to represent a person's gender (1 for male and 2 for female) and the remaining nine digits the person's Social Security number, the ten digits would mean something else. Numerous other interpretations are possible, but without a context it is impossible to say what the digits represent. However, when framed in a specific context (such as being told that the first digit represents a person's gender and the remaining digits the Social Security number), the data is transformed into information.

Metadata, in a database environment, is data that describes the properties of data. It contains a complete definition or description of database structure (i.e., the file structure, data type, and storage format of each data item), and other constraints on the stored data. For example, when the structure of the ten digits 1713445232 is revealed, the ten digits become information, such as a phone number. Metadata defines this structure. In other words, through the lens of metadata, data takes on specific meaning and yields information.[1] Table 1.1 contains metadata for the data associated with a manufacturing plant. Later in this chapter we will see that in a database environment, metadata is recorded in what is called a data dictionary.

Table 1.1   Some metadata for a manufacturing plant

Record Type	Data Element	Data Type	Size	Source	Role	Domain
PLANT	Pl_name	Alphabetic	30	Stored	Non-key	
PLANT	Pl_number	Numeric	2	Stored	Key	Integer values from 10 to 20
PLANT	Budget	Numeric	7	Stored	Non-key	
PLANT	Building	Alphabetic	20	Stored	Non-key	
PLANT	No_of_employees	Numeric	4	Derived	Non-key	

As reflected in Table 1.1, the smallest unit of data is called a **data element**. A group of related data elements treated as a unit (such as Pl_name, Pl_number, Budget, Building, and No_of_employees) is called a **record type**. A set of values for the data elements constituting a record type is called a record instance or simply a **record**. A **file** is collection of records. A file is sometimes referred to as a **data set**. A company with ten plants would have a PLANT file or a PLANT data set that contains ten records.

1.2 Data Management

This book focuses strictly on management of data, as opposed to the management of human resources. Data management involves four actions: (a) data creation, (b) data retrieval, (c) data modification or updating, and (d) data deletion. Two data management functions support these four actions: data must be accessed and, for ease of access, data must be organized.

Despite today's sophisticated information technologies, there are still only two primary approaches for accessing data. One is **sequential access**, where in order to get to the *nth* record in a data set it is necessary to pass through the previous *n-1* records in the data

---

[1] With the advent of the data warehouse, the term metadata assumes a more comprehensive meaning to include business and technical metadata, which is outside the scope of the current discussion.

set. The second approach is **direct access**, where it is possible to get to the *nth* record without having to pass through the previous *n-1* records. While direct access is useful for *ad hoc* querying of information, sequential access remains essential for transaction processing applications such as payroll, generating grade reports, and generating utility bills.

In order to access data, the data must be organized. For sequential access, this means that all records in a file must be stored (organized) ordered on some unique identifier, such as employee number, inventory number, flight number, account number, or stock symbol. This is called sequential organization. A serial (unordered) collection of records, also known as a heap file, cannot provide sequential access. For direct access, the records in a file can be stored serially and be organized either randomly or using an external index. A randomly organized file is one in which the value of a unique identifier is processed by some sort of transformation routine (often called a hashing algorithm) that computes the location of records within the file (relative record numbers). An indexed file makes use of an index external to the data set similar in nature to the one found at the back of this book to identify the location where a record is physically stored.

As will be seen in Section 1.5, a database takes advantage of software called a database management system (DBMS) that sits on top of a set of files physically organized as sequential files and/or as some form of direct access files. A DBMS facilitates data access in a database without burdening a user with the details of how the data is physically organized.

1.3 Limitations of File-Processing Systems

Computer applications in the 1960s and 1970s focused primarily on automating clerical tasks. These applications made use of records stored in separate files and thus were called file-processing systems. Although file-processing systems for information systems applications have been useful for many years, database technology has rendered them obsolete except for their use in a few legacy systems such as some payroll and customer billing systems. Nonetheless, understanding their limitations provides insight into the development of and justification for database systems.

Figure 1.1 shows three file-processing systems for a hypothetical university. One processes data for students, another processes data for faculty and staff, and a third processes data for alumni. In such an environment, each file-processing system has its own collection of private files and programs that access these files.

While an improvement over the manual systems that preceded them, file-processing systems suffer from a number of limitations:

- *Lack of data integrity*: **Data integrity** ensures that data values are correct, consistent, complete, and current. Duplication of data in isolated file-processing systems leads to the possibility of inconsistent data. Then it is difficult to identify which of these duplicate data is correct, complete, and/or current. This creates data integrity problems. For example, if an employee who is also a student and an alumnus changes his or her mailing address, files that contain the mailing address in three different file-processing systems require updating

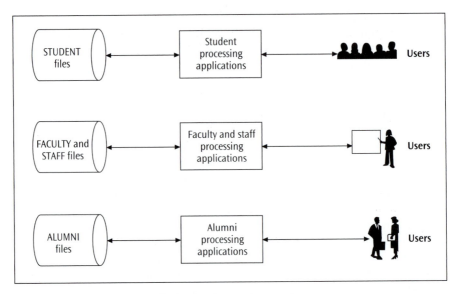

Figure 1.1   An example of a file-processing environment

to ensure consistency of information across the board. Data redundancy across the three file-processing systems not only creates maintenance inefficiencies, but also leads to the problem of not knowing which is the current, correct, and /or complete address of the person.

- *Lack of standards*: Organizations with file-processing systems often lack or find it difficult to enforce standards for naming data items as well as for accessing, updating, and protecting data. The absence of such standards can lead to unauthorized access and accidental or intentional damage to or destruction of data. In essence, security and confidentiality of information may be compromised.

- *Lack of flexibility/maintainability*: Information systems make it possible for end users to develop information requirements that they had never envisioned previously. This inevitably leads to a substantial increase in requests for new queries and reports. However, file-processing systems are dependent upon a programmer who either has to write or modify program code to meet these information requirements from isolated data. This can bring about information requests that are not satisfied or programs that are inefficiently written, poorly documented, and difficult to maintain.

These limitations are actually symptoms resulting from two fundamental problems: lack of integration of related data and lack of program-data independence.

- *Lack of data integration*: Data is separated and isolated, and ownership of data is compartmentalized, resulting in limited data sharing. For example, to produce a list of employees who are students and alumni at the same time, data from multiple files must be accessed. This process can be quite complex and time consuming since a program has to access and perform logical comparisons across independent files containing employee, student, and alumni data. In short, lack of integration of data contributes to all of the problems listed above as symptoms.

- *Lack of program-data independence*: In a file-processing environment, the structural layout of each file is embedded in the application programs. That is, the metadata of a file is fully coded in each application program that uses the particular file. Perhaps the most often-cited example of the program-data dependence problem occurred during the file-processing era, when it was common for an organization to expand the zip code field from five digits to nine digits. In order to implement this change, every program in the employee, student, and alumni file-processing systems containing the zip code field had to be identified (often a time-consuming process itself) and then modified to conform to the new file structure. This not only required modification of each program and its documentation but also recompiling and retesting of the program. Likewise, if a decision was made to change the organization of a file from indexed to random, since the structure of the file was mapped into every program using the file, every program using the file had to be modified. Identifying all the affected programs for corrective action was not a simple task, either. Thus, because of lack of program-data independence, file-processing systems lack flexibility since they are not amenable to structural changes in data. Program-data dependence also exacerbates data security and confidentiality problems.

It is only through attacking the problems of lack of program-data independence and lack of integration of related data that the limitations of file-processing systems can be eliminated. If a way is found to deal with these problems so as to establish centralized control of data, then unnecessary redundancy can be reduced, data can be shared, standards can be enforced, security restrictions can be applied, and integrity can be maintained. One of the objectives of database systems is to integrate data without programmer intervention in a way that eliminates data redundancy. The other objective of database systems is to establish program-data independence, so that programs that access the data are immune to changes in storage structure (how the data is physically organized) and access technique.

1.4 The ANSI/SPARC Three-Schema Architecture

In the 1970s the Standards Planning and Requirements Committee (SPARC) of the American National Standards Institute (ANSI) offered a solution to these problems by proposing what came to be known as the **ANSI/SPARC three-schema architecture**.[2] The ANSI/SPARC three-schema architecture, as illustrated in Figure 1.2, consists of three perspectives of metadata in a database. The conceptual schema is the nucleus of the three-schema architecture. Located between the external schema and internal schema, the **conceptual schema** represents the global conceptual view of the structure of the entire database for

[2] In a database context, the word "schema" stands for description of metadata.

a community of users. By insulating applications/programs from changes in physical storage structure and data access strategy, the conceptual schema achieves program-data independence in a database environment.

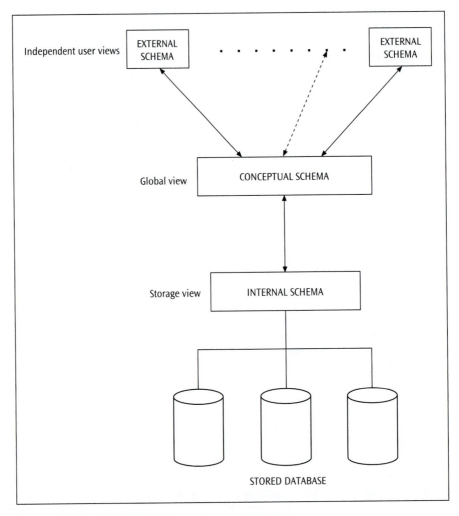

Figure 1.2 The ANSI/SPARC three-schema architecture

The **external schema** consists of a number of different user **views**[3] or subschemas,[4] each describing portions of the database of interest to a particular user or group of users. The conceptual schema represents the global view of the structure of the entire database for a community of users. The conceptual schema is the consolidation of user views. The data specification (metadata) for the entire database is captured by the conceptual schema. The **internal schema** describes the physical structure of the stored data (how the data is actually laid out on storage devices) and the mechanism used to implement the access strategies (indexes, hashed addresses, and so on). The internal schema is concerned with efficiency of data storage and access mechanisms in the database. Thus, the internal schema is technology-dependent while the conceptual schema and external schema are technology-independent. In principle, user views are generated on demand using the logical names of the data items in the conceptual schema independent of the logical or physical structure of the data. Thus, *the conceptual schema insulates user views from changes in the physical storage structure of the data*. This property is referred to as **data independence**. For example, if direct access to data ordered by zip code is required, the change is recorded as "direct access" in the conceptual schema and the indexing technique may be employed in the internal schema. However, the user views are completely shielded from even the knowledge of this change in the internal schema. That is, the specification and implementation of an index on zip code does not require any modification and testing of the application programs that use the views containing zip code. In short, the presence of the conceptual schema effectively hides the physical storage structures and access techniques from the external schema. Sometimes, data independence is also referred to as **physical data independence**. Redefinition of logical structures of a data model (such as adding or restructuring tables in a relational database) may sometimes be in order. Since the user views (external schema) are generated exclusively by logical references, the user views are immune to such logical design changes in the conceptual schema. This property is called **logical data independence**. Logical data independence enables a user view to be immune to changes in the other user views.

A file-processing system, in contrast, may be viewed as a two-schema architecture consisting of the internal schema and the programmer's view (external schema), as shown in Figure 1.3. Here, the programmer's view corresponds to the physical structure of the data, meaning that the physical structure of data (internal schema) is fully mapped (incorporated) into the application program. The file-processing system lacks program-data independence because any modification to the storage structure or access strategy in the internal schema necessitates changes to application programs and subsequent recompilation and testing. In the absence of a conceptual schema, the internal schema structures are necessarily mapped directly to external views (or subschemas). Consequently, changes in the internal schema require appropriate changes in the external schema; therefore, data independence is lost. Because changes to the internal schema such as incorporating new user requirements and accommodating technological enhancements are expected in a typical application environment, absence of a conceptual schema essentially sacrifices data

---

[3] Informally, a view is a term that describes the information of interest to a user or a group of users where a user can be either an end user or a programmer. See Chapter 6 (Section 6.4) for a more precise definition of a view.

[4] While an external schema is technically a collection of external subschemas or views, the term "external schema" is used here in the context of either an individual user view or a collection of different user views.

independence. In short, file-processing systems lack data independence because they employ what amounts to a two-schema architecture.

The three-schema architecture described in this section is required to achieve data independence. This book uses the entity-relationship modeling grammar to specify the conceptual schema.

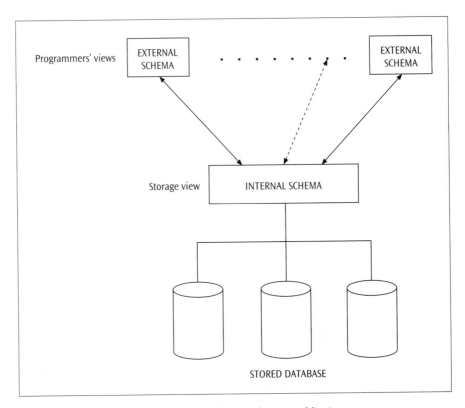

Figure 1.3   The file-processing system: A two-schema architecture

1.5 Characteristics of Database Systems

Database systems seek to overcome the two root causes of the limitations that plague file-processing systems by creating a single integrated set of files that can be accessed by all users. This integrated set of files is known as a **database**. A database management system (typically referred to as a DBMS) is a collection of general-purpose software that facilitates the processes of defining, constructing, and manipulating a database for various applications. Figure 1.4 provides a layman's view of the difference between a database and a database management system. This illustration shows how neither a user nor a programmer is able to access data in the database without going through the database management system software. Whether a program is written in Java, C, COBOL, or some other language, the program must "ask" the DBMS for the data, and the DBMS will fetch the data. SQL (Structured Query Language) has been established as the language for accessing data in a database by the International Organization for Standardization (ISO) and the American National Standards Institute (ANSI). Accordingly, any application program that seeks to access a database must do so via embedded SQL statements.

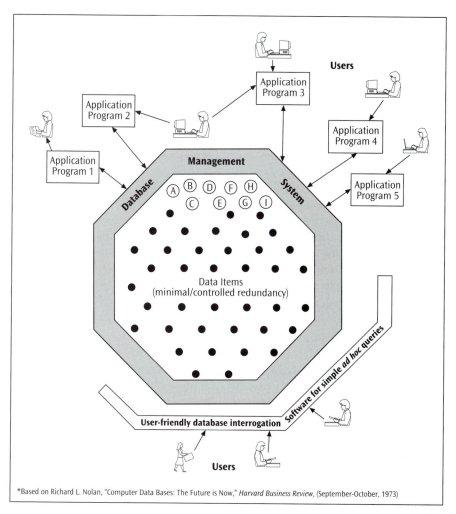

Figure 1.4 An early view of a database system

An important purpose of this book is to discuss how to organize the data items conceptualized in Figure 1.4. In reality, data items do not exist in one big pool surrounded by the database management system. Several different architectures exist for organizing this data. One is a hierarchical organization, another is a network organization, and a third is relational. Although each of the ways to organize data is discussed in this book, the relational approach is emphasized.[5] While the data items that collectively comprise the database at the physical level are stored as sequential, indexed, and random files, the DBMS is a layer on top of these files that frees the user and application programs from the burden of knowing the structures of the physical files (unlike a file-processing system).

Next, let us look more closely at what constitutes a database, a database management system, and finally a database system.

1.5.1 What Is a Database System?

A system is generally defined as a set of interrelated components working together for some purpose. A **database system** is a self-describing collection of interrelated data. A database system includes data and metadata. Properties of a database system are:

- Data consists of recorded facts that have implicit meaning.
- Viewed through the lens of metadata, the meaning of recorded data becomes explicit.
- A database is self-describing in that the metadata is recorded within the database, not in application programs.
- A database is a collection of files whose records are logically related to one another. In contrast with that of a file-processing system, integration of data as needed is the responsibility of the DBMS software instead of the programmer.
- Embedded pointers and various forms of indexes exist in the database system to facilitate access to the data.

A database system may be classified as single-user or multi-user. A single-user database system supports only one user at a time. In other words, if user A is using the database, users B and C must wait until user A has completed his or her database work. When a single-user database runs on a personal computer, it is also called a desktop database system. In contrast, a multi-user database system supports multiple users concurrently. If the multi-user database supports a relatively small number of users (usually fewer than 50) or a specific workgroup within an organization, it is called a workgroup database system. If the database is used by the entire organization and supports many users (more than 50, usually hundreds) across many locations, the database is known as an enterprise database system.

The term "enterprise database system" is somewhat misleading. In the early days of database processing, the goal was to have a single database for the entire organization. While this type of database is possible in a small organization, in large organizations multiple databases exist that are indeed used enterprise-wide. For example, large oil companies have databases organized by function: one database for exploration, one database for refining, another for marketing, a fourth for royalty payments, and so on. On the other hand, a consumer-products company might have several product databases. Each one of these is an enterprise database system since its use extends enterprise-wide.

---

[5] Two relatively new data modeling architectures (the object-oriented data model and the object-relational model) also exist. Appendix B discusses each of these architectures. Appendix A reviews architectures based on the hierarchical and network organizations.

A natural extension to the enterprise database system is the concept of distributed data-base systems. With the tremendous strides made in network and data communication technology in the last two decades, distribution of databases over a wide geographic area has become highly feasible. A **distributed database (DDB)** is a collection of multiple logi-cally interrelated databases that may be geographically dispersed over a computer network. A **distributed database management system (DDBMS)** essentially manages a distributed database while rendering the geographical distribution of data transparent to the user community. The advent of DDBs facilitated replacement of the large, centralized mono-lithic databases of the 1980s with decentralized autonomous database systems that are interrelated via a computer network.

Another important trend that emerged in the 1990s is the development of data warehouses. The distinguishing characteristic of a **data warehouse** is that it is mainly intended for decision-support applications used by knowledge workers. As a conse-quence, data warehouses are optimized for information retrieval rather than transaction processing. By definition, a data warehouse is subject-oriented, integrated, nonvolatile, and time-variant. Since its relatively recent inception, data warehousing has evolved rapidly in large corporations to support business intelligence, data mining, decision analytics, and customer relations management (CRM).

1.5.2 What Is a Database Management System?

Figure 1.5 illustrates the components of a database system, consisting of the DBMS, data-base, data dictionary, and data repository. A **database management system (DBMS)** is a collection of general-purpose software that facilitates the processes of defining, construct-ing, and manipulating a database. The major components of a DBMS include one or more query languages; tools for generating reports; facilities for providing security, integrity, backup, and recovery; a data manipulation language for accessing the database; and a data definition language used to define the structure of data. As shown in Figure 1.5, Struc-tured Query Language (SQL) plays an integral role in each of these components. SQL is used in the **data definition language (DDL)** for creating the structure of database objects such as tables, views, and synonyms. SQL statements are also generated by program-ming languages used to build reports in order to access data from the database. In addi-tion, people involved in the data administration function use **data control languages (DCLs)** that make use of SQL statements to (a) control the resource locking required in a multi-user environment, (b) facilitate backup and recovery from failures, and (c) pro-vide the security required to ensure that users access only the data that they are autho-rized to use.

Data manipulation languages (DMLs) facilitate the retrieval, insertion, deletion, and modification of data in a database. SQL is the most well-known nonprocedural[6] DML and can be used to specify many complex database operations in a concise manner. Most DBMS products also include procedural language extensions to supplement the capabilities of SQL, such as Oracle PL/SQL. Other examples of procedural language extensions are lan-guages such as C, Java, Visual Basic, and COBOL, in which precompilers extract data

---

[6] SQL is known to be a nonprocedural language since it only specifies what data to retrieve as opposed to specifying how actually to retrieve it. A procedural language specifies how to retrieve data in addition to what data to retrieve.

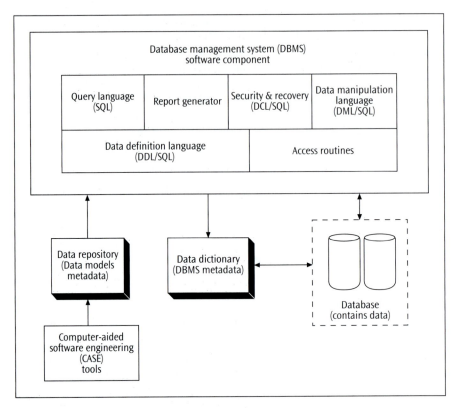

Figure 1.5 Components of a database system

manipulation commands written in SQL from a program and send them to a DML compiler for compilation into object code for subsequent database access by the run-time subsystem.[7] Finally, the access routines handle database access at run time by passing requests to the file manager of the operating system to retrieve data from the physical files of the database.

Much as a dictionary is a reference book that provides information about the form, origin, function, meaning, and syntax of words, a **data dictionary** in a DBMS environment stores metadata that provides such information as the definitions of the data items and their relationships, authorizations, and usage statistics. The DBMS makes use of the data dictionary to look up the required data component structures and relationships, thus relieving application developers (end users and programmers) from having to incorporate data structures and relationships in their applications. In addition, any changes made to the physical structure of the database are automatically recorded in the data dictionary. This removes the need to modify application programs that access the modified structure.

---

[7] The run-time subsystem of a database management system processes applications created by the various design tools at run time.

While not a component of the DBMS per se, the data repository has become an integral part of the data management suite of tools. The **data repository** is a collection of metadata about data models and application program interfaces. CASE (computer-aided software engineering) tools such as Oracle Designer and ERWIN that are used for developing a conceptual/logical schema[8] interact with the data repository and are independent of the database and the DBMS.

1.5.3 Advantages of Database Systems

A database system is comprised of a database, general-purpose software called the database management system that creates and manipulates the database, a data dictionary, and appropriate hardware. While an organization could write its own special-purpose database management system software to create and manipulate the database, a database management system is usually purchased from a software vendor such as IBM, Microsoft, or Oracle. Figure 1.6 illustrates how a database system for the hypothetical university introduced earlier might be structured.

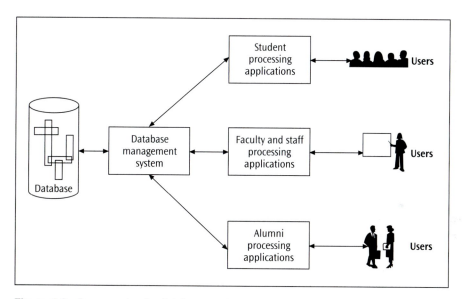

Figure 1.6 An example of a database system

As mentioned previously, in file-processing systems access to information is achieved through programs written in conventional programming languages by application programmers. In contrast, database systems contain a variety of tools such as query languages and report writers that make it possible for end users to access and make use of much of the information needed for analysis and decision-making.

Data integrity ensures that data is consistent, accurate, and reliable. A properly designed database system allows data integrity to be achieved through controlling

---

[8] The data modeling activity includes the development of the conceptual and logical schema. The role of the conceptual schema in database design is discussed in Section 1.6.

redundancy. As we will see throughout this book, while database systems do not elimi-nate redundancy (that is, storing the same piece of data in more than one place), they can—in fact they must, through adherence to business rules that take the form of constraints—control unnecessary and harmful redundancies that often accompany file-processing systems.

Recall that file-processing systems are plagued by lack of integration of data and lack of program-data independence. The reason for program-data dependence is that data structures used to physically store data must be referred to directly by the application program. Database systems do not suffer from this limitation as data structures are stored, along with the data, in the database as opposed to being recorded in the applica-tion program. The DBMS accesses the database and supplies data appropriate to the needs of individual application programs. The DBMS accomplishes this using the data dictio-nary where the metadata is recorded. Thus the data dictionary in a database environ-ment insulates application programs from changes to the structure of data.

The notion of data relatability involves the creation of logical relationships between dif-ferent types of records, usually in different files. In a file-processing environment, infor-mation often cannot be generated without a programmer writing or at least modifying an application program to consolidate the files. With the advent of database systems, all that is necessary is for one to specify the data to be combined; the DBMS will perform the nec-essary operations to accomplish the task.

1.6 Data Models

A model is a simplified expression of observed or unobservable reality used to perceive relationships in the real world. Models are used in many disciplines and can take a vari-ety of forms. Some are small-scale physical representations of reality, like model air-craft in a wind tunnel. Others take the form of mathematical models that use a set of mathematical symbols and logic to describe the behavior of a system—such as economet-ric models, optimization models, and statistical models. In the mathematical computing field, directed graphs (digraphs) are used to model relationships among variables and essen-tially represent an analog model. Likewise, in the database design arena, a **data model** is used to represent real-world phenomena. Data modeling is considered the most important part of the database design process.

A drawing of a chair is essentially a model of the real-world object "chair." A chair con-structed using Legos is also a model of the real-world object "chair." However, modeling an intangible real-world object that cannot be physically grasped, such as a department, requires a different approach—one that makes use of descriptors. For example, a depart-ment can be described by specifying that it has a name, can have multiple locations, has a number, has a budget, has a certain number of employees, and the like. This is an example of data model that forms a conceptual expression of the intangible real-world object "department" using descriptors. Objects, tangible as well as intangible, such as employ-ees, inventory, sales, and projects, can be modeled in a similar manner.

As simplified abstractions of reality, data models enable better understanding of data speci-fications such as data types, relationships, and constraints gathered from user requirements. Data models serve as the blueprints for designing databases. Just as one would not build a house from incomplete blueprints with known errors, it is imperative that a data model also be a com-plete and accurate reflection of the data requirements of a database system.

1.6.1 Data Models and Database Design

Figure 1.7 illustrates the role of data models in the phases of database design. A database represents some aspect of the real world called the **universe of interest**. For a small organization, the universe of interest may be all functionalities of the company (marketing, finance, accounting, production, human resources, and so on). A large oil company may have islands of interest centered around functions such as exploration, refining, and marketing, and a separate database designed for each function. A large consumer-products company might be oriented around product groups and have separate databases for different product groups.

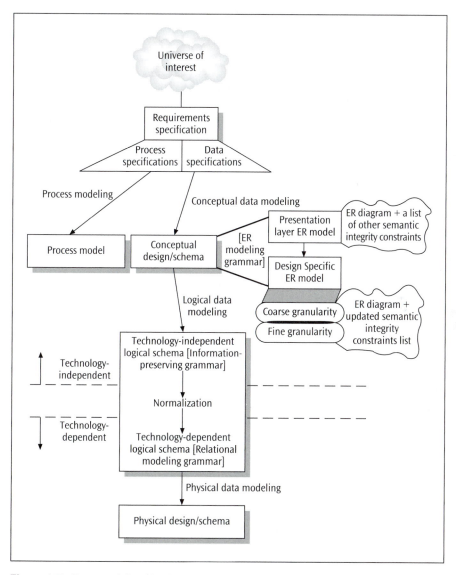

Figure 1.7   Data modeling/database design life cycle

The initial step in the design process is the **requirements specification**. During this step, systems analysts review existing documents and systems and interview prospective users in an effort to identify the objectives to be supported by the database system. The output of the requirements specification activity is a set of data and process specifications. This is essentially an organized conglomeration of user-specified restrictions on the organization's activities (business processes) that must be reflected in the database and/or database applications. Such restrictions are commonly referred to as **business rules**.

In order to define the data requirements, one needs to know the process requirements, that is, what is going to be done with the data. For example, suppose a company is going to sell a product. What processes are involved? When a company sells a product it bills the customers who purchase the product. Then, shipping has to be notified to dispatch the product to the customer. Shipping also has to check the inventory and make sure that inventory levels are adjusted as a result of sales. The inventory system must make sure that inventory levels are optimal and, accordingly, replenish inventory periodically.

Data is required in order to accomplish processes such as those mentioned above. Customers' names, addresses, and telephone numbers are needed for billing purposes. For shipping, a shipping address is required. In the inventory system one needs to know in which warehouse and in which bin a particular product is located. One also needs to know quantity on hand, quantity on order, and the lead time required by the supplier to fill an order. In short, data and process requirements go hand-in-hand. This book assumes that the requirements specification has been done, and separates out the data requirements from the process requirements.[9] The data requirements will be narrated in the form of stories and vignettes. For example, Chapters 3, 4, and 5 contain stories that provide the data requirements for two hypothetical organizations: Bearcat Incorporated and Cougar Medical Associates.

1.6.2 The Database Design Life Cycle

Database design follows a life cycle that includes three tiers: conceptual data modeling, logical data modeling, and physical data modeling. This book is organized to follow this progression. As shown in Figure 1.7, conceptual design is done first.

1.6.2.1 Conceptual Data Modeling

The conceptual data model describes the structure of the data to be stored in the database without specifying how and where it will be physically stored or the methods used to retrieve it. The conceptual design activity is technology-independent. During the conceptual design, the focus should be on capturing the user-specified business rules in all their richness, unconstrained by the boundaries of the anticipated technology or DBMS product that will be used for implementation. The product of the conceptual design activity is the conceptual schema.

---

[9] Process requirements are often expressed in the form of data flow diagrams. An examination of approaches for identifying process requirements is usually part of a systems analysis and design textbook and thus is not discussed here.

Several conceptual data modeling methods exist (for example, ER modeling and NIAM modeling[10]), each with its own specific grammar. This book uses the ER modeling technique due to its significant popularity in the database design sphere. The conceptual model is portrayed in two progressive layers: the Presentation Layer ER model and the Design-Specific ER model. The Presentation Layer ER model is intended for user-analyst communication and entails the database analyst working with the user community and developing a script (a Presentation Layer ER diagram) accompanied by a set of semantic integrity constraints.[11] Next, the Presentation Layer ER model (both the ER diagram and semantic integrity constraints) is mapped (translated) to the Design-Specific ER model. The Design-Specific ER model is expressed in terms of two levels of granularity—the coarse level and the fine level. These two levels of the Design-Specific ER model progressively incorporate technical details needed for the final database design but not quite necessary for interactions with the user community. Chapters 3, 4, and 5 discuss these ER models in detail.

A conceptual data model is supposed to capture all business rules implicitly or explicitly present in the requirements specification. Simply going through a detailed process of representing the requirements in a conceptual model does not assure that the resulting conceptual design captures all business rules. Therefore, it is imperative that a validation step is included as the concluding activity in the conceptual modeling phase to systematically verify that the conceptual design developed indeed answers all the questions stated implicitly or explicitly in the requirements specification. Where the validation fails, the model should be refined accordingly. Since the ER modeling grammar is used for the conceptual design activity in this book, validation of an ER model must be addressed. Validating an ER model is discussed at the end of Chapter 5.

1.6.2.2 Logical Data Modeling

As a consequence of the technology-independent orientation of the conceptual design activity, it is possible that the conceptual schema may contain constructs not directly compatible with the technology intended for implementation. It is also possible that some of the design may require refinement to eliminate data redundancy problems. Transforming a conceptual schema to a schema more compatible with the implementation technology of choice becomes necessary. Thus, the second tier of the data modeling activity is called logical design. The product of the logical design activity is the logical schema, which evolves from a technology-independent representation to a technology-dependent representation. The latter is typically modeled using the hierarchical, network, or relational architecture. The relational model forms the basis for most of the material in this textbook. Discussion of characteristics of the other models is confined to the appendices. Logical design and the role of normalization are discussed in detail in Chapters 6 through 9.

---

[10] Entity relationship (ER) modeling is a "design by analysis" modeling approach and is top down in nature while NIAM (Nijssen Information Analysis Methodology) modeling is a "design by synthesis" approach and is bottom up in nature.

[11] Semantic integrity constraints are business rules that are not captured in the ER diagram.

1.6.2.3 Physical Data Modeling

Once a logical schema has been developed, the process moves on to the physical design tier. Here the internal storage structure and access strategies are specified. As shown in Figure 1.7, the physical design activity is fully technology-dependent. Physical design involves using the tools of a particular DBMS product to create the database and to design and develop applications that address the high-level requirements of the universe of interest.

The ANSI/SPARC three-schema architecture in Figure 1.3 forms the basis for the data modeling/database design life cycle shown in Figure 1.7. Starting with the conceptual design activity and progressing through the logical design and physical design activities mirrors the nucleus of interest of the three-schema architecture, that is, the conceptual schema. This figure is replicated at the beginning of Parts I, II, III, and IV of the book to serve as a roadmap through the topics covered in the chapters of each part.

Chapter Summary

Data consists of raw facts, that is, facts that have not yet been organized or processed to reveal their meaning. Information is data in context—that is, data that has been organized into a specific context that has meaning and value. Metadata describes the properties of data. It is through the lens of metadata that data becomes information.

Data management involves four actions: creating, retrieving, updating, and deleting data. Two data management functions support these actions: organizing data and accessing data. Two primary forms of access are sequential access and direct access.

Database systems have been successful because they overcome the problems associated with the lack of integration of data and program-data dependence that plague their predecessors, file-processing systems. Database management system (DBMS) software has been the vehicle that has allowed many organizations to move from a file-processing environment to a database system environment. Among the components of a DBMS are tools for (a) retrieving and analyzing data in a database, (b) creating reports, (c) creating the structure of database objects, (d) protecting the database from unauthorized use, and (e) facilitating recovery from various types of failures. In a DBMS environment, the data dictionary (metadata that describes characteristics of data) functions as the lens through which data in the database is viewed.

The ANSI/SPARC architecture divides a database system into three levels or tiers. The external level is closest to the users and is concerned with the way data is used or viewed by individual users. The conceptual level is technology-independent and represents the global or community view of the entire database. The internal level is the one closest to physical storage and is concerned with the way the data is physically stored. As such, the internal level is technology-dependent. The data as perceived at each level are described by a schema (or subschemas in the case of the external level). A file-processing system is essentially a two-tier architecture with only external and internal levels. Without the conceptual schema, the internal schema must be mapped directly into external views. Thus changes in the internal schema require appropriate changes in the external subschemas; this is how data independence is lost.

Data models play a crucial role in database design. Data models describe the database structure. The approach described in this book begins with the creation of a conceptual schema that describes the structure of the data to be stored in the database without specifying how and where it will be stored or the methods used to retrieve it. The conceptual schema takes the form of Presentation Layer and Design-Specific ER models and, once appropriately validated, serves as input to the logical design activity. During logical design, the technology-independent conceptual schema evolves into a technology-dependent logical schema. This technology-dependent logical schema is subsequently used during the physical design activity.

Exercises

1. What is the difference between data, metadata, and information?

2. Demonstrate your understanding of data, metadata, and information using an example.

3. Describe the four actions involved in data management.

4. Distinguish between sequential access and direct access. Give an example of a type of application for which each is particularly appropriate.

5. Identify a common task in a payroll system for which sequential access is more appropriate than direct access and explain why this is so.

6. What is the difference between a serial collection of data and a sequential collection of data? Which can be used for direct access?

7. What is the purpose of an external index?

8. What is data integrity, and what is the significance of a lack of data integrity?

9. Describe the limitations of file-processing systems. How do database systems make it possible to overcome these limitations?

10. Using the Internet, trace the history of ANSI and ISO and their relevance to the information systems discipline. Write a summary of your findings.

11. Describe the structure of the ANSI/SPARC three-schema architecture. Compare this structure with that of the two-schema architecture inherent in a file-processing system.

12. Explain why a file-processing system may be referred to as belonging to a two-schema architecture.

13. Define data independence.

14. What is the difference between logical and physical data independence? Why is the distinction between the two important?

15. What is the difference between a database and a database management system?

16. Since ANSI and ISO have adopted SQL as the standard language for database access, explore via the Internet the history and features of SQL and its appropriateness for database access. Write a summary of your findings.

17. Write a short essay (one or two pages) about distributed databases using information available from Internet resources.

18. Write a short essay (one or two pages) about data warehousing using information available from Internet resources.

19. Oil companies have functional databases, and the consumer-product industry tends to have product databases. How do financial institutions and the airline industry classify their enterprise database systems? Use Internet resources to find the answer, and record your findings.

20. Find out and describe briefly what a CASE tool is, using Internet sources.

21. Distinguish between a model and a data model.

22. What is the role of data models in database design?

23. Write a short essay (one or two pages) summarizing the content of the *Harvard Business Review* article on databases cited in Figure 1.4.

Selected Bibliography

Connolly, T. M. and Begg, C. E. (2002) *Database Systems: A Practical Approach to Design, Implementation, and Management*, Third Edition, Addison-Wesley.

Courtney, J. F. and Paradice, D. B. (1988) *Database Systems for Management*, Times Mirror/ Mosby College Publishing.

Elmasri, R. and Navathe, S. B. (2003) *Fundamentals of Database Systems*, Fourth Edition, Addison-Wesley.

Hansen, G. W. and Hansen, J. V. (1992) *Database Management and Design*, Prentice Hall.

Hoffer, J. A.; Prescott, M. B.; and McFadden, F. R. (2002) *Modern Database Management*, Sixth Edition, Prentice Hall.

Kim, Y. and March, S. T. (1995) "Comparing Data Modeling Formalisms," *Communications of the ACM*, June, Vol. 38 No. 6, pp. 103-112.

Kroenke, D. M. (2004) *Database Processing: Fundamentals, Design, and Implementation*, Ninth Edition, Prentice Hall.

Martin, J. (1976) *Principles of Data-Base Management, Prentice Hall.*

Nolan, R. L. (1973) "Computer Data Bases: The Future is Now," *Harvard Business Review*, September-October, Vol. 51 No. 4, pp. 98-114.

Pratt, P. J. and Adamski, J. J. (1994) *Database Systems Management and Design*, Third Edition, Boyd & Fraser Publishing Company.

Ramakrishnan, R. and Gehrke, J. (2000) *Database Management Systems*, McGraw Hill.

Rob, P. and Coronel, C. (2006) *Database Systems: Design, Implementation, and Management* Seventh Edition, Course Technology.

Shepherd, J. C. (1990) *Database Management: Theory and Application*, Richard D. Irwin, Inc.

Turban, E.; McLean, E.; and Wetherbe, J. (1996) *Information Technology for Management*, John Wiley & Sons, Inc.

PART I

CONCEPTUAL DATA MODELING

INTRODUCTION

Database systems typically have a very limited understanding of what the data in the database *actually mean*. Therefore, **semantic modeling** (the overall activity of attempting to represent meaning) can be a valuable precursor to the database design process to capture at least some of the meaning conveyed by the users' business rules. The term conceptual modeling is often used as a synonym for semantic modeling. In this book, **conceptual modeling** refers to only data specifications (not process specifications) of the user requirements, at a high level of abstraction. It has been argued that conceptual modeling in database design is justified because it encourages user participation in the design process, allows the model to be more DBMS-independent, facilitates understanding of how the database fits into the organization as a whole, and eases maintenance of schemas and applications in the long run (Batini, Ceri, and Navathe, 1992). Sophisticated interpretations of the meaning of data are still left to the user (Date, 2004).

Conceptual modeling entails constructing a representation of selected phenomena in a certain application domain, for example, an airline management system, a patient care system, or a university enrollment system. A conceptual model portrays data specifications at the highest level of abstraction in the data modeling hierarchy. The input source for this modeling phase is the business rules culled out from the requirements specification supplied by the user community. A conceptual model includes:

- A *context,* a setting in which the phenomenon being modeled transpires; the universe of interest (application domain) being modeled represents the context
- A *grammar* that defines a set of constructs and rules for modeling the phenomenon
- A *method* that describes how to use the grammar

The product of conceptual modeling is a conceptual modeling *script* such as an ER model, NIAM model, or semantic object model for the phenomenon being modeled. This framework for conceptual modeling is based on Wand and Weber (2002), and is illustrated in Figure I.1.

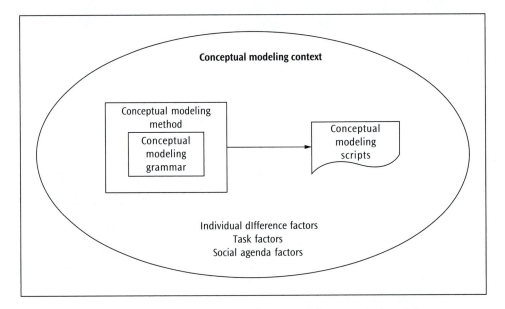

Figure I.1 Wand and Weber (2002) framework for research in conceptual modeling

The chapters in Part I introduce and elaborate upon conceptual data modeling using the Wand and Weber framework. Chapter 2 introduces the fundamental constructs and rules for the ER modeling grammar, which is used in this book as the tool for conceptual data modeling. The constructs pertain to inter-entity class binary relationships. Chapter 3 employs a comprehensive case to illustrate the method component of the Wand and Weber framework. In this chapter, the method to use the entity-relationship (ER) modeling grammar is explicated in progressive steps leading to the emergence of a specific script (i.e., an ER diagram and a semantic integrity constraints list) for the case in question. Chapter 4 introduces newer constructs that enhance the ER (EER) modeling grammar with means to model intra-entity class relationships. An extension to the case from Chapter 3 incorporating a story line that requires application of EER constructs is used here to demonstrate the method pertaining to the use of the EER constructs. Chapter 5 presents higher-order relationships, namely relationships of degree 3 and beyond, innovative use of the grammar, and a few additional ER constructs (e.g., cluster entity type, interrelationship dependency). A second comprehensive case is used to illustrate the method component of the modeling framework.

Figure I.2 replicates Figure 1.7 in Chapter 1, showing a "road map" that provides an overview of the process of data modeling and database design and indicates how the topics in Part I fit into the overall picture.

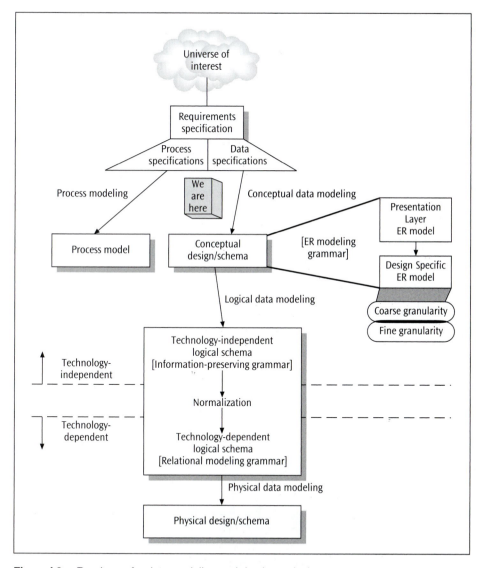

Figure I.2   Road map for data modeling and database design

CHAPTER **2**

Foundation Concepts

This chapter introduces the fundamental constructs and rules for conceptual data modeling using the entity-relationship (ER) modeling grammar as the modeling tool. The basic units of the ER model, that is, entity type, entity class, attribute, unique identifier, and relationship type, are treated in detail. The chapter also includes three vignettes designed to illustrate a few data modeling errors that can occur when the ER modeling grammar is not applied properly or the requirements stated by the users are misinterpreted and thus improperly specified.

2.1 A Conceptual Modeling Framework

There are many modeling schemes that attempt to represent the data semantics of a database application, and in fact many of these schemes bear a strong family resemblance to one another and often use graphical representations. The entity-relationship (ER) modeling grammar originally proposed by Peter Chen in 1976 and further refined by others over the years is probably the most widely accepted data modeling tool for conceptual design and was chosen by ANSI in 1988 as the standard model for Information Resource Directory Systems (IRDSs) (Batini, et al., 1992, p. 30). It aspires to capture the overall data semantics of an application in a concise manner in terms appropriate for subsequent mapping to specific database models. While criticized by some for its insufficiency for the completion of a database design (Date, 2004), the ER modeling grammar is an effective tool for communicating technical information in the development of large database applications. Thus, the ER model is used in this book to illustrate conceptual data modeling.

2.2 ER Modeling Primitives

Table 2.1 contains a series of terms (referred to here as **primitives**) that correlate the real world to the conceptual world and serve as a foundation for the entity-relationship (ER) modeling grammar. The real world consists of (a) tangible objects of an **object type** (such as a specific student, a specific piece of furniture, or a specific building), or (b) intangible objects of an object type (a specific department, a particular project, or a certain course).

Table 2.1 Equivalence between real world primitives and conceptual primitives

Real World Primitive	Conceptual Primitive
Object {type}	Entity(ies) {type}
Object (occurrence)	Entity(ies) {instance}
Property	Attribute
Fact	Value

Table 2.1 Equivalence between real world primitives and conceptual primitives (continued)

Real World Primitive	Conceptual Primitive
Property value set	Domain
Association	Relationship
Object class	Entity class

In the conceptual world, an object type is referred to as an **entity type**. **Objects** belonging to an object type are considered to be **entities** or **entity instances** of the corresponding entity type. The concept of an entity is the most fundamental concept of the ER modeling grammar and serves as the foundation for other concepts (Chen, 1993). Instead of the actual person or object, Anna Li, the entity representing Anna Li takes the form of a record in a STUDENT data set. Thus actual students are student objects and are referred to in terms of a STUDENT object type, whereas a representation of the STUDENT object type is called the STUDENT entity type.

In the real world there can be many occurrences of a particular object type. For example, there can be 35,000 students enrolled and taking courses in a university during a semester and thus 35,000 student objects. In this case, there would be 35,000 student entities represented by 35,000 records in a STUDENT data set. The collection of these 35,000 student entities is referred to as an **entity set**.

An object type can have many properties. For example, a STUDENT object type has properties such as student number, date of birth, gender, and so on. Correspondingly, an entity type is said to have **attributes**.

An entity or entity instance is created when a value is supplied for some attribute(s). Thus a STUDENT data set with 35,000 records would contain values that represent the facts associated with the 35,000 student objects. In addition, in the real world, a fact is drawn from a **property value set**. In the conceptual world, the value of an attribute comes from a **domain** of possible values. For example, the domain for the attribute **Gender** can be (Male, Female) whereas the domain associated with the set of two-character U.S. postal codes is (AK, AL, AR, ..., WY). A domain can be either explicit or implicit. The domain for **Gender** is an example of an explicit domain consisting of a set of only two possible values. The domain for the attribute **Salary** would be an example of an implicit domain, because it would not be practical to explicitly list the set of all possible salaries between say $10,000 and $2,000,000.

In the real world, **associations** between objects of different object types exist. For example, students enroll in courses, suppliers supply parts, and salespersons process orders. In the conceptual world, these associations are referred to as **relationships**.

Two other terms need to be introduced as part of our foundation concepts: object class and entity class. In the real world an **object class** is a generalization of related object types that have shared properties, whereas in the conceptual world an **entity class** is a generalization of different related entity types that have shared attributes. For example, the entity type chair and the entity type table would both refer to the entity class furniture. Likewise, student, faculty, and customer entity types can be said to belong to an entity class called human beings.

2.3 Foundations of the ER Modeling Grammar

The basic building blocks of ER modeling include entity types, attributes, and relationship types. A set of attributes makes up or gives structure to an entity type. A rule in the grammar specifies that entity types can only be associated via a relationship type and a relationship type can only show association between entity types.

2.3.1 Entity Types and Attributes

As described in Section 2.2, an entity type is a set of related attributes that comprise a conceptual representation of an object type. Typically an entity type takes the form of a singular noun (a person, thing, place, or concept). Attributes shape an entity type. An entity type can participate in one or more relationships with other entity types. Table 2.2 lists a number of characteristics possessed by an attribute. First, each attribute has a name that generally conforms to a standardized naming convention. For example, Company A may use the attribute **Ename** to represent the name of an employee in the EMPLOYEE entity type, but require a different attribute name to be used for the name of the employee who manages a particular department (**Empname**) in the DEPARTMENT entity type. Company B, on the other hand, may require that each employee name attribute in the database make use of the same attribute name **Ename**. Then, the employee name in the EMPLOYEE entity type is referred to as **EMPLOYEE.Ename** whereas the employee name in the DEPARTMENT entity type would be referred to as **DEPARTMENT.Ename**.

Table 2.2   Characteristics of attributes

Attribute	Characteristics
Name	Standardized naming convention
Type	Numeric, alphabetic, alphanumeric, logical, date/time, etc.
Classification	Atomic or composite/molecular
Category	Single-valued or multi-valued
Source	Stored (real) or derived (virtual)
Domain*	Property value set—implicit or explicit
Value	Conceptual representation of a fact about a property
Optionality	Optional value or mandatory value
Role	Key (unique identifier) or non-key

* In a stricter sense, other characteristics of an attribute (e.g., type, value) are also viewed as part of the domain of an attribute.

A variety of **data types** can be associated with attributes. A *numeric* data type is used when an attribute's value can consist of positive and negative numbers, and is often used in arithmetic operations. Numeric attributes can further be constrained so as to allow only integer values, decimal values, and so on. The *alphabetic* data type permits an attribute to consist of only letters and spaces, while an *alphanumeric* data type allows the value of an attribute to consist of text, numbers (telephone numbers, postal codes, account numbers, and so on), and certain special characters. Alphabetic and alphanumeric data types should not be used for attributes involved in arithmetic operations. Likewise, an attribute not involved in an arithmetic operation should not be defined as a numeric data type even if it contains only numbers (telephone number, Social Security number) to enable textual manipulations. A *logical* data type is associated with an attribute whose value can be either true or false. Attributes with a *date* data type occur frequently in database applications, for example date of birth, date hired, or flight date.

A particularly important characteristic of an attribute is its classification—that is, whether it is an atomic attribute or a composite attribute. An attribute that has a discrete factual value and cannot be meaningfully subdivided is called an **atomic or simple attribute**. On the other hand, a **composite or molecular attribute** can be meaningfully subdivided into smaller subparts (i.e., atomic attributes) with independent meaning. **Salary** is an atomic attribute because it cannot be meaningfully divided further. Depending on the user's specification, **Name** can be an atomic attribute or a composite attribute made up of **First name, Middle initial**, and **Last name**. Figure 2.1 illustrates how an address might be modeled in the form of a hierarchy of composite attributes.

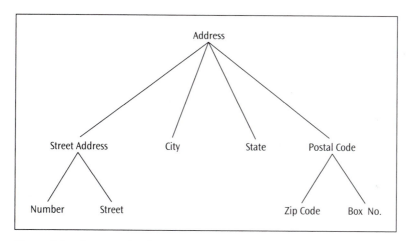

Figure 2.1   An example of a composite attribute hierarchy

An attribute can be either a **stored attribute** or a **derived attribute**. In some cases two or more attributes are related in the sense that the value of one can be calculated or derived from the values of the other(s). For example, if **Flight time** is calculated as the difference between the arrival time at the destination and departure time at the point of origin, then **Flight time** in this case can be a derived attribute since it need not be stored.

Most attributes have a single value for a particular entity and are referred to as **single-valued attributes**. For example, **Date of birth** is a single-valued attribute of an employee. There are attributes, however, that can have more than one value. For example, a programmer may be skilled in several programming languages, thus making the attribute **Skill** a **multi-valued attribute**.

For each entity of an entity type, some attributes *must* be assigned a value. Such attributes are referred to as **mandatory attributes**. On the other hand, attributes that need not be assigned a value for each entity are referred to as **optional attributes**. For example, it is possible that the attribute **Salary** might be classified as an optional attribute if the salary of each employee need not necessarily be provided or is unknown. Another attribute **Commission** might be classified as an optional attribute because a commission is not necessarily meaningful for job types other than perhaps the job type of salesperson. Attributes classified as optional are assigned a special value called *null* when their value is not available or is unknown.[1]

Composite and/or multi-valued attributes that are nested as a meaningful cluster are called **complex attributes**.[2] As an example, let's consider the medical profile for a patient as shown below.

Medical_profile (Blood (Type, Cholesterol (HDL, LDL, Triglyceride), Sugar), Height, Weight, {Allergy (Code, Name, Intensity)})

Observe that composite attributes are enclosed in parentheses ( ) while multi-valued attributes are enclosed in braces { }. The medical profiles for two patients appear in Figure 2.2.

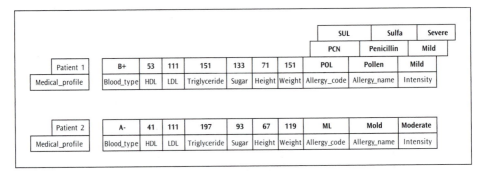

Figure 2.2   Medical profile of patients

2.3.2 Entity and Attribute-Level Data Integrity Constraints

Data integrity constraints, generally referred to as just **integrity constraints**, are rules that govern the behavior of data at all times in a database. The source of integrity constraints are the business rules that emerge from the user requirements specification for a database application system. In other words, integrity constraints are technical expressions of business rules. To that extent, the integrity constraints prevail and thus must be preserved across all three tiers of data modeling—conceptual, logical, and physical.

A **business rule**, in a data modeling context, is a short statement of a specific condition or procedure relevant to the universe of interest (application domain) being modeled expressed in a precise, unambiguous manner. Business rules often are implicitly interleaved in the requirements specification and are culled out as explicit expressions. While this process of developing business rules from the requirements specification is not quite

---

[1] "Value unknown" and "value exists but is not available" constitute what is typically called "missing data."

[2] A complex attribute is most likely an entity type instead of an attribute. Nonetheless, the purpose here is to explain the existence of an attribute called a complex attribute.

scientific, it certainly can be systematic. A step-by-step analysis of the requirements specification will enable an analyst/modeler to extract specific conditions and procedures inherent in the universe of interest as business rules. Further, such a systematic analysis will also facilitate identification of ambiguities which, when clarified by the user community, will yield additional business rules and also facilitate correction of other business rules. Business rules not only facilitate development of data models, but also aid in validating them.

Consider the following example. The universe of interest for modeling is a university academic program. A short extract from a comprehensive requirements specification for this application domain is narrated here to illustrate development of business rules.

> *There are several colleges in the university. A college offers many courses and a college term is divided into four quarters—Fall, Winter, Spring, and Summer—during which one or more of these courses may be offered. Let us even say that every quarter at least 23 courses are offered. The college also has several instructors. Instructors teach; that is why they are called instructors. Often, not all instructors teach during all quarters. Further, instructors are capable of teaching a variety of courses that the college offers.*

At the outset, four entity types—COLLEGE, COURSE, QUARTER, and INSTRUCTOR—may emerge from this narrative. Some of the business rules that can be extracted from the narrative are:

- There are four quarters; specifically, they are, Fall, Winter, Spring, and Summer.
- A college offers many courses.
- At least 23 courses are offered during every quarter.
- A college has several instructors.
- An instructor need not teach in a given quarter.
- An instructor is capable of teaching a variety of courses offered by the college.
- *An instructor must teach in some quarter.*
- *An instructor must be capable of teaching at least one course that the college offers.*

Observe that the first six business rules are obvious from the narrative, while the last two are essentially inferred and precisely specified from the requirements specification as a whole.

A systematic study of the narrative reveals ambiguities that require clarification by the user community. A few examples are:

- Is a particular course offered in more than one quarter?
- Are there courses that are just in the books but are never offered?
- Can an instructor teach for more than one college in the university?

Answers to such questions will sharpen the requirements specification by way of additional business rules. In short, business rules are indispensable in the process of translating a requirements specification to integrity constraints.

In general, integrity constraints are considered to be part of the schema (the description of the metadata) in that they are declared along with the structural design of the data model (conceptual, logical, and physical) and hold for all valid states of a database that correctly model an application (Ullman and Widom, 1997). While it is possible to specify all the integrity constraints as a part of the modeling process, some cannot be expressed explicitly or implicitly in the schema of the data model. For instance, an ER diagram (conceptual schema) is not capable of expressing domain constraints of attributes. Likewise, there are other constraints that a logical schema (relational schema) is not capable of expressing (this topic is covered in more detail in Chapter 6). As a consequence, such constraints are carried forward through the data modeling tiers in textual form and are often referred to as **semantic integrity constraints**.

At the conceptual tier of data modeling, two types of data integrity constraints pertaining to entity types and attributes are specified:

- The **domain constraint** imposed on an attribute to ensure that its observed value is not outside the defined domain.
- The **uniqueness constraint** that requires entities of an entity type to be uniquely identifiable. This is also sometimes referred to as the *key constraint*.

2.3.2.1 Domain Constraint

The concept of domain was briefly introduced in Section 2.2. The characteristics of an attribute discussed in Section 2.3.1 (name, type, class, and domain) are in themselves data integrity constraints on the attribute. The fact that the attribute **Name** is defined as 30 positions alphanumeric is an example of a constraint on the attribute **Name**. Similarly the fact that the attribute **Salary** is eight positions numeric is a constraint on the attribute **Salary**. However, the requirement that the attribute **Salary** must be in the range between $60,000 and $80,000 is an example of an implicit domain constraint.[3] While constraints are intended to make sure that the integrity of data is maintained, a domain constraint cannot ensure total data integrity. For example, if the actual salary of an employee is $65,000 but is erroneously entered as $56,000, the domain constraint will result in the identification of the error. On the other hand, if the erroneous entry is $67,000, that is, within the domain, the error cannot be identified. An example for an explicit specification of a domain constraint can be: A **Student_type** takes a value from the set {FR, SO, JR, SR, GR}. Another example for an explicit domain constraint is: **Music_skill** takes a value from the set {Rock, Jazz, Classical}.

2.3.2.2 Unique Identifier of an Entity Type

It is important to be able to uniquely identify entities in the entity set of the entity type. An atomic or composite attribute whose values are distinct for each entity in the entity set is the **unique identifier**[4] of the entity type. This attribute, then, can be used to identify an entity distinctly from other entities in the entity set. Note that specifying a unique identifier for the entity type signifies the imposition of the uniqueness constraint on the entity

---

[3] Strictly speaking, data type and size of an attribute can also be construed as domain constraints on the attribute.

[4] Another popular term, primary key, is deliberately avoided here because a strict definition of the term is possible only in the context of a relation schema discussed in Chapter 6.

type. It is also important to recognize that the idea of the uniqueness constraint is not limited to finding one unique identifier for the entity type, but extends to identifying all irreducible unique identifiers of the entity type.[5]

Every attribute plays only one of three roles in an entity type. It is a key attribute, a non-key attribute, or a unique identifier. Any attribute that is a constituent part of a unique identifier is a **key attribute**. An attribute that is not a constituent part of a unique identifier is a **non-key attribute**.[6]

A graphical representation of an entity type along with its attributes is portrayed in Figure 2.3. Here, an entity type PATIENT is portrayed by the rectangle, and attributes are shown as circles. Various characteristics of an attribute are exemplified in the figure as follows:

- Date of birth and address are optional atomic attributes (shown as empty circles).
- Phone is an optional multi-valued attribute (shown as an empty double circle).
- Age is an optional atomic attribute that is derived by subtracting date of birth from the current date (shown as a dotted empty circle).
- SSN is a mandatory atomic attribute (shown as a dark circle) that serves as a unique identifier. Note that attribute names of unique identifiers are underlined.
- Pat_Name is a mandatory composite attribute composed of the atomic attributes first name, middle initial, and last name. While first name and last name are mandatory atomic attributes, middle initial is an optional atomic attribute.
- Pat_ID is a mandatory composite attribute and serves as a second unique identifier.
- Medical Profile is a complex attribute.

Observe that in ER diagrams, a rectangular box is used to represent an entity type whereas a circle symbolizes an attribute. The atomic attributes that make up a composite attribute are attached to the circle that represents the composite attribute. Dark circles represent mandatory attributes while empty circles signify optional attributes. The name of the attribute appears adjacent to the circle. A multi-valued attribute (e.g., **Phone** and **Allergy**) is shown by a double circle and a derived attribute (e.g., **Age**) is shown by a dotted circle.

2.3.3 Relationship Types

A **relationship type** is a meaningful association among entity types. The **degree** of a relationship type is defined as the number of entity types participating in that relationship type. A relationship type is said to be binary (or of degree two) when two entity types are involved. Relationship types that involve three entity types (of degree three) are defined as

---

[5] A unique identifier is irreducible when none of its proper subsets are unique identifiers. The concept of irreducibility is discussed in detail in Chapter 6.

[6] In formal terms, a key attribute is a proper subset of a unique identifier, and a non-key attribute is any attribute that is not a subset of a unique identifier.

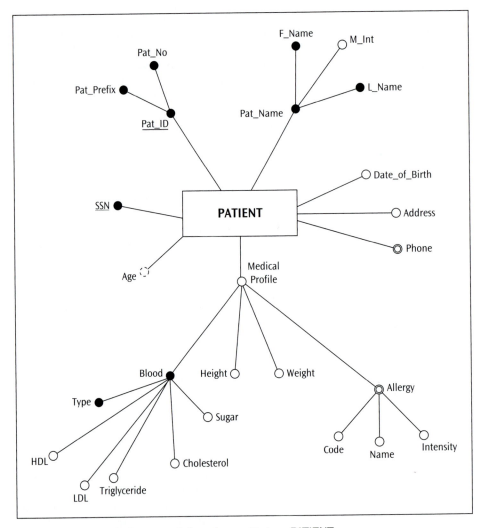

Figure 2.3 A graphical representation of an entity type PATIENT

ternary relationships. While a relationship of degree four can be referred to as a quaternary relationship, a more generalized term beyond ternary relationship is *n-ary* relationship where the degree of the relationship type is *n*. An entity type related to itself is termed a recursive relationship type.

Figures 2.4 through 2.8 illustrate binary, ternary, quaternary, and recursive relationship types respectively where a diamond signifies a relationship type. Figure 2.4 illustrates a binary relationship type *Flies* between Pilot and FLIGHT. A particular pilot flying a specific flight is an instance of the relationship type *Flies*. This relationship instance is often referred to as a relationship. The set of all relationship instances involving pilots and flights is defined as a **relationship set**.

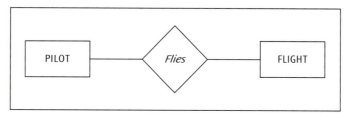

Figure 2.4 A binary relationship

Figure 2.5 illustrates a ternary relationship type *Teaches* among PROFESSOR, SUBJECT, and COURSE. An instance of the relationship type *Teaches* involves a particular Professor teaching a certain Course in a specific Subject. Another relationship instance of *Teaches* could involve the same Professor teaching another Course in the same Subject area. A third instance of the relationship type *Teaches* involves the same Professor teaching a Course in a different Subject area. A fourth relationship instance of *Teaches* involves another Professor teaching a different Course in the previous Subject area. Examples of the four relationship instances described above are as follows.

- Example 1: Professor Einstein teaches the course Optics in Physics.
- Example 2: Professor Einstein teaches the course Mechanics in Physics.
- Example 3: Professor Einstein teaches the course Calculus in Mathematics.
- Example 4: Professor Chu teaches the course Algebra in Mathematics.

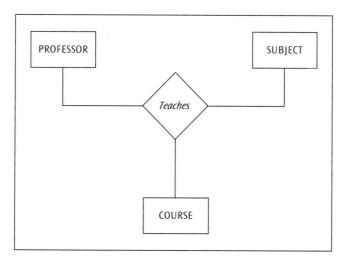

Figure 2.5 A ternary relationship

Diagrammatic representation of the relationship among entity types in terms of relationship instances among the instances of the participant entity types as enumerated in the example above is called an **instance diagram**. Figure 2.6 uses an instance diagram in the form of a vertical ellipse to illustrate these relationship instances. Ternary relationship types are discussed in greater detail in Chapter 5.

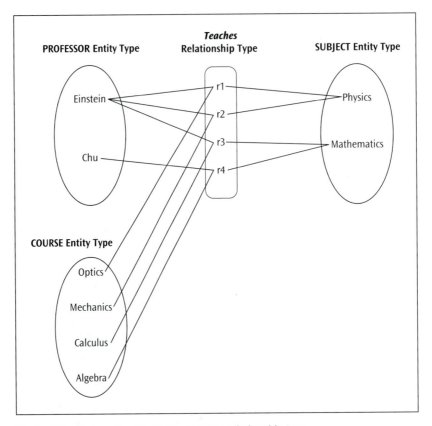

Figure 2.6 Four instances of the *Teaches* relationship type

Figure 2.7 illustrates a quaternary relationship type *Prescribes* among PHYSICIAN, PATIENT, MEDICATION, and ILLNESS. An instance of the relationship type *Prescribes* involves a particular Physician prescribing a specific Medication to treat a particular Patient for a certain Illness. A second instance of the relationship type *Prescribes* might involve the same Physician prescribing the same Medication to treat a different Patient for the same Illness. Examples of these two relationship instances follow.

- Example 1: Dr. Fields prescribes Advil to treat Sharon Moore for a headache.
- Example 2: Dr. Fields prescribes Advil to treat Michelle Li for a headache.

Quaternary relationship types are discussed in greater detail in Chapter 5. Figure 2.8 illustrates a recursive relationship type where a NURSE can act as a supervisor of other nurses. Two instances of this relationship type might involve the same Nurse (e.g., Florence Nightingale) supervising two different Nurses (e.g., Jean Warren and Michael Evans).

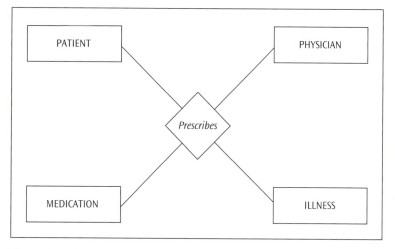

Figure 2.7   A quaternary relationship

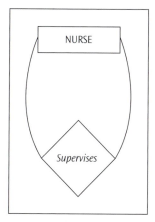

Figure 2.8   A recursive relationship

The participation of an entity type in a relationship type can be indicated by its **role name**. When used in recursive relationship types, role names describe the function of each participation. The use of role names to describe the *Supervises* relationship type that appears in Figure 2.8 is given in Figure 2.9. One participation of NURSE in the *Supervises* relationship is given the role name of Supervisor to reflect the possibility that a nurse may supervise other nurses, and the second participation is given the role name of Supervisee to indicate the possibility that a nurse may be supervised by another nurse.

Role names may also be used when two entity types are associated through more than one relationship type. For example, consider the possibility that the PILOT and FLIGHT entity types are associated through the two relationship types *Flies* and *Scheduled_for*. *Flies* exists in order to indicate the specific pilot in charge of the flight, whereas as *Scheduled_ for* indicates all pilots and co-pilots who are members of the flight crew. As shown in Figure 2.10, the use of role names can clarify the purpose of each relationship.

Foundation Concepts

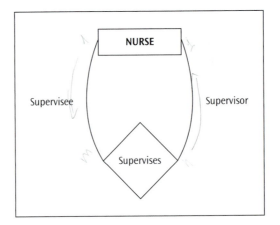

Figure 2.9 Role names in a recursive relationship

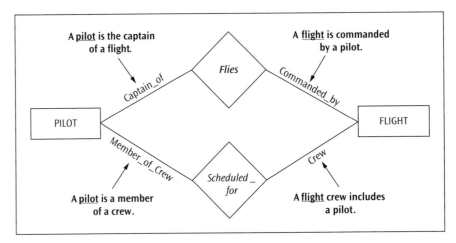

Figure 2.10 Role names in binary relationships

Except for the previous two kinds of relationships, role names are often unnecessary when the relationship specification is unambiguous.[7]

2.3.4 Structural Constraints of a Relationship Type

The data integrity constraints pertaining to relationship types specified in an ER diagram are referred to as the **structural constraints of a relationship type**.[8] Two independent structural constraints together define a relationship type: cardinality constraint (also

---

[7] Often in the specification of a relationship type, role names replace the relationship symbol in commercial software engineering tools.

[8] Sometimes (for example, in UML—Unified Modeling Language) the term "multiplicity" is used instead of structural constraints.

known as mapping cardinality or connectivity) and participation constraint. A relationship type is not fully specified until both structural constraints are explicitly imposed in the model. The context of the binary and recursive relationship types are used here to introduce these structural constraints.

The **cardinality constraint** for a binary relationship type is a constraint that specifies the maximum number of entities of an entity type to which another entity can be associated through a specific relationship set expressed as a ratio. For example, a binary relationship between the two entity types, SALESPERSON and VEHICLE, may possess a 1:n cardinality constraint, meaning that (a) each salesperson entity can be related to many vehicle entities (up to n), and (b) each vehicle entity is related to at most one salesperson entity (see Figure 2.11).

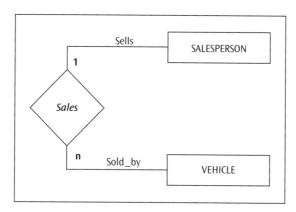

Figure 2.11   Cardinality ratio: An introduction

The relationship type *Sales* between SALESPERSON and VEHICLE is reflected by the *1* on the edge connecting SALESPERSON to *Sales* and the *n* on the edge connecting VEHICLE to *Sales*, indicating that *one* Salesperson sells *many* Vehicles and *one* Vehicle is sold by *one* Salesperson. A relationship of this type is sometimes referred to as a **parent-child relationship (PCR)**.[9]

For a binary relationship between entity types A and B, four mapping cardinalities are possible:

- **m:n**. An entity in A is associated with any number (zero or more) of entities in B and *vice versa*. This is the general form of a cardinality constraint in a binary relationship. An example of an m:n cardinality constraint involves the two entity types EMPLOYEE and CERTIFICATION where each Employee could possess many different Certifications and each Certification could be possessed by many different Employees (see Figure 2.12).
- **1:n**. An entity in A is associated with any number (zero or more) of entities in B; an entity in B, however, is associated with no more than 1 entity of A.

[9] The original source of this acronym is said to be the practitioners' world. Bachmann used it in the data structure diagram that predates the ER model. The term was also prevalent in hierarchial and network data models before the advent of relational data model.

When m takes a value of 1, the general form, m:n, becomes 1:n. The *Sales* relationship type enumerated above is reproduced in Figure 2.13 as an example of 1:n cardinality constraint.

- **n:1**. An entity in A is associated with no more than 1 entity of B. An entity in B, however, is associated with any number (zero or more) of entities in A. This is just a reverse expression of the cardinality constraint, 1:n. The same *Sales* relationship type presented above also serves as an example of n:1 cardinality constraint (see Figure 2.13).

- **1:1**. An entity in A is associated with no more than 1 entity of B; and an entity in B is associated with no more than 1 entity of A. When both m and n take a value of 1, the general form, m:n, becomes 1:1. A 1:1 cardinality constraint would exist between the two entity types EMPLOYEE and COMPUTER if each Employee is assigned no more than a single Computer and each Computer is assigned to no more than a single Employee (see Figure 2.14).

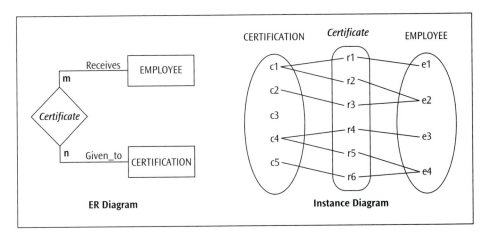

Figure 2.12 An illustration of cardinality ratio of m:n

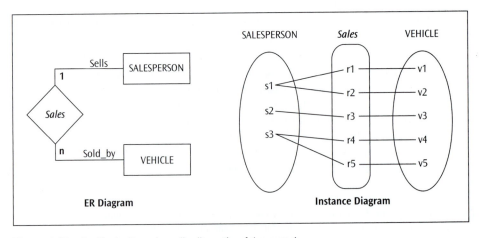

Figure 2.13 An illustration of cardinality ratio of 1:n or n:1

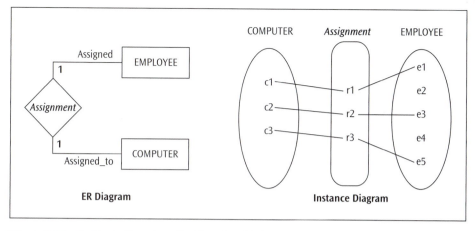

Figure 2.14 An illustration of cardinality ratio of 1:1

The cardinality constraint reflects the **maximum cardinality** of the entity types participating in the binary relationship type. The maximum cardinality indicates the maximum number of relationship instances in which an entity participates. For example, in the *Sales* relationship type in Figure 2.13, a Salesperson entity is connected to a maximum of n *Sales* relationship instances, while a Vehicle entity is connected to a maximum of 1 *Sales* relationship instance.

The **participation constraint** for an entity type in a binary relationship type is based on whether, in order to exist, an entity of that entity type needs to be related to an entity of the other entity type through this relationship type. Participation can be total or partial. If, in order to exist, every entity must participate in the relationship, then participation of the entity type in that relationship type is **total participation**. On the other hand, if an entity can exist without participating in the relationship, then participation of the entity type in that relationship type is **partial participation**. Total and partial participation are also commonly referred to as **mandatory** and **optional** participation, respectively. For example, if every Salesperson must have sold at least one Vehicle, then there is total participation of SALESPERSON in the *Sales* relationship type. Instead, if a Salesperson need not have sold any Vehicle, then there is partial participation of a SALESPERSON in the *Sales* relationship type. Since the participation constraint specifies the minimum number of relationship instances in which each entity can participate, it reflects the **minimum cardinality** of an entity type's participation in a relationship type.

Figure 2.15 contains several versions of the participation constraint involving SALESPERSON and VEHICLE. In Figure 2.15a, the bar next to the VEHICLE entity type indicates that each SALESPERSON must sell at least one VEHICLE, while in Figure 2.15b the oval next to the VEHICLE entity type indicates that a SALESPERSON may or may not sell a vehicle. In summary, the participation of the entity type SALESPERSON in the *Sales* relationship type is conveyed by looking across the *Sales* relationship to the VEHICLE side. What about the participation of VEHICLE in the *Sales* relationship? In Figure 2.15a, a VEHICLE *must* be sold by a SALESPERSON, while in Figure 2.15b it is possible for a VEHICLE to exist without having been sold by a SALESPERSON.

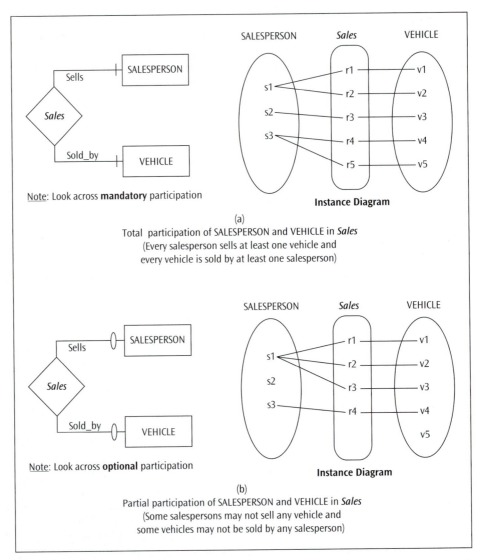

Note: Look across **mandatory** participation

(a)
Total participation of SALESPERSON and VEHICLE in *Sales*
(Every salesperson sells at least one vehicle and
every vehicle is sold by at least one salesperson)

Note: Look across **optional** participation

(b)
Partial participation of SALESPERSON and VEHICLE in *Sales*
(Some salespersons may not sell any vehicle and
some vehicles may not be sold by any salesperson)

Figure 2.15   Examples of the participation constraint

Total participation of an entity type in a relationship type is also called **existence dependency** of that entity type in that relationship type. Accordingly, in Figure 2.15a SALESPERSON has existence dependency on the *Sales* relationship type, whereas in Figure 2.15b SALESPERSON does not have existence dependency on the *Sales* relationship type.

The structural constraints for the relationship between the SALESPERSON and VEHICLE appear in Figure 2.16. In each of Figures 2.16a through 2.16d, a SALESPERSON can sell many VEHICLEs. In Figure 2.16a, a SALESPERSON must sell at least one VEHICLE and a VEHICLE must be sold by a SALESPERSON. In Figure 2.16b a SALESPERSON may or may not sell a VEHICLE and a VEHICLE need not be sold by a SALESPERSON. Figure 2.16c illustrates the requirement that a SALESPERSON must sell at least one VEHICLE, while a VEHICLE need not be sold by a SALESPERSON. Finally, in Figure 2.16d a SALESPERSON may or may not sell a VEHICLE but a VEHICLE must be sold by a SALESPERSON.

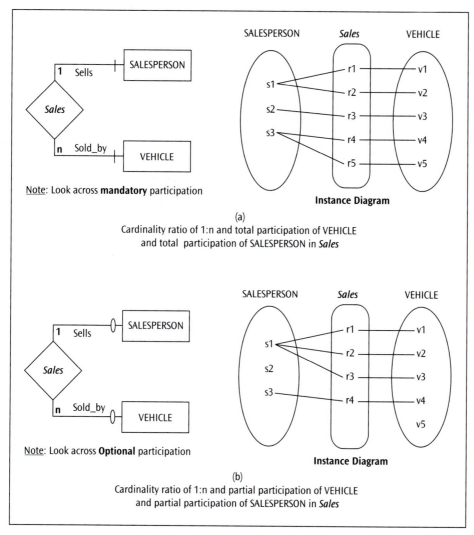

Figure 2.16 Cardinality ratio and participation constraints for a relationship

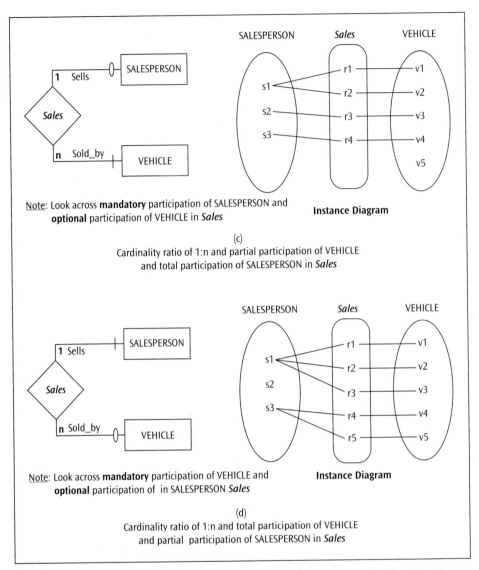

Figure 2.16 Cardinality ratio and participation constraints for a relationship (continued)

The structural constraints for the recursive relationship involving the NURSE entity type appear in Figures 2.17a and 2.17b. These examples reflect a cardinality ratio of 1:n, meaning that a NURSE supervises many other NURSEs—at most n. Based on the directional nature of the cardinality ratio, this *automatically* implies that a NURSE is *supervised* by no more than one NURSE. The oval and the hash (ǀ) introduced in Figure 2.15 are used to describe the participation constraints. In Figure 2.17a, the oval on the right-hand side indicates that a NURSE may or may not supervise other NURSEs, while the oval on the left-hand side indicates that a NURSE need not be supervised (i.e., there are NURSEs who are not supervised by another NURSE). The instance diagram in Figure 2.17a illustrates the partial participation of NURSE as Supervisor and as Supervisee in the *Supervises* relationship type. For example, observe how Nurse n1 is the supervisor of Nurses n2 and n3 but is not supervised by another Nurse. On the other hand, Nurse n2 is a supervisee of Nurse n1 (i.e., is supervised by Nurse n1) but is not a supervisor of any Nurses.

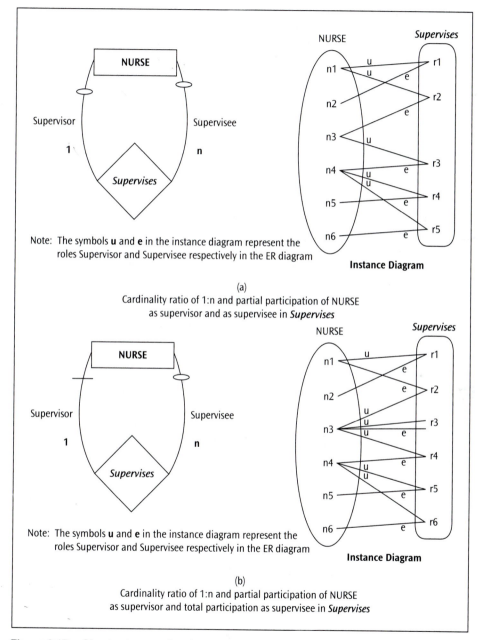

Figure 2.17 Structural constraints for recursive relationships: cardinality ratio of 1:n

In Figure 2.17b, the oval on the right-hand side indicates that a NURSE may or may not supervise other NURSEs. However, the hash on the left-hand side indicates that every NURSE must be supervised by another NURSE. Observe that in the instance diagram, Nurse n3 is the supervisor of Nurses n1, n3, and n4 (i.e., Nurse n3 acts as his/her own supervisor). In addition, note how each of the six relationship instances pertains to a different one of the six nurses—indicating that each nurse is supervised.

A recursive relationship type with an m:n cardinality ratio appears in Figure 2.18. In this relationship, a Course may not only serve as a prerequisite for many other courses but may also have many other courses as its prerequisites. The instance diagram in Figure 2.18a illustrates this *Prerequisite* relationship type by showing a duplicate copy of the COURSE entity type. Observe how Course c2 has Courses c1, c3, and c4 as its prerequisites. In addition, note that Course c3 is a prerequisite of Course c2 as well as of Course c5, while Course c2 is not a prerequisite for any other courses. Moreover, note that Courses c1 and c3 have no prerequisites. The instance diagram in Figure 2.18b illustrates the same relationships among courses in a truly recursive sense through the use of only one COURSE entity type.

The attributes described in Section 2.3.1 can also be assigned to relationship types. For example, consider the 1:1 relationship type *Heads* between PROFESSOR and DEPARTMENT (see Figure 2.19). In a semantic sense, the attribute **Start_dt** belongs to the relationship type *Heads* because the value of this attribute is determined based on when a particular Professor assumed the duties of the Head of a particular Department.

In a technical sense, the attribute **Start_dt** can be shown either in PROFESSOR or in DEPARTMENT because the cardinality of the relationship is 1:1.

In a 1:n relationship type, attributes of the relationship can alternatively be shown only as attributes of the child entity type in the relationship and not of the parent entity type. Consider the example shown in Figure 2.20. Here **Rent** is shown as an attribute of the relationship type *Occupies* in order to convey the semantics of the relationship. However, in a technical sense the attribute **Rent** belongs to the entity type STUDENT (Figure 2.21). If **Rent** is included as an attribute of DORMITORY, first of all, it will have to be a multivalued attribute; further, there will be no way to identify the rent paid by each individual student living in the dormitory.

Attributes of m:n relationship types cannot be shown anywhere other than as an attribute of the relationship type. For example, consider the attribute **Cost** in the *Imports* relationship type in Figure 2.22. Since the cost incurred by a vendor to import a product is determined by a vendor-product combination, the cost can only be specified as an attribute of *Imports* and not as an attribute of either VENDOR or PRODUCT.

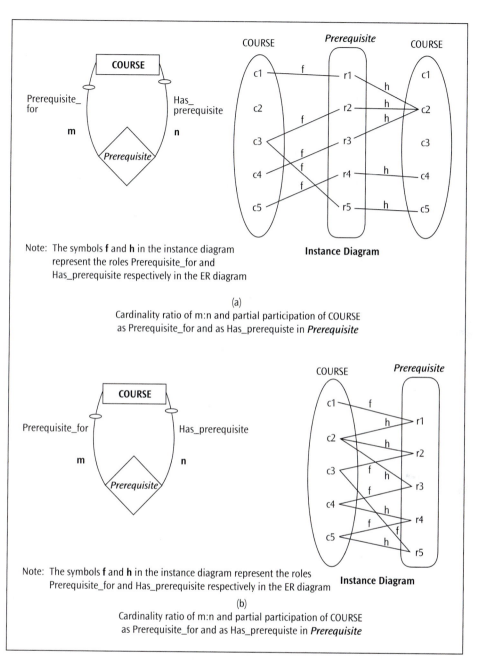

Note: The symbols **f** and **h** in the instance diagram represent the roles Prerequisite_for and Has_prerequisite respectively in the ER diagram

(a)

Cardinality ratio of m:n and partial participation of COURSE as Prerequisite_for and as Has_prerequiste in *Prerequisite*

Note: The symbols **f** and **h** in the instance diagram represent the roles Prerequisite_for and Has_prerequisite respectively in the ER diagram

(b)

Cardinality ratio of m:n and partial participation of COURSE as Prerequisite_for and as Has_prerequiste in *Prerequisite*

Figure 2.18 Structural constraints for recursive relationships: cardinality ratio of m:n

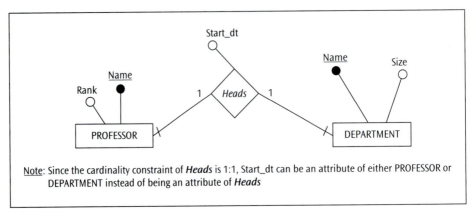

Note: Since the cardinality constraint of **Heads** is 1:1, Start_dt can be an attribute of either PROFESSOR or DEPARTMENT instead of being an attribute of **Heads**

Figure 2.19 An attribute of a relationship in a 1:1 relationship type

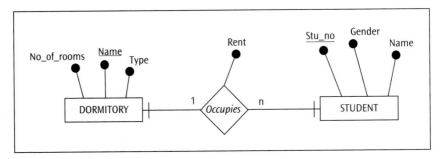

Figure 2.20 An attribute of a relationship in a 1:n relationship type

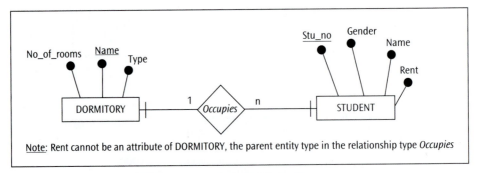

Note: Rent cannot be an attribute of DORMITORY, the parent entity type in the relationship type *Occupies*

Figure 2.21 **Rent** as an attribute of the entity type STUDENT instead of being an attribute of the relationship type *Occupies* as in Figure 2.20

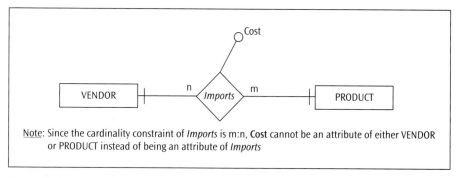

Note: Since the cardinality constraint of *Imports* is m:n, **Cost** cannot be an attribute of either VENDOR or PRODUCT instead of being an attribute of *Imports*

Figure 2.22   An attribute of a relationship in an m:n relationship type

2.3.5 Base Entity Types and Weak Entity Types

As a brief review of the relationship between entities and attributes, recall that a set of attributes gives structure to an entity type. When each attribute is given a value (at least one attribute should have a non-null value), an entity of this entity type is created. A second set of values for each of the attributes that constitute the entity type results in a second entity of this entity type. A collection of all entities of this entity type becomes an entity set of this entity type.

An entity type where the entities have independent existence (that is, each entity is unique) is referred to as a **base (or strong) entity type**. For example, a set of attributes **Part#**, **Part_name**, **Color**, **Cost**, **Price**, and so on constitute a *base* entity type INVENTORY if there are no duplicate entities in that entity set. Independent existence of each entity of a base entity type is accomplished via the uniqueness of value for an attribute, atomic or composite, in the entity set. INVENTORY, for example, is a base entity type because **Part#** is different for each entity. No two entities in the INVENTORY entity set have the same **Part#** even though any two or more entities may have the same part name and/or color, and/or cost, and/or price. In other words, **Part#** is a unique identifier of INVENTORY. In another base entity type, the identification of the independent existence of entities in the entity set may require a composite attribute. For example, **Survey#**, **Lot#**, **Block#**, and **Plot#** together may be required for establishing the independent existence of entities in a base entity type, PROPERTY. In short, the property of independent existence in a base entity type implies presence of a unique identifier in the entity type. Incidentally, some base entity types may have more than one unique identifier. For example, the composite attribute **[State, License_plate#]** as well as the atomic attribute **Vehicle_id#** are unique identifiers of an automobile.

ER modeling grammar allows for the conceptualization of an entity type that does not have independent existence, that is, an entity type that does not have its own unique identifier. Such an entity type is called a **weak entity type**. Presence of duplicate entity instances of this entity type in the entity set is legal. Independent existence of a weak entity can be achieved only by borrowing part or all of its unique identifier from another entity (or other entities) through one or more of a special type of relationship called an *identifying relationship*. Such a dependence of a weak entity on other entities is called *identification dependency*. For example, several apartments in an apartment complex may have exactly the same properties (e.g., size, number of bedrooms, number of bathrooms, etc.)

including the apartment number. That is, the same apartment number may appear in different buildings of the complex. Therefore, the unique identification of an apartment is impossible without associating it with a building, assuming that the buildings have independent existence. Semantically this means that an apartment cannot exist apart from the building in which it is located. Figure 2.23 illustrates the relationship between a BUILDING entity type and an APARTMENT entity type and shows their respective data sets. BUILDING is a base entity type because it has a unique identifier, **Bldg_no**. APARTMENT, however, is a weak entity type because it does not have a unique identifier of its own. In order to distinguish a base entity type from a weak entity type, in Figure 2.23, the weak entity type is shown as a double rectangular box. To signify the identification dependency of APARTMENT on BUILDING, a double diamond is used to portray the identifying relationship type, *Contains*. Recall that a single diamond is used to represent a "regular" relationship between entity types.

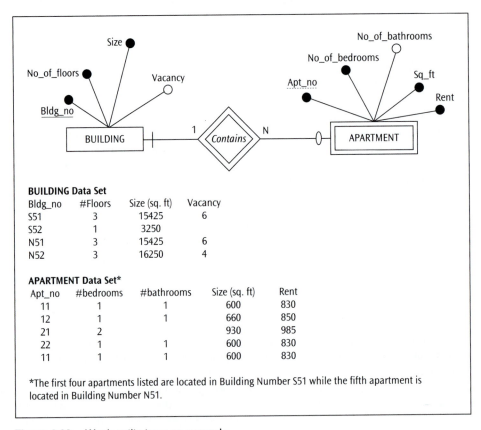

BUILDING Data Set

Bldg_no	#Floors	Size (sq. ft)	Vacancy
S51	3	15425	6
S52	1	3250	
N51	3	15425	6
N52	3	16250	4

APARTMENT Data Set*

Apt_no	#bedrooms	#bathrooms	Size (sq. ft)	Rent
11	1	1	600	830
12	1	1	660	850
21	2		930	985
22	1	1	600	830
11	1	1	600	830

*The first four apartments listed are located in Building Number S51 while the fifth apartment is located in Building Number N51.

Figure 2.23   Weak entity type: an example

An attribute, atomic or composite, in a weak entity type (which in conjunction with a unique identifier of the parent entity type in the identifying relationship type uniquely identifies weak entities), is called the **partial key** of the weak entity type and is denoted figuratively by a dotted underline. The partial key of a weak entity type is sometimes referred to as a **discriminator**. The sample data shown in the BUILDING and APART-MENT data sets in Figure 2.23 illustrate how **Apt_no** is unique for those apartments within Building S51 but is not unique within the APARTMENT data set. In order to identify each apartment in the apartment complex, the unique identifier of the BUILDING in which the apartment is located (**Bldg_no**) must be concatenated with the partial key of APARTMENT (**Apt_no**).

Recall that existence dependency of an entity type in a relationship type implies mandatory participation of all entities in the entity set of the said entity type in that relationship (see Figures 2.15 and 2.16 for examples). By virtue of being identification-dependent on the identifying parent entity type, a weak entity type is always also existent-dependent on the corresponding identifying relationship type. In the previous example, APARTMENT has existence dependency on *Contains* because APARTMENT has identification dependency on BUILDING through the identifying relationship type, *Contains*.

Additional examples of weak entity type appear in Figures 2.24 through 2.27. Observe in Figure 2.24, that the weak entity type INTERNSHIP has a collective identification dependency on both COMPANY and STUDENT via two independent identifying relationships. Sample data for Figure 2.24 is shown in Figure 2.25.

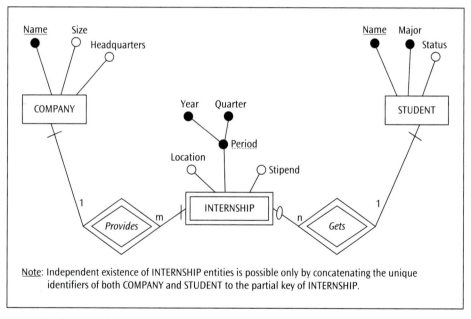

Note: Independent existence of INTERNSHIP entities is possible only by concatenating the unique identifiers of both COMPANY and STUDENT to the partial key of INTERNSHIP.

Figure 2.24   A weak entity type with multiple identifying parents: an example

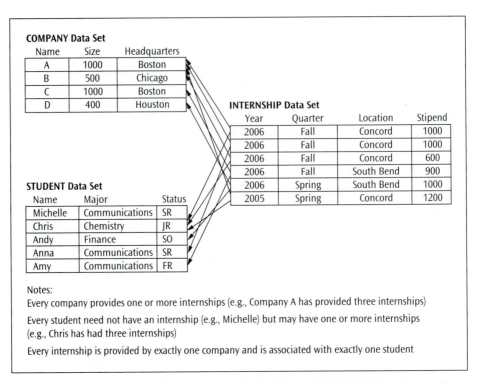

COMPANY Data Set

Name	Size	Headquarters
A	1000	Boston
B	500	Chicago
C	1000	Boston
D	400	Houston

INTERNSHIP Data Set

Year	Quarter	Location	Stipend
2006	Fall	Concord	1000
2006	Fall	Concord	1000
2006	Fall	Concord	600
2006	Fall	South Bend	900
2006	Spring	South Bend	1000
2005	Spring	Concord	1200

STUDENT Data Set

Name	Major	Status
Michelle	Communications	SR
Chris	Chemistry	JR
Andy	Finance	SO
Anna	Communications	SR
Amy	Communications	FR

Notes:

Every company provides one or more internships (e.g., Company A has provided three internships)

Every student need not have an internship (e.g., Michelle) but may have one or more internships (e.g., Chris has had three internships)

Every internship is provided by exactly one company and is associated with exactly one student

Figure 2.25 Sample data for the COMPANY, STUDENT, and INTERNSHIP entity types in Figure 2.24

As a second example, Figure 2.26 portrays a weak entity type, TRAINING_PROGRAM, that is identification-dependent on another weak entity type, INTERNSHIP. Sample data sets are shown in Figure 2.27. Note that in both cases, the identification-dependent weak entity type is also existent-dependent on the identifying relationship type.

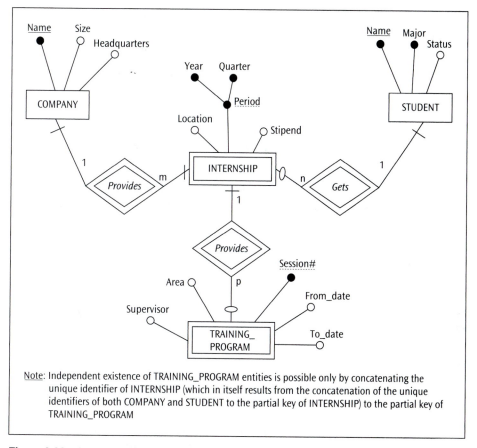

Figure 2.26 A weak entity type TRAINING_PROGRAM identity-dependent on another weak entity type, INTERNSHIP

INTERNSHIP Data Set

Year	Quarter	Location	Stipend
2006	Fall	Concord	1000
2006	Fall	Concord	1000
2006	Fall	Concord	600
2006	Fall	South Bend	900
2006	Spring	South Bend	1000
2005	Spring	Concord	1200

TRAINING_PROGRAM Data Set

Session#	Supervisor	Area	From_date	To_date
3	Ken	Microsoft Office	12/2/2006	12/7/2006
3	Ken	Open Office	12/10/2006	12/13/2006
2	Ken	Microsoft Office	12/2/2006	12/7/2006
3	John	Microsoft Project	12/10/2006	12/12/2006

Figure 2.27 Sample data for the INTERNSHIP and TRAINING_PROGRAM entity types of Figure 2.26

2.4 Data Modeling Errors

Research in conceptual modeling indicates that modeling errors are not uncommon, especially among novice data modelers (Batra and Davis, 1992). This section presents a few common data modeling errors using short vignettes. These errors can be of two kinds: **semantic errors** that arise from misinterpretations of the requirements specification, and **syntactic errors** that violate the grammar of the modeling language. The former entails incorrect imposition of a business rule implicit in the requirements specification and is more difficult to identify since conceptual modeling is not an exact science. Attention to details, repeated practice and experience over the long term are means by which one can minimize semantic errors in conceptual modeling. The latter (syntactic errors), are relatively easy to avoid by simply knowing the rules of the grammar of the modeling language—in this case the ER modeling grammar. How to avoid semantic errors in conceptual modeling is a topic addressed across all chapters of Part I of this book. The following examples articulate some possible syntactic and semantic errors in conceptual modeling using the ER modeling grammar, so that a student can learn to avoid similar modeling errors.

2.4.1 Vignette 1

This vignette is a slightly expanded version of the example about a university's academic program presented in Section 2.3.2.

There are several colleges in the university. Each college has a name, location, and size. A college offers many courses over four college terms or quarters—Fall, Winter, Spring, and Summer—during which one or more of these courses are offered. Course#, name, and credit hours describe a course. No two courses in any college have the same course#; likewise no two courses have the same name. Terms are identified by year and quarter, and contain enrollment numbers. Courses are offered during every term. The college also has several instructors. Instructors teach; that is why they are called instructors. Often, not all instructors are scheduled to teach during all terms; but every term has some instructors teaching. Also, the same course is never taught by more than one instructor in a specific term. Further, instructors are capable of teaching a variety of courses offered by the college. Instructors have a unique employee ID and their name, qualification, and experience are also recorded.

To begin with, COLLEGE may be modeled as an entity type since a collection of attributes, namely **Name**, **Location**, **Size**, **Course**, and **Instructor**, seem to cluster under this title. Figure 2.28a portrays the COLLEGE entity type using the ER modeling grammar. The ER diagram at this point is syntactically correct. However, there are a couple of problems with reference to the semantics conveyed by this entity type. COLLEGE in Figure 2.28a is modeled as an entity type with attributes that indicate that every college has one name, one location, one size, one course, and one instructor. Of course, the attributes **Course** and **Instructor** are correctly shown as composite attributes with their appropriate content of atomic attributes. Nonetheless, this is not quite correct according to the semantics conveyed in the requirements specification. A college indeed offers many courses and also has several instructors. Therefore, these two attributes (**Course** and **Instructor**), must be modeled as multi-valued attributes, as shown in Figure 2.28b. Also, the attribute

Name is duplicated in COLLEGE, one referring to the name of a college and the other referring to the name of a course. Duplicate attribute names within an entity type are semantically incorrect since the only reference point for the attributes is the entity type. Thus one of the attributes labeled **Name** is changed to **Course_name**, as shown in Figure 2.28b. The ERD at this point is both semantically and syntactically correct.

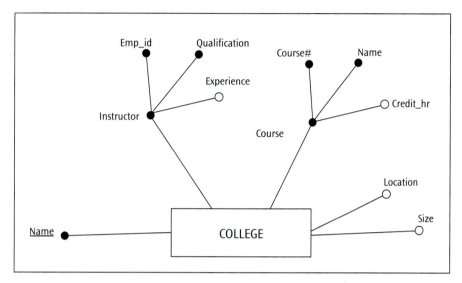

Figure 2.28a   The entity type COLLEGE

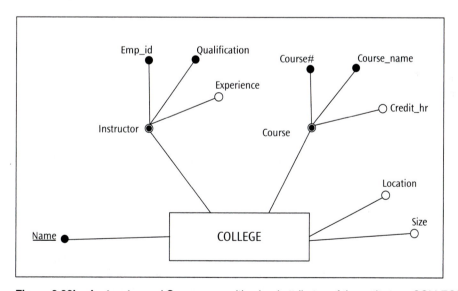

Figure 2.28b   **Instructor** and **Course** as multi-valued attributes of the entity type COLLEGE

Next we notice that instructors are capable of teaching a variety of courses. This indicates a relationship between instructors and courses. Given the current portrayal of the COLLEGE entity type, the relationship between instructors and courses can be modeled as shown in Figure 2.28c. While the data model shown in Figure 2.28c correctly conveys the intended semantics, the model violates a syntactic rule of the ER modeling grammar. While all attributes of an entity type are implicitly related to each other, an explicit relationship between attributes of an entity type independent of the entity type is not permitted in the ER modeling grammar. This syntactic error can be corrected only by modeling COURSE and INSTRUCTOR as independent entity types related to the COLLEGE entity type, and then establishing the relationship between the INSTRUCTOR and COURSE entity types. The ER diagram corrected accordingly is shown in Figure 2.28d. Since both COURSE and INSTRUCTOR have unique identifiers, they both are modeled as base entity types. Further, since courses are offered every term and there are four terms, **Term** is also modeled here as an optional multivalued attribute of COURSE; the optional property of the attribute allows for the possibility that some courses are not offered at all even though they may still be on the books. However, when **Term** has a value, it must have the quarter made up of **Year** and **Qtr#**.

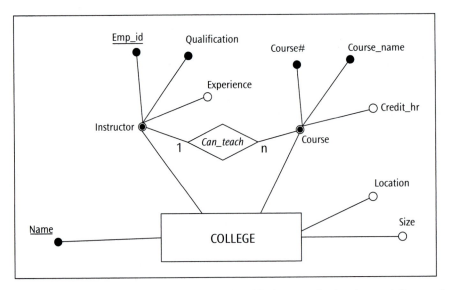

Figure 2.28c A syntactically incorrect relationship between **Instructor** and **Course** in COLLEGE

Next, **Course_name**, according to the requirements specification, is also a unique identifier of COURSE and is not defined so in Figure 2.28d. The underlining of **Course_name** as in Figure 2.28e corrects this error. We further notice that instructors are scheduled to teach during specific terms. The ER diagram in Figure 2.28e portrays this relationship. Once again, while semantically correct in expressing the notion of a relationship between terms and instructors, the ER modeling grammar does not permit modeling a relationship between an entity type and an attribute of an entity type. This grammatical error in the model script can be corrected by expressing TERM as a base entity type. The ER diagram revised accordingly is shown in Figure 2.28f.

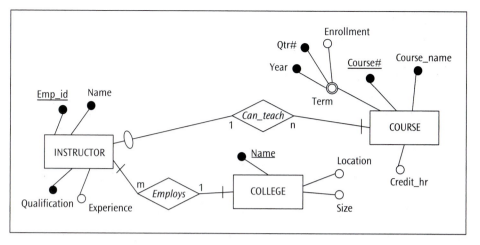

Figure 2.28d A syntactically correct relationship between INSTRUCTOR and COURSE

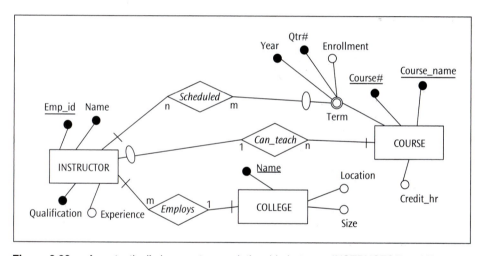

Figure 2.28e A syntactically incorrect m:n relationship between INSTRUCTOR and **Term**

The conceptual model at this point captures the fact that instructors are scheduled to teach in the four terms and the instructors are capable of teaching several courses. However, the fact that courses are offered over the four terms and in each term one or more of the courses are offered is yet to be incorporated in the ER diagram. More importantly, the business rule that *the same course is never taught by more than one instructor in a specific term* is not incorporated in the conceptual data model either. An m:n relationship between COURSE and TERM will capture the semantics of the first statement. This is shown in Figure 2.28g. On another note, the unique identifier of TERM is defined as the combination of **Year** and **Qtr#**. However, in Figure 2.28f these two attributes are shown as two independent unique identifiers of the entity type TERM. This is a syntactic error and is corrected in Figure 2.28g by first constructing a composite attribute **Quarter** whose atomic components are **Year** and **Qtr#** and then labeling (underlining) **Quarter** as the unique identifier of TERM.

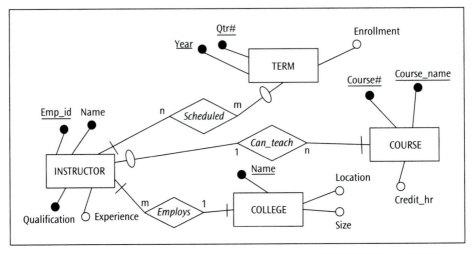

Figure 2.28f A syntactically correct m:n relationship between INSTRUCTOR and TERM

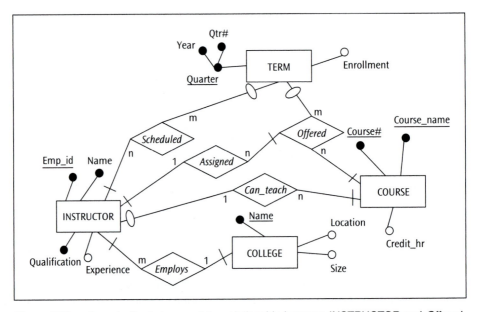

Figure 2.28g A syntactically incorrect 1:n relationship between INSTRUCTOR and *Offered*

In addition, an attempt is made to convey the business rule that *the same course is never taught by more than one instructor in a specific term* by establishing a relationship between INSTRUCTOR and *Offered* (see Figure 2.28g). This relationship does indeed correctly convey the semantics that a course offered in a term is assigned to only one instructor, while any instructor may teach several courses in the same term, and the same and other courses in other terms too. However, there is a violation of the ER modeling grammar in the expression of the relationship type *Assigned* since it relates an entity type, INSTRUCTOR, with a relationship type, *Offered*—i.e., a diamond connecting to another diamond. The solution for this syntactic error is not immediately obvious. To get a better understanding of this scenario, let us review the ER diagram that appears in Figure 2.24. Here, it is possible to conceptualize INTERNSHIP as an m:n relationship type between COMPANY and STUDENT. However, since the objective of this example is to demonstrate that a weak entity type can have multiple identifying parents, INTERNSHIP is modeled as a weak entity type. Similarly, the m:n *Imports* relationship type shown in Figure 2.22 could have been modeled as a weak entity type IMPORTS with VENDOR and PRODUCT as its identifying parents. The transformation of an m:n relationship to a weak entity type with multiple identifying parents is more formally explained in the next chapter (see Section 3.2.4). Following this line of logic, the m:n relationship type *Offered* may be modeled as a weak entity type with TERM and COURSE as its identifying parents. Then the expression of *Assigned* as a relationship type relating the base entity type INSTRUCTOR and the weak entity type OFFERING will accomplish expression of the intended semantics without violating the rules of the ER modeling grammar. The revised ER diagram is shown in Figure 2.28h.

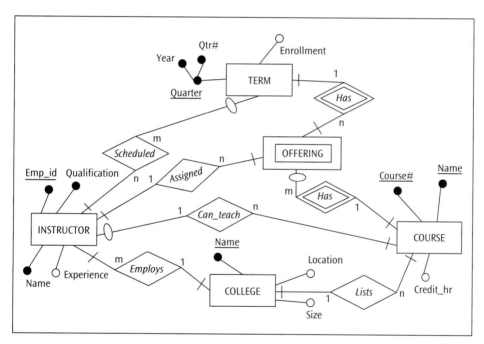

Figure 2.28h A syntactically correct m:n relationship between INSTRUCTOR and OFFERING

As a final note, the scenario described in the vignette is not quite complete in its specification. The reader may find it a good exercise to first identify all business rules implicitly expressed in the vignette and also then list the ambiguities present in the description of the scenario due to incomplete specification.

2.4.2 Vignette 2

Let us revisit one of the scenarios modeled as a weak entity type INTERNSHIP in Section 2.3.5 (see Figure 2.24) and construct the story underlying the scenario. Companies have work and students want work experience. Internships for students in companies emerge from the confluence of these two needs. Thus, the concept of INTERNSHIP in this scenario is not conducive for modeling as an independent base entity type. Instead, internships can be viewed as a relationship between companies and students. All companies in our scenario offer internship opportunities; the companies that do not will not be part of our scenario. However, all students need not get internships—some may not qualify for internships for various reasons. Thus, *Internship*, can be modeled as an m:n relationship between COMPANY and STUDENT with total participation of COMPANY and partial participation of STUDENT in the relationship type *Internship*, as shown in Figure 2.29a. Dependent existence of internships on companies and students can also be captured by modeling INTERNSHIP as a weak entity type with both COMPANY and STUDENT as its identifying parents. This design alternative, shown earlier in Figure 2.24, enabled explication of a weak entity type with multiple identifying parents in Section 2.3.5. The two designs are semantically equivalent and syntactically correct. In Chapter 3, the former conceptualization (*Internship* as a relationship type) is suggested as more palatable for user interaction, while the latter (INTERNSHIP as a weak entity type) is presented as a designer's view.

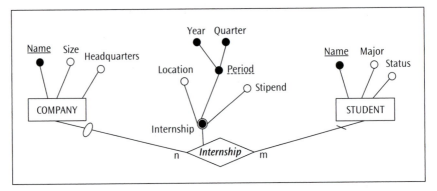

Figure 2.29a *Internship* (Figure 2.24) modeled as a relationship type between COMPANY and STUDENT

Next, suppose that some of the internships include various training sessions. If training programs already exist for employees, customers, etc., independent of internships exclusively for the students, it is quite conceivable to model TRAINING_PROGRAM as a base entity type. Instead, let us limit the scope of our scenario to training programs developed exclusively for internships. This is reflected in Figure 2.26 where TRAINING_PROGRAM is modeled as a weak entity type identification dependent on INTERNSHIP. The objective in Section 2.3.5 was to demonstrate how a weak entity type can have another weak entity type as its identifying parent. How would we model this scenario when *Internship* is depicted as a relationship type, as in Figure 2.29a? Figure 2.29b presents a design that captures the semantics of this scenario. However, this design violates the ER modeling grammar rules, namely, a relationship can exist only between entities—i.e., a diamond cannot connect anything other than rectangles. Therefore, the design depicted in Figure 2.29b is syntactically incorrect according to the ER modeling grammar. The only correct solution for this scenario is the model depicted in Figure 2.26. A more elegant solution is presented in Chapter 5.

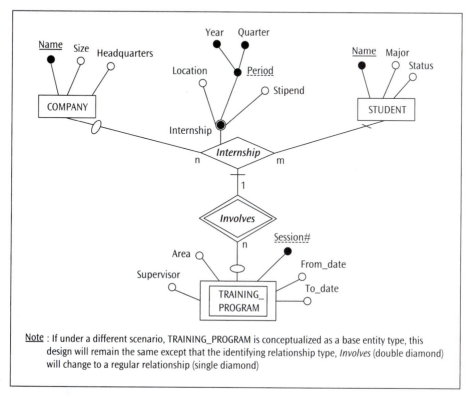

Figure 2.29b Syntactically incorrect relationship between *Internship* and TRAINING_ PROGRAM

2.4.3 Vignette 3

Widget USA is a manufacturer of Widgets located in Whitefield, IN. A widget is an intricate assembly of numerous parts. The assembly and the manufacture of a few intricate parts are done by Widget USA at its small, but highly sophisticated plant in Whitefield, IN. All the other parts of the widget are outsourced to various vendors. Some of the vendors supply more than one part; but a part is supplied by only one vendor. A part has a unique part number, and a unique name. Therefore, it is enough if a value for one of these two attributes is present for a given part. Other attributes of parts available are: size, weight, color, design, and quality standard. Manufactured parts have a cost and raw material, and follow a production plan, while purchased parts have a price and delivery schedule. Production plans have machine sequence, timetable, and capacity, and are identified by a unique plan number. Vendors are identified by a vendor name. Other information available on a vendor includes its address, phone number, and vendor rating.

At the outset, it is possible to identify VENDOR, PART, and PRODUCTION_PLAN as base entity types and establish relationships among them, as shown in Figure 2.30a. The cardinality constraint between VENDOR and PART is 1:n, indicating that a vendor supplies many parts and a part is supplied by only one vendor. Also, the participation constraints (look across) indicate that every vendor must be a supplier of part(s); however, all parts are not supplied by vendors since some are manufactured. A similar relationship exists between PRODUCTION_PLAN and PART since some but not all parts are manufactured by Widget USA in its own plant. The ER diagram is syntactically correct. There are two commonly committed semantic errors seeded in the ER diagram in Figure 2.30a in order to point out that careful scrutiny of the details present in the requirements specification is crucial to accurate data modeling. First, note that a vendor has telephone numbers (plural) indicating that **Phone#** in the VENDOR entity type should be a multi-valued attribute. Also, every part has a unique part number and a unique name implying that PART has two unique identifiers. Concatenating **Part#** and **Name** to form a single composite attribute and labeling that as a single unique identifier of PART is incorrect. These errors are corrected in Figure 2.30b. The fact that an attribute called **Name** occurs in the entity type VENDOR as well as the entity type PART is not an error.

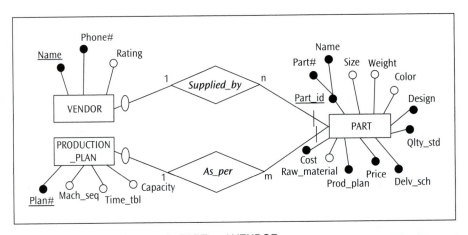

Figure 2.30a Semantic errors in PART and VENDOR

The conceptual model in Figure 2.30b is still semantically incomplete. What is missing here can be discovered only when the scenario described in the vignette is systematically analyzed for all explicit and implicit business rules conveyed by the story. For instance, a business rule that *a manufactured part is not purchased and vice versa* is implicitly stated in the story. The ER diagram in Figure 2.30b does not capture this business rule.

It must be noted that the ER modeling grammar is not rich enough always to capture all specified data requirements. Under these circumstances, the business rules that cannot be implemented in the ER diagram will have to be carried forward in the design in some other form—e.g., text form. A more formal method to address this issue is discussed in

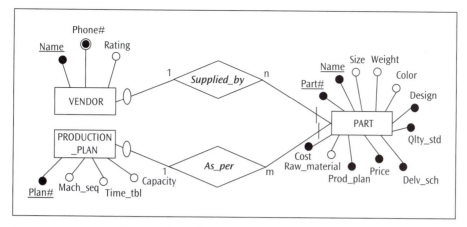

Figure 2.30b Semantic errors in PART and VENDOR in Figure 2.30a corrected

Chapter 3. That said, this particular business rule that *a manufactured part is not pur-chased and vice versa* can indeed be modeled in an ER diagram in several different ways. Suppose we split PART into two different entity types, namely MANUFACTURED_PART and PURCHASED_PART, as shown in Figure 2.30c. Does this solve the problem? This design certainly offers an opportunity to separate manufactured parts from purchased parts and relate them to production plans and vendors respectively. However, the data model does not explicitly prohibit a manufactured part from also being a purchased part and vice versa. In other words, inclusion of the same part in both entity sets, even if done inadvertently, will not be considered an error by this design. We also notice that a significant number of attributes are duplicated across MANUFACTURED_PART and PURCHASED_PART. Does this mean data redundancy? The answer is "No." As long as the manufactured parts are included in the MANUFACTURED_PART entity set and purchased parts are present in the PURCHASED_PART entity set, mere duplication of attribute names does not result in data redundancy. Also, observe that the participation constraint of the entity type MANUFACTURED_PART in *As_per* relationship type (look across) and the participation constraint of the entity type MANUFACTURED_PART in *Supplied_by* relationship type in this design are mandatory—different from the design depicted in Figure 2.30b.

In short, the semantic incompleteness of both designs (Figure 2.30b and 2.30c) with respect to the business rule that *a manufactured part is not purchased and vice versa* is the same. In both cases, the business rule should be carried forward in the design pro-cess in, say, a text form. Additional ER modeling grammar constructs introduced in Chap-ters 3 and 4 will enable unambiguous incorporation of this business rule in the ER diagram in two different ways. The discussion of these solutions is deferred to Chapters 3 and 4, respectively. Let us now explore the business rule: *It is enough if a value for one of the two attributes, Part# or Name, is present for a given part.* In Figure 2.30b, both **Part#** and **Name** are designated mandatory (dark circle), implying that a value must be present for both these unique identifiers (underlined attributes) in all entities of the PART entity set. Thus the design does not reflect the above stated business rule. Designating both the unique identifiers as optional (empty circle) will not accomplish the stated objective either. The solution then is to specify these two unique identifiers as optional and include the business rule in the conceptual model in text form, as in Figure 2.30c. Can an optional attribute be a unique identifier? The answer to this question is a qualified "Yes." If an entity type has only one unique identifier, then the attribute, atomic or composite, designated as the unique identifier must be a mandatory attribute. Instead, if there are multiple unique identifiers for the entity type, it is enough if one of these unique identifiers is designated as a mandatory attribute (observe entity type MANUFACTURED_PART in Figure 2.30c). Alternatively, it is possible to specify all the unique identifiers as optional attributes followed by a business rule of the form: *In every entity of this entity set any one of these attributes must have a value—not all.* This is reflected in the entity type PURCHASED_PART in Figure 2.30c accompanied by the note at the bottom of the figure. Some data modeling scholars may disagree with this design option because at some point in the design cycle one of the unique identifiers must be designated as the primary means of identifying entities of the entity set and that unique identifier must be a mandatory attribute. However, at the conceptual level of data modeling, there is no reason why all the richness of the scenario not be captured.

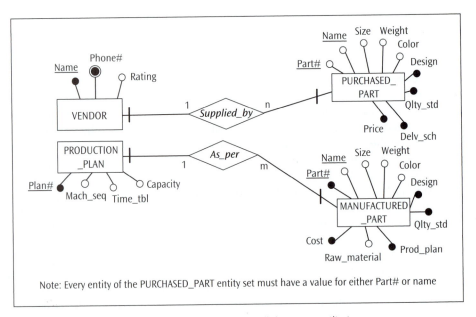

Figure 2.30c   Entity type PART in Figure 2.30b split into two entity types

At this point, let us tweak the scenario of this vignette to incorporate the fact that a widget is an intricate assembly of numerous parts. Some of the parts are actually subassemblies in that they have many subparts. However, a part can be a subpart of only one part—i.e., a part can be a subpart in only one subassembly.

Since we have not been advised to the contrary, the practical assumption to make is that any subpart of a part can be either a manufactured part or a purchased part. Therefore, the design depicted in Figure 2.30c may not be readily conducive for incorporating the new twist to the story. It is possible to work from the design shown in Figure 2.30b. Let us, instead, explore a different alternative based on Figure 2.30c. Since manufactured parts and purchased parts have several common characteristics (e.g., the attributes **Part#, Name, Size, Weight, Color, Design**, and **Qlty_std** are common to both), it is possible to model the collection of these common attributes as an entity type, PART, while the entity types MANUFACTURED_PART and PURCHASED_PART retain only the attributes specific to each of them. Then, it is necessary that both MANUFACTURED_PART and PURCHASED_PART are independently related to PART—a 1:1 relationship with partial participation of PART, and total participation of MANUFACTURED_PART and PURCHASED_PART in the relationships *Manufactured* and *Purchased* respectively, as reflected in Figure 2.30d. Now that the PART entity set includes all the manufactured and purchased parts, a part containing subparts can be modeled by establishing a relationship between PART and SUB_PART, a mirror image of the entity type PART. The ER diagram in Figure 2.30d reflecting this relationship is syntactically correct. Is it also semantically correct? The answer is 'No.' Since any part can also be a subpart of another part, duplication of several part entities in parts and sub-parts entity set is imminent. This is data redundancy and can create semantic problems of data consistency, currency, and correctness in addition to storage inefficiencies during database implementation.

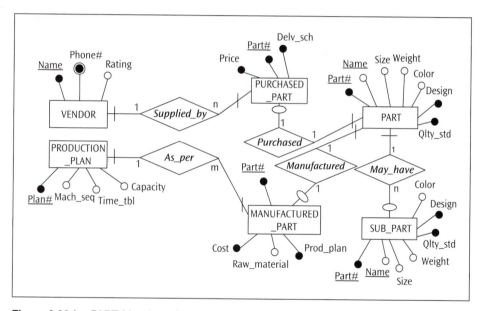

Figure 2.30d   PART-*May_have*-SUB_PART modeled

The correct solution for this scenario is to model *Sub_part* as a recursive relationship, as shown in Figure 2.30e. An alternative design built along the lines of Figure 2.30b is presented in Figure 2.30f. Observe that in this design the composite attributes **Manufactured** and **Purchased** are designated optional—correctly so, because for part entities that are manufactured the attribute **Purchased** and the constituent atomic attributes will not have any value. Likewise, for part entities that are purchased the attribute

Manufactured and its constituent atomic attributes will not have any values. When a composite attribute is designated optional, is it semantically correct for it to contain a mandatory atomic attribute as constituent elements? Yes. The semantics conveyed here is that when the composite attribute does not have any value the constituent atomic attributes (optional or mandatory) also do not have any values. However, when the composite attribute has a value, only the constituent mandatory atomic attribute need have values. For instance, in a manufactured part entity, the attributes **Price** and **Delv_sch** and therefore the composite attribute **Purchased** will be empty. On the other hand, for this entity, **Cost** and **Prod_plan** must not be empty while **Raw_material** can be empty. Collectively, then the composite attribute **Manufactured** is not empty.

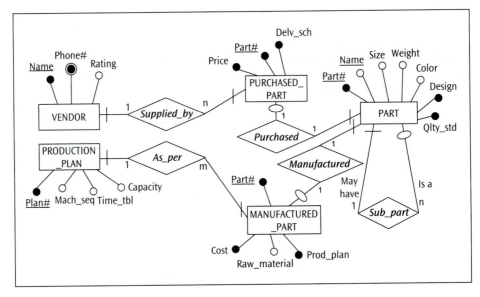

Figure 2.30e   PART-*Sub_part* as a recursive relationship

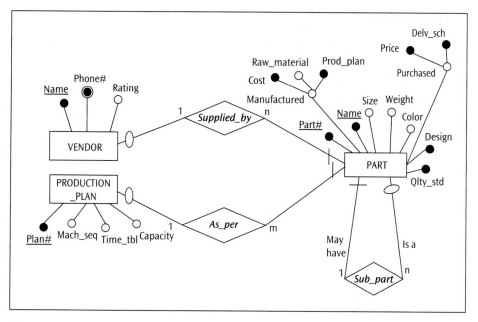

Figure 2.30f   PART-*Sub_part* as a recursive relationship – an alternative design

Chapter Summary

The fundamental constructs and rules for conceptual data modeling can be understood using the Wand and Weber (2002) framework for research in conceptual modeling. The entity-relationship (ER) modeling grammar is a popular tool for conceptual data modeling.

The entity-relationship (ER) model was developed by Peter Chen in 1976. Using this model, object types are conceptualized as entity types. Objects belonging to an object type are considered to be entities of the corresponding entity type. Properties of an object type are represented as attributes of an entity type. An entity is created when a value is supplied for each attribute. Some, but not all, attribute values can be null. Associations exist between objects of different object types. In conceptual modeling, these associations are referred to as relationships between entity types.

An attribute possesses a number of characteristics. These include a name, a data type, and a class (atomic or composite). Furthermore, an attribute can be stored or derived, single or multi-valued, and mandatory or optional. An atomic attribute or collection of atomic attributes (that is, a composite attribute) can serve as a unique identifier of an entity type. Every attribute plays only one of three roles in an entity type. It is a key attribute, a non-key attribute, or a unique identifier. Any attribute that is a constituent part of a unique identifier is a key attribute. An attribute that is not a constituent part of a unique identifier is a non-key attribute.

Business rules supplied by the users expressed in terms of constraints allow data integrity to be achieved. At the conceptual tier of data modeling, two types of data integrity constraints must be specified: (a) the domain constraint imposed on an attribute to ensure that its observed value is not outside the defined domain, and (b) the key (or uniqueness) constraint, which requires entities of an entity type to be uniquely identifiable.

A relationship type is a meaningful association among entity types. The degree of a relationship is defined as the number of entity types participating in a relationship type. A relationship type is said to be binary or of degree two when two entity types are involved. Relationship types that involve three entity types (of degree three) are defined as ternary relationships, while relationships that involve four or more entity types are referred to as n-ary relationships. An entity type related to itself is termed a recursive relationship type. A relationship type is not fully specified until two structural constraints are explicitly imposed. These are the cardinality constraint and the participation constraint. Role names are used to indicate the participation of entity types in relationship types. In addition, a relationship type can have attributes.

An entity type where the entities have independent existence (that is, where each entity is unique) is referred to as a base or strong entity type. An entity type that does not have independent existence (where some entities in the entity set may be identical) is known as a weak entity type. An attribute, atomic or composite, in a weak entity type, which in conjunction with a unique identifier of the parent entity type in the identifying relationship type uniquely identifies weak entities, is called the partial key or the discriminator of the weak entity type.

Exercises

1. What is the difference between the conceptual world and the real world? Is it possible for a conceptual model to represent reality in total? Why or why not?

2. Use examples to distinguish between:
   a. An object type and an entity type
   b. An object and an entity
   c. A property and an attribute
   d. An entity and an entity instance
   e. An association and a relationship
   f. An object class and an entity class

3. Describe various data types associated with attributes.

4. What is the difference between a stored attribute and a derived attribute?

5. What would be the domain of the attribute **County_name** in the state of Texas?

6. Distinguish between a simple attribute, a single-valued attribute, a composite attribute, a multi-valued attribute, and a complex attribute. Develop an example similar to Figure 2.3 that illustrates the difference between each type of attribute.

7. What is a unique identifier of an entity type? Is it possible for there to be more than one unique identifier for an entity type?

8. What is the difference between a key attribute and a non-key attribute?

9. Consider the EMPLOYEE entity type given below.
   a. List all key and non-key attributes.
   b. What is (are) the unique identifier(s)?
   c. Which attribute(s) is (are) derived attributes?
   d. Using the following figure as a guide, develop sample data for four employees that illustrate the nature of the various mandatory and optional attributes in the EMPLOYEE entity type. Be sure to illustrate the various ways the **Name** attribute might appear.

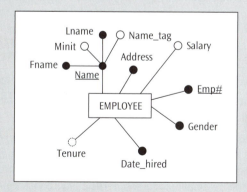

10. Discuss how to distinguish between an entity type and an attribute.

11. Give an example of three entity types and accompanying attributes that might be associated with a database for a car rental agency.

12. What is a relationship type? How does a relationship type differ from a relationship instance?

13. What is meant by the "degree" of a relationship?

14. What is the value of using role names to describe the participation of an entity type in a relationship type?

15. What is the difference between a binary relationship that exhibits a 1:1 cardinality constraint and a binary relationship that exhibits a 1:n cardinality constraint?

16. Describe how *Married_to* can be modeled as a recursive relationship.

17. Create an example of a recursive relationship with an m:n cardinality constraint.

18. Distinguish between a participation constraint and minimum cardinality.

19. Why can total participation of an entity type in a relationship type also be referred to as existence dependency of that entity type in that relationship type?

20. How do cardinality constraints and participation constraints relate to the notions of total and partial participation?

21. Discuss the difference between existence dependency and identification dependency.

22. Give an example of a relationship type between two entity types where an attribute can be assigned to the relationship type instead of to one of the two entity types.

23. What is the difference between a base entity type and a weak entity type? When is a weak entity type used in data modeling?

24. Define the term partial key.

25. This is a narrative about a small university in Kodai, CA. There are several colleges in the university. Each college has a name, location, and size. A college offers many courses over four college terms or quarters—Fall, Winter, Spring, and Summer—during which one or more of these courses are offered. Course#, name, and credit hours describe a course. No two courses in any college have the same course#; likewise no two courses have the same name. Terms are identified by year and quarter, and contain numbers. Courses are offered during every term. The college also has several instructors. Instructors teach; that is why they are called instructors. Often, not all instructors are scheduled to teach during all terms; but every term has some instructors teaching. Also, the same course is never taught by more than one instructor in a specific term. Further, instructors are capable of teaching a variety of courses offered by the college. Instructors have a unique employee ID and their name, qualification, and experience are also recorded.

    a. List the business rules explicitly stated and implicitly indicated in the narrative.

    b. Study the narrative carefully and identify the missing information required for developing a semantically complete conceptual data model.

26. The instance diagram shown on the next page illustrates the relationship between Sullivan Insurance Agency's agents and clients. Using this instance diagram, write the narrative that describes the relationship between agents and clients. Your narrative should include a description of both the cardinality ratio and participation constraints implied in the instance diagram. In addition, draw the ER diagram that fully describes the relationship between the company's agents and clients.

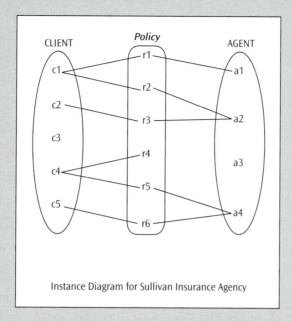

Instance Diagram for Sullivan Insurance Agency

27. Revise the ER diagram drawn in the previous exercise to include the following mandatory attributes: CLIENT—ID number, name, address (city, state, zip), phone number(s), birthdate; AGENT—agent number, name, phone number, area; and commission received by an agent for selling a *Policy* to a client.

28. Use the instance diagram depicting the ternary relationship *Orders* shown on the next page to answer the following questions.

    a. Which customers order pens from the Galveston warehouse?

    b. Which items are ordered by customers from both warehouses?

    c. Which warehouse fills one or more orders of items from both customers?

    d. Describe orders filled from both warehouses.

    e. What changes must be made to the instance diagram for order r1 to involve both pencils and pens?

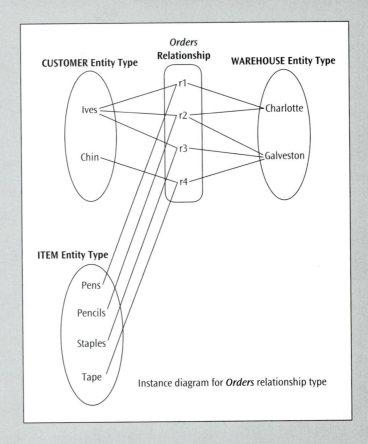

Instance diagram for *Orders* relationship type

29. Adams, Ives, and Scott Incorporated is an agency that specializes in representing clients in the fields of sports and entertainment. Given the nature of the business, some employees are given a company car to drive and each company car must be assigned to an employee. Each employee has a unique employee number, plus an address and set of certifications. Not all employees have earned one or more certifications. Company cars are identified by their vehicle ID and also contain a license plate number, make, model, and year. Employees represent clients. Not all employees represent clients, while some employees represent many clients. Each client is represented by one and only one employee. Sometimes clients refer one another to use Adams, Ives, and Scott to represent them. A given client can refer one or more other clients. A client may or may not have been referred to Adams, Ives, and Scott by another client, but a client may be referred by only one other client. Each client is assigned a unique client number. Additional attributes recorded for each client are name, address, and date of birth.

Draw an ER diagram that shows the entity types and relationship types for Adams, Ives, and Scott. While you must name each relationship type and define its structural constraints, it is not necessary that you supply role names.

Selected Bibliography

Batini, C.; Ceri, S.; and Navathe, S. B. (1992) *Conceptual Database Design: An Entity-Relationship Approach*, Benjamin/Cummings.

Chen, P. (1976) "The Entity Relationship Model: Towards a Unified View of Data," *ACM Transactions on Database Systems*, 1 (March) 9-36.

Chen, P. (1993) "The Entity-Relationship Approach," in [R. Y. Wang, Ed.] *Information Technology in Action: Trends and Perspectives*, Prentice Hall.

Connolly, T. M.; Begg, C. E. (2005) *Database Systems: A Practical Approach to Design, Implementation, and Management*, Fourth Edition, Addison-Wesley.

Date, C. J. (2004) *An Introduction to Database Systems*, Eighth Edition, Addison-Wesley.

Elmasri, R.; and Navathe, S. B. (2003) *Fundamentals of Database Systems*, Fourth Edition, Addison-Wesley.

Navathe, S. B. (1992) "Evolution of Data Modeling For Databases," *Communications of the ACM*, 35, 9 (September) 112-123.

Teory, T. J. (1999) *Database Modeling and Design*, Third Edition, Morgan Kaufman.

Wand, Y.; and Weber, R. (2002) "Research Commentary: Information Systems and Conceptual Modeling – A Research Agenda," *Information Systems Research*, 13, 4 (December) 363-376.

Entity-Relationship Modeling

Chapter 3 presents the application of ER modeling grammar to conceptual data modeling. The whole chapter expounds the "Method" component of the Wand and Weber framework for conceptual modeling research presented at the beginning of Part I of this book (see Figure I.1). The discussion is framed in the context of Bearcat Incorporated, a manufacturing company with several plants located in the northeastern part of the United States. This example database application is also used in subsequent chapters.

Section 3.1 provides the narrative data requirements for Bearcat Incorporated. Section 3.2 applies the ER modeling grammar developed in Chapter 2 to a full-scale conceptual modeling process of the Bearcat Incorporated requirements. The two basic layers of the ER model are introduced, progressing from an end-user communication tool (the Presentation Layer ER model) to a database design oriented product (the Design-Specific ER model). The Design-Specific ER model is then developed systematically through two stages, resulting in the coarse granularity layer and the fine granularity layer. The Fine-granular Design-Specific ER model is then ready for direct mapping to the logical tier.

3.1 Bearcat Incorporated: A Case Study

The case narrated in this section may initially appear overwhelming. However, it is not unusual to find a multitude of data requirements or business rules like these in the real world of database design. The intent of this chapter is to present an array of requirements, then to show how the ER modeling grammar can be applied to systematically dissect the case content into manageable scenarios. We will see how ER diagrams are developed for each scenario, and how these scenarios are synthesized into a full-fledged ER model.

Here are some user requirements or business rules for Bearcat Incorporated that serve as a starting point.

Bearcat Incorporated is a manufacturing company with several plants in the northeastern United States. These plants are responsible for leading different projects that the company might undertake, depending on a plant's function. A certain plant might even be associated with several projects, but a project is always under the control of just one plant. Some plants do not undertake any projects at all. If a plant is closed down, the projects undertaken by that plant cannot be canceled. The project assignments from a closed plant must be temporarily removed in order to allow the project to be transferred to another plant.

Employees work in these plants, and each employee works in only one plant. A plant may employ many employees but must have at least 100 employees in order to exist. A plant with employees cannot be closed down. Every plant is managed by an employee who works in the same plant; but every employee is not a plant manager nor can an employee manage more than one plant. Company policy dictates that every plant must have a manager. Therefore, an employee currently managing a plant cannot be deleted from the database. If a plant is closed down, the employee no longer manages the plant but becomes an employee of another plant.

Some employees are assigned to work on projects and in some cases might even be assigned to work on several projects simultaneously. For a project to exist, it must have at least one employee assigned to it. Projects might need several employees depending on their size and scope. As long as an employee is assigned to a project, his or her record cannot be removed from the database. However, once a project ends it is removed from the database and all assignments of employees to that project must be removed.

Some employees also supervise other employees, but all employees need not be supervised—the employees that are supervised are supervised by just one employee. An employee may be a supervisor of several employees, but no more than 20. The Human Resources Department uses a designated default employee number to replace a supervisor who leaves the company. It is not possible for an employee to be his or her own supervisor.

Some employees may have several dependents. Bearcat Incorporated does not allow both husband and wife to be an employee of the company. Also, a dependent can only be a dependent of one employee at any time.

Bearcat Incorporated offers credit union facilities as a service to its employees and to their dependents. An employee is not required to become a member of Bearcat Credit Union (BCU). However, most employees and some of their dependents have accounts in BCU. Some BCU accounts are individual accounts, and others are joint accounts between an employee and his or her dependent(s). Every BCU account must belong to at least an employee or a dependent. Each joint account must involve no more than one employee and no more than one of his or her dependents. If an employee leaves the company, all dependents and BCU accounts of the employee must be removed. In addition, as long as a dependent has a BCU account, deletion of the dependent is not permitted.

Bearcat Incorporated sponsors recreational opportunities for the dependents of employees in order to nurture the hobbies of the dependents. Dependents need not have a hobby, but it is possible that some dependents may have several hobbies. Because some hobbies are not as popular as others, every hobby need not have participants. If a dependent is no longer in the database, all records of the participation of that dependent in hobbies should not exist in the database either. Finally, as long as at least one dependent participates in a hobby, that hobby should continue to exist.

All plants of Bearcat Incorporated have a plant name, number, budget, and building. A plant has three or more buildings. Each plant can be identified by either its name or number. Bearcat Incorporated operates seven plants and these plants are numbered 10 through 20. However, either the plant number or plant name must always be recorded.

The name of an employee of Bearcat Incorporated consists of the first name, middle initial, last name, and a nametag. Employee number is used to identify employees in the company. However, name can also be used as an identifier. Both employee number and employee name must be recorded. Where two or more employees have the same name, a one-position numeric nametag is used so that up to ten otherwise duplicate names can be distinguished from one another. Sometimes the middle initial of an employee may not be available.

While the address, gender (male or female), and date hired of each employee must be recorded, salary information is optional. Salaries at Bearcat Incorporated range from $35,000 to $90,000. Also, the salary of an employee cannot exceed the salary of the employee's supervisor.

The start date of an employee as a manager of a plant should also be gathered. In addition, there is the requirement that the number of employees working in each plant be available; this can be computed. Information about the dependents related to each employee, such as the dependent's name, relationship to the employee, birth date, and gender, should also be captured. The name of the dependent along with how the dependent is related to an employee is mandatory. A mother or daughter must be a female, a father or son must be a male, and a spouse can be either male or female. Since a dependent cannot exist independently of an employee, the dependent name and relationship to the employee, in conjunction with either the employee name or the employee number, is used to identify the dependents of an employee. There is also the requirement that the number of dependents of each employee be captured, but this, like the number of employees working in each plant, can be computed.

Projects have unique names and numbers and their location must be specified. Every project must have a project number but sometimes may not have a project name. Project numbers range from 1 to 40; Bearcat Incorporated's projects are located in the cities of Stafford, Bellaire, Sugarland, Blue Ash, and Mason. The amount of time an employee has been assigned to a particular project should be recorded for accounting purposes.

BCU accounts are identified by a unique account ID, composed of an account number and an account type (C—checking account, S—savings account, I—investment account). For each account, the account balance is recorded. Only the account ID is required for every account. Account numbers contain a maximum of six alphanumeric characters.

Hobbies are identified by a unique hobby name and include one code that indicates whether the hobby is an indoor or outdoor activity, and another code that indicates if the hobby is a group or individual activity. The time spent by a dependent per week on each hobby and the associated annual cost are also captured.

3.2 Applying the ER Modeling Grammar to the Conceptual Modeling Process

The ER modeling grammar for conceptual modeling serves two major purposes:

- As a communication/presentation device used by an analyst to interact with the end-user community (the **Presentation Layer ER model/schema**)
- As a design tool at the highest level of abstraction to convey a deeper level understanding to the database designer (the **Design-Specific ER model/schema**)

An ER model includes (1) an **ER diagram (ERD)** portraying entity types, attributes for each entity type, and relationships among entity types, and (2) **semantic integrity constraints** that reflect the business rules about data not captured in the ER diagram.

Figure 3.1 illustrates how the Presentation Layer and the Design-Specific ER models fit into the conceptual modeling activity. The development of the Presentation Layer ER model entails the analyst working with the user community and culminates in the generation of a script, that is, a Presentation Layer ER diagram, accompanied by a list of semantic integrity constraints. Next, the Presentation Layer ER model (including the ER diagram and semantic integrity constraints) is mapped or translated to the Design-Specific ER model. The Design-Specific ER model is expressed in terms of two levels of granularity—*the* coarse level and *the fine level*. Section 3.2.1 describes the method for the development of the Presentation Layer ER model for Bearcat Incorporated, while Sections 3.2.2 and 3.2.3 focus on the methods for the development of the coarse and fine granularity Design-Specific ER models respectively for Bearcat Incorporated.

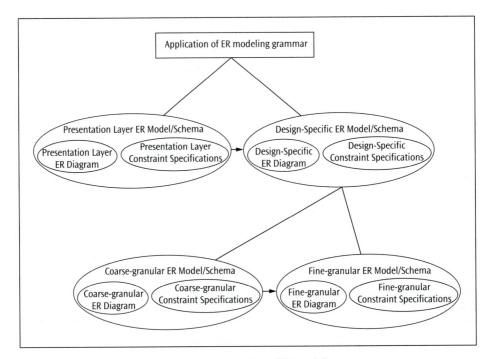

Figure 3.1 Conceptual modeling method using the ER modeling grammar

3.2.1 The Presentation Layer ER Model

As the name implies, this layer of the ER model serves the principal purpose of communication with the end-user community. The Presentation Layer ER diagram is a surface-level expression of the application domain, and the semantic integrity constraints are a reiteration of the business rules that are not captured in the Presentation Layer ER diagram.

While a variety of notational variations for entity-relationship diagrams appear in the conceptual data modeling literature, only a handful are relatively popular and are also used in commercial computer-aided software engineering (CASE) tools. Often the notational variations occur in the expression of the *properties* of a relationship type. In all these notations, an entity type is expressed by a rectangular box while a weak entity type appears as a double rectangular box. In order to contrast it from a weak entity type, the entity type is often referred to as a base (or strong) entity type. A relationship type is shown as a diamond whereas an identifying relationship type (a relationship type that connects a weak entity type to its identifying parent entity type) is shown as a double diamond. The CASE tools tend to avoid the use of the diamond for a relationship type, instead labeling the edges connecting the entity types to capture the semantics of the relationship. There are just a couple of different ways to express attributes graphically. This text uses the convention where the optional/mandatory property can be explicitly expressed. Thus, an attribute is shown by a circle with the name of the attribute written adjacent to the circle. A dark circle represents an attribute with a mandatory value (also known as a mandatory attribute) while an empty circle indicates an attribute with an optional value (also known as an optional attribute). Component attributes constituting a composite attribute are attached to the circle that represents the composite attribute. A multi-valued attribute is shown by a double circle while a derived attribute is shown by a dotted circle. An attribute that serves as a unique identifier of a base entity type is underlined with a solid line, whereas the partial key of a weak entity type (also known as a discriminator) is underlined with a dotted line. Note that an entity type can and often does have multiple unique identifiers and each identifier can be an atomic or composite attribute.

Figure 3.2 summarizes the notational scheme used for the Presentation Layer ER diagrams in this book. Although it is based on Peter Chen's original notation (1976), it also incorporates a few desirable features of other commonly used notational schemes. A relationship is shown through the use of edges (lines) that connect the relationship type to the participating entity types. Both the cardinality ratio and the participation constraint are expressed via a "look across" approach. The cardinality ratio is placed on the connectors adjacent to the relationship type. The oval adjacent to E2 in Figure 3.2 indicates that E1 is optionally related to E2, that is there can be entities (e_{11}, e_{12}, ...,. e_{1x}, ... e_{1n}) of E1 not related to any entity of E2. As stated in Section 2.3.4, this is known as partial participation of E1 in R. The bar (|) adjacent to E1 signifies that E2 is mandatorily related to E1.[1] As discussed in Section 2.3.4, this is known as total participation of E2 in R. This also implies that E2 has existence dependency on R. That is, in order for an entity e_{21} of E2 to exist, it must participate in a relationship r_1 with an entity e_{1x} of E1. The Crow's Foot notation to specify the relationship properties is also popular and essentially replaces the **m** and **n** in the Chen scheme with fork-like symbols. The Crow's Foot notation, originally introduced by Everest (1986) for the Knowledgeware software, follows the same "look-across" strategy to specify both cardinality ratios and participation constraints as is done in the Chen scheme. The meaning of the exclusive, inclusive and noninclusive arc notations is discussed later in the chapter.

As a high-level diagrammatic portrayal of the application domain, an ER diagram is not capable of capturing some of the finer business rules that are part of the data requirements. The specification of constraints is the mechanism to record the business rules not captured in the ER diagram. Together the ER diagram and the semantic integrity constraints must preserve all the information conveyed in the data requirements of the application. The ER diagram coupled with the semantic integrity constraints represents the conceptual schema for the application, and because the ER modeling grammar in this case expresses the conceptual schema, the resulting script is referred to as the ER model/schema.

---

[1] Mandatory participation in Chen's original notational scheme is implicitly indicated by the absence of the oval (symbol for partial participation). We, however, employ an explicit indicator—that is, the bar (|).

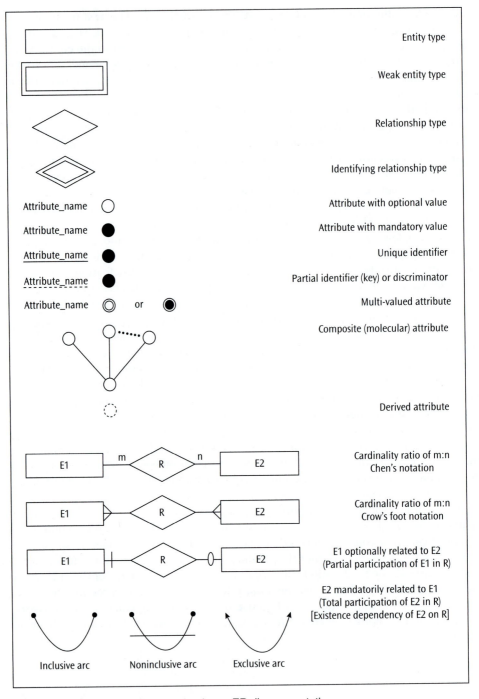

Figure 3.2 Summary of Presentation Layer ER diagram notation

3.2.2 The Presentation Layer ER Model for Bearcat Incorporated

The development of an ER diagram often begins by examining the data requirements specification for the application domain in an effort to extract possible entity types, attributes, and relationships among entities. Conceptual modeling is a heuristic process, as opposed to a scientific process, and is therefore intuitive and iterative in nature. For example, at first glance, it may appear that Company should be an entity type in the ER diagram. However, a careful reading of the data requirements in Section 3.1 indicates that if we specify an entity type for Company, it will have exactly one entity, Bearcat Incorporated. In other words, the scope of the application domain here is just one company, not a number of companies. Thus, an entity type for Company is not necessary. In fact, notice that in the narrative no attributes have been specified for Company and therefore an entity type Company simply cannot exist unless we make up some attributes. Recall that attributes give shape to an entity type and therefore an entity type cannot exist without at least one attribute. Alternatively, one could be futuristic and speculate on the possibility of Bearcat Incorporated acquiring other companies, or simply deciding to organize itself into several autonomous business units (divisions). Then, the Company entity type in the data model would add value. The sole purpose of single-instance entity types is to provide an avenue for future expansion of the database. However, it is possible for a designer to speculate numerous such situations which can lead to inefficient design based on unsubstantiated expansion of the scope of the desired database application. Any expansion of the scope of a design based on anticipated future needs should be done in consultation with the user community. For instance, in order to choose this second alternative, the designer should elicit a list of attributes from the user for a Company entity type.

One way to arrive at the first cut of an ER diagram is to begin by listing all discernable data elements from the narrative, treat them all as attributes, and then attempt to group these attributes based on apparent commonalities among them. This should lead to the identification of different clusters of attributes, each of which may now be designated as an entity type and labeled with an appropriate name, culled from the narrative. A review of the leftover data elements in the list should facilitate recognition of some of them as possible links among the entity types previously identified. Thus, relationship types can be generated. Note once again that an entity type cannot exist without having one or more attributes. In addition, while a standalone entity type is not technically prohibited in an ERD, an entity type, in general, does not exist in an ERD without being related to at least one other entity type. The short list of remaining data elements, if any, can be reconciled as either nonexistent in the data requirements or as part of an entity type or relationship type created thus far or as other new entity types. The first cut of the ER diagram at this point is ready for recursive refinement. This approach is usually labeled the **synthesis approach**.

Alternatively, one could consider an **analysis approach** for discovering entity types from the narrative. This approach begins with a search for things that can be labeled by singular nouns (a person, place, thing, or concept) which are modeled as entity types of the ER diagram. This is followed by gathering properties that appear to belong to individual entity types. These properties, nouns in themselves, belonging to each entity type are labeled as attributes of that entity type. As was the case in the synthesis approach, throughout this process the identification of relationships among various entity types must also be recognized.

Under both the synthesis and analysis approaches caution should be exercised to ensure that elements outside the scope of the narrative are not brought into the modeling process based on individual whims and fancies.

The excerpts that follow are taken from the narrative for Bearcat Incorporated in Section 3.1 to illustrate the application of the analysis approach to identify possible entity types and their attributes. Capitalized nouns constitute the entity types while nouns in *italics* are the attributes of the respective entity types.

- Bearcat Incorporated is a manufacturing company with several PLANTs .... All plants of Bearcat Incorporated have a *plant name, number, budget,* and *building*....
- These plants are responsible for leading different PROJECTs.... Projects have ... *names* and *numbers* and their *location* must be specified....
- EMPLOYEEs work in these plants.... The *name* of an employee ... consists of the *first name, middle initial, last name,* and a *nametag.* While the *address, gender,* ... and *date hired* ... must be recorded, *salary* information is optional.... The *start date* of an employee as a manager of a plant should also be gathered. There is also the requirement that the *number of employees* working in each plant be available.... Company policy dictates that every plant must have a MANAGER....
- Some employees may have several DEPENDENTs.... Information about the dependents related to each employee such as the dependent's *name, relationship to the employee, birth date,* and *gender* should also be captured.... There is also the requirement that the *number of dependents* of each employee be captured....
- Bearcat Incorporated offers CREDIT UNION facilities as a service to its employees and to their dependents.... BCU accounts are identified by a unique *account ID,* composed of an *account number* and an *account type* (C—checking account, S—savings account, I—investment account). For each account, the *account balance* is recorded.
- Bearcat Incorporated sponsors recreational opportunities for the dependents of employees in order to nurture the HOBBY(ies) of the dependents. Hobbies are identified by a unique *hobby name* and include one *code* that indicates whether the hobby is an indoor or outdoor activity, and another *code* that indicates if the hobby is a group or individual activity. The *time spent by a dependent per week* on each hobby and the associated *annual cost* are also captured.

A series of eight boxes is used to illustrate how the relationships among the entity types for Bearcat Incorporated can be detected. Some of these boxes contain an excerpt from the Bearcat Incorporated data requirements in Section 3.1 plus the ER diagram for the italicized sentences that appear in the excerpt. The use of information in sentences not italicized in a particular box will be described later in the chapter. Note that in order to enhance the clarity of the presentation, attributes are not shown in these ER diagrams.

Bearcat Incorporated is a manufacturing company that has several plants in the northeastern part of the United States. *These plants are responsible for leading different projects that a company might undertake, depending on a plant's function. A certain plant might even be associated with several projects but a project is always under the control of just one plant. Some plants do not undertake any projects at all.* If a plant is closed down, the projects undertaken by that plant cannot be canceled. The project assignments from a closed plant must be temporarily removed in order to allow the project to be transferred to another plant.

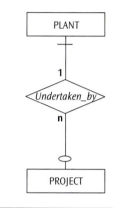

Box 1

In Box 1, observe that the cardinality ratio of the relationship type *(Undertaken_by)* is shown as 1:n by looking across from PLANT to PROJECT. This is because a plant can be associated with several projects but a project is always under the control of a single plant. Also, since the data requirements explicitly specify that *"Some plants do not undertake any projects at all,"* one can infer that a plant may or may not undertake/control a project (that is, a plant optionally controls a project). This is indicated in the diagram in Box 1 by looking across from the PLANT to the PROJECT and placing the oval optionality marker just above the PROJECT entity type. Accordingly, the participation constraint of PLANT in this relationship is said to have the value "partial." Likewise, since every project must be controlled by a plant, a look across from the PROJECT to the PLANT signifies the mandatory participation of the PROJECT in the relationship through the bar (l) just below the PLANT.

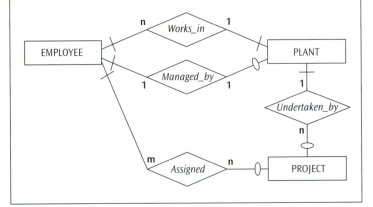

Box 2

In Box 2, the *Works_in* relationship type between the PLANT and the EMPLOYEE entity types indicates that every employee must work in a plant and only in one plant (look across from EMPLOYEE to PLANT) and that a plant contains many employees. However, note that the requirement that a plant must have at least 100 employees in order to exist is not reflected in the ER diagram. Instead, the ER diagram in Box 2 only indicates that a plant must have at least one employee and may have more than one employee. A second relationship type, *Managed_by*, also exists between the PLANT and EMPLOYEE entity types in order to show that (a) each plant must be managed by one and no more than one employee, and (b) an employee may manage one plant but all employees are not managers. Observe that earlier in this section, MANAGER was identified as a possible base entity type. The basis for the design decision portrayed here in Box 2 is discussed at the end of this section. The relationship type *(Assigned)* between the EMPLOYEE and PROJECT entity types exhibits an m:n cardinality ratio. The oval next to PROJECT indicates that every employee need not be assigned to a project (looking across from EMPLOYEE), and the bar next to EMPLOYEE indicates that every project has at least one employee assigned to it (looking across from PROJECT).

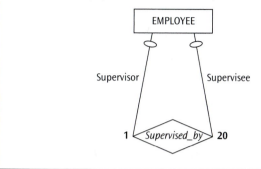

Some employees also supervise other employees but all employees need not be supervised – the employees that are supervised are supervised by just one employee. An employee may be a supervisor of several employees, but no more than 20. The Human Resources Department uses a designated default employee number to replace a supervisor who leaves the company. It is not possible for an employee to be his or her own supervisor.

Box 3

The *Supervised_by* relationship type in Box 3 is a recursive relationship and indicates that a given employee *may* be supervised—but there may be only one supervisor per employee. At the same time, an employee may supervise many other employees – no more than 20 though. Finally, an employee need neither be supervised nor be a supervisor.

86

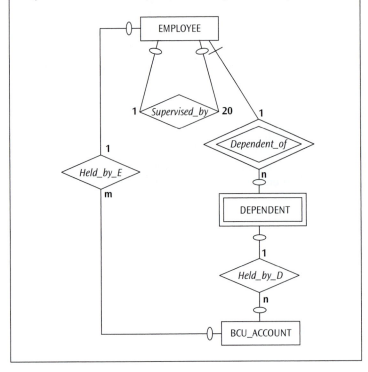

Some employees may have several dependents. Bearcat Incorporated does not allow both husband and wife to be an employee of the company. Also, a dependent can only be a dependent of one employee at any time. Bearcat Incorporated offers credit union facilities as a service to its employees and to their dependents. An employee is not required to become a member of Bearcat Credit Union (BCU). However, most employees and some of their dependents have accounts in BCU. Some BCU accounts are individual accounts and others are joint accounts between an employee and his or her dependent(s). Every BCU account must belong to at least an employee or a dependent. Each joint account must involve no more than one employee and no more than one of his or her dependents. If an employee leaves the company, all dependents and BCU accounts of the employee must be removed. In addition, as long as a dependent has a BCU account, deletion of the dependent is not permitted.

Box 4

The DEPENDENT entity type in Box 4 is shown as a weak entity type, and the *Dependent_of* relationship type is shown as an identifying relationship type. How do we know that DEPENDENT is a weak entity type? The answer is "We do not know" from what is stated in Box 4. The statement " . . . *a dependent can only be a dependent of one employee at any time"* only indicates total participation of DEPENDENT in the relationship. Any entity type, base or weak, has existence dependency in a relationship type if its participation in the relationship is total. The reason we know that DEPENDENT is a weak entity type participating as a child in an identifying relationship type with EMPLOYEE is because later in the narrative it is stated, "Since a dependent cannot exist independently of an

Chapter 3

employee, the dependent name and relationship to the employee, in conjunction with either the employee name or the employee number, is used to identify the dependents of an employee." Observe that while not all employees have dependents, each dependent is related to one and only one employee.

The BCU_ACCOUNT entity type participates in two relationships. The *Held_by_E* relationship type reflects the fact that an employee may have a BCU account while not all BCU accounts are held by employees. In addition, no more than one employee can be associated with a BCU account. The *Held_by_D* relationship type reflects a similar relationship between the BCU_ACCOUNT entity type and the DEPENDENT entity type. Here, while a dependent may have several BCU accounts, he or she need not have a BCU account. Likewise, a BCU account need not be associated with a dependent, but if it is, then there can be no more than one dependent per BCU account. The business rules that every BCU account must belong to at least an employee or a dependent, and if held jointly, must belong to an employee and his or her dependent—not an employee nor the dependent of *any* other employee—are not reflected in the ERD in Box 4. These business rules that cannot be captured in the ERD must be included in the list of semantic integrity constraints that accompanies the ERD as part of the ER model.

What if joint accounts between employees and dependents are not permitted? This means that the same BCU account entity cannot be related to an employee entity as well as a dependent entity. That is, the relationship types *Held_by_E* and *Held_by_D* are mutually exclusive. An **exclusive arc**, as shown in Box 5, represents this constraint. Likewise, if, for any reason, the user requirement specifies that all BCU accounts must be jointly held between an employee and a dependent, then an **inclusive arc** is used to express such a business rule as a constraint in the ERD, as shown in Box 6.

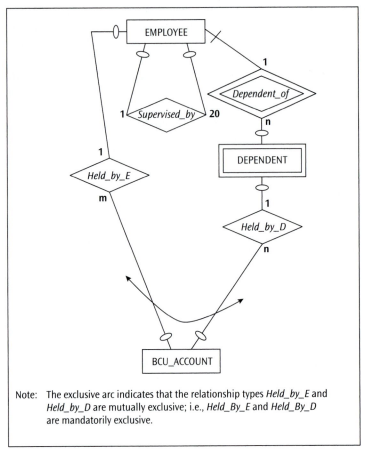

Note: The exclusive arc indicates that the relationship types *Held_by_E* and *Held_by_D* are mutually exclusive; i.e., *Held_By_E* and *Held_By_D* are mandatorily exclusive.

Box 5

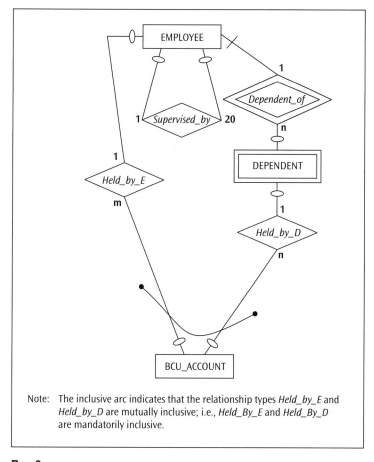

Note: The inclusive arc indicates that the relationship types *Held_by_E* and *Held_by_D* are mutually inclusive; i.e., *Held_By_E* and *Held_By_D* are mandatorily inclusive.

Box 6

Observe that the business rule, "*Every BCU account must belong to at least an employee or a dependent,*" cannot be captured in the ER diagram using either the inclusive arc or the exclusive arc. Use of an inclusive arc here would indicate that every BCU account *must* always belong jointly to an employee and a dependent. An exclusive arc, on the other hand, would mean that a BCU account *must never* belong jointly to an employee and a dependent. The business rule, however, indicates that a BCU account *must* belong to *either* an employee *or* a dependent, or sometimes an employee and a dependent together. The noninclusive arc in Box 7 is designed to capture this semantic, namely, *optional inclusiveness*. Accordingly, some BCU accounts are held independently by employees and dependents while others are held jointly between employees and dependents.

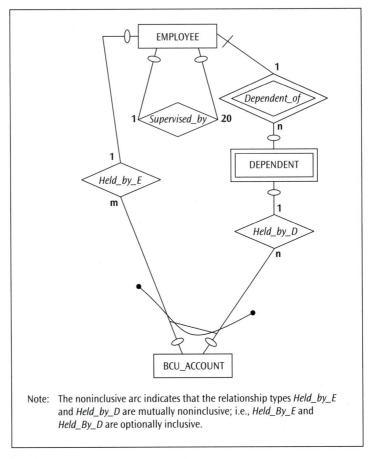

Box 7

Note: The noninclusive arc indicates that the relationship types *Held_by_E* and *Held_by_D* are mutually noninclusive; i.e., *Held_By_E* and *Held_By_D* are optionally inclusive.

The relationship type *(Participates)* between the DEPENDENT entity type and the HOBBY entity type reflects an m:n cardinality ratio. Observe that a dependent may optionally have many hobbies (up to n), and up to m dependents may participate in a hobby (Box 8). Also, a hobby need not have any participant. Observe that while DEPENDENT is a weak entity type, its relationship with HOBBY is not an identifying relationship type.

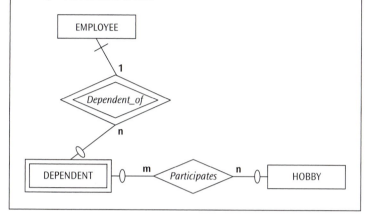

Bearcat Incorporated sponsors recreational opportunities for the dependents of employees in order to nurture the hobbies of the dependents. Dependents need not have a hobby, but it is possible that some dependents may have several hobbies. Because some hobbies are not as popular as others, every hobby need not have participants. If a dependent is no longer in the database, all records of the participation of that dependent in hobbies should not exist in the database either. Finally, as long as at least one dependent participates in a hobby, that hobby should continue to exist.

Box 8

A review of the data requirements for Bearcat Incorporated in Section 3.1 also makes it possible to identify the attributes of each of the entity types. Figure 3.3[2] is a collective representation of Boxes 1, 2, 3, 7, and 8, and includes the attributes. Although the attribute names in the ERD are arbitrary, as much as possible, meaningful abbreviated attribute names have been used to identify the attributes. The following comments elaborate on selected attributes in the ER diagram.

- *Mandatory attributes.* The data requirements in Section 3.1 indicate that a number of attributes are mandatory, that is, they cannot have missing values. These attributes are shown as dark circles. Attributes shown as empty circles are *optional attributes* (can have missing values).
- *Multi-valued attributes.* Since a plant must have three or more buildings, in addition to a plant name, number, and budget, **Building** is shown as a mandatory multi-valued attribute.
- *Derived attributes.* A derived attribute is shown as a dotted circle. Since the data requirements indicate that the number of employees working in each plant and the number of dependents of each employee can be calculated, the attributes **No_of_employees** and **No_of_dependents** appear as derived attributes. Recall that derived attributes are not stored in the database.

---

[2] All ER diagrams that appear in this book were created using Microsoft Visio.

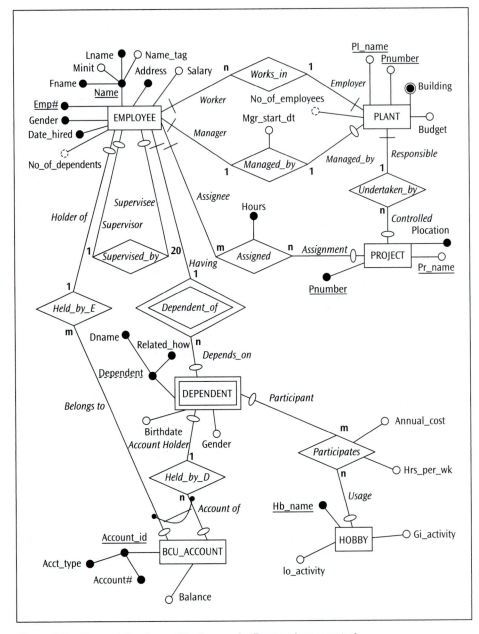

Figure 3.3   Presentation Layer ER diagram for Bearcat Incorporated

- *Composite attributes.* The definitions of both the employee name and the account ID in the data requirements indicate that both can be subdivided into smaller subparts composed of atomic attributes. In the ER diagram, the composite attribute **Name** of the EMPLOYEE entity type is shown as consisting of the atomic attributes **Fname, Minit, Lname**, and **Name_tag** since *"...the name of an employee of Bearcat Incorporated consists of the first name, middle initial, last name and a name tag."* In addition, *"... BCU accounts are identified by a unique account ID, composed of an account number and an account type..."* which gives rise to the definition of **Account_id** as a composite attribute consisting of the atomic attributes **Account#** and **Acct_type**.

- *Attributes of relationship types.* The data requirements suggest that certain relationship types can also have attributes. For example, the optional attribute **Mgr_start_dt** of the *Managed_by* relationship type is used to record the date on which an employee begins managing a plant. Since **Mgr_start_dt** is associated with the 1:1 relationship type *Managed_by,* this attribute could have been assigned to either the EMPLOYEE or PLANT entity type. A second attribute of a relationship type is the mandatory attribute **Hours** of the *Assigned* relationship type reflecting the number of hours an employee works on a project. Note that the placement of this attribute in either the EMPLOYEE or PROJECT entity types will not convey the intended meaning. The same notion applies to the two optional attributes, **Annual_cost** and **Hrs_per_wk**, of the *Participates* relationship type which indicate the annual cost and hours per week that a dependent spends on a particular hobby.

- *Unique identifiers.* Embedded within the data requirements is the definition of certain attributes as unique identifiers. The names of these attributes (for example, **Emp#** of EMPLOYEE, **Pnumber** and **Pl_name** of PLANT,[3] and **Account_id** of BCU_ACCOUNT) appear underlined in the ER diagram.

- *Partial keys.* Since the combination of the dependent name and the dependent's relationship to the employee is unique for a given employee, the composite attribute **Dependent**, a composite attribute, consisting of the atomic attributes **Dname** and **Related_how**, appears as a partial key and is underlined with a dotted line in the ER diagram.[4]

A variety of other business rules were introduced in the data requirements in order to maintain consistency among associated objects in the Bearcat Incorporated case. Since these business rules are not reflected in the Presentation Layer ER diagram (Figure 3.3), they are

---

[3] Note that **Pnumber** and **Pl_name** are shown as optional attributes in Figure 3.3. Both are also designated as unique identifiers (underlined). This may initially appear contradictory because we expect unique identifiers to be mandatory attributes. However, at the conceptual level, the task is to simply identify all possible unique identifiers of an entity type. The implication here is that in any entity of this entity type (PLANT) it is enough if either **Pnumber** or **Pl_name** has a value. This requirement is specified as a semantic integrity constraint (see Table 3.1). Later, at the logical level, only one of these unique identifiers (**Pnumber** or **Pl_name**) will be chosen to serve as the primary unique identifier. At that point, the primary unique identifier is made mandatory.

[4] It is conceivable that a DEPENDENT entity type can have a naturally occurring unique identifier (e.g., Social Security number) in which case it should be modeled as a base entity type. Since the requirements specification of Bearcat Incorporated does not include such an attribute, DEPENDENT is modeled here as a weak entity type identification-dependent on the identifying parent EMPLOYEE.

Entity-Relationship Modeling

expressed as the semantic integrity constraints shown in Table 3.1 in order to preserve all the information conveyed in the data requirements. These semantic integrity constraints are grouped into three categories: attribute-level business rules, entity-level business rules, and business rules governing entity deletion. Constraints that do not fall into one of these three categories have been listed under miscellaneous business rules. For example, the business rule requiring gender to be either male or female cannot be captured in an ER diagram and therefore must be recorded as a semantic integrity constraint. In addition, recall that while each plant can be identified either by its name or number, either the plant number or plant name must always be recorded. A requirement of this type also cannot be reflected in an ER diagram but can be defined as an entity-level business rule since the constraint involves multiple attributes. As a third example, the business rule that the salary of an employee cannot exceed the salary of the employee's supervisor cannot be captured as a constraint in the ER diagram *per se* and therefore must be captured as a semantic integrity constraint that takes the form of a miscellaneous business rule.

The deletion constraints describe business rules that define the specific action(s) allowed when an attempt is made to delete an instance of an entity type. These constraints apply to possible actions when the parent in a parent-child (binary) relationship (PCR) is deleted. Note that the deletion of an instance of the child entity type in a PCR does not entail any constraint on the relationship. For example, consider the possible impact on the database of an employee leaving the company. If the employee entity is arbitrarily deleted without specifying the consequences of this deletion on other related entity types, the integrity of the database will be compromised. Though this may at first glance appear to be an implementation level detail, the business rules for such constraints must still come from the user community; therefore it is imperative that this information be captured during the initial stages of database design, and preserved across the Presentation Layer ER model and both the coarse and fine levels of the Design-Specific ER model.

Often it can be difficult to determine if a data element is an attribute or something that should be modeled as an entity type or a relationship type. For instance, shouldn't MANAGER of a plant be modeled as an entity type in Figure 3.3? If it is not, where will the start date of the manager of a plant be recorded? Although creating a MANAGER entity type appears reasonable, a closer look at the data requirements described in the narrative in Section 3.1 reveals that the only attribute of such an entity type would be manager start date. Further, if a unique identifier for MANAGER is not obvious, one might, as a first step, consider the possibility of modeling plant manager as a weak entity type child of either the EMPLOYEE entity type or the PLANT entity type. Alternatively, given its relationship to both EMPLOYEE and PLANT, one might approach modeling the presence of a plant manager as a relationship between EMPLOYEE and PLANT (see Figure 3.3), and capture **Mgr_start_date** as an attribute of the relationship type *Managed_by,* where the EMPLOYEE entity type plays the role Manager, and the PLANT entity type plays the role Managed_by.

Also, if the **Building** attribute of the PLANT entity type can be modeled as a multivalued attribute, why can't the dependents of an employee be modeled as a multi-valued attribute of the EMPLOYEE entity type? The answer is that this is possible except for the fact that DEPENDENT is involved in two "external" relationships (one called *Held_by_D* with BCU_ ACCOUNT, another called *Participates* with HOBBY). In other words, if DEPENDENT is modeled as a multi-valued attribute of EMPLOYEE instead of the weak entity child of EMPLOYEE, the *Held_by_D* and *Participates* relationships could not be shown. The fact that the ER modeling grammar does not permit specification of a relationship between an attribute and an entity type is illustrated by Vignette 1 in Section 2.4.1 of Chapter 2.

Attribute-Level Business Rules

1. Each plant has a plant number that ranges from 10 to 20.
2. The gender of each employee or dependent is either male or female.
3. Project numbers range from 1 to 40.
4. Project locations are confined to the cities of Bellaire, Blue Ash, Mason, Stafford, and Sugarland.
5. Account types are coded as C—checking account, S—savings account, I—investment account.
6. A hobby can be either I—indoor activity or O—outdoor activity.
7. A hobby can be a G—group activity or I—individual activity.

Entity-Level Business Rules

1. A mother or daughter dependent must be a female, a father or son dependent must be a male, and a spouse dependent can be either male or female.
2. An employee cannot be his or her own supervisor.
3. A dependent may have a joint account only with an employee of Bearcat Incorporated to whom he or she is related.
4. Either plant number or plant name must be present.
5. Every plant is managed by an employee who works in the same plant.

Business Rules Governing Entity Deletion

1. A plant with employees cannot be closed down.
2. If an employee is deleted, all BCU accounts of that particular employee must be deleted.
3. *If a plant is closed down, the projects undertaken by that plant cannot be canceled. The project assignments from a closed plant must be temporarily removed in order to allow the project to be transferred to another plant.
4. The Human Resources Department uses a designated default employee number to replace a supervisor who leaves the company.
5. An employee currently managing a plant cannot be deleted from the database.
6. If a plant is closed down, the employee no longer manages the plant but becomes an employee of another plant.
7. **If an employee leaves the company, all dependents and BCU accounts of the employee must be removed.
8. **As long as a dependent has a BCU account, deletion of the dependent is prohibited.
9. As long as an employee is assigned to a project, his or her record cannot be removed from the database.
10. If a project is deleted, all assignments of employees to that project must be deleted.
11. If a dependent is deleted, all records of the participation of that dependent in hobbies must be deleted.
12. A hobby with at least one dependent participating in it cannot be deleted.

*Honoring this rule entails relaxation of the requirement at the beginning of Section 3.1 that a project is always under control of a plant.

**Rule 7 cannot be honored for dependents who have bank accounts because Rule 8 prohibits deletion of such dependents. This is resolved in favor of Rule 7 by letting the bank account be deleted when a dependent is deleted. Additional discussion of this conflict resolution appears in Section 3.2.3.

Miscellaneous Business Rules

1. Each plant has at least three buildings.
2. Each plant must have at least 100 employees.
3. The salary of an employee cannot exceed the salary of the employee's supervisor.

Table 3.1 Semantic integrity constraints for the Presentation Layer ER model

3.2.3 The Coarse-Granular Design-Specific ER Model

While the Presentation Layer ER model serves as a vehicle for the analyst to interact with the user community, additional information is required for the ultimate design and implementation of the database. For instance, additional details about the characteristics of attributes must be obtained from the users in order to further define the nature of each attribute. For each attribute defined in the data requirements, Table 3.2 lists an entity type name, abbreviated attribute name, data type, and size. While domain values for attributes for which a property value set is available in the Presentation Layer ER model are mapped into this table, an additional domain constraint on **Emp#** was also secured from the user(s) at this time. The entity type names are obtained from the Presentation Layer ER diagram.

Entity/Relationship Type Name	Attribute Name	Data Type	Size	Domain Constraint
PLANT	Pl_name	Alphabetic	30	
PLANT	Pnumber	Numeric	2	Integer values from 10 to 20.
PLANT	Budget	Numeric	7	
PLANT	Building	Alphabetic	20	
PLANT	No_of_employees	Numeric		
EMPLOYEE	Emp#	Alphanumeric	6	ANNNNN, where A can be letter and NNNNN is a combination of five digits
EMPLOYEE	Fname	Alphabetic	15	
EMPLOYEE	Minit	Alphabetic	1	
EMPLOYEE	Lname	Alphabetic	15	
EMPLOYEE	Name_tag	Numeric	1	Integer values from 1 to 9.
EMPLOYEE	Address	Alphanumeric	50	
EMPLOYEE	Gender	Alphabetic	1	M or F
EMPLOYEE	Date_hired	Date	8	
EMPLOYEE	Salary	Numeric	6	Ranges from $35,000 to $90,000
EMPLOYEE	No_of_dependents	Numeric	2	
Managed_by	Mgr_start_dt	Date		
DEPENDENT	Dname	Alphabetic	15	
DEPENDENT	Related_how	Alphabetic	12	
DEPENDENT	Birthdate	Date	8	
PROJECT	Pr_name	Alphabetic	20	
PROJECT	Pnumber	Numeric	2	Integer values from 1 to 40.
PROJECT	Plocation	Alphabetic	15	Bellaire, Blue Ash, Mason, Stafford, Sugarland
Assigned	Hours	Numeric	3	
BCU_ACCOUNT	Balance	Numeric	(8.2)*	
BCU_ACCOUNT	Account#	Alphanumeric	6	
BCU_ACCOUNT	Acct_type	Alphabetic	1	C=Checking Acct., S=Savings Acct., I=Investment Acct.
HOBBY	Hb_name	Alphabetic	20	
HOBBY	Io_activity	Alphabetic	1	I=Indoor Activity, O=Outdoor Activity
HOBBY	Gi_activity	Alphabetic	1	G=Group Activity, I=Individual Activity
Participates	Hrs_per_wk	Numeric	(2.1)*	
Participates	Annual_cost	Numeric	6	

*$(n_1.n_2)$ is used to indicate n_1 places to the left of the decimal point and n_2 places to the right of the decimal point

Entity-Level Business Rules
1. A mother or daughter dependent must be a female, a father or son dependent must be a male, and a spouse dependent can be either male or female.
2. An employee cannot be his or her own supervisor.
3. A dependent may have a joint account only with an employee of Bearcat Incorporated to whom he or she is related.
4. Either PLANT.Pnumber or Pl_name must be present.
5. Every plant is managed by an employee who works in the same plant.

Miscellaneous Business Rules
1. Each plant has at least three buildings.
2. The salary of an employee cannot exceed the salary of the employee's supervisor.

Table 3.2   Semantic integrity constraints for the Coarse-granular Design-Specific ER model

As noted in Section 3.2, a handful of popular notational schemes for the ER diagram exist. In this section and throughout the book, the **(min, max) notation**, originally prescribed by Abrial (1974) for specifying the structural constraints of a relationship, is used. Here, *min* depicts the minimum cardinality of an entity type's participation in a relationship type—the participation constraint—and *max* indicates the maximum cardinality of an entity type's participation in a relationship, thus reflecting the cardinality ratio. This notation is in general more precise and particularly more expressive for specifying relationships of higher degrees beyond binary (see Section 5.1.1). In this notation, a pair of finite whole numbers (min, max) is used with each *participation* of an entity type E in a relationship type R without any reference to other entity types participating in the relationship R where $0 \leq min \leq max$ and $max > 0$. The meaning conveyed here is that each entity e in E participates in at least "min" and at most "max" relationships (r_1, r_2, ...) in R. In this notation, min = 0 implies partial or optional participation of E in R, and min = or > 1 implies total or mandatory participation of E in R. The mapping of the structural constraints of a relationship from the presentation layer to the design-specific layer—in other words, converting Chen's notation to the (min, max) notation—is demonstrated in Figure 3.4. Figure 3.4a indicates that an employee may have several BCU accounts; but need not have any. This is done by "looking across" from EMPLOYEE to BCU_ACCOUNT through the relationship *Held_by_E* (Chen's notation). On the other hand, a BCU account belongs to no more than one employee and some BCU accounts need not belong to any employee. This is inferred by "looking across" from BCU_ACCOUNT to EMPLOYEE in the relationship *Held_by_E*. The same metadata is reflected in Figure 3.4b using the Crow's Foot notation. In the (min, max) notation shown in Figure 3.4c, the cardinality ratio (maximum cardinality) and participation (minimum cardinality) constraint in the *Held_by_E* relationship type are captured in terms of the *participation* of EMPLOYEE in the *Held_by_E* relationship type independent of the *participation* of BCU_ACCOUNT in the *Held_by_E* relationship type. An employee participates in at least zero (0) (optional participation) and at most m *Held_by_E* relationships, meaning an employee need not have a BCU account, but may have many BCU accounts. Likewise, a BCU account participates in at least zero (0) (optional participation) and at most one (1) *Held_by_E* relationship, meaning a BCU account need not belong to any employee, but can belong to a maximum of one employee. Note that the (min, max) notation employs a "look here" approach as opposed to the "look across" approach used in the Presentation Layer ER diagram.

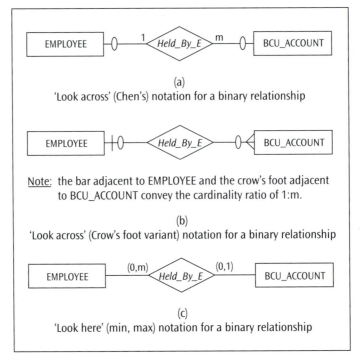

Figure 3.4   Introduction of (min, max) notation for a binary relationship

The mapping of the structural constraints described in the Presentation Layer ER diagram in Figure 3.3 to the (min, max) notation appears in the initial version of the coarse granularity level of the Design-Specific ER diagram in Figure 3.5. In contrast to what is shown in Figure 3.3, *observe how the (min, max) notation allows the requirement that a plant must have at least 100 employees to be explicitly specified*. Also, note that a weak entity type participating as a child in an identifying relationship type, since it also has existence dependency on the identifying parent, will always have a (min, max) value of (1,1) in that relationship.

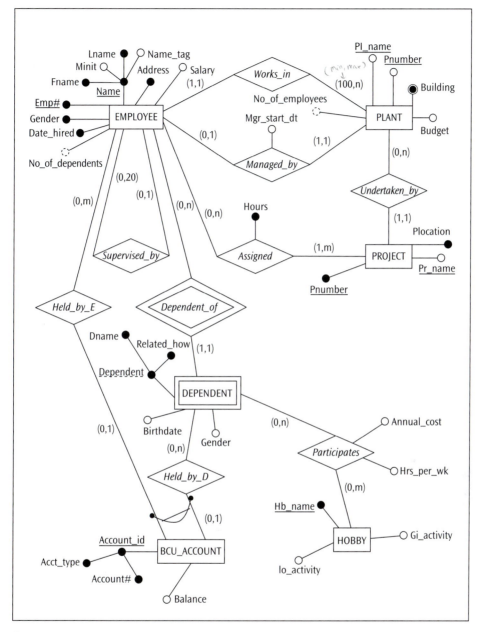

Figure 3.5   Coarse-granular Design-Specific ER diagram for Bearcat Incorporated – Stage 1

The next step is to incorporate the **deletion constraints** in the Design-Specific ER model. In the context of every relationship type in an ER model, the deletion of an entity from the parent entity type requires specific action either in the parent entity set or in the child entity set in order to maintain consistency of the relationships in the database. There are four rules that are applicable to deletion constraints: the restrict rule, the cascade rule, the "set null" rule, and the "set default" rule.[5]

- When a parent entity in a relationship is deleted, if all child entities related to this parent in this relationship should not be deleted, then the deletion of the parent should be disallowed. This is called the **restrict rule**.
- When a parent entity in a relationship is deleted, if all child entities related to this parent in this relationship should also be deleted, the **cascade rule** applies.
- When a parent entity in a relationship is deleted, if all child entities related to this parent in this relationship should be retained but no longer referenced to this parent, the **"set null" rule** applies.
- When a parent entity in a relationship is deleted, if all child entities related to this parent in this relationship should be retained but no longer referenced to this parent but should be referenced to a predefined default parent, the **"set default" rule** applies.

For example, Figure 3.6 illustrates these deletion rules in the context of a relationship between the entity type FACULTY_MEMBER and the entity type PHD_STUDENT. In Figure 3.6a, the restrict rule (R) prohibits the deletion of a faculty member serving as the dissertation chair of one or more Ph.D. Students. On the other hand, in Figure 3.6b, the cascade rule (C) implies that the deletion of a faculty member leads to the deletion of all Ph.D. students for whom the faculty member serves as dissertation chair. The "set null" rule (N) allows a Ph.D. student to exist without a dissertation chair by simply nullifying the relationship of the Ph.D. student with the faculty member, should the faculty member be deleted (see Figure 3.6c). Observe how total participation (i.e., min = 1) in Figure 3.6c is incompatible with the "set null" rule and must therefore be corrected to permit partial participation of PHD_STUDENT in the *Diss_chair* relationship (i.e., min = 0). Finally, the "set default" rule (D) portrayed in Figure 3.6d is somewhat similar to the "set null" rule. Here, instead of nullifying the relationship, the Ph.D. student is linked to a predetermined (default) dissertation chair, should the student's current dissertation chair be deleted. Conventionally, when a deletion constraint is not specified, the restrict rule is implied.

When the cardinality ratio in a binary relationship is 1:n, one of the entity types in the relationship is unambiguously the parent and the other is the child. On the other hand, in m:n and 1:1 relationship types it is not possible to unequivocally assign the parent/child role to either one of the participating entity types. Under these circumstances, the deletion rules will be predicated on all entity types participating in the relationship under assessment.

Also, note that the restrict rule always entails an action on the parent entity in a relationship whereas the other three rules always result in actions on the child.

It is also important to understand that deletion rules cannot be arbitrarily applied in a relationship type. For instance, in an identifying relationship type, imposition of the "set null" rule on a weak entity type is invalid because the weak entity type has existence

---

[5] Some argue that the deletion constraints belong in physical database design. It is our view that the semantics for the deletion constraints also emerge from user-specified business rules and ought to be captured, modeled, and passed through the data modeling tiers. A similar constraint is necessary when occasionally the value of the unique identifier of the parent entity is changed. This type of constraint is called an update constraint. This is discussed in Section 10.1.1.1 of Chapter 10.

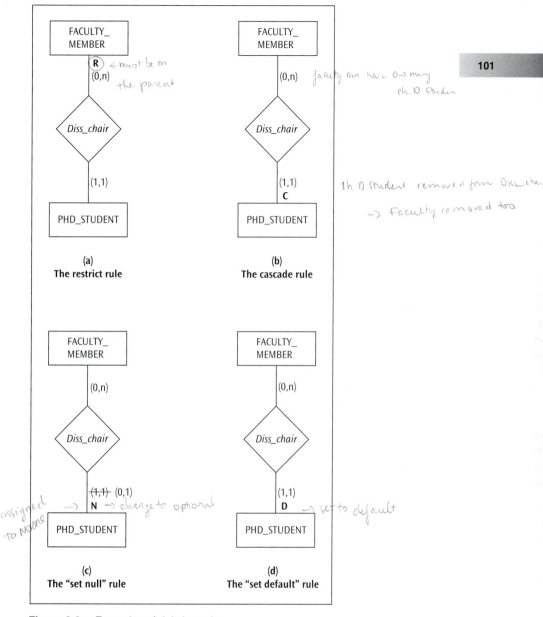

The following handwritten annotations appear on the figure:

(a) R ← must be on the parent

(b) faculty can have 0→many ph.D students

Ph.D student removed from Diss_chair → Faculty removed too

(c) (1,1) → N → change to optional; assigned to No one

(d) D → set to default

Figure 3.6 Examples of deletion rules

dependency in the identifying relationship. Likewise, specification of the "set null" rule is incompatible with an m:n binary relationship type. In the case of an 1:1 binary relationship type, specification of the restrict rule on both the participating entity types either explicitly or by default is incorrect since there is an opportunity for mutually referencing entities of the two entity types to be hung in a deadlock. In other words, deletion of either one may get restricted by the other; as a consequence, neither can ever be deleted.

Entity-Relationship Modeling

Similar conflicts may arise in a hierarchy of relationships where the cascade rule at one level may contradict a restrict rule specified at the next lower level in the hierarchy.

A total of 12 deletion constraints were identified during the development of the Presentation Layer ER model and listed in Table 3.1. Figure 3.7 illustrates how the representation of these constraints can be incorporated into the coarse granularity level of the Design-Specific ER diagram. Each of the 12 deletion constraints repeated below is followed *in italics* by a description of how the relevant deletion rule is shown in Figure 3.7. Note that Figure 3.7 contains two planted errors.

1. A plant with employees cannot be closed down.

    *The R adjacent to the PLANT entity type indicates the **restriction** of the deletion of a plant if the plant has one or more employees. However, the deletion of a plant without employees is permitted.*

2. If an employee is deleted, all BCU accounts of that particular employee must be deleted.

    *The C immediately above the BCU_ACCOUNT entity type under the relationship Held_by_E indicates that the deletion of an employee should **cascade** through to delete all BCU accounts associated with this employee.*

3. If a plant is closed down, the projects undertaken by that plant cannot be canceled. The project assignments from a closed plant must be temporarily removed in order to allow the project to be transferred to another plant.

    The N (representing the "set null" rule) immediately above the PROJECT entity type indicates that a project's relationship with a plant can be temporarily removed resulting in the project not being undertaken by (i.e., under the control of) any plant.

Observe that the specification of the "set null" rule here conflicts with the minimum cardinality (participation constraint) of PROJECT in *Undertaken_by*. The total participation constraint requiring that a project must be controlled by a plant conflicts with the 'set null' rule that allows a project to be temporarily removed from a plant. As a consequence, one or the other of these two constraints must be changed. Figure 3.8 incorporates this change by making the participation constraint partial (min = 0)—that is, optional participation of PROJECT in *Undertaken_by*—instead of total (mandatory participation of PROJECT in *Undertaken_by*).

4. The Human Resources Department uses a designated default employee number to replace a supervisor who leaves the company.

    The D (representing the "set default" rule) shown immediately below the EMPLOYEE entity type indicates that if a supervisor is deleted all supervisees under that supervisor are reassigned to the designated default supervisor.

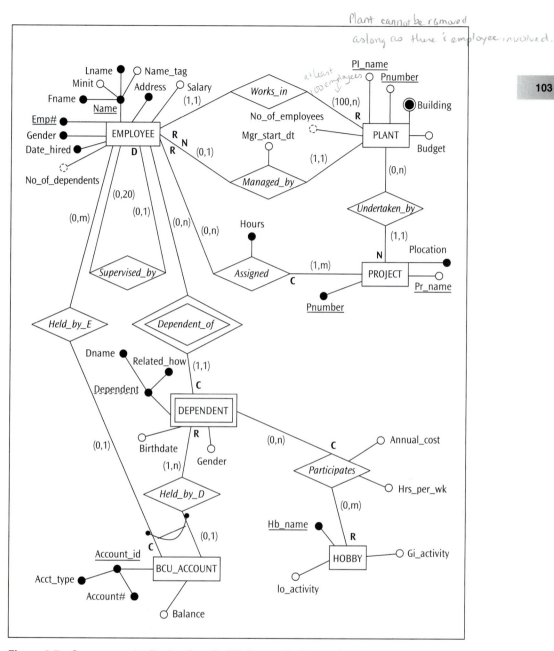

Figure 3.7 Coarse-granular Design-Specific ER diagram for Bearcat Incorporated – Stage 2
{deletion constraints added}

5. An employee currently managing a plant cannot be deleted from the database.

   The R adjacent to the EMPLOYEE entity type indicates the **restriction** *of a deletion of an employee if the employee manages a plant since company policy dictates that every plant must have a manager.*

6. If a plant is closed down, the employee no longer manages the plant but becomes an employee of another plant.

   The N adjacent to the EMPLOYEE entity type indicates that an employee's relationship with a plant as a manager can be removed resulting in the employee no longer playing the role of a manager anymore.

7. If an employee leaves the company, all dependents and BCU accounts of the employee must be removed.

   The C immediately above the DEPENDENT entity type indicates that the deletion of an employee should **cascade** *through to delete all dependents associated with this employee Similarly, the C immediately above the BCU ACCOUNT entity type indicates that the deletion of an employee should* **cascade** *through to delete all BCU accounts associated with this employee.*

8. As long as a dependent has a BCU account, deletion of the dependent is not permitted.

   The R immediately below the DEPENDENT entity type indicates the **restriction** *of the deletion of a dependent if the dependent has one or more BCU accounts. However, the deletion of a dependent without a BCU account is permitted.*

Observe that constraints 7 and 8 create a conflict when an attempt is made to delete an employee who has a dependent with a BCU account. Note that while constraint 7 requires that all dependents of the employee be deleted, constraint 8 restricts (i.e., prohibits) the deletion of a dependent with a BCU account. In cases such as this, the analyst must meet with the user(s) and modify the business rule(s) that gave rise to the conflicting constraints. In this situation, either (a) the deletion of a dependent must cascade through to delete all of the dependent's BCU accounts, or (b) the deletion of an employee who has dependents with BCU accounts must be restricted. The constraint requiring that the deletion of a dependent must cascade through to delete all of the dependent's BCU accounts is shown in Figure 3.8.

Note that the two planted errors present in Figure 3.7 have been corrected in Figure 3.8.

9. As long as an employee is assigned to a project, his or her record cannot be removed from the database.

   The R adjacent to the EMPLOYEE entity type indicates the **restriction** *of a deletion of an employee who is assigned to one or more projects. Once the assignments of the employee have been deleted, then the employee can be deleted.*

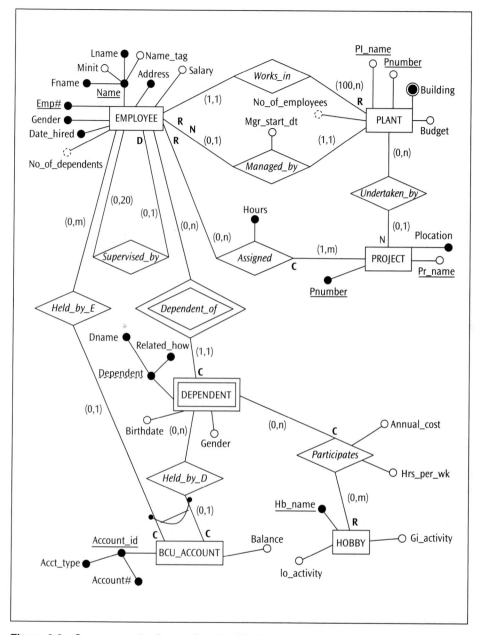

Figure 3.8 Coarse-granular Design-Specific ER diagram for Bearcat Incorporated – Final

10. If a project is deleted, all assignments of employees to that project must be deleted.

   The C adjacent to the Assigned relationship type indicates that the deletion of a project should **cascade** *through to delete all related assignments of employees to that project.*

11. If a dependent is deleted, all records of the participation of that dependent in hobbies must be deleted.

    The C above the Participates relationship type indicates that the deletion of a dependent should **cascade** *through to delete all related participations of the dependent in hobbies.*

12. A hobby with at least one dependent participating in it cannot be deleted.

    The R immediately above the HOBBY entity type indicates the **restriction** *of a deletion of a hobby if the hobby has one or more participants. Once the participation of all dependents in a particular hobby has been deleted, then the hobby can be deleted.*

The other business rules not mapped to the Coarse-granular Design-Specific ER diagram are included with the semantic integrity constraints for the Design-Specific ER model in Table 3.2. Thus, the Design-Specific ER model comprising the ER diagram in Figure 3.8 and constraint specifications in Table 3.2 fully preserve all constructs and constraints reflected in the Presentation Layer ER model.

Now we have seen how the user-oriented Presentation Layer ER model is translated to a database design orientation. To this end, three specific steps were taken to develop a Design-Specific ER model (coarse granularity):

1. Collection of a few characteristics for attributes (for example, data type and size)
2. Introduction of the technically more precise (min, max) notation for the specification of relationships in the ER diagram
3. Mapping of the deletion rules to the ER diagram

The next section presents the second stage of this line of enquiry in order to render the conceptual schema amenable to direct mapping to the logical level.

3.2.4 The Fine-granular Design-Specific ER Model

The Coarse-granular Design-Specific ER model incorporates data modeling constructs in the conceptual model beyond the Presentation Layer model. However, the Design-Specific coarse layer, while capturing the conceptual design reflected in the Presentation Layer ER model in all its richness, contains constructs that cannot be directly mapped to certain logical data models. It would be ideal to be able to portray all the specified constructs and constraints—inherent, implicit, and explicit—in the ER diagram itself so that an integrated view of the database design is available in one place. The incremental contributions of the Fine-granular Design-Specific ER model are:

- Mapping the attribute characteristics to the ER diagram
- Appropriately decomposing constructs in the coarse layer that cannot be directly mapped to a logical schema—i.e., preparing a Fine-granular Design-Specific ER model for direct mapping to a logical schema

At the same time, the richness captured in the coarse layer is fully preserved in this transformation to the Fine-granular Design-Specific ER model, even though all the constructs and constraints are not fully incorporated exclusively in the ER diagram. Table 3.3 contains the translated semantic integrity constraints not captured in the Fine-granular Design-Specific ER diagram (shown later in Figure 3.11). The ER diagram and the constraint specifications in Table 3.3 together are fully information preserving.

> Constraint	PLANT.Pnumber	IN (10 through 20)
> Constraint	Nametag	IN (1 through 9)
> Constraint	Gender	IN ('M', 'F')
> Constraint	Salary	IN (35000 through 90000)
> Constraint	PROJECT.Pnumber	IN (1 through 40)
> Constraint	Plocation	IN ('Bellaire', 'Blue Ash', 'Mason', 'Stafford', 'Sugarland')
> Constraint	Acct_type	IN ('C', 'S', 'I')
> Constraint	Io_activity	IN ('I', 'O')
> Constraint	Gi_activity	IN ('G', 'I')
> Constraint	Related_how Related_how Related_how	IN ('Spouse') OR IN (('Mother', 'Daughter') AND Gender IN ('F')) OR IN (('Father', 'Son') AND Gender IN ('M'))
> Constraint	Building	COUNT (not < 3)
> Constraint	No_of_employees	NOT < 100

Constraints Carried Forward to Logical Design

1. An employee cannot be his or her own supervisor.
2. A dependent can have a joint account only with an employee of Bearcat Incorporated with whom he or she is related.
3. The salary of an employee cannot exceed the salary of the employee's supervisor.
4. Either PLANT.Pnumber or Pl_name must have a value.
5. Every plant is managed by an employee who works in the same plant.

Table 3.3   Semantic integrity constraints for the Fine-granular Design-Specific ER model

Of the various constructs and constraints supported by the ER modeling grammar, multi-valued attributes and binary relationships of cardinality ratio m:n, while adding to the expressive power of a conceptual schema, cannot be directly implemented in a logical data model. As a consequence, these two constructs will have to be decomposed to lower-level constructs that can be directly mapped to a logical schema.

3.2.4.1 Resolution of multi-valued attributes

A multi-valued attribute in an entity type can be transformed to a single-valued attribute by simply including the attribute as part of the unique identifier of the entity type. Figure 3.9a shows the resolution of the multi-valued attribute **Building** in the entity type PLANT for Bearcat Incorporated. Note that the presence of two unique identifiers in PLANT makes the solution rather interesting in the sense that the attribute **Building** is appended to either **Pnumber** or **Plname** to form a single new unique identifier, **Pl_id**. Table 3.4a shows sample data for each variation.

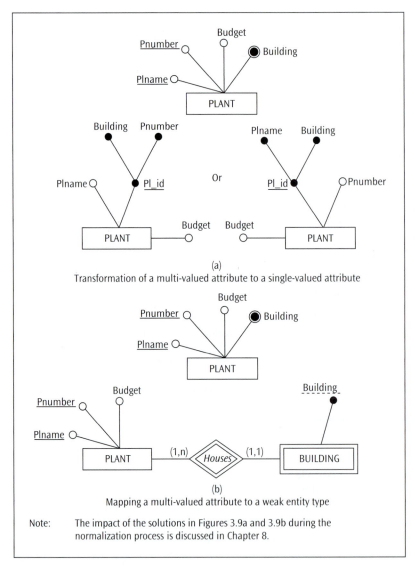

(a)
Transformation of a multi-valued attribute to a single-valued attribute

(b)
Mapping a multi-valued attribute to a weak entity type

Note: The impact of the solutions in Figures 3.9a and 3.9b during the normalization process is discussed in Chapter 8.

Figure 3.9   Two methods for the resolution of a multi-valued attribute

Original PLANT Data Set

Pnumber	Plname	Budget	Building
10	Underwood	3000000	1
			2
			3
11	Garnett	3000000	1
			2
12	Belmont	3500000	1
13	Vanderbilt	3500000	1
			2

Revised PLANT Data Set- Variation 1

Pnumber	Building	Plname	Budget
10	1	Underwood	3000000
10	2	Underwood	3000000
10	3	Underwood	3000000
11	1	Garnett	3000000
11	2	Garnett	3000000
12	1	Belmont	3500000
13	1	Vanderbilt	3500000
13	2	Vanderbilt	3500000

Revised PLANT Data Set- Variation 2

Pnumber	Building	Plname	Budget
10	1	Underwood	3000000
10	2	Underwood	3000000
10	3	Underwood	3000000
11	1	Garnett	3000000
11	2	Garnett	3000000
12	1	Belmont	3500000
13	1	Vanderbilt	3500000
13	2	Vanderbilt	3500000

(a)

Sample data illustrating the transformation of a multi-valued attribute to a single-valued attribute (see Figure 3.9a)

Original PLANT Data Set

Pnumber	Plname	Budget	Building
10	Underwood	3000000	1
			2
			3
11	Garnett	3000000	1
			2
12	Belmont	3500000	1
13	Vanderbilt	3500000	1
			2

Revised PLANT Data Set

Pnumber	Plname	Budget
10	Underwood	3000000
11	Garnett	3000000
12	Belmont	3500000
13	Vanderbilt	3500000

Revised BUILDING Data Set

Building
1
2
3
1
2
1
1
2

(b)

Sample data illustrating the transformation of a multi-valued attribute to a weak entity type (see Figure 3.9b)

Table 3.4　Sample data sets for Figure 3.9

Entity-Relationship Modeling

An alternative solution to decomposing a multi-valued attribute to a single-valued attribute is to transform the multi-valued attribute to an entity type. In this case, BUILDING will become a weak entity type child in an identifying relationship with PLANT and the attribute **Building** will serve the role of the partial key of BUILDING (see Figure 3.9b and Table 3.4b).

The first solution (Figure 3.9a), apparently simpler and more efficient than the alternate solution (Figure 3.9b), may pose a data redundancy problem that will be resolved during normalization.[6] Interestingly, the resolution in most cases will result in a schema equivalent to the alternate solution proposed in Figure 3.9b.

3.2.4.2 Resolution of m:n relationship types

A binary relationship with a cardinality ratio of m:n also poses a problem in that it cannot be represented in a logical data model as is (that is, as an m:n relationship). The solution for this problem is to decompose the m:n relationship to a 1:n and a 1:m relationship with a newly created entity type serving as a bridge between the two base entity types in the binary relationship. Since such a bridging entity type at the intersection of the two base entity types is artificially created, it does not have its own natural unique identifier. As a consequence, the entity type thus created is in essence a weak entity type with two identifying parents. This is called a **gerund entity type**.[7] A gerund entity type is sometimes referred to as an **associative entity type**, a **composite entity type**, or a **bridge entity type**.

Any attributes of the m:n relationship type now become the attributes of the gerund entity type. The example in Figure 3.10 illustrates the resolution of an m:n cardinality ratio in a binary relationship type. Observe that the relationship type *Assigned* in Figure 3.10a is expressed as a weak entity type ASSIGNMENT in the decomposition shown in Figure 3.10b. In essence, a new (gerund) entity type has been created in place of the relationship type depicting the m:n relationship between EMPLOYEE and PROJECT. The direct relationship between EMPLOYEE and PROJECT no longer exists. The new (gerund) entity type, ASSIGNMENT, is not only related to the EMPLOYEE as well as the PROJECT but also both the relationships are identifying relationship types. This means that ASSIGNMENT has two identifying parents and has identification dependency on both parents (EMPLOYEE and PROJECT). As a result, the m:n cardinality ratio has been decomposed to a 1:n (*Uses*) and 1:m (*Belongs_to*) relationship.

---

[6] Normalization is covered in Chapter 8.

[7] This issue is not unique to binary relationships. Any n-ary relationship (ternary, quaternary, etc.) poses the same problem and has a similar solution. This issue is discussed further in Chapter 5.

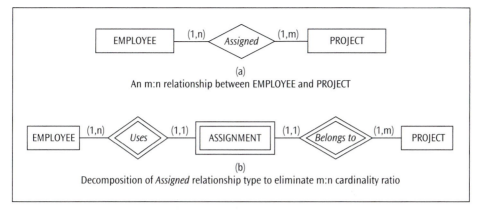

Figure 3.10   Resolution of the m:n cardinality ratio

3.2.4.3 The Fine-granular Design-Specific ER model

The final transformation of the Design-Specific ER model from coarse granularity to fine granularity for Bearcat Incorporated is shown in Figure 3.11. In addition to the transformations discussed above (that is, multi-valued attribute, and m:n relationship), this figure introduces a notation that allows the data type and size associated with each atomic attribute listed in Table 3.2 to be incorporated into an ER diagram. Enclosed in square brackets [ ] adjacent to the name of each atomic attribute is its data type (**A**—Alphabetic data type, **N**—Numeric data type, **X**—Alphanumeric data type, and **Dt**—Date data type) followed by its size. When a numeric attribute represents a decimal value, its size is represented by $n_1.n_2$, where n_1 is the scale (number of positions to the left of the decimal point) and n_2 is the precision (number of positions to the right of the decimal point).

Note that the domain constraint on the **Emp#** in Table 3.2 specifies a numeric and alphabetic component for **Emp#**. This requirement is expressed in Figure 3.11 by making **Emp#** a composite attribute composed of the two atomic attributes **Emp_n** and **Emp_a**.

The integrity constraints not mapped to the Fine-granular Design-Specific ER diagram are included with the semantic integrity constraints for the Design-Specific ER model in Table 3.3. Thus, the Fine-granular Design-Specific ER model comprising the ER diagram in Figure 3.11 and constraint specifications in Table 3.3 fully preserves all constructs and constraints reflected in the Coarse-granular Design-Specific ER model.

Entity-Relationship Modeling

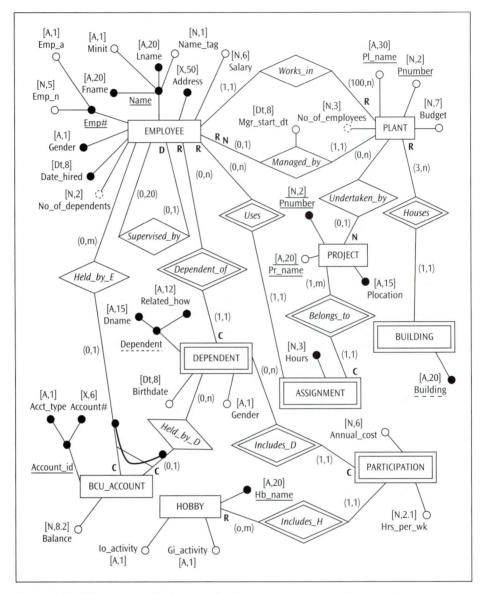

Figure 3.11 Fine-granular Design-Specific ER diagram for Bearcat Incorporated

Chapter Summary

The ER modeling grammar for the conceptual data modeling activity includes an ER diagram as well as the specification of semantic integrity constraints not captured in the ER diagram. The ER modeling framework presented in this book contains two basic layers: a Presentation Layer and a Design-Specific Layer. The Design-Specific ER model in turn is made up of two layers—the coarse and fine granularities.

The case of Bearcat Incorporated, a manufacturing company with several plants located in the northeastern part of the United States, is used to illustrate the ER modeling framework, proceeding from the Presentation Layer ER model to the Design-Specific ER model.

The Presentation Layer ER model is the principal vehicle for communicating with the end-user community. The Presentation Layer ER diagram contains the initial definition of attributes, entity types, and relationship types. Business rules and data requirements not incorporated into the Presentation Layer ER diagram are expressed as semantic integrity constraints (domain constraints on an attribute or collection of attributes, deletion constraints, and miscellaneous constraints).

Transformation of the end-user oriented Presentation Layer ER model to the database designer oriented Design-Specific ER model incorporates additional details about the characteristics of attributes obtained from the users. The Design-Specific ER diagram makes use of the (min, max) notation to represent the structural constraints and allows for the deletion constraints to be explicitly shown at the coarse-granular level. Next, the ER model with the greatest detail, the Fine-granular Design-Specific ER model, maps the attribute characteristics to the ER diagram and decomposes specific constructs (multi-valued attributes and m:n relationship types) to prepare the conceptual schema for direct mapping to a logical schema. The richness of the original requirements specification is preserved all the way through the conceptual modeling process.

Exercises

1. What is information preservation and why is it important?
2. Describe what constitutes an ER model.
3. What is the focus of a Presentation Layer ER model?
4. Crow's Foot notation and IDEF1X notation are two other popular notations for the ER modeling grammar. Investigate these two grammars and compare them with the Chen's notation used in this chapter.
5. Examine the CASE tools, ERWin and Oracle/Designer, and discuss the ER modeling grammar supported by each of them.
6. Give examples of types of business rules that are not reflected in a Presentation Layer ER diagram.
7. Consider the following Presentation Layer ER diagram.
   a. Identify the base and weak entity types. What is (are) the unique identifier(s) of each entity type? Which unique identifiers are composite attributes?
   b. Identify the partial key(s).

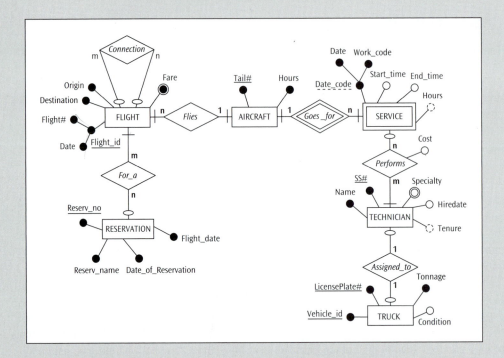

c. Identify the optional attributes. Which of these are multi-valued attributes?

d. Identify the derived attributes. What is it about these attributes that allow them to be considered "derived" attributes?

e. Identify the recursive and binary relationship types.

f. Which relationship types exhibit (a) total participation of each entity type, (b) partial participation of each entity type, and (c) total participation of one entity type and partial participation of the second entity type?

g. Describe the nature of each relationship type with (a) 1:1 cardinality ratio, (b) a 1:n cardinality ratio, and (c) an m:n cardinality ratio.

h. What is the significance of the use of the double diamond in naming the *Goes_for* relationship type versus the use of a single diamond to name the *Performs* relationship type?

i. Describe an obvious business rule that would be associated with the **Date_of_ Reservation** attribute in the RESERVATION entity type.

j. What does the assignment of the attribute **Cost** to the *Performs* relationship type mean?

8. What is the difference between an exclusive arc and an inclusive arc?

9. The figure below is a reproduction of Figure 2.30b in Section 2.4.3 of Chapter 2 pertaining to Vignette 3. We were unable to portray the business rule: "*a manufactured part is not purchased and vice versa*" in that ER diagram and so chose to carry forward the business rule as an item in a semantic integrity constraints list. Incorporate the business rule in the ER diagram now and explain how the revised ER diagram portrays this business rule.

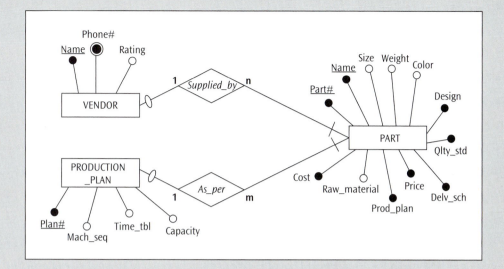

10. Suppose in the previous exercise the business rule is changed to indicate that, due to insufficient production capacity, some of the manufactured parts are also purchased. Revise the ER diagram to capture this revised business rule and explain how the revised ER diagram portrays this business rule.

11. What is a deletion constraint?

12. What four deletion rules are applicable to deletion constraints? Which rule(s) refer to an action on the parent and which rule(s) refer to an action on the child?

13. Is the use of the "set null" rule applicable to an identifying relationship type? If yes, explain. If no, definitely explain.

14. What must be done to develop a Coarse-granular Design-Specific ER model from a Presentation Layer ER model?

15. When used in the context of data modeling, what is meant by the use of the term "mapping"?

16. What constructs in a Coarse-granular Design-Specific ER diagram cannot be directly mapped to a logical schema? What is required to represent these constructs in a Fine-granular Design-Specific ER diagram?

17. Assume that data are maintained on airports around the country for a company that offers a flight chartering service for college basketball teams. They gather information from a wide variety of sources, but have had considerable difficulty in the past obtaining data on runway surfaces at airports located in small college towns. Company pilots need access to this information when they land, either at airports such as these or in occasional emergencies when they must land at other small airports that are not located in college towns. As a result, as shown here, they have created a RUNWAY_SURFACE entity type separate from a RUNWAY entity type.

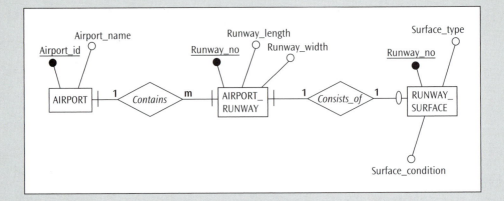

a. What does the cardinality ratio and participation constraint suggest about some airport runways in the *Consists_of* relationship?

b. How might the Presentation Layer ER diagram be revised to make the relationship between an airport runway and runway surface more efficient?

18. Transform the Presentation Layer ER diagram for Exercise 7 to a Coarse-granular Design-Specific ER diagram that makes use of the (min, max) notation for specifying the structural constraints of all relationships.

19. Which ER constructs in the Coarse-granular Design-Specific ER diagram in Exercise 7 need to be decomposed preparatory to logical model mapping? Transform the Coarse-granular Design-Specific ER diagram to a Fine-granular Design-Specific ER diagram.

20. Nilakanta, Inc., a factory manufacturing miscellaneous spare parts, wants to implement a computer-based information system to manage their operation. Recently a team of systems analysts worked with the various end-users in the factory and developed the requirements specification for this system. Since they know that in most large companies database development and application development are done by different groups in the IS department, the team wrote separate specifications for data requirements and process requirements. The data requirements for the system are described below. Your task is to develop a conceptual design that captures the data specifications using the ER modeling grammar.

The factory has several departments. A department may have many employees but at least three. Every employee works for one and only one department. Every department has a manager—only one manager per department. Clearly, a manager is an employee of the company, but all employees are not managers. A department may have many machines and every machine is assigned to a specific department. A machine may go for maintenance numerous times. Maintenance is done on a machine only once on a given day. Some machines are so new that they may not have gone for maintenance yet.

Products are produced on machines. A product can be an assembly of several different components or a single piece. Every product/component goes through one or more machines for appropriate production operations. Likewise, several products may go through a particular machine for some standard process. Designers design the products. Some well-qualified designers may design more than one product. All designers are employees of the factory. Operators, who are also employees of the factory, operate these machines. Due to multiple shifts, several operators operate the same machine. All operators are routinely assigned to work on only one machine and no operator is kept idle. A machine is never kept idle either, except when it is out for maintenance.

The database should capture employee's name, which would include first name, last name and middle initial. It must also capture gender, address and salary. An employee's salary cannot exceed his/her manager's salary. A Social Security number (SSN) uniquely identifies an employee. Likewise, department number, department name, type, and location must be captured. The department number and department name are both unique identifiers of a department. Every machine will have a unique machine number. It will also consist of other details like name of machine, type, and vendor's name. When a machine goes for maintenance, the maintenance date for that machine needs to be captured. A product is identified by its component ID. The Component name and description must also be recorded. It should be possible to compute the number of components in a product. When a component goes through machining, the starting time, completion time, and hours of machining for each product on every machine must be captured. The information about a designer includes his/her qualification, specialization field, and experience in years. Operators, who are responsible for operating the machines, belong to a labor union and have certain skill sets associated with them.

Develop a Presentation Layer ER model for Nilakanta, Inc. The ERD should be fully specified with the unique identifiers, other attributes for each entity type, and the relationship types that exist among the various entity types. All business rules that can be captured in the ERD must be present in the ERD. Any business rule that cannot be captured in the ERD should be specified as part of a list of semantic integrity constraints.

21. The NCAA (National Collegiate Athletic Association) wants to develop a database to keep track of information about college basketball. Each university team belongs to only one conference (e.g., the University of Houston belongs to Conference USA; the University of Cincinnati belongs to the Big East Conference, etc.), but a few teams may not belong to any conference. A conference has several teams; no conference has less than five (5) teams. Each team can have a maximum of twenty (20) players and a minimum of thirteen (13) players. Each player can play for only one team. Each team has from three (3) to seven (7) coaches on its coaching staff and a coach works for only one team. Lots of games are played in each university location every year; but a game between any two universities is played at a given location only one time a year. Three referees from a larger pool of referees are assigned to each game. A referee can work several games; however some referees may not be assigned to any game. Players are called players because they play in games; in fact several games. A game involves at least ten players. It is possible that some players simply sit on the bench and do not play in any game. Player performance statistics (i.e., points scored, rebounds, assists, minutes played, and personal fouls committed) are recorded for each player for every game. Information collected about a game includes the final score, the attendance, and the date of the game. During the summer months, some of the players serve as counselors in summer youth basketball camps. These camps are identified by their unique campsite location (e.g., Mason, Bellaire, Kenwood, League City, etc.). Each camp has at least three (3) players who serve as counselors, and a player serving as a counselor may work in a number of camps.

A player can be identified by student number (i.e., Social Security number) only. The other attributes for a player include name, major, and grade point average. For a coach, relevant attributes include name, title (e.g., head coach, assistant coach), salary, address, and telephone number. Attributes for a referee include name, salary, years of experience, address, telephone number, and certifications. Both coaches and referees are identified by their personal NCAA identification number. A team is identified by the name of the university (i.e., team). Other team attributes include current ranking, capacity of home court, and number of players. Each conference has a unique name, number of teams, and an annual budget. For the basketball camps, data is available on the campsite (i.e., location) and the number of courts.

Develop a Presentation Layer ER model for the NCAA database. The ERD should be fully specified with the unique identifiers, other attributes for each entity type, and the relationship types that exist among the various entity types. All business rules that can be captured in the ERD must be present in the ERD. Any business rule that cannot be captured in the ERD should be specified as part of a list of semantic integrity constraints.

Hint: No more than seven entity types are needed to complete this design.

22. This exercise contains additional information in the form of deletion rules that will enable us to develop a Design-Specific ER diagram for the NCAA database in Exercise 21.

When a referee retires, all links to the games handled by that referee should be removed. Likewise, if a game is cancelled, all links to the referees for that game should be dropped. Although it is does not happen often, a university may sometimes leave the conference of which it is a member. Naturally, we want to keep the team in the database since the university could decide to join another conference at a later date. However, if a team (university) leaves the NCAA altogether, all players and coaches of that team should be removed from the database along with the team. In all other relationships that exist in the database, the default value of "Restriction of Deletion" should be explicitly indicated.

a. Develop a Coarse-granular Design-Specific ER diagram for the NCAA database.

b. Transform the design in (a) to a Fine-granular ER diagram. Note that attribute characteristics are not provided and thus need not appear in the diagram.

Selected Bibliography

Abrial, J. (1974) "Data Semantics," in Klimbie and Koffeman (eds.): Data Base Management, North-Holland.

Elmasri, R.; and Navathe, S. B. (2003) *Fundamentals of Database Systems*, Fourth Edition, Addison-Wesley.

Fahrner, C.; and Vossen, G. (1995) "A Survey of Database Design Transformations Based on the Entity-Relationship Model," *Data & Knowledge Engineering*, 15, 213–250.

Navathe, S. B. (1992) "Evolution of Data Modeling for Databases," *Communications of the ACM*, 35, 9 (September) 112–123.

Song, I.; Evans, M.; and Park, E. K. (1995) "A Comparative Analysis of Entity-Relationship Diagrams," *Journal of Computer & Software Engineering*, 3, 4, 427–459.

Teorey, T. J. (1999) *Database Modeling and Design*, Third Edition, Morgan Kaufman.

Teorey, T. J.; Yang, D.; and Fry, J. P. (1986) "A Logical Design Methodology for Relational Databases Using the Extended Entity-Relationship Model," *Computing Surveys*, 18, 2 (June) 197–222.

## CHAPTER 4

Enhanced Entity-Relationship (EER) Modeling

Enhanced entity-relationship (EER) modeling is an extension to ER modeling that incorporates additional constructs to enhance the capability of the ER modeling grammar for conceptual data modeling. Sometimes the EER model is also referred to as the extended entity-relationship model. The additional semantic modeling constructs in the EER model were developed in response to the demands of more complex database applications in the 1980s and the inadequacy of the ER model to fulfill these requirements. Some of these constructs/concepts were also independently developed in related areas such as software engineering (object modeling) and artificial intelligence (knowledge representation).

This chapter introduces the Superclass/subclass (SC/sc) relationship as the basis for the EER modeling specialization/generalization and categorization constructs. Section 4.1.1 presents a scenario that exposes the inadequacy in the expressive power of ER modeling constructs. Section 4.1.2 introduces the concept of the intra-entity class relationship and the SC/sc relationship as the building block to model such intra-entity class relationships. Sections 4.1.3 through 4.1.8 follow with descriptions of EER constructs that extend the power of the ER model. Section 4.1.3 elaborates on the general properties of an SC/sc relationship, while Section 4.1.4 engages in an in-depth coverage of specialization/generalization. Section 4.1.5 pertains to the development of the specialization/generalization construct in a hierarchy and network of relationships. The categorization construct is discussed in Section 4.1.6. Specialization/generalization and categorization essentially serve two different modeling needs. Section 4.1.7 gives some guidelines on how to determine which construct to use in a business database design scenario. While aggregation is often discussed as a construct in the object-oriented paradigm, a presentation of this construct appears in Section 4.1.8 because aggregation adds to the expressive power of the EER modeling grammar. Section 4.2 describes the process of converting the Presentation Layer EER diagram to the Design-Specific EER diagram as a preliminary step for mapping the conceptual schema to a logical schema. Section 4.3 begins with an extension to the Bearcat Incorporated story that incorporates business rules which require the use of EER modeling constructs. Finally, Section 4.4 takes the reader through the conceptual data modeling layers for Bearcat Incorporated.

4.1 Superclass/subclass Relationship

Vignette 1 presents a scenario to set the stage for introducing the building block of enhanced ER modeling constructs, namely the Superclass/subclass relationship, as an extension to the ER modeling grammar.

4.1.1 Vignette 1

This vignette is about Division I collegiate sports where universities provide academic support in the form of academic advising, tutoring, career counseling, and so on to student-athletes who participate in the sports programs sponsored by the university. In particular, a university provides this support to numerous student-athletes while each student-athlete receives such support from only the university he or she attends. Every student-athlete participates in one or more sponsored sports. Football, basketball, and baseball are among the many sponsored sports. Attributes common to all student-athletes, regardless of sport, include student# (a unique identifier), name, major, grade point average, eligibility, sport (a multi-valued attribute), weight, and height. For student-athletes participating in the sports of football, basketball, and baseball, attributes specific to each sport are also collected. These include:

- *For football players: touchdowns, position, speed, and uniform#.*
- *For basketball players: uniform# (a unique identifier), position (a multi-valued attribute), points scored per game, assists per game, and rebounds per game.*
- *For baseball players: uniform# (a unique identifier), position (a multi-valued attribute), batting average, home runs, and errors.*

For other sports, such sport-specific attributes are not available. Also, note that uniform# is not a unique identifier of football players because it is possible for two football players to have the same uniform number in squads with more than 100 players (one may be an offensive player and the other a defensive player).

Figure 4.1 models a student-athlete who participates in one or more sports including football, basketball, or baseball by creating the entity type STUDENT_ATHLETE with three optional composite attributes **Football_player**, **Basketball_player**, and **Baseball_player**. This design is semantically and syntactically correct in the context of this vignette. Now, suppose that student-athletes who participate in basketball may also serve as counselors in summer basketball camps. Since the ER modeling grammar does not permit modeling a relationship between an entity type and an attribute of an entity type (see Vignette 1 of Section 2.4.1 in Chapter 2 for an illustration), a relationship between BASKETBALL_CAMP (an entity type) and **Basketball_player** (an attribute) is syntactically incorrect.

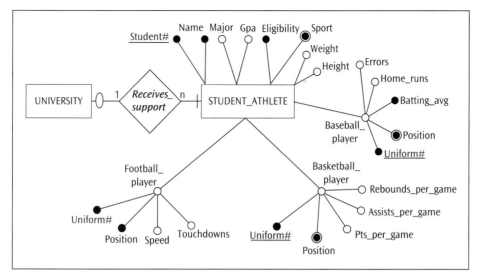

Figure 4.1 Modeling STUDENT_ATHLETE with optional composite attributes

Figure 4.2 replaces the three composite attributes with the three entity types FOOTBALL_PLAYER, BASKETBALL_PLAYER, and BASEBALL_PLAYER, each of which contains those attributes specific to that sport. Creating three 1:1 relationships between the STUDENT_ATHLETE entity type and the FOOTBALL_PLAYER, BASKETBALL_PLAYER, and BASEBALL_PLAYER entity types permits modeling a relationship between BASKETBALL_PLAYER and BASKETBALL_CAMP. While the possibility that a student-athlete may participate in more than one of the sports of football, basketball, and baseball is implicit in the ERD shown in Figure 4.2, one may also express it explicitly as a semantic integrity constraint associated with the ERD.

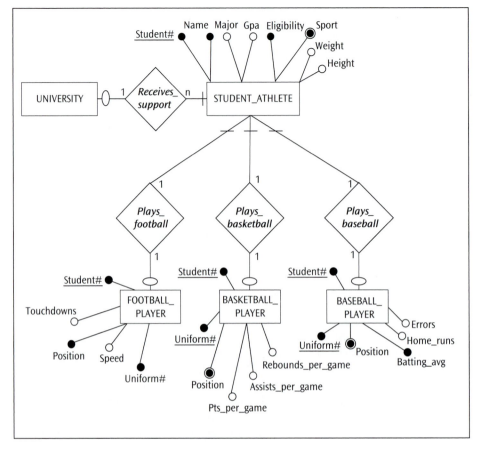

Figure 4.2 Modeling STUDENT_ATHLETE using inter-entity class relationships

Suppose a student-athlete is required to participate in at least one of the three sports of football, basketball, or baseball. The current design shown in Figure 4.2, while permitting a student-athlete to participate in one or more of these sports, does not enforce any requirement that the student-athlete *must* participate in at least one of these three sports. That is, as per this design, some student-athletes need not participate in any of these three sports.

At first glance, one may consider the use of an inclusive arc to allow for the possibility that a student-athlete may participate in more than one of the sports of football, basketball, and baseball. However, such a model would be semantically incorrect because an inclusive arc implies that a student-athlete who participates in one of these sports *must* participate in the other two as well. On the other hand, a noninclusive arc conveys the concept of optionality in inclusiveness. The design shown in Figure 4.3 makes it possible to show that a student-athlete is required to participate in at least one of these three sports.

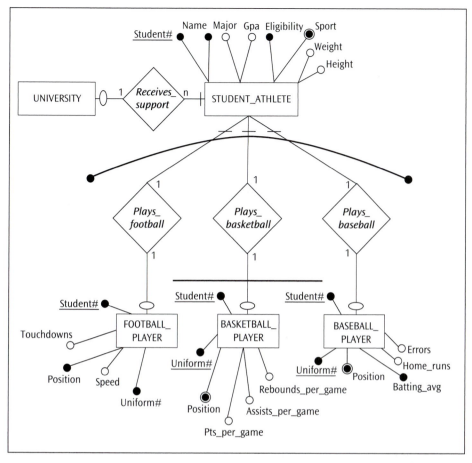

Figure 4.3 Modeling STUDENT_ATHLETE using an optionally inclusive noninclusive arc

While the design presented in Figure 4.3 is syntactically correct, the accuracy of the semantics conveyed by this design is questionable. For instance, the design indicates that each student-athlete may be related to at most one football player, basketball player, or baseball player as if the student-athlete is different from a football, basketball, or baseball player. The idea that football players, basketball players, and baseball players are indeed student-athletes and possess all the attributes of a STUDENT_ATHLETE is not explicitly conveyed in the design. In other words, this design is at best an awkward expression of the semantics enunciated in the vignette.

In the following sections of this chapter, enhancements to the ER modeling grammar by way of additional constructs to express sharper relationships among entity types, especially the ones that belong to the same entity class, are discussed. The building block for these modeling constructs is the fundamental construct called the Superclass/subclass relationship.

4.1.2 A Motivating Exemplar

Section 2.2 introduced the entity class as a construct that conceptualizes an object class. For example, stores selling furniture can be modeled as a relationship between a STORE entity type and a FURNITURE entity type. Here STORE and FURNITURE represent entity types of different generic entity classes because they represent different independent object types and the relationship between the two is a binary relationship that represents an **inter-entity class relationship**. Since STORE *has* a relationship with FURNITURE, an inter-entity class relationship is often called a *Has-a relationship*.

Now consider the fact that FURNITURE can be chairs, tables, desks, sofas, beds, and so on. Although FURNITURE and STORE do not share any common properties (attributes), chairs, tables, and sofas are all furniture and possess some common properties. In this case, CHAIR, TABLE, and SOFA are referred to as entity types that belong to the generic entity class, FURNITURE, and a relationship between CHAIR and FURNITURE, or TABLE and FURNITURE, or SOFA and FURNITURE is called an **intra-entity class relationship** (see Figure 4.4). As CHAIR (or TABLE or SOFA) *is* FURNITURE, this relationship is also referred to as an *Is-a relationship* (see Figure 4.4).

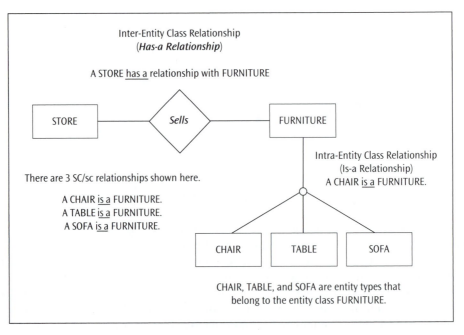

Figure 4.4   Inter-entity and intra-entity class relationships

Since FURNITURE represents the generic class of entity that includes one or more entity type occurrences (CHAIR, TABLE, SOFA), FURNITURE is labeled a **superclass** entity type or just **SC**. Since CHAIR (or TABLE or SOFA) represents an entity type that is a subgroup of FURNITURE, it is labeled a **subclass** entity type or just **sc**. The relationship between a superclass and any one of the subclasses is called an **SC/sc relationship**. There are three SC/sc relationships present in the intra-entity class relationship shown in Figure 4.4. Note that FURNITURE and CHAIR (or TABLE or SOFA) are separate entity types, though they are not independent like FURNITURE and STORE are. Figure 4.5 shows that entities that belong to the CHAIR, TABLE, and SOFA subclasses also belong to the FURNITURE superclass. Note that an entity that belongs to a subclass represents the same entity that is connected to it from the FURNITURE superclass.

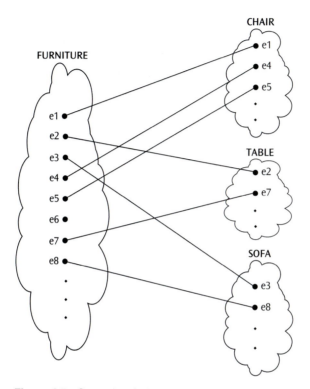

Figure 4.5 Superclass/subclass entity instances

It is not necessary that CHAIR, TABLE, and SOFA be abstracted to an SC called FURNITURE. There may be situations where direct relationships between CHAIR and STORE, TABLE and STORE, and SOFA and STORE may be modeled. Likewise, instead of creating subclasses, the generic entity type FURNITURE may incorporate all the attributes of CHAIR, TABLE, and SOFA in a single entity type. These modeling variations are context dependent.

4.1.3 General Properties of a Superclass/subclass Relationship

There are two basic kinds of SC/sc relationships: Specialization/generalization and categorization. Specialization/generalization is discussed in Sections 4.1.4 and 4.1.5, and categorization in Section 4.1.6.

The general properties of an SC/sc relationship are as follows:

- One superclass (SC) is related to one or more subclasses (sc) (specialization/generalization) or one subclass (sc) is related to one or more superclasses (SC) (categorization).
- An entity[1] that exists in a subclass can be associated with only one superclass entity.
- An entity cannot exist in the database merely by being a member of a subclass; it must also be a member of an associated superclass.
- An entity that is a member of a superclass can be optionally included as a member of any number of its subclasses.
- It is not required that every entity (member) of a superclass be a member of a subclass.
- A subclass inherits all the attributes of the superclass to which it is related. In addition, it also inherits all the relationship types in which the superclass participates. This is known as the **type inheritance property**.
- A subclass may also possess its own *specific attributes*[2] in addition to the attributes inherited from the superclass to which it is related. Likewise, a subclass may also have its own specific relationship(s) with other entity type(s) (that is, it may have its own inter-entity class relationships).

In Section 2.3.4, we asserted that a relationship type is not fully specified until both structural constraints, that is, cardinality ratio and participation constraints, are specified. While that rule applies to both of the EER constructs above, the cardinality ratio of any SC/sc relationship in both is always 1:1. Likewise, the participation of a subclass in an SC/sc relationship is always total.

In order to represent the EER constructs diagrammatically, the ER diagramming notation set is expanded with a few additional symbols, shown in Figure 4.6. Use of these symbols is illustrated in the remainder of Section 4.1.

4.1.4 Specialization and Generalization

Generating subgroups ("sc"s) of a generic entity class (SC) by specifying the distinguishing properties (attributes) of the subgroups is called **specialization**. **Generalization**, on the other hand, crystallizes the common properties (attributes) shared by a set of entity types ("sc"s) into a generic entity type (SC). In other words, generalization is a reverse process of specialization where the differences among a set of entity types are suppressed and the common features are "generalized" into a single superclass of which the original source entity types become subclasses. In short, specialization is a top-down approach to viewing an SC/sc relationship while generalization is a bottom-up approach to describe the same relationship. Specialization and generalization can be thought of as two sides of the same coin. Therefore all discussions about specialization apply equally well to a generalization and vice versa.

At this point, let us revisit Vignette 1 and see if the design reflected in Figure 4.3 can be better expressed using the EER construct, specialization. Since a football player *is a*

---

[1] Note that in Section 2.2, the term *entity* has been defined to mean an instance (occurrence) of an entity type.

[2] These attributes need not be unique to the subclass.

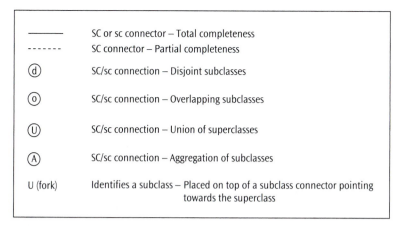

——————	SC or sc connector – Total completeness
- - - - - - -	SC connector – Partial completeness
(d)	SC/sc connection – Disjoint subclasses
(o)	SC/sc connection – Overlapping subclasses
(U)	SC/sc connection – Union of superclasses
(A)	SC/sc connection – Aggregation of subclasses
U (fork)	Identifies a subclass – Placed on top of a subclass connector pointing towards the superclass

Figure 4.6   EER diagram notation

student-athlete (the same is true for a basketball player and a baseball player), the design shown in Figure 4.3 can indeed be modeled as a specialization. Figure 4.7 is a diagrammatic depiction of this scenario and takes the form of a specialization. The subclasses that participate in the specialization, FOOTBALL_PLAYER, BASKETBALL_PLAYER, and BASEBALL_PLAYER, are represented as base entity types (as was the case in Figure 4.2) and are collectively connected to the parent superclass, STUDENT_ATHLETE, via a circle symbol that signifies the relationship of specialization. A fork (U) symbol placed on each subclass connector (the line connecting the subclass to the circular specialization symbol) opens towards the superclass. A solid line flowing from a subclass to the back of the fork denotes total participation of that subclass in that particular specialization. The cardinality ratio (1:1) is inherent to an SC/sc relationship, and so is not overtly indicated in the diagram.

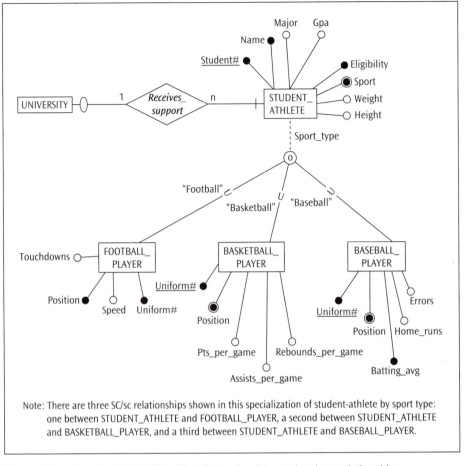

Figure 4.7 Modeling STUDENT_ATHLETE using intra-entity class relationships

The participation of the superclass in the specialization is referred to as the **completeness constraint** and can assume one of two values: total or partial. **Total specialization** means that every entity of the superclass must participate in this specialization relationship. This is indicated in the diagram by drawing a solid line from the superclass to the circular specialization symbol. A dotted line, on the other hand, connotes **partial specialization** meaning that there may be entities present in the superclass that do not participate in this specialization. Each subclass may have its own specific attributes that the superclass does not have. For example, in Figure 4.7, FOOTBALL_PLAYER has the attributes **Speed**, **Position**, **Touchdowns**, and **Uniform#**; BASKETBALL_PLAYER has the attributes **Uniform#** (unique identifier), **Pts_per_game**, **Position** (multi-valued), **Assists_per_game**, and **Rebounds_per_game**; and BASEBALL_PLAYER has the attributes **Position** (multi-valued), **Uniform#** (unique identifier), **Batting_avg**, **Home_runs**, and **Errors**. Other subclasses may or may not have similar attributes. Notice that **Uniform#** is not a unique identifier in FOOTBALL_PLAYER because in squads that have more than 100 players, the same uniform number will be assigned to more than one player. That each subclass inherits all the attributes of the superclass and all the relationship types in which the superclass participates is implicit in this diagram. Using the sample data sets in Table 4.1, each subclass (FOOTBALL_PLAYER, BASKETBALL_PLAYER, BASEBALL_PLAYER) inherits the attributes **Student#**, **Name**, **Major**, **Gpa**, **Eligibility**, **Sport** (multi-valued), **Height**, and **Weight** from the STUDENT_ATHLETE superclass. Furthermore, each of the three subclasses in Figure 4.7 also implicitly inherit the inter-entity class relationship type *Provides_support* with the entity type UNIVERSITY.

STUDENT_ATHLETE Data Set

Student#	Name	Major	Gpa	Eligibility	Sport	Weight	Height
345212	Routt	Sociology	3.25	2	Football Track	185	72
672333	Evans	Communications	2.68	2	Football	205	70
502123	Gettys	Business	3.65	0	Basketball Baseball	215	78
230543	Francis	Psychology	2.58	4	Basketball	200	79
902341	Pakilana	Communications	3.72	0	Swimming	115	65
324543	Oliver	Communications	3.12	1	Basketball Soccer	125	70

FOOTBALL_PLAYER Data Set

Student#	Uniform#	Touchdowns	Speed	Position
345212	6	0	4.23	Cornerback
672333	6	12	4.39	Halfback

BASKETBALL_PLAYER Data Set

Student#	Uniform#	Position	Pts/Game	Ast/Game	Reb/Game
502123	10	Pt Guard	12	9	6
230543	1	Sh Guard Sm Forward	7	2	5
324543	44	Sh Guard	8	4	6

BASEBALL_PLAYER Data Set

Student#	Uniform#	Position	Average	HomeRuns	Errors
502123	10	First Base Outfield	.320	6	0

Note: The shaded Student# in the FOOTBALL_PLAYER, BASKETBALL_PLAYER, and BASEBALL_PLAYER data sets is used to illustrate which student-athletes participate in the sports of football, basketball, and baseball. Observe that Routt and Oliver participate in these sports plus other sports as well. Pakilana, on the other hand, is a student-athlete who participates in a sport other than football, basketball, and baseball and thus data for Pakilana appears only in the STUDENT_ATHLETE data set.

Table 4.1 Sample data sets for Figure 4.7

In the example shown in Figure 4.7, the dotted line for the **superclass connector** indicates that there are student-athletes who are neither football players, nor basketball players, nor baseball players (they participate in other sports). Had this been a solid line, it would have conveyed the meaning that there are no student-athletes playing anything other than football, basketball, or baseball. Another constraint pertaining to SC/sc participation in the specialization is called the **disjointness constraint**, which is used to specify that the subclasses of a specialization must be disjoint and is done by placing the letter "d" in the circular specialization symbol. This means that an entity of the superclass cannot be a member of more than one subclass of the specialization. If, however, the subclasses in a specialization are not constrained to be "disjoint," their sets of entities may *overlap* across the subclasses in this specialization. This is indicated by placing the letter "o" in the circular specialization symbol. In Figure 4.7, the value "o" for the disjointness constraint says that a football player can also be a basketball player and/or baseball player. Had this value been a "d," it would prohibit a student-athlete from participating in more than one of these three sports. It is important to note that the completeness constraint and the disjointness constraint are *two independent, complementary constraints* and a specialization is not fully specified until *both* of these constraints are specified.

Frequently the membership of an entity in a subclass can be exactly determined based on a specified condition reflected by the value of some attribute(s) in the superclass of the specialization. In this case, the subclasses in the specialization are called **predicate-defined** (or **condition-defined**) subclasses. The condition itself is called the **defining predicate** or **defining condition** and is shown beside the superclass connector, while the values of the defining predicate are coded next to the subclass connectors. In Figure 4.7, *Sport_type* is the defining predicate[3] and can have the values "Football," "Basketball," and/or "Baseball." This models the semantics that all student-athlete entities in the superclass STUDENT_ ATHLETE marked by the defining predicate "Baseball" will also be members of the subclass BASEBALL_PLAYER.

Often attribute(s) in the superclass are used to define the predicate. If all values of the predicate are determined by the same attribute in the superclass, this attribute is referred to as the **defining attribute**[4] and the specialization itself is labeled **attribute-defined specialization**. Figure 4.7 presents an interesting variation of this constraint in that **Sport** in STUDENT_ATHLETE is a mandatory multi-valued attribute. The overlapping property indicated by the disjointness constraint value of the specialization relationship requires that **Sport** be a multi-valued attribute in order to permit multiple values for the defining predicate *Sport_type* for the same superclass entity.

There are times when the membership of a superclass entity in the subclass(es) of its particular specialization cannot be exactly determined by a specified condition that can be evaluated automatically. Often in these cases the membership of the entity is determined by the end-user individually for each entity, based on some not necessarily well-defined procedure. The specialization then is called a **user-defined** (or **procedure-defined**) **specialization**.

Had the process of generalization been used to develop Figure 4.7, the designer would have begun by observing a number of common attributes[5] (**Student#**, **Name**, **Major**, **Gpa**, **Eligibility**, **Sport**, **Weight**, and **Height**) in the process of defining each of the individual entity types FOOTBALL_PLAYER, BASKETBALL_PLAYER, and BASEBALL_PLAYER. These common attributes suggest that each of the three entity types really belongs to a more general entity type. Thus the superclass STUDENT_ATHLETE emerges. The result enables the

---

[3] It should be noted that a defining predicate is not an attribute; it is a condition based on the value(s) of one or more attributes in the superclass.

[4] Sometimes the term "subclass discriminator" is used instead.

[5] Commonality implies that *all* characteristics of the individual attributes across the entity types are the same.

attributes common to the three subclass entity types to be highlighted in the superclass while at the same time preserving the attributes that are specific to each of the subclasses. A closer inspection of Figure 4.7 may suggest that the attributes **Uniform#** and **Position** should be included among the attributes of the STUDENT_ATHLETE entity type instead of each of the individual subclasses. However, there are a variety of factors that prohibit adding these attributes to STUDENT_ATHLETE. First, the intent is to show **Uniform#** as a unique identifier in two of the subclasses. But it cannot be shown as a unique identifier of STUDENT_ATHLETE since two student-athletes in baseball and basketball respectively may have the same uniform number. Then the requirement that **Uniform#** is a unique identifier in BASKETBALL_PLAYER and BASEBALL_PLAYER would not be preserved. In addition, designation of **Uniform#** as a unique identifier in STUDENT_ATHLETE will propagate the same property to FOOTBALL_PLAYER, which would also be incorrect. The multi-valued attribute **Position** also poses a similar problem because a student-athlete participating in multiple sports may play the same position in more than one sport—a center in football could also be a center in basketball.

Vignette 1 can be extended to demonstrate two other properties of a specialization/ generalization: (1) multiple specialization of the same entity type (SC), and (2) specific relationship of a subclass entity type. For example, suppose a student-athlete, in addition to participating in various sport programs, may also serve as a captain of a team and participate as either a varsity-level player or intramural-level player. In addition, let us also say that an intramural-level player may be sponsored by an organization (e.g., fraternity, sorority, club, etc.).

Figure 4.8 illustrates how it is also possible for an entity type to participate as a superclass in multiple independent specializations. That is, STUDENT_ATHLETE specialized as the subclass set {FOOTBALL_PLAYER, BASKETBALL_PLAYER, BASEBALL_PLAYER} can also be independently specialized as a TEAM_CAPTAIN and the subclass set {VARSITY_ PLAYER, INTRAMURAL_PLAYER}. In essence, STUDENT_ATHLETE here is participating as a superclass in three independent specializations—the first one with three subclasses, the second having just one subclass, and the third with two subclasses. Note that since TEAM_CAPTAIN is the only subclass in that specialization, the disjoint property is irrelevant. Since not all student-athletes are team captains, the dotted line emanating from STUDENT_ATHLETE in this specialization indicates partial specialization (i.e., the completeness constraint in this case is 'partial'). Likewise, the {VARSITY_PLAYER, INTRAMURAL_PLAYER} specialization of STUDENT_ATHLETE is a disjoint specialization as indicated by the **d** in the circular specification circle, meaning that a student-athlete can either be a varsity player or an intramural player, but not both. In addition, the solid line emanating from STUDENT_ATHLETE in this specialization denotes total specialization (i.e., the completeness constraint is 'total')—meaning that every student-athlete *must* either be a varsity player or intramural player.

Also, in Figure 4.8, the inter-entity class relationship depicted by *Sponsored_by* between INTRAMURAL_PLAYER and ORGANIZATION illustrates that an entity type participating in a specialization as a subclass may also have specific (external) relationship(s) beyond the specialization. Sample data representing this extension of Vignette 1 appears in Table 4.2. Observe both in Figure 4.8 and Table 4.2 that the subclasses INTRAMURAL_PLAYER and VARSITY_PLAYER also have their own specific attributes.

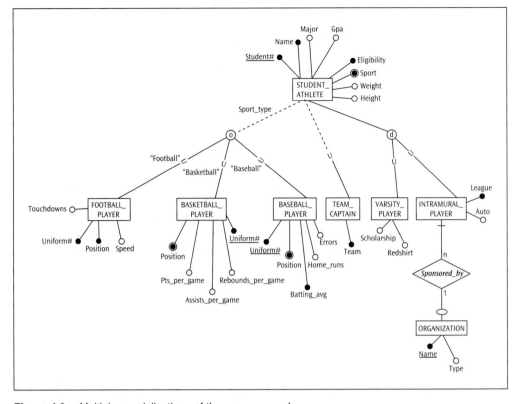

Figure 4.8 Multiple specializations of the same superclass

STUDENT_ATHLETE Data Set

Student#	Name	Major	Gpa	Eligibility	Sport	Weight	Height
345212	Routt	Sociology	3.25	2	Football Track	185	72
672333	Evans	Communications	2.68	2	Football	205	70
502123	Gettys	Business	3.65	0	Basketball Baseball	215	78
230543	Francis	Psychology	2.58	4	Basketball	200	79
902341	Pakilana	Communications	3.72	0	Swimming	115	65
324543	Oliver	Communications	3.12	1	Basketball Soccer	125	70
454212	Newman			2	Baseball		

FOOTBALL_PLAYER Data Set

Student#	Uniform#	Touchdowns	Speed	Position
345212	6	0	4.23	Cornerback
672333	6	12	4.39	Halfback

BASKETBALL_PLAYER Data Set

Student#	Uniform#	Position	Pts/Game	Ast/Game	Reb/Game
502123	10	Pt Guard	12	9	6
230543	1	Sh Guard Sm Forward	7	2	5
324543	44	Sh Guard	8	4	6

BASEBALL_PLAYER Data Set

Student#	Uniform#	Position	Average	HomeRuns	Errors
502123	10	First Base Outfield	.320	6	0
454212	19	Pitcher	.535		

TEAM_CAPTAIN Data Set

Student#	Team
502123	Baseball

VARSITY_PLAYER Data Set

Student#	Scholarsp	Redshirt
345212	Y	N
672333	Y	N
502123	Y	N
230543	N	N
902341	Y	N
324543	N	N

INTRAMURAL_PLAYER Data Set

Student#	League	Auto
454212	Baseball	Camry

Note: Two student-athletes (Francis and Oliver) are non-scholarship varsity players. Francis participates in basketball while Oliver partcipates in basketball and soccer. Newman participates in baseball but only as an intramural player.

Table 4.2 Sample data sets for Figure 4.8

4.1.5 Specialization Hierarchy and Specialization Lattice

Just as an entity type can be a superclass in multiple specializations, an entity type participating as a subclass in a specialization can also serve as a superclass of another specialization. When this happens there is a hierarchy of specializations. Often, a **specialization hierarchy** will have tiers of specialization cascading through multiple levels, as shown in Figure 4.9. Vignette 2 describes a situation to which a specialization hierarchy can be applied.

4.1.5.1 Vignette 2

Vignette 2 is a continuation of Vignette 1 introduced in Section 4.1.1. It is common today for a football player to specialize in either offense or defense. Attributes collected for offensive players include receptions and yards gained. Attributes collected for defensive players include number of tackles and number of interceptions. Likewise, some baseball players are also pitchers. Attributes collected about pitchers include earned run average (ERA), pitching speed, innings pitched, strikeouts, and walks.

The structure of a specialization hierarchy is constrained to an inverted tree in that an entity type must not participate as a child in more than one specialization. In informal terms, a child entity type cannot have more than one parent. Notice that Figure 4.9 exhibits a three-level specialization hierarchy [STUDENT → STUDENT_ATHLETE → FOOTBALL_PLAYER → {DEFENSIVE_PLAYER, OFFENSIVE_PLAYER}].[6] There are no entity types in this structure that participate as a subclass in more than one specialization. Clearly, in such a hierarchy, the type inheritance rule means that a subclass inherits the attributes and relationship types of not just the immediate parent, but also of the predecessor superclasses in the hierarchy all the way up to the root of the specialization hierarchy.

4.1.5.2 Vignette 3

Vignette 3 describes a scenario suitable for modeling as a specialization lattice. Each member of the high school staff at Homer Hanna High School is either a full-time or part-time employee. At the same time, a staff member may be either a trainer or coach and a member of the teaching staff or support staff. A part-time employee may be a part-time trainer, and each part-time trainer is also a trainer. Finally, a member of the athletic staff is a full-time employee, a coach, and a teacher.

This scenario describes a situation where an entity type can participate as a subclass in more than one specialization; in simple terms a child can have more than one parent. Such a specialization is called a **specialization lattice**. Since a specialization can involve only one superclass, each parent in a specialization lattice comes from a different specialization. The subclass itself inherits all the attributes and relationship types from the superclasses of all the specializations participating in the specialization lattice and the predecessor hierarchy of all these superclasses. This is called **multiple type inheritance**, and the subclass in the specialization lattice is referred to as a **shared subclass** since it is participating as a subclass in multiple specializations. As a rule, any attribute or relationship type inherited more than once via different paths in the specialization lattice is not duplicated in the shared subclass.

Figure 4.10 illustrates a specialization lattice. Here the entity type ATHLETIC_STAFF_MEMBER is a shared subclass in three distinct specializations: [FULL_TIME_EMPLOYEE → ATHLETIC_STAFF_MEMBER], [COACH → ATHLETIC_STAFF_MEMBER], and [TEACHER → ATHLETIC_STAFF_MEMBER] and inherits the specific attributes of FULL_TIME_EMPLOYEE, COACH, and TEACHER. In addition, ATHLETIC_STAFF_MEMBER also inherits the attributes of HIGH_SCHOOL_STAFF, but only once even though the inheritance itself occurs via three paths. It should be noted that the relationship types *Scheduled* between TEACHER and COURSE and *Coaches* between COACH and TEAM are also inherited by ATHLETIC _STAFF_MEMBER. At the same

---

[6] The symbol → is used here to convey the hierarchical predecessor and successor.

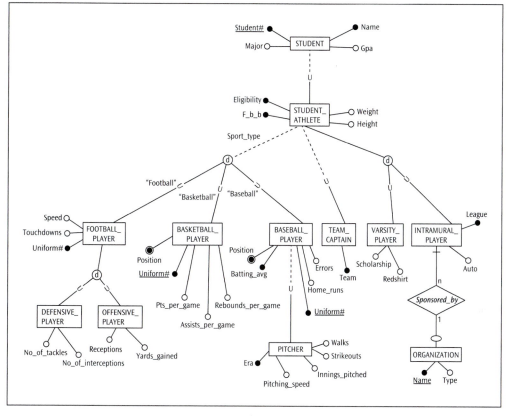

Figure 4.9 A specialization/generalization hierarchy

time, the specific attributes of ATHLETIC_STAFF_MEMBER and the specific relationship type *Recruits* belong only to ATHLETIC_STAFF_MEMBER.

Finally, while specialization is used as the basis for all discussions in Sections 4.1.4 and 4.1.5, all the concepts covered apply equally to generalization as these two constructs portray the same EER modeling features viewed from opposite ends.

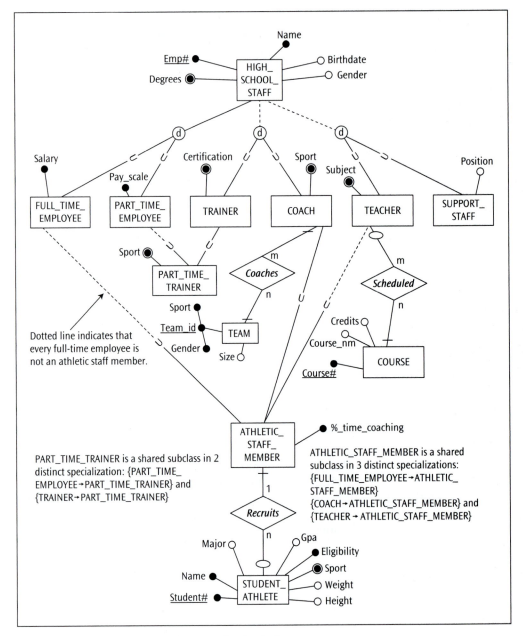

Figure 4.10 Two occurrences of a specialization lattice

4.1.6 Categorization

A fundamental characteristic of the specialization/generalization construct is that there can be only one superclass in the construct. For this reason, there are intra-entity class (*Is-a*) relationships that cannot be modeled as a specialization/generalization. For example, donors that make donations to a university can be individuals, companies, or foundations.

Since INDIVIDUAL, COMPANY, and FOUNDATION are entity types of different entity classes and any one or more of individual, company, and foundation can be a donor, this relationship requires a construct that involves three superclasses and a subclass in a *single* relationship type. Therefore, this relationship cannot be modeled as a specialization/ generalization. In other words, the donor set is a collection of entities that is a subset of the union of the three superclass entity sets. Thus, this relationship construct captures a concept different from a specialization/generalization and is called **categorization**. The participating subclass (in this case DONOR) is labeled a **category**. *An entity that is a member of the category (subclass) must exist in only one of the superclasses in the categorization relationship.*

Diagrammatically, as shown in Figure 4.11, categorization is represented similarly to specialization except that the circle symbol that signifies the categorization relationship contains the letter "U" to indicate that the membership in the subclass set results from the *union* of multiple superclasses. Also, one or more superclass connectors radiate from the relationship (circle symbol) connecting the participating superclasses. A characteristic of categorization is that there is only one subclass in each categorization. The cardinality ratio in categorization is 1:1 within and across the SC/sc relationships, and the participation of the subclass in the categorization is always total (solid line from the category to the back of the fork). Once again, these properties are inherent to the categorization construct and so are not overtly indicated in the diagram. Each superclass in a categorization may exhibit total participation (solid line from the entity type to the circle symbol) or partial participation (dotted line from the entity type to the circle symbol). In the example shown in Figure 4.11, the participation of COMPANY and FOUNDATION are total (every company and foundation must be a donor) while the participation of INDIVIDUAL is partial (some individuals are not donors). On the other hand, a donor is either a company, or a foundation, or an individual. If the completeness constraint for all superclasses in the categorization exhibits total participation, then the category (subclass) itself is called a **total category**. That is, the category set is a union of all the superclasses entities. Likewise, if the completeness constraint is partial, the category itself is referred to as a **partial category** (i.e., the category set is a proper subset of the union of all the superclass entities). Finally, type inheritance in categorization is selective. That is, members of the category (subclass) selectively inherit attributes and relationships of the superclass entity based on the SC/sc relationship in the categorization in which the member participates. This is often referred to as the **selective type inheritance** property of a category. Observe that this is diametrically opposite to the property of multiple type inheritance exhibited by a shared subclass in a specialization lattice.

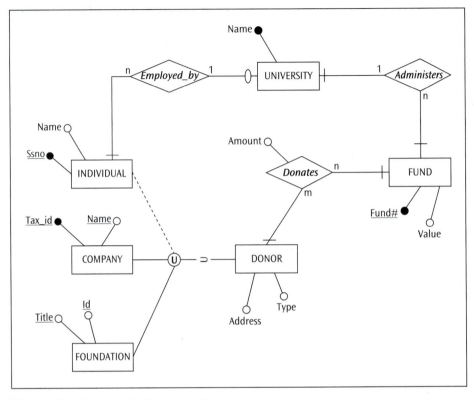

Figure 4.11 An example of categorization

Sometimes a category may not have a unique identifier. For instance, DONOR does not have a unique identifier. While type inheritance, even when selective, will furnish the category (in this case, DONOR) with a unique identifier, the properties of the unique identifier will vary across category instances depending on the superclass from which the attributes are inherited. This situation is often simplified by specifying a "manufactured" surrogate key for the category to serve the role of unique identifier.

The sample data shown in Figure 4.12 illustrates the relationship among INDIVIDUAL, COMPANY, FOUNDATION, DONOR, and FUND shown in Figure 4.11. Observe that DONOR is a partial category because, while the participation of COMPANY and FOUNDATION is total, the participation of INDIVIDUAL is partial (that is, the Jim Jones with Social Security number 456456456 is not a donor). In Figure 4.12, the surrogate key "manufactured" for each donor begins with a one-character code that represents the donor category (F = Foundation, C = Company, I = Individual) followed by the donor number within that category. In addition, the DONATION data reflects the fact that each donor makes a donation to at least one fund and that each fund has at least one donation from a donor.

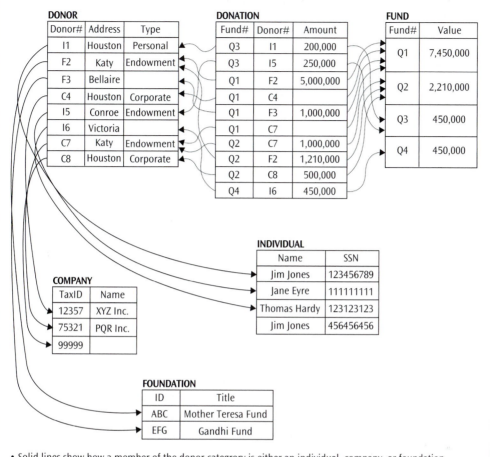

DONOR				DONATION				FUND	
Donor#	Address	Type		Fund#	Donor#	Amount		Fund#	Value
I1	Houston	Personal		Q3	I1	200,000		Q1	7,450,000
F2	Katy	Endowment		Q3	I5	250,000			
F3	Bellaire			Q1	F2	5,000,000		Q2	2,210,000
C4	Houston	Corporate		Q1	C4				
I5	Conroe	Endowment		Q1	F3	1,000,000		Q3	450,000
I6	Victoria			Q1	C7				
C7	Katy	Endowment		Q2	C7	1,000,000		Q4	450,000
C8	Houston	Corporate		Q2	F2	1,210,000			
				Q2	C8	500,000			
				Q4	I6	450,000			

INDIVIDUAL

Name	SSN
Jim Jones	123456789
Jane Eyre	111111111
Thomas Hardy	123123123
Jim Jones	456456456

COMPANY

TaxID	Name
12357	XYZ Inc.
75321	PQR Inc.
99999	

FOUNDATION

ID	Title
ABC	Mother Teresa Fund
EFG	Gandhi Fund

- Solid lines show how a member of the donor categrory is either an individual, company, or foundation.
- Dotted lines are not part of the categorization construct but used to show how detailed information about the fund and donor of each donation can be obtained.

Figure 4.12 Sample data sets for the categorization example in Figure 4.11

4.1.7 Choosing the Appropriate EER Construct

It should be clear at this point that specialization/generalization and categorization serve different modeling needs. Are there any rules of thumb that will help determine which to use in a given situation? Yes—and the principles can best be illustrated with an example.

Consider a scenario of an automobile dealership in the state of Texas. The dealership typically stocks cars, trucks, vans, and sport utility vehicles (SUVs). Figure 4.13 displays the entity types CAR, TRUCK, VAN, and SUV. The generalization of these entity types to the entity type VEHICLE that contains the attributes common to cars, trucks, vans, and SUVs appears in Figure 4.14. The result is a generalization/specialization where VEHICLE represents the entity class (superclass) to which the entity types (subclasses) CAR, TRUCK, VAN, and SUV belong. Notice that the completeness constraint indicates that

Enhanced Entity-Relationship (EER) Modeling

the dealership has other vehicles not captured explicitly in this generalization and the only data captured on these vehicles is that which is common to all vehicles. It is also important to realize that an entity *cannot* be a member of one of the subclasses {CAR, TRUCK, VAN, SUV} unless it exists in the superclass VEHICLE.

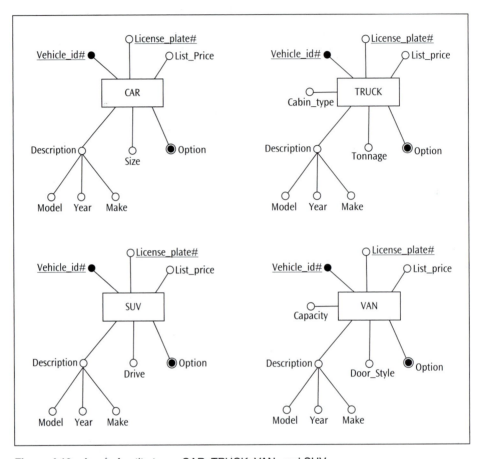

Figure 4.13 A set of entity types CAR, TRUCK, VAN, and SUV

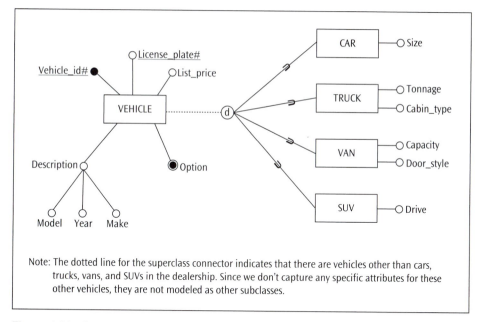

Note: The dotted line for the superclass connector indicates that there are vehicles other than cars, trucks, vans, and SUVs in the dealership. Since we don't capture any specific attributes for these other vehicles, they are not modeled as other subclasses.

Figure 4.14 Generalization of subclasses CAR, TRUCK, VAN, and SUV to a VEHICLE superclass

Not all vehicles in the dealership are registered vehicles, though. In other words, the dealership has in its inventory lots of vehicles not yet registered; they do not have an assigned license plate number. How would you model REGISTERED_VEHICLE? The crucial issue here is that the model should allow for the possibility of some unregistered cars, trucks, vans, and/or SUVs to be present. This property cannot be captured in the generalization/specialization construct, shown in Figure 4.14. One method to model this situation involves the use of the categorization construct. Figure 4.15 presents this scenario. Note that REGISTERED_VEHICLE is the subclass in this categorization while CAR, TRUCK, VAN, and SUV are superclasses of the categorization. Participation of CAR and VAN in this relationship is partial, while TRUCK and SUV exhibit total participation. This is indicated in Figure 4.15 by the dotted lines and solid lines for the completeness constraint. That there are other vehicles in the lot and some of them may be registered vehicles is incorporated via the superclass OTHER with a partial participation in the categorization. Also, since the category REGISTERED_VEHICLE has a unique identifier **License_plate#**, there is no need to create a surrogate key for REGISTERED_VEHICLE.

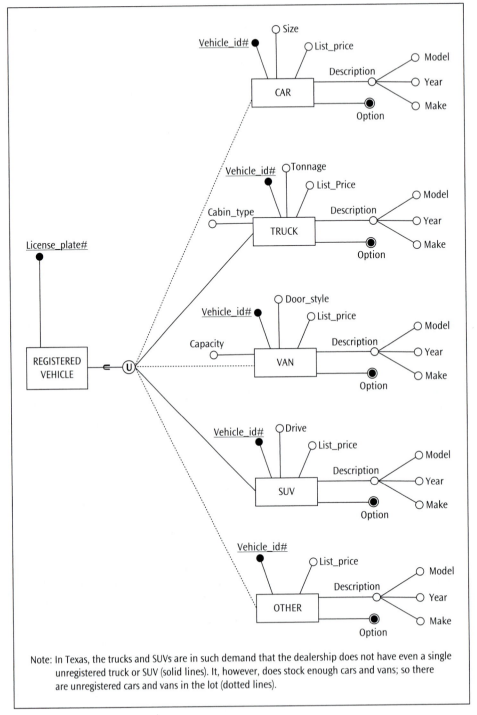

Figure 4.15 Categorization of superclasses CAR, TRUCK, VAN, and SUV to a REGISTERED_VEHICLE subclass

Note that if REGISTERED_VEHICLE is a "total category" (that is, if the participation of all the superclasses in the categorization is total), then a generalization/specialization can replace the categorization. But then, in this case there is no difference between VEHICLE (Figure 4.14) and REGISTERED_VEHICLE (Figure 4.15). In essence, *with a total category, categorization and generalization/specialization are mutually substitutable constructs.* If REGISTERED_VEHICLE is a "partial category," an alternative design of the following form is possible. You could portray the registered subset of CAR, TRUCK, VAN, SUV, and OTHER as subclasses REGISTERED_CAR, REGISTERED_TRUCK, REGISTERED_VAN, REGISTERED_SUV, and REGISTERED_OTHER of the respective vehicle types CAR, TRUCK, VAN, SUV, and OTHER. CAR, TRUCK, VAN, SUV, and OTHER can participate as subclasses of a generalization where VEHICLE serves as the superclass, thus creating a generalization hierarchy which includes both vehicles and registered vehicles, as shown in Figure 4.16.

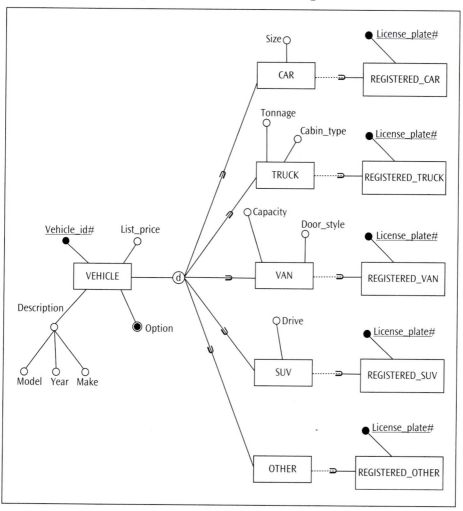

Figure 4.16 The partial category REGISTERED_VEHICLE in Figure 4.15 expressed as a specialization

Enhanced Entity-Relationship (EER) Modeling

In general, it may not be efficient to replace a partial category with a generalization/specialization construct. A total category, on the other hand, can always be alternatively modeled as a generalization/specialization—total or partial. Although the choice is subjective and context specific, if the superclasses belong to the same semantic class of entity, that is, if they share numerous attributes that are common including sometimes unique identifiers with the same properties, generalization/specialization may be a more appropriate modeling construct to adopt; otherwise one must resort to categorization as the modeling construct.

4.1.8 Aggregation[7,8]

While categorization is capable of expressing a modeling variation that generalization/specialization cannot incorporate, there are other constraints categorization imposes in order to sharpen its expressive power. In categorization, an entity that is a member of a category (subclass) must exist in only one of its superclasses. With aggregation, this constraint is relaxed. In fact, this condition is not only relaxed but also reversed in the aggregation construct, thereby enriching the capabilities of an EER modeling domain.

Aggregation allows modeling a "whole/part" association as an *"Is-a-part-of"* relationship between a superclass and a subclass. An aggregate here is a subclass that is a subset of the aggregation of the superclasses in the relationship. In other words, an entity in the aggregate *contains* superclass entities from *all* SC/sc relationships in which it participates. Therefore, in this case the type inheritance is collective, as opposed to categorization where it is selective. **Collective type inheritance** connotes inheritance of attributes and relationships from all superclass entities contained in the specific aggregation.

A diagrammatic representation of the aggregation construct in the EER diagram is shown in Figure 4.17. The notation is similar to that of categorization except that the union indicator (U) is replaced by the aggregation indicator (A). In the example in Figure 4.17, the subclass PERSONAL_COMPUTER is the aggregate of which HARDWARE and OPERATING_SYSTEM are parts. While a category can be total or partial, an aggregate can never be partial (no connector is a dotted line). That is, all hardware and operating system entities are "part of" some personal computer. Further, a hardware entity or an operating system entity can belong to only one personal computer entity. Figure 4.18 portrays an aggregation hierarchy.

---

[7] Unified modeling language (UML) for object-oriented modeling distinguishes between simple aggregation, which is entirely conceptual, and composition, which is classified as a variation of simple aggregation and does add some valuable semantics (Booch, Rumbaugh, and Jacobson, 1999). Composition changes the meaning of navigation across the association between the whole and its parts and links the lifetimes of the whole and its parts. Aggregation, as described in this section, corresponds to UML's composition construct.

[8] The term aggregation is also used in inter-entity class relationships to indicate a cluster of related entity types which is referred to as an aggregate entity type or a cluster entity type. This will be addressed in Chapter 5.

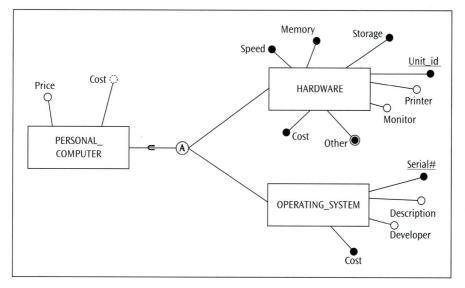

Figure 4.17 An example of the EER aggregation construct

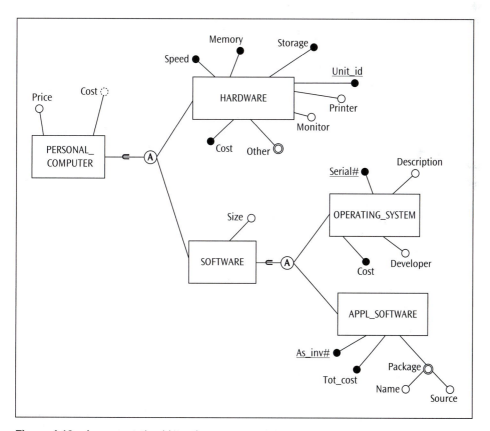

Figure 4.18 An aggregation hierarchy

Figure 4.19 depicts an aggregation and a categorization involving the same set of entity types. Every taxable property is a lot or a house (not a lot *and* a house together as a single taxable property). Some lots and some houses are not taxable properties (superclass connectors are dotted lines, denoting partial completeness). Each lot and each house is recorded as a separate taxable property. On the contrary, a lot and a house are "parts of" a home—that is, a HOME entity includes both a LOT entity and a HOUSE entity. All lots and houses participate in the aggregation. A house and a lot cannot belong to more than one home.

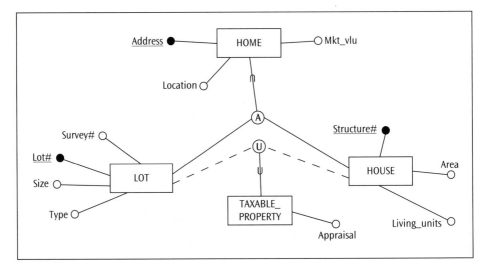

Figure 4.19 A category and an aggregate contrasted

4.2 Converting from the Presentation Layer to a Design-Specific EER Diagram

Recall that the Design-Specific ER model transforms the user-oriented Presentation Layer ER model to a database designer orientation, while at the same time fully preserving all constructs and constraints that are inherent, implicit, and explicit in the Presentation Layer ER model. Figure 4.20 illustrates the representation of the Presentation Layer ER diagram of Figure 4.9 as a Fine-granular Design-Specific ER diagram. Observe that the converted diagram contains (a) the data type and size of each atomic attribute, (b) the transformation of the multi-valued attributes in the Presentation Layer ER diagram into weak entity types, and (c) deletion constraints that govern inter-entity class relationships.[9]

[9] Since SC/sc relationship types, by definition, have a cardinality ratio of 1:1, the deletion constraints associated with SC/sc relationship types will exhibit the same properties as a 1:1 inter-entity class relationship type. However, this topic is not covered in this book.

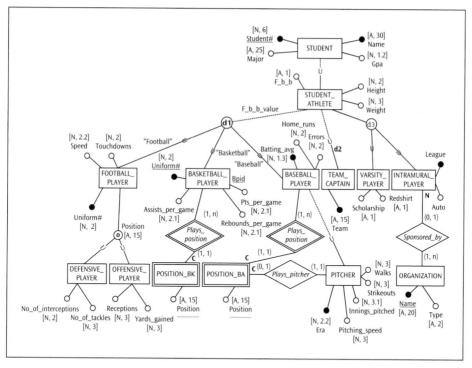

Figure 4.20 A Fine-granular Design-Specific ER diagram

In the Presentation Layer ER diagram shown in Figure 4.9, the multi-valued attribute **Position** in BASKETBALL_PLAYER and BASEBALL_PLAYER indicates that both baseball and basketball players can play more than one position within that sport. Observe in Figure 4.20 the multi-valued attribute **Position** in both BASKETBALL_PLAYER and BASEBALL_PLAYER have each been replaced by the weak entity types POSITION_BK (for BASKETBALL_PLAYER) and POSITION_BA (for BASEBALL_PLAYER) with the attribute **Position** serving as its partial key in each weak entity type. In BASEBALL_PLAYER, replacing the multi-valued attribute **Position** with the weak entity type POSITION_BA also requires that a relationship *Plays_pitcher* between the POSITION_BA and PITCHER be established, where each POSITION_BA entity is associated with at most one PITCHER entity and each PITCHER entity is associated with exactly one POSITION_BA entity. The SC/sc relationship between the BASEBALL_PLAYER and PITCHER remains intact.

The deletion constraints associated with the inter-entity class relationship types in Figure 4.20 are listed here followed by a description, *in italics*, of how the deletion constraint is shown in the figure.

- If an organization stops sponsoring intramural players, then intramural players sponsored by that organization are retained in the database for possible future affiliation with another organization. *The N (representing the set null rule) located immediately below INTRAMURAL_PLAYER indicates that an intramural player entity is retained when the related organization is deleted.*

Enhanced Entity-Relationship (EER) Modeling

- When a basketball player does not play any longer, all records of the positions s/he plays are removed along with the player entity. *The C (representing the cascade rule) located immediately above the weak entity type POSITION_BK specifies the deletion of related entities in the POSITION_BK entity set when a basketball player is deleted.*

- When a baseball player does not play any longer, all records of positions s/he plays are removed along with the player entity. *The C (representing the cascade rule) located immediately above the weak entity type POSITION_BA specifies the deletion of related entities in the POSITION_BA entity set when a baseball player is deleted.*

- When a pitcher is no longer in service, the associated position entity in POSITION_BA is removed along with the pitcher entity. *The C (representing the cascade rule) located immediately next to the weak entity type POSITION_BA specifies the deletion of the related entity in the POSITION_BA entity set when a pitcher is deleted.*

4.3 Bearcat Incorporated Data Requirements Revisited

Recall that the discussion of the ER model in Chapter 3 is framed in the context of data requirements for Bearcat Incorporated, a manufacturing company with several plants located in the northeastern part of the United States. This section introduces additional requirements that will exemplify the incorporation of EER modeling constructs to the conceptual model of Bearcat Incorporated.

Bearcat Incorporated provides significant support to recreational opportunities for the dependents of employees. However, in order to provide a comprehensive set of recreational opportunities for dependents, the employee association of Bearcat Incorporated aggressively solicits sponsors for various hobbies. The sponsor of a hobby can be one or more of an individual, a school, or a church. Although each hobby need not have a sponsor, each individual, school, and church is involved in sponsoring one or more hobbies. Social Security Number is used to identify each individual sponsor. Other data captured about individuals include name, address, and phone number. Schools and churches are each identified by their name. For a church its denomination and pastor are recorded, while for a school its size and the name of the principal are recorded. The church denominations when available are Catholic, Baptist, Methodist, or Lutheran.

Many of the sponsors of hobbies are not-for-profit-organizations and for these organizations the type, exempt ID, and annual operating budget are recorded. Some of the schools are public schools and therefore are also classified as not-for-profit-organizations. For the public schools, the name of the school district and its tax base are recorded.

A plant's project may be done in-house or outsourced to one or more contracted vendors. However, a vendor can participate in only one outsourced project at a time. A plant employee is assigned to an in-house project, and an in-house project involves one or more employees. An employee is involved in no more than seven in-house projects but need not be involved in any project. For both in-house and outsourced projects, a description of the project is gathered. Data gathered about each vendor include a vendor name, address, phone number, and contact person. Vendor name is used to identify each vendor.

Since the same vendors are often contracted for future projects, when an outsourced project is removed from the system, the vendor information should be retained for future use. If a hobby is removed from the recreation portfolio of Bearcat Incorporated its relationship with any sponsor is removed as well. Likewise, when a sponsor is removed from the recreation portfolio of Bearcat Incorporated its relationship with any hobby is removed.

4.4 ER Model for the Revised Story

Figure 4.21 and Table 4.3 incorporate the additional requirements described in Section 4.3 into the Presentation Layer ER model for Bearcat Incorporated. Observe that PROJECT has become a superclass entity type with IN_HOUSE_PROJECT and OUTSOURCED_PROJECT as subclasses. Since a project may be done either in-house or outsourced, the circular specialization symbol contains the letter "d". Further, since vendors are contracted to do any outsourced project, the inter-entity relationship type *Contracted_to* was created between OUTSOURCED_PROJECT and VENDOR. Figure 4.21 also illustrates two intra-entity relationships that involve SPONSOR. SPONSOR is a category with the three superclasses: CHURCH, SCHOOL, and INDIVIDUAL. While each sponsor is either a church, school, or individual (as reflected by the category), the fact that each sponsor can sponsor one or more hobbies is reflected by the inter-entity relationship type *Supports* between SPONSOR and HOBBY. In addition, SPONSOR is also the superclass in a partial specialization where the subclass is NOT_FOR_PROFIT_ORGANIZATION. A specialization lattice involving SCHOOL, NOT_FOR_PROFIT_ORGANIZATION, and PUBLIC_SCHOOL also exists in Figure 4.21 because a public school is both a school and a not-for-profit organization. Table 4.3 lists the semantic integrity constraints (statements which are not shaded) introduced by the requirements added to the Bearcat Incorporated scenario in Section 4.3 that are not reflected in the Presentation Layer ER diagram in Figure 4.21.

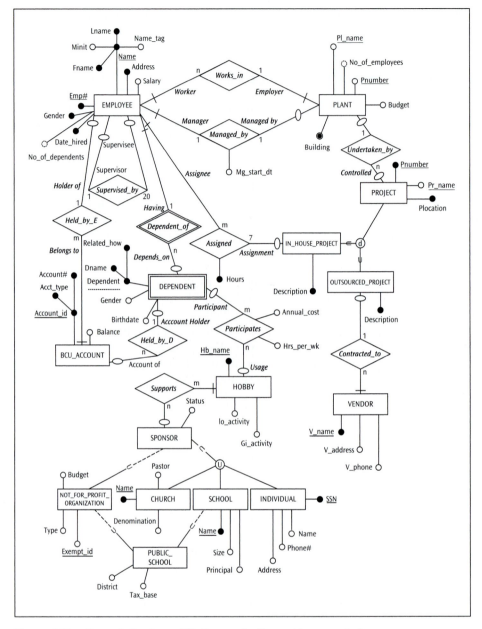

Figure 4.21   Presentation Layer ER diagram for Bearcat Incorporated

Attribute Level Business Rules

1. Each plant has a plant number that ranges from 10 to 20.
2. The gender of each employee or dependent is either male or female.
3. Project numbers range from 1 to 40.
4. Project locations are confined to the cities of Bellaire, Blue Ash, Mason, Stafford, and Sugarland.
5. Account types are coded as C – checking account, S – savings account, I – investment account.
6. A hobby can be either I – indoor activity or O – outdoor activity.
7. A hobby can be a G – group activity or I – individual activity.
8. Church denomination, if available, is confined to Baptist, Catholic, Lutheran, or Methodist.
9. The status of a sponsor can be either A – active or I – inactive.

Entity Level Business Rules

1. A mother or daughter dependent must be a female, a father or son dependent must be a male, and a spouse dependent can be either a male or female.
2. An employee cannot be his or her own supervisor.
3. A dependent may have a joint account only with an employee of Bearcat Incorporated to whom he or she is related.

Business Rules Governing Entity Deletion

1. A plant with employees cannot be closed down.
2. If an employee is deleted, all BCU accounts of that particular employee must be deleted.
3. *If a plant is closed down, the projects undertaken by that plant cannot be canceled. The project assignments from a closed plant must be temporarily removed in order to allow the project to be transferred to another plant.
4. The Human Resources Department uses a designated default employee number to replace a supervisor who leaves the company.
5. An employee currently managing a plant cannot be deleted from the database.
6. If a plant is closed down, the employee no longer manages the plant but becomes an employee of another plant.
7. **If an employee leaves the company, all dependents and BCU accounts of the employee must be removed.
8. As long as an employee is assigned to a project, his or her record cannot be removed from the database.
9. If a project is deleted, all assignments of employees to that project must be deleted.
10. If a dependent is deleted, all records of the participation of that dependent in hobbies must be deleted.
11. A hobby with at least one dependent participating in it cannot be deleted.
12. When an outsourced project is deleted, information about the vendor should be retained.
13. When a sponsor is deleted, the sponsor's association with any hobby is also removed.
14. If a hobby is removed from the recreation portfolio of Bearcat Incorporated, its relationship with any sponsor is also removed.

*In order to honor this rule, the requirement here has been relaxed.
**Rule 7 reflects the modified version of the rule after its conflict with Rule 8 in Table 3.1 has been resolved in favor of Rule.7

Miscellaneous Business Rules

1. Each plant has at least three buildings.
2. Each plant must have at least 100 employees.
3. The salary of an employee cannot exceed the salary of the employee's supervisor.

Table 4.3 Presentation Layer semantic integrity constraints for expanded Bearcat Incorporated scenario
(Note: Shaded statements represent constraints originally established in Chapter 3)

Recall that the Design-Specific ER model consists of two levels of granularity—coarse and fine. Tables 4.4 and 4.5 contain the expression of the semantic integrity constraints introduced by the requirements added to the Bearcat Incorporated scenario (see the unshaded areas) for each of these models. Figure 4.22 contains the ER diagram for the Design-Specific ER model. In the interest of brevity, Figure 4.22 depicts the Fine-granular Design-Specific ER diagram, since a fine-granularity ER diagram subsumes the features of a coarse-granular diagram.

Entity/Relationship Type Name	Attribute Name	Data Type	Size	Domain Constraint
PLANT	Pl_name	Alphabetic	30	
PLANT	Pnumber	Numeric	2	Integer values from 10 to 20.
PLANT	Budget	Numeric	7	
PLANT	Building	Alphabetic	20	
PLANT	No_of_employees	Numeric		
EMPLOYEE	Emp#	Alphanumeric	6	ANNNNN, where A can be letter and NNNNN is a combination of five digits
EMPLOYEE	Fname	Alphabetic	15	
EMPLOYEE	Minit	Alphabetic	1	
EMPLOYEE	Lname	Alphabetic	15	
EMPLOYEE	Name_tag	Numeric	1	Integer values from 1 to 9.
EMPLOYEE	Address	Alphanumeric	50	
EMPLOYEE	Gender	Alphabetic	1	M or F
EMPLOYEE	Date_hired	Date	8	
EMPLOYEE	Salary	Numeric	6	Ranges from $35,000 to $90,000
EMPLOYEE	No_of_dependents	Numeric	2	
Managed_by	Mgr_start_dt	Date		
DEPENDENT	Dname	Alphabetic	15	
DEPENDENT	Related_how	Alphabetic	12	
DEPENDENT	Birthdate	Date	8	
PROJECT	Pr_name	Alphabetic	20	
PROJECT	Pnumber	Numeric	2	Integer values from 1 to 40.
PROJECT	Plocation	Alphabetic	15	Bellaire, Blue Ash, Mason, Stafford, Sugarland
Assigned	Hours	Numeric	3	
BANK ACCOUNT	Bank#	Numeric	2	Integer values from 10 to 90.
BANK ACCOUNT	Account#	Alphanumeric	6	
BANK ACCOUNT	Acc_type	Alphabetic	1	C=Checking Acct., S=Savings Acct., I=Investment Acct.
HOBBY	Hb_name	Alphabetic	20	
HOBBY	Io_activity	Alphabetic	1	I=Indoor Activity, O=Outdoor Activity
HOBBY	Gi_activity	Alphabetic	1	G=Group Activity, I=Individual Activity
Participates	Hrs_per_wk	Numeric	(2.1)*	
Participates	Annual_cost	Numeric	6	
IN_HOUSE_COMPONENT	Proj_pct	Numeric	(3.1)	
IN_HOUSE_COMPONENT	Component	Alphanumeric	50	
OUTSOURCED_COMPONENT	Proj_pct	Numeric	(3.1)	
OUTSOURCED_COMPONENT	Component	Alphanumeric	50	
VENDOR	V_name	Alphabetic	30	
VENDOR	V_address	Alphanumeric	50	
VENDOR	V_phone#	Alphanumeric	10	
SPONSOR	Status	Alphabetic	1	A = Active, I = Inactive
NOT_FOR_PROFIT_ORGANIZATION	Exempt_id	Alphanumeric	9	

Table 4.4 Semantic integrity constraints for Coarse-granular Design-Specific ER model for expanded Bearcat Incorporated scenario
(Note: Shaded areas represent constraints originally established in Chapter 3)

Entity/Relationship Type Name	Attribute Name	Data Type	Size	Domain Constraint
NOT_FOR_PROFIT_ORGANIZATION	Budget	Numeric	(10,0)	
NOT_FOR_PROFIT_ORGANIZATION	Type	Alphabetic	1	
CHURCH	Name	Alphabetic	30	
CHURCH	Pastor	Alphabetic	30	
CHURCH	Denomination	Alphabetic	20	Baptist, Catholic, Lutheran, Methodist
SCHOOL	Name	Alphanumeric	40	
SCHOOL	Size	Numeric	4	
SCHOOL	Principal	Alphabetic	30	
INDIVIDUAL	Ssno	Numeric	9	
INDIVIDUAL	Address	Alphanumeric	50	
INDIVIDUAL	Name	Alphabetic	30	
INDIVIDUAL	Phone#	Alphanumeric	10	
PUBLIC_SCHOOL	District	Alphabetic	30	
PUBLIC_SCHOOL	Tax_base	Numeric	(10,0)	

*(n.m) is used to indicate n places to the left of the decimal point and m places to the right of the decimal point

Entity Level Domain Constraints
1. A mother or daughter dependent must be a female, a father or son dependent must be a male, and a spouse dependent can be either a male or female.
2. An employee cannot be his or her own supervisor.
3. A dependent can have a joint account with only an employee of Bearcat Incorporated with whom he or she is related.

Remaining Miscellaneous Constraints
1. Each plant has at least three buildings.
2. The salary of an employee cannot exceed the salary of the employee's supervisor.

Table 4.4 Semantic integrity constraints for Coarse-granular Design-Specific ER model for expanded Bearcat Incorporated scenario (continued)
(Note: Shaded areas represent constraints originally established in Chapter 3)

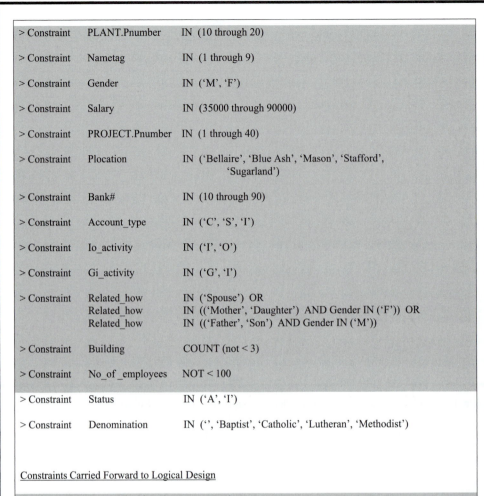

> Constraint	PLANT.Pnumber	IN (10 through 20)
> Constraint	Nametag	IN (1 through 9)
> Constraint	Gender	IN ('M', 'F')
> Constraint	Salary	IN (35000 through 90000)
> Constraint	PROJECT.Pnumber	IN (1 through 40)
> Constraint	Plocation	IN ('Bellaire', 'Blue Ash', 'Mason', 'Stafford', 'Sugarland')
> Constraint	Bank#	IN (10 through 90)
> Constraint	Account_type	IN ('C', 'S', 'I')
> Constraint	Io_activity	IN ('I', 'O')
> Constraint	Gi_activity	IN ('G', 'I')
> Constraint	Related_how Related_how Related_how	IN ('Spouse') OR IN (('Mother', 'Daughter') AND Gender IN ('F')) OR IN (('Father', 'Son') AND Gender IN ('M'))
> Constraint	Building	COUNT (not < 3)
> Constraint	No_of_employees	NOT < 100
> Constraint	Status	IN ('A', 'I')
> Constraint	Denomination	IN ('', 'Baptist', 'Catholic', 'Lutheran', 'Methodist')

Constraints Carried Forward to Logical Design

1. An employee cannot be his or her own supervisor.
2. A dependent can have a joint account only with an employee of Bearcat Incorporated with whom he or she is related.
3. The salary of an employee cannot exceed the salary of the employee's supervisor.

Table 4.5 Semantic integrity constraints for Fine-granular Design-Specific ER model for expanded Bearcat Incorporated scenario
(Note: Shaded areas represent constraints originally established in Chapter 3)

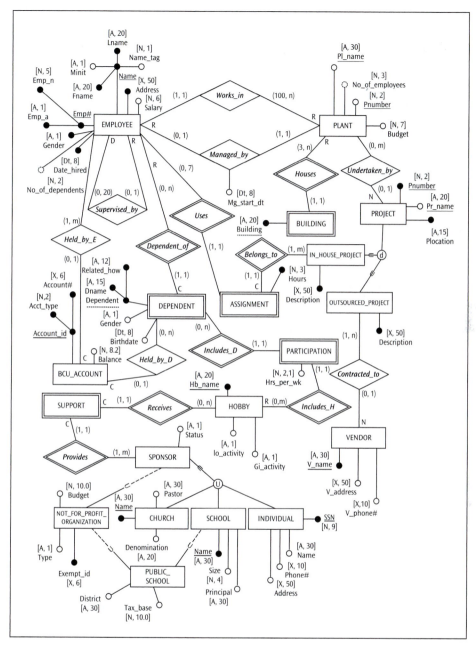

Figure 4.22 Fine-granular Design-specific ER diagram for Bearcat Incorporated

Similar to the gerund entity type ASSIGNMENT created as a result of the m:n relationship between EMPLOYEE and IN_HOUSE_PROJECT, a gerund entity type SUPPORT appears in Figure 4.22 as a result of the decomposition of the m:n relationship between HOBBY and SPONSOR. This makes it possible to tie the support of a certain sponsor of a specific hobby to its two identifying parents (SPONSOR and HOBBY). Observe that the three additional deletion rules (12 – 14) listed in Table 4.3 are now incorporated into Figure 4.22. Further, that an employee cannot be assigned to more than seven project components is captured by the maximum cardinality shown on the edge connecting employee to the *Uses* relationship. Table 4.5 records the remaining constraints in order to render the transformation fully information preserving.

Chapter Summary

With the advent of complex database applications, the constructs available in the entity-relationship (ER) modeling grammar were found inadequate to fully capture the richness of the conceptual design. The enhanced entity-relationship (EER) modeling grammar extends the original entity-relationship modeling grammar to include a few new constructs: specialization/generalization, specialization hierarchies and lattices, categorization, and aggregation. Each of these new constructs can be displayed in an EER diagram.

A fundamental unit of these intra-entity class relationships is the superclass/subclass (SC/sc) relationship where the superclass represents a generic entity type for a group of entity types (subclasses). Since the generic entity type subsumes the group, it is also referred to as an entity class.

Specialization and generalization can be viewed as two sides of the same coin. Specialization involves the generation of subgroups (i.e., subclasses) of a generic entity class by specifying the distinguishing attributes of the subgroups, whereas generalization consolidates common attributes shared by a set of entity types into a generic entity type. Two main constraints apply to specialization/generalization: the completeness constraint which can be total or partial, and the disjointness constraint which can be disjoint or overlapping.

A specialization/generalization can take the form of a hierarchy or a lattice. In a specialization hierarchy, an entity type can participate as a subclass in only one specialization, whereas in a specialization lattice an entity type can participate as a subclass in more than one specialization. A category allows for the modeling of a situation where a subclass can be a subset of the union of several superclasses. A constraint imposed by the use of a category is that an entity that is a member of a category (subclass) must exist in only one of its superclasses. On the other hand, aggregation allows for the relaxation of this constraint and requires that an entity that is a member of an aggregate must exist in all of its superclasses.

Exercises

1. What is the difference between an inter-entity class relationship and an intra-entity class relationship?

2. What is a subclass and when is a subclass of use in data modeling?

3. Under what circumstances in a specialization is it possible for one superclass to be related to more than one subclasses and one subclass to be related to one or more superclasses?

4. What is the type inheritance property?

5. What is the difference between specialization and generalization? Why is this difference not reflected in ER diagrams?

6. What is the difference between total specialization and partial specialization and how is each reflected in an ER diagram?

7. What is the difference between a specialization hierarchy and a specialization lattice?

8. How does categorization differ from specialization?

9. What is the difference between a total category and a partial category?

10. In categorization, what is meant by the property of selective type inheritance?

11. Explain how a total category and specialization are mutually substitutable constructs.

12. Contrast categorization and aggregation.

13. Consider the following Presentation Layer ER diagram.

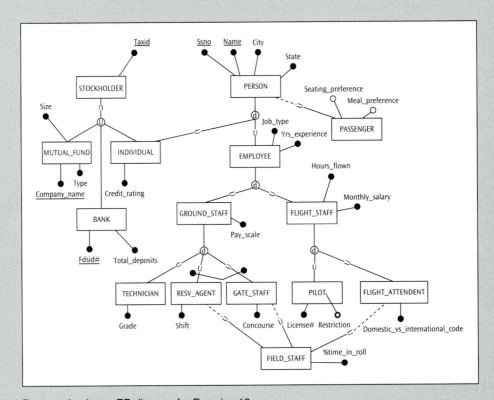

Presentation Layer ER diagram for Exercise 13

a. Identify all entity types which at one time or another function as a superclass.

b. List the superclass entity type and subclass entity type(s) in each distinct specialization.

c. Identify an entity type that functions as a shared subclass? What type of specialization does this entity type reflect?

d. Which entity types comprise the entire specialization hierarchy? For each level in this specialization hierarchy, define which entity types serve as a superclass and which serve as a subclass.

e. Which entity type functions as a superclass in more than one specialization?

f. How many Superclass/subclass(SC/sc) relationships appear in the ER diagram?

g. How many specializations appear in the ER diagram?

h. Which entity types in the diagram form a category?

i. What is the meaning of the arc in the diagram?

j. List the attributes inherited by the entity type RESV_AGENT.

k. List the attributes inherited by the entity type INDIVIDUAL.

l. List the attributes inherited by FIELD_STAFF.

m. Should all members of the GATE_STAFF also be FIELD_STAFF? If the answer is no, what modification must be made to Figure 4.23 to enforce such a business rule?

n. List the attributes inherited by MUTUAL_FUND.

o. Which, if any, entity type(s) function as both a superclass and a category?

p. What is the identifier of the entity type MUTUAL_FUND?

q. Which entity type(s) *require* the use of a surrogate key as an identifier?

r. Which entity types(s) function as a superclass in one SC/sc relationship and as a subclass in another SC/sc relationship?

s. Is it possible for an individual stockholder to be an employee of the airline company? If yes, explain. If no, explain.

t. Suppose it is possible for an employee to own shares in a mutual fund that is an owner of the airline. Modify the ER diagram to incorporate this enhancement.

14. Revise Figure 4.11 to allow (i.e., require) each individual to be a donor. Show how your revision allows the categorization construct to be replaced by the specialization/generalization construct. Explain why is it not possible to make a similar revision to the original version of Figure 4.11.

15. This exercise is based on the three Presentation Layer ER diagrams that follow.

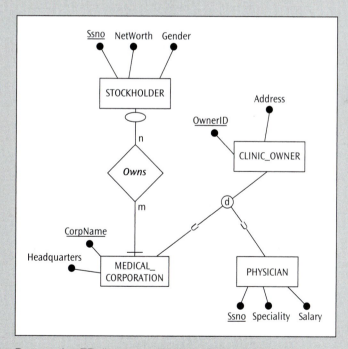

Presentation ER diagram number 1 for Exercise 15

Enhanced Entity-Relationship (EER) Modeling

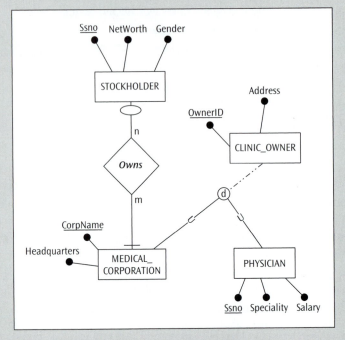

Presentation ER diagram number 2 for Exercise 15

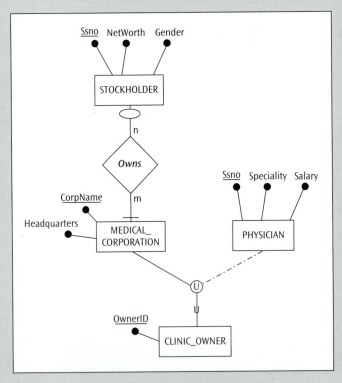

Presentation ER diagram number 3 for Exercise 15

a. Describe what is reflected by the entity types and relationship types in Presentation Layer ER diagram number 1. In other words, please tell the basic facts of the story. An explanation of the various attributes need not be part of your story.

b. How do the basic facts of the story change with the change noted in Presentation Layer ER diagram number 2?

c. How many data sets would be required to illustrate the nature of the information requirements in Presentation Layer ER diagram number 1?

d. What would be the name and attributes associated with each data set identified in question (c)?

e. Using Presentation Layer ER diagram number 1, assume there are four clinic owners, at least one of which is a physician and at least one of which is a medical corporation. How many entities (instance of each entity type) would appear in the MEDICAL_CORPORATION, PHYSICIAN, and CLINIC_OWNER data sets? It is possible that there may be more than one correct answer to this question.

f. How would your answer to question (e) change if it were based on illustrating the difference between Presentation Layer ER diagram number 1 and Presentation Layer ER diagram number 2? Please justify your answer.

g. Discuss how the basic facts of the story change if the story is based on Presentation Layer ER diagram number 3.

16. Consider the Presentation Layer ER diagram that appears in Figure 4.21.

a. What makes SCHOOL part of the specialization lattice involving SCHOOL, NOT_FOR_PROFIT_ORGANIZATION, and PUBLIC_SCHOOL and, at the same time, part of the SPONSOR category that involves CHURCH, SCHOOL, and INDIVIDUAL? What role (i.e., subclass, superclass, category, aggregate) do SCHOOL and NOT_FOR_PROFIT_ORGANIZATION serve in each of these structures?

b. Does SPONSOR take the form of a total category or a partial category?

c. Is it possible to redraw the SPONSOR category as a specialization and still retain the participation of SCHOOL in the specialization? If the answer is "yes," redraw this portion of Figure 4.21.

17. This exercise contains additional information over and above that given in Exercises 17 and 18 in Chapter 3 in order to give you an opportunity to work with various types of enhanced ER modeling constructs.

- Both coaches and referees are basketball professionals having chosen different careers within this profession.

- Players no longer serve as counselors in summer youth basketball camps. Instead some of the players and all of the coaches do voluntary service as trainers in the summer youth basketball camps.

Incorporate this additional information into the Presentation Layer ER diagram that you developed for Exercise 17 in Chapter 3.

Enhanced Entity-Relationship (EER) Modeling

Selected Bibliography

Booch, G.; Rumbaugh, J.; and Jacobson, I. (1998) *The Unified Modeling Language User Guide*, Addison-Wesley.

Connolly, T. M.; and Begg, C. E. (2002) *Database Systems: A Practical Approach to Design, Implementation, and Management*, Third Edition, Addison-Wesley.

Elmasri, R.; and Navathe, S. B. (2003) *Fundamentals of Database Systems*, Fourth Edition, Addison-Wesley.

Elmasri, R.; Weeldreyer, J.; and Hevner, A. (1985) "The Category Concept: An Extension to the Entity-Relationship Model," *International Journal on Data and Knowledge Engineering*, 1, 1 (May).

Smith, J. M.; and Smith, D. C. P. (1977) "Data Abstractions: Aggregation and Generalization," *ACM Transactions on Database Systems*, 2, 2 (June) 105-133.

Modeling Complex Relationships

The ER modeling grammar for conceptual data modeling presented in Chapters 2, 3, and 4 can portray many rich circumstances that represent business scenarios. Yet these designs are insufficient to fully express some of the intricate yet real data relationships in an enterprise. Although it is practically impossible to express *all* business rules of an application in an ER diagram (ERD), a few additional ER modeling constructs are available and can be used to model more business rules. In addition, existing constructs can be used in interesting ways to incorporate certain business rules in the ERD instead of relegating the business rules to additional textual semantic constraints of the ER model.[1] This chapter presents several application scenarios as vignettes and demonstrates how the business rules pertaining to these scenarios can be modeled in an ERD using advanced modeling techniques.

This chapter begins with a presentation of the ternary relationship type. Two vignettes depicting real-world examples are used in Section 5.1 to describe the need for and use of relationships of degree three. Section 5.2 extends this discussion to examine relationship types beyond degree three. This is done by imposing additional constraints on the scenario of the first two vignettes along with vignettes 3 and 4. The weak relationship type and its practical applications are introduced in Section 5.3. In Section 5.4, innovative use of weak relationship types is presented through composites of weak relationship types that reflect inclusion and exclusion dependencies pertaining to relationship types. Following the format of Chapters 3 and 4, Section 5.5 discusses how to decompose some of the complex relationship types preparatory to logical schema mapping in the next step. Chapter 5 concludes with two important sections. Section 5.6 addresses the need for the conceptual modeling process to include a careful validation of the semantics captured in an ERD. Using revisions to vignettes introduced previously in the chapter, semantic errors caused by the misinterpretation of relationships are identified. In cases where semantic errors are of significance in the context of the requirements specification, alternatives for restructuring the ERD are discussed. Section 5.7 begins with a narrative of a new story for a small medical clinic, once again seeking to model a real-world scenario. The rest of the section is devoted to the development of a Presentation Layer ER diagram as well as a Design-Specific ER diagram for the medical clinic, Cougar Medical Associates.

---

[1] Textual expression of the business rules as semantic integrity constraints is always available as a last resort.

5.1 The Ternary Relationship Type

In Chapter 2 we learned that the degree of a relationship is the number of entity types that participate in a relationship type. Accordingly, a **binary relationship** and a **ternary relationship** indicate a degree of two and three, respectively.

It is sometimes argued that relationship types beyond degree two (binary) are rarely found in real-world applications and thus are not crucial for data modeling and database design. This argument is far from the realities of the business world. Costly errors of expression may occur in database design when a genuinely ternary relationship is inadvertently expressed as a set of binary relationships among the three entity types taken two at a time. Likewise, combining independent binary relationships to a ternary (or n-ary) relationship is also problematic. Contemporary CASE tools tend to sacrifice expressive power in database design by not providing for relationships beyond degree two. As a consequence, designers often are not afforded an opportunity to consider relationships beyond degree two during the data modeling and database design phases. In this section, several examples are presented to illustrate the value of ternary relationship types in conceptual data modeling.

5.1.1 Vignette 1—Madeira College

Madeira College offers many courses, and a college term is divided into four quarters—fall, winter, spring, and summer—during which one or more of these courses may be offered. Every quarter at least one course is offered. The college also has several instructors. Often, not all instructors teach during all quarters. Further, instructors are capable of teaching a variety of courses that the college offers. Likewise, at least one, but often more than one instructor can teach a specific course. A course may be offered during some or all quarters; some courses may not be offered at all. Finer specifications such as the minimum number of instructors teaching in a quarter, the minimum number of quarters in which an instructor teaches, and so on could be stated. For now, let us just say that at least one instructor teaches per quarter and that some instructors may not teach any course in any quarter—they may be doing research.

A representation of this vignette as a Presentation Layer ER diagram is shown in Figure 5.1, along with sample data sets.

Is it possible to infer from this ER diagram which instructor teaches what course during which quarter? The answer is "no" from a semantic perspective. Semantically, the information on what an instructor "can teach" can never be used to infer what an instructor actually teaches. Just because an instructor is capable of teaching a course does not mean that he or she indeed teaches that particular course. For example, using the data sets for *Can_teach*, *Teaches_during*, and *Offered_during*, since both Pezman and Fite can teach EE812, both teach during the Fall and Spring quarters, and EE812 is offered during the Fall and Spring quarters, the inference is that both Pezman and Fite teach EE812 during the Fall and Spring quarters. However, suppose, as per the vignette's story line, Pezman actually teaches EE812 only during the Spring quarter while Fite teaches EE812 only during the Fall quarter.

Now, what if we add another binary relationship *Teaches* between INSTRUCTOR and COURSE that captures the courses an instructor actually teaches irrespective of what he or she is capable of teaching (see Figure 5.2). Is it now possible to infer which instructor

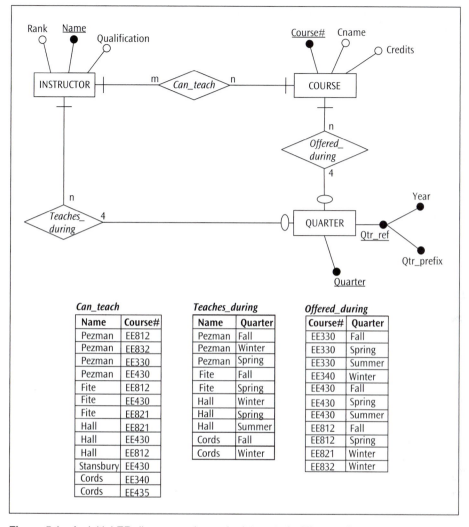

Figure 5.1 An initial ER diagram and sample data sets for Vignette 1

teaches what course during which quarter? The answer now is "not always." For example, using the data sets for *Teaches*, *Teaches_during*, and *Offered_during* in Figure 5.2, it is still not possible to determine that Pezman actually teaches EE812 *only* during the Spring quarter whereas Fite teaches the course *only* during the Fall quarter. There are times, however, when one may be able to infer the course/teaching schedule with certainty from the three data sets, *but not always*. For instance, had there been a business rule that requires that an instructor who teaches a course must be teaching that course if it is offered during the quarter when he or she teaches, then both Pezman and Fite will be scheduled to teach EE812 in both Fall and Spring. This result can, in fact, be derived from the three binary relationship types *Teaches, Teaches_during,* and *Offered_during* shown in Figure 5.2. Absent such a business rule, it is possible, as stated in the story line, for Pezman to teach EE812 in the Spring quarter and for Fite to teach EE812 in the Fall quarter; this can never be derived from the three binary relationship types *Teaches, Teaches_during,* and *Offered_during*, shown in Figure 5.2.

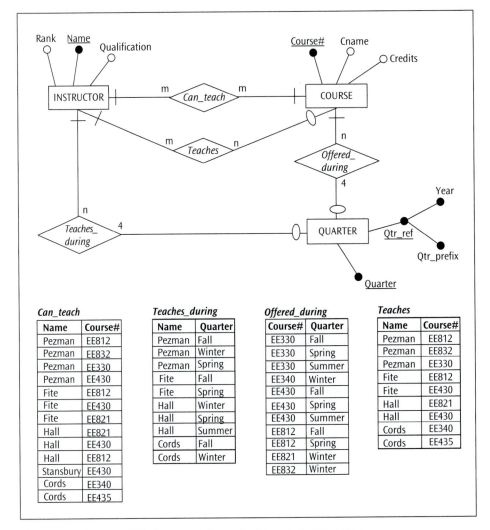

Figure 5.2 A second ER diagram and sample data sets for Vignette 1

What is the solution to this problem? When can one unequivocally infer who teaches what and when? It is possible to capture this condition precisely via a ternary relationship type among INSTRUCTOR, COURSE, and QUARTER? This relationship type is shown as *Schedule* in Figure 5.3 with a supporting sample data set.

Note that ER modeling grammars that use the "look across" notation (Chen's notation employed in the Presentation Layer ER diagram) to express a cardinality ratio cannot capture the cardinality ratio of any relationship type beyond degree two accurately. In a binary relationship type, there is only one entity type present when looking across the

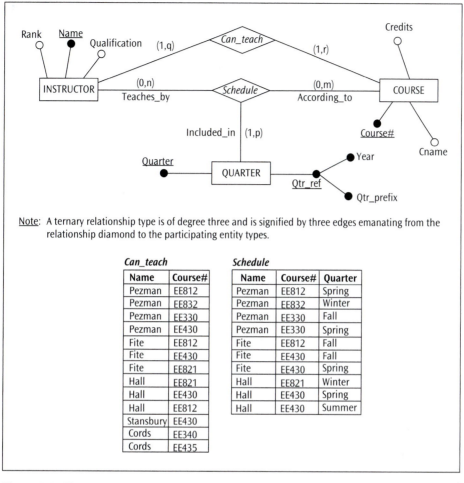

Note: A ternary relationship type is of degree three and is signified by three edges emanating from the relationship diamond to the participating entity types.

Can_teach

Name	Course#
Pezman	EE812
Pezman	EE832
Pezman	EE330
Pezman	EE430
Fite	EE812
Fite	EE430
Fite	EE821
Hall	EE821
Hall	EE430
Hall	EE812
Stansbury	EE430
Cords	EE340
Cords	EE435

Schedule

Name	Course#	Quarter
Pezman	EE812	Spring
Pezman	EE832	Winter
Pezman	EE330	Fall
Pezman	EE330	Spring
Fite	EE812	Fall
Fite	EE430	Fall
Fite	EE430	Spring
Hall	EE821	Winter
Hall	EE430	Spring
Hall	EE430	Summer

Figure 5.3 The ternary relationship type *Schedule* and associated sample data set

relationship type. For instance, in Figure 5.1 it is possible to state unambiguously that one instructor entity is related to a maximum of n course entities. However, looking across from the INSTRUCTOR in a ternary relationship type *Schedule* (Figure 5.3), we see both COURSE and QUARTER, and the maximum cardinality should actually reflect the maximum combination of courses and quarters per instructor. The (min, max) notation, which uses the "look here" approach to express the structural constraints of a relationship type, is able to express the maximum cardinality precisely in a relationship type of any degree. For this reason the (min, max) notation is used here to specify the structural constraints of relationship types beyond degree two; in the interest of consistency, the same notation will be used for recursive and binary relationship types henceforth.

In the ER diagram that appears in Figure 5.3, the (0, n) on the edge labeled "Teaches_ by" indicates that an instructor need not teach any course in any quarter (0 for the min value) and an instructor may teach up to n {course, quarter} pairs (n for the max value). For example, using the data sets in Figure 5.3, while Stansbury can teach EE430 (see the *Can_ teach* data set), the data set *Schedule* indicates that he is not scheduled to teach a course during any of the four quarters. Likewise, Cords can teach EE435 but is not scheduled to teach a course during any of the four quarters either. Hall, on the other hand, who can teach three different courses, is scheduled to teach two of them. Looking at the (1, p) on the "Included_in" edge associated with the ternary relationship type *Schedule* reveals that a quarter must be related to at least one {instructor, course} pair (1 for the min value) and that a quarter may be related to up to p {instructor, course} pairs (p for the max value). The data set *Schedule* in Figure 5.3 indicates that there is at least one {instructor, course} pair related to each quarter and that each quarter has one or more instructors teaching one or more courses.

One can also argue that *Schedule* can be conceptualized as a base entity type and, in effect, preempt the formulation of a ternary relationship type as unnecessary. The ER diagram in Figure 5.4 models this viewpoint. This is a valid argument since SCHEDULE in Figure 5.4 lends itself to conceptualization as a base entity type with its own unique identifier (**Call#**).

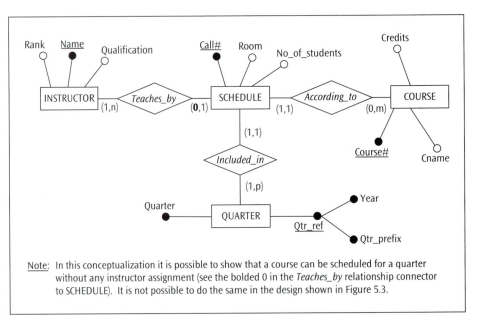

Note: In this conceptualization it is possible to show that a course can be scheduled for a quarter without any instructor assignment (see the bolded 0 in the *Teaches_by* relationship connector to SCHEDULE). It is not possible to do the same in the design shown in Figure 5.3.

Figure 5.4 Modeling the ternary relationship type *Schedule* of Figure 5.3 as a base entity type

However, there are situations where the conceptualization of a ternary relationship as a base entity type is not necessarily semantically obvious. For instance, consider possible relationships among INSTRUCTOR, COURSE, and BOOK. It is quite conceivable that one may need to know which instructor uses what book for which course. After all, one instructor may use a certain book for a course and a different instructor may teach the same

168

Chapter 5

course using a different book, and so on. While binary relationships among the three entity types taken two at a time are meaningful, they cannot even collectively capture the semantics stated above. In this case, a ternary relationship type *Uses* among INSTRUCTOR, COURSE, and BOOK becomes genuinely necessary. However, a base entity type to pre-empt this ternary relationship is not semantically obvious (see Figure 5.5), essentially demonstrating that a ternary relationship type is not just a syntactically valid construct, but a semantically valuable construct in the ER modeling grammar.

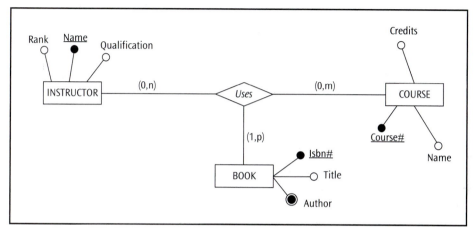

Figure 5.5   The ternary relationship type *Uses*

Let us now examine another scenario to signify the utility of a ternary relationship.

5.1.2 Vignette 2—Get Well Pharmacists, Inc.

Get Well Pharmacists, Inc. has numerous pharmacies across the state of Ohio. A pharmacy dispenses medication to patients. It is imperative that the records at Get Well, Inc. should always have the data on which of its pharmacies dispensed what medication to which patient. In addition, every pharmacy stocks numerous different medicines and the same medicine is carried in several pharmacies. A patient often takes one or more medicines, and the fact that the same medicine may be used by at least one and often many patients can also be a meaningful relationship. Finally, a pharmacy typically has one or more patients as customers and some patients use one or more pharmacies. To make the story a little more interesting, let us impose another business rule (V2R1) that a particular physician prescribes a certain medication to a specific patient.

At the outset, it is seen that a series of binary relationships as well as a ternary relationship exist among the PHARMACY, PATIENT, and MEDICATION entity types. It is not always possible to capture the import of the business rule V2R1 by specifying binary relationships among PATIENT, MEDICATION, and PHYSICIAN taken two at a time. But it is always possible with certainty for a ternary relationship among the three to precisely frame this relationship (V2R1). Figure 5.6 shows two ternary relationships: *Dispenses* among PHARMACY—PATIENT—MEDICATION and *Prescribes* among PHYSICIAN—PATIENT—MEDICATION. Once again, it is conceivable that the ternary relationship *Prescribes* lends itself to be conceptualized as a base entity type with the unique identifier **Prescription#**, thus possibly preempting the need for the ternary relationship. However, the same is not true for *Dispenses*. This alternative appears in Figure 5.7.

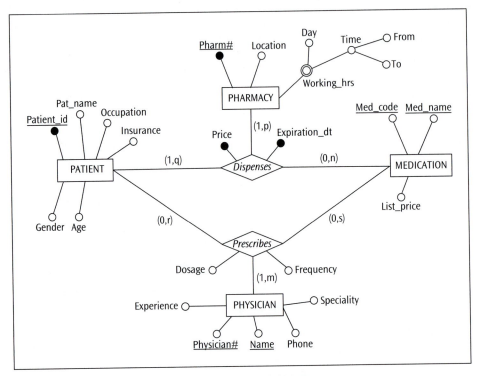

Figure 5.6 Two ternary relationship types *Dispenses* and *Prescribes*

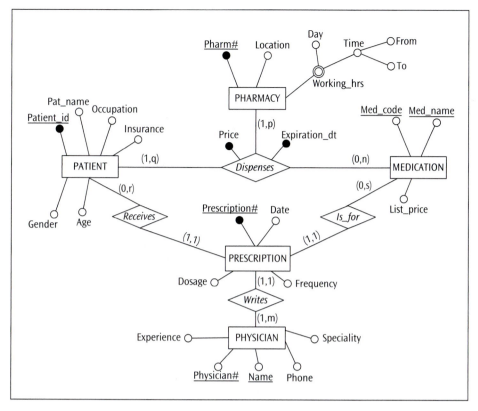

Figure 5.7 Representing the *Prescribes* relationship type as the PRESCRIPTION base entity type

5.2 Beyond the Ternary Relationship Type

Sometimes business rules defining real-world occurrences spontaneously convey relationships even beyond degree three. Though the frequency of such occurrences may not be high, a data modeler/database designer must be sensitive to these possibilities lest such occurrences go unnoticed and compromise the richness of expression of a data model. The question of implementability is a different issue and should not erode the expressive power of a conceptual data model, which is supposed to be technology-independent.

5.2.1 The Case for a Cluster Entity Type

Entity clustering as an ER modeling construct is a useful abstraction to present an ER diagram at a broader level of conceptualization. The technique incorporates object clustering ideas to produce a bottom-up consolidation of natural groupings of entity types. A **cluster entity type** literally emerges as a result of a grouping operation on a collection of entity types and relationship(s) among them. Some database design applications lend themselves to a repeated roll-up to higher levels of abstraction through clustering and provide an opportunity to conceptualize a layered set of ER diagrams. This can be especially useful for large database design projects where the different layers within the

Presentation Layer ER diagram can be used to inform the end-user hierarchy (for example, executives, managers, and staff). Even for an office staff working at the detail level of a business, a layered presentation can facilitate quicker understanding of the semantics captured by the overall design when presented through a cascading set of ER diagrams. The example in the next section serves as a simple illustration of the concept of the cluster entity type.

5.2.2 Vignette 3—More on Madeira College

Madeira College, originally introduced in Vignette 1, has students as well. Students must declare a major field of study and in fact can enroll in up to three different majors. Every major has some students. The advising office of the college has staff specially trained in advising in each major. An advisor is restricted to two majors by training. Every major has exactly two trained advisors. A student can have multiple advisors for each major he or she enrolls in but must have at least one advisor per major. Therefore it is imperative that information about which advisor advises which student for what major is recorded.

Notice in the ER diagram of Figure 5.8a that in order to capture all aspects of the above scenario, a ternary relationship type *Advising* among ADVISOR, STUDENT, and MAJOR is necessary, but not sufficient. This is why the additional two binary relationship types *Trained_in* between ADVISOR and MAJOR as well as *Enrolls* between STUDENT and MAJOR are included. There is no story line that requires the presence of a binary relationship type between ADVISOR and STUDENT, nor can such a relationship type collectively with the other two binary relationship types replace the ternary relationship type *Advising* as demonstrated through the previous two examples. The data set in Figure 5.8b contains representative data for the *Advising* relationship.

Suppose we introduce a new business rule (Rule V3R1) in Vignette 3: *An advisor may advise a student for only one major.*

Will this be accomplished if the maximum cardinality of the edge "Advisee" is changed from a **p** to a **1**? The answer here is "yes"; but a "yes" with an unintended side effect. The maximum cardinality of 1 on the edge "Advisee" (see Figure 5.9) does indeed limit an advisor to advising a student for only one major; but it also limits the relationship of a student entity to no more than one {major, advisor} pair. This essentially implies that a student may have up to three majors as indicated on the edge "Enrollee" but no advisor for more than one major, and in fact no more than one advisor, period. The data set in Figure 5.9 shows the effect of restricting a student to one {major, advisor} pair. The reader should note that changing the maximum cardinality of the "Advisor" edge or the "Advised_for" edge to a **1** instead of the "Advisee" edge as discussed above does not yield a correct solution either to the implementation of Rule V3R1 in the ER diagram.

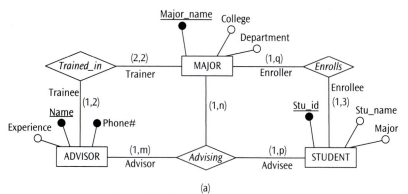

(a)
ER diagram for the scenario described in Vignette 3

Note: The "Advisee" edge of the *Advising* relationship type indicates that a student must be related to at least 1 {major, advisor} pair and may be related to up to p {major, advisor} pairs (e.g., SID 123 in Figure 5.8b has three different major/advisor combinations and SID 345 also has three different major/advisor combinations but only two different majors and two different advisors).

Advising

Stu_id	Major_name	Name
123	Physics	Hawking
123	Music	Mahler
123	Chemistry	Kouri
456	Literature	Michener
789	Music	Bach
678	Physics	Hawking
456	Music	Bach
345	Physics	Kouri
345	Physics	Hawking
345	Chemistry	Kouri

(b)
Data set for the *Advising* relationship in the Figure 5.8a

Figure 5.8 The ternary relationship type *Advising* plus two binary relationship types *Trained_in* and *Enrolls*

Modeling Complex Relationships

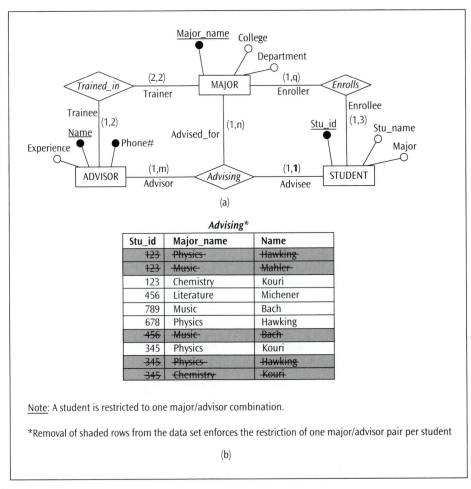

(a)

Stu_id	Major_name	Name
~~123~~	~~Physics~~	~~Hawking~~
~~123~~	~~Music~~	~~Mahler~~
123	Chemistry	Kouri
456	Literature	Michener
789	Music	Bach
678	Physics	Hawking
~~456~~	~~Music~~	~~Bach~~
345	Physics	Kouri
~~345~~	~~Physics~~	~~Hawking~~
~~345~~	~~Chemistry~~	~~Kouri~~

Advising*

Note: A student is restricted to one major/advisor combination.

*Removal of shaded rows from the data set enforces the restriction of one major/advisor pair per student

(b)

Figure 5.9 Changing the maximum cardinality of the "Advisee" edge in Figure 5.8a from p to 1

Rule V3R1 requires that while a student may have multiple advisors and also multiple majors, the student is not permitted to have the same advisor for more than one major. This essentially means that a {student, advisor} pair must be related to no more than one major, while a major may be related to many such pairs. This constraint is reflected in the revised *Advising* data set in Table 5.1.

Advising Data Set*

SID	Major	Advisor
123	Physics	Hawking
123	Music	Mahler
123	Chemistry	Kouri
456	Literature	Michener
789	Music	Bach
678	Physics	Hawking
456	Music	Bach
345	Physics	Kouri
345	Physics	Hawking
~~345~~	~~Chemistry~~	~~Kouri~~

Note: An advisor is limited to advising a student for only one major.

*Removal of the shaded row enables a constraint that limits an advisor to advising a student in only one major.

Table 5.1 Revised *Advising* data set

As discussed above, Rule V3R1 cannot be incorporated in the ER diagram by changing the structural constraints of the ternary relationship type *Advising* in Figure 5.8. The implication of this rule is that one {student, advisor} pair can be related to only one major. This does not prohibit the presence of several {student, advisor} pairs nor does it preclude a major being related to several {student, advisor} pairs. Rule V3R1 can then be interpreted as a relationship between MAJOR and an emergent entity type, say, COUNSELING that results from the cluster {ADVISOR—*Counseling*—STUDENT}, as shown in Figure 5.10. COUNSELING here is called a cluster entity type,[2] indicated in the ERD by the dotted rectangle. Note that each cluster entity "COUNSELING" represents an interrelated {Student, Advisor} pair. Rule V3R1 is implemented in the ER diagram by replacing the ternary relationship type *Advising* in Figure 5.8 with the cluster entity type COUNSELING and its relationship with MAJOR (Figure 5.10). The maximum cardinality of **1** on the edge labeled "Takes_charge" is also required to specify the constraint implied by the Rule V3R1.

---

[2] Aggregate entity type is also a popular term for this. Sometimes this is referred to as a composite (molecular) object. However, in order to distinguish this from an aggregate arising from an aggregation construct (see Section 4.1.6) the term cluster entity type is used. Notice that this entity type is essentially virtual (not real) and does not have a real existence like a base or weak entity type.

The other two relationship types, *Trained_in* and *Enrolls*, shown in Figure 5.8, also persist in the ER diagram. They are not shown in Figure 5.10 in order to enhance the clarity of expression of modeling this particular business rule.

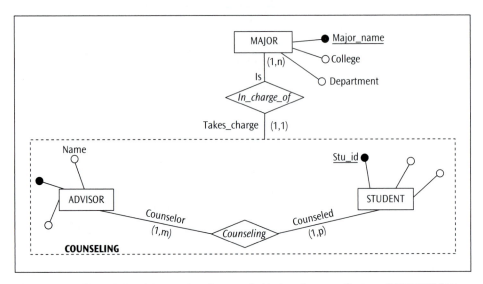

Figure 5.10   The relationship type *In_charge_of* with the cluster entity type COUNSELING

Finally, consider the addition of one more business rule (V3R2) to Vignette 3: *In order to minimize advising snafus, the college mandates that no more than one advisor can advise the same student for the same major.*

The *Advising* data set in Table 5.2 reflects Rule V3R2. This rule can be interpreted as one advisor per {student, major} pair triggering the emergence of another cluster entity type ENROLLMENT as the product of the relationship type *Enrolls* between STUDENT and MAJOR. ENROLLMENT then is related to ADVISOR.

Figure 5.11 portrays the cluster entity type ENROLLMENT and its relationship type *Assignment* with ADVISOR. Once again, note that the maximum cardinality on the edge labeled "Assigned_to" indicates a **1** in order to enforce the constraint of no more than one advisor per {student, major} pair in the ER diagram. Figure 5.12 is a consolidated view of the two cluster entity types ENROLLMENT and COUNSELING, along with the rest of the scenario from the beginning of the episode in Vignette 1 and the embellishment added in Vignette 3.

5.2.3 Vignette 4—A More Complex Entity Clustering

Vignette 3 demonstrated the use of a new ER modeling grammar construct—namely, the cluster entity type—using a simple example. In the vignette that follows, a slightly more complex entity clustering is demonstrated.

Surgeons perform surgeries on patients to treat illnesses. Also, a surgery event pertains to a certain surgery type and there can be numerous surgeries of a certain surgery type. Suppose there is a need to keep track of which surgeons perform what surgery(ies) on which

Advising Data Set*

SID	Major	Advisor
123	Physics	Hawking
123	Music	Mahler
123	Chemistry	Kouri
456	Literature	Michener
789	Music	Bach
678	Physics	Hawking
456	Music	Bach
345	Physics	Kouri
~~345~~	~~Physics~~	~~Hawking~~
345	Chemistry	Kouri

Note: An student is limited to only one advisor in one major.

*Row shaded allows the restriction that limits a student to only one advisor in each major to be enforced.

Table 5.2 *Advising* data: Next revision

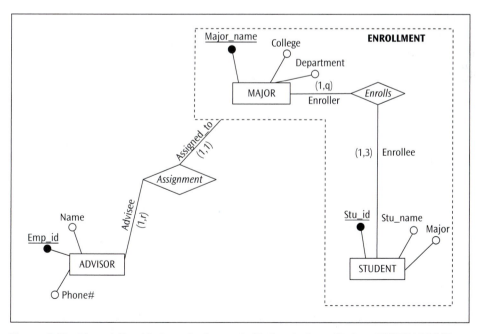

Figure 5.11 The relationship type *Assignment* with the cluster entity type ENROLLMENT

Modeling Complex Relationships

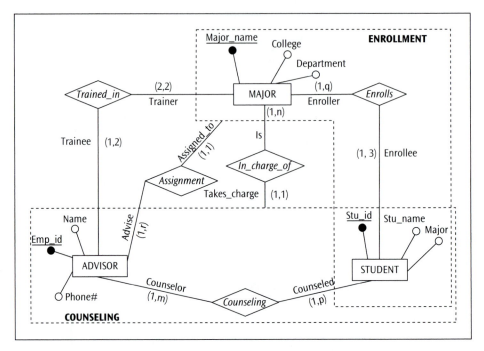

Figure 5.12 Two cluster entity types: COUNSELING and ENROLLMENT

patient(s) to treat what illness(es) for insurance purposes. In fact, the primary insurance covers such treatments. However, all treatments are not necessarily covered by insurance.

How do we model this scenario? This story line naturally lends itself to being modeled as a **quaternary relationship** type (a relationship of degree four) among PATIENT, SURGEON, SURGERY, and ILLNESS in addition to other possible meaningful lower-degree relationships among some of these entity types not indicated in this vignette. Notice that the relationship type *Performs* does not spontaneously lend itself to be modeled as a base entity type, thus inducing the emergence of a quaternary relationship type. In addition, SURGERY can be modeled as a weak entity type identity-dependent on SURGERY_TYPE. Figure 5.13a depicts the quaternary relationship type *Performs* along with the other relevant relationship types specified in the vignette. As a matter of convenience, the attributes in the ERD as well as the various participation constraints in the relationship types are arbitrarily created since they are not the primary focus of this vignette.

Observe that the insurance does not cover the PATIENT or SURGERY or ILLNESS individually. The insurance coverage pertains to the surgery performed on the patient by a surgeon to treat an illness. Thus the cluster entity emerging from the *Performs* relationship is the one that is covered by insurance. The ERD in Figure 5.13a captures this by creating the cluster entity type TREATMENT. This cluster entity type TREATMENT exhibits a relationship with an entity type PRIMARY_INSURANCE outside the cluster. In Figure 5.13b, the roll-up of the cluster entity type TREATMENT to a base entity type is demonstrated.

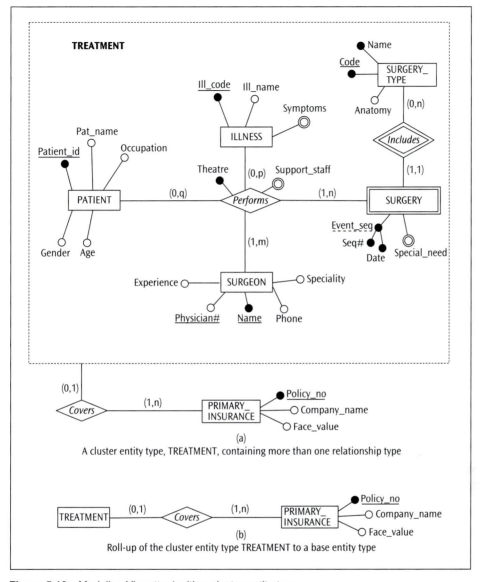

Figure 5.13   Modeling Vignette 4 with a cluster entity type

5.2.4 Cluster Entity Type—Additional Examples

Let us now evaluate a few nuances that can be teased out of the Madeira College story (see Vignette 1). Suppose a course is offered by a specific instructor during a certain quarter at a particular time slot. This enables the same course to be offered at the same time slot during the same quarter by more than one instructor. This story line is modeled as a quaternary relationship type *Offering* among COURSE, INSTRUCTOR, QUARTER, and TIME_SLOT, as shown in Figure 5.14. On closer inspection, one may see an entity type emerging as a product of the quaternary relationship type *Offering*. In other words, the cluster of entity types involved in the quaternary relationship type *Offering* lends itself to the state of an entity type called SECTION. SECTION then is a cluster entity type.

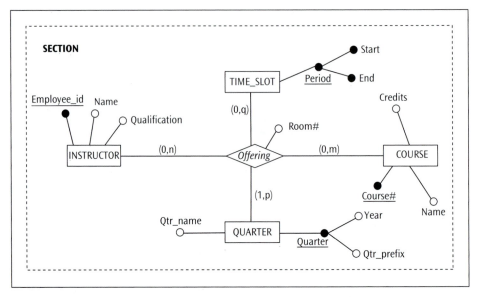

Figure 5.14   An example of the cluster entity type SECTION emerging as a product of a quaternary relationship type

If TIME_SLOT has no attributes other than **Period**, one can question if it is necessary to model it as a base entity type. Accordingly, an alternative way of modeling the cluster entity type SECTION is to treat **Time_slot** as a multi-valued attribute of the *Offering* relationship type. This conceptualization is shown in Figure 5.15.[3]

Suppose we model CLASS_ROOM also as a component of the cluster entity type SECTION. This is accomplished by including CLASS_ROOM as an entity type in the *Offering* relationship type. *Offering* now becomes a **quintary relationship**, a relationship of degree five, illustrated in Figure 5.16. This relationship can be alternatively modeled similar to Figure 5.15 with **Location** and **Time_slot** becoming part of the multi-valued attribute of the *Offering* relationship type, as long as the entity types CLASS_ROOM and TIME_SLOT need not be represented as entity types. This design appears in Figure 5.17.

Observe that the cluster entity type SECTION (ER diagram in Figure 5.16 or 5.17) in its current form implies that "team teaching" of a course section is possible—that is, a particular course in a certain time slot in a given room during a specific quarter can be offered

---

[3] While specification of attributes for a relationship type in the ER modeling grammar is an accepted practice, designation of a multi-valued attribute for a relationship type is somewhat uncommon. An alternate solution essentially resulting from the decomposition of this multi-valued attribute of a relationship type is presented in Section 5.5.2.

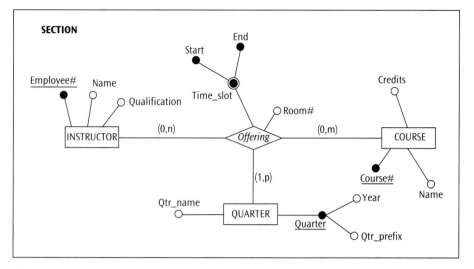

Figure 5.15 An alternate representation of the cluster entity type SECTION (TIME_SLOT reduced to a multi-valued attribute of *Offering*)
Note: The semantics of the cluster entity type have changed.

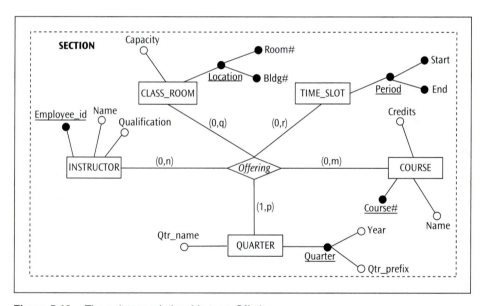

Figure 5.16 The quintary relationship type *Offering*

Modeling Complex Relationships

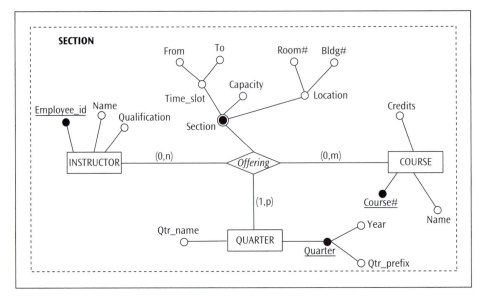

Figure 5.17   Reducing *Offering* to a ternary relationship type with a multi-valued attribute
<u>Note</u>: The semantics of the cluster entity type have changed.

by more than one instructor. Observe that the current design also permits other semantically impractical states; for example, an instructor (or another instructor) may teach a different course in the same room at the same time slot during a specific quarter. Nonetheless, let us presently focus on the "team teaching" issue because the purpose of the design in Figure 5.16 is limited to exemplifying the possibility of a relationship type of degree five.

5.2.5 Madeira College—The Rest of the Story

Let us entertain, at this time, imposition of a series of additional business rules on the Madeira College scenario and observe the unfolding of the ER modeling variations to implement these business rules in the design.

To begin with, consider a new business rule (V4R1) that *team teaching is not practiced in Madeira College.* In other words, no two instructors can teach the same course, at the same time slot, in the same room during a quarter. This implies that INSTRUCTOR is no longer a component of the cluster entity type SECTION, shown in Figure 5.16. Instead there is a 1:n relationship between INSTRUCTOR and SECTION. In addition, **Time_slot**, while correct, is rather a clumsy composite attribute to work with. Thus, from a practical perspective, one may choose to make up a **Section#** to serve as a surrogate for **Time_slot** as the partial key in Offering, in which case **Time_slot** need not be a mandatory attribute anymore. This design variation is shown in Figure 5.18.

Note, however, that the ERD in Figure 5.18a is incorrect because in the ER modeling grammar a relationship type cannot be directly related to another relationship type (i.e., *Teaching* cannot be related to *Offering*).[4] The correct rendition of this design appears in

---

[4] The ER modeling grammar does allow a weak relationship type to be involved in a relationship with another relationship type. A weak relationship type is discussed in Section 5.3.

Figure 5.18b. Observe that the edge "Taught_by" emanating from the *Offering* relationship type in Figure 5.18a has been replaced by the edge "Taught_by" in Figure 5.18b emerging from the cluster entity type SECTION. Also, in Figure 5.18b, in order to enforce the "no team teaching" business rule, the maximum cardinality reflected on the edge labeled "Taught_by" in the

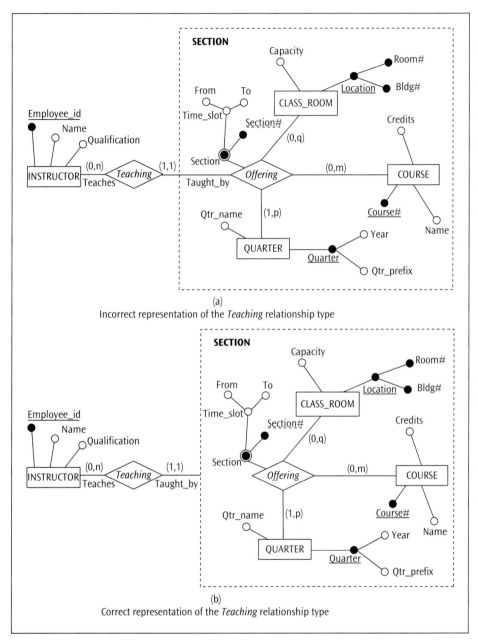

(a)
Incorrect representation of the *Teaching* relationship type

(b)
Correct representation of the *Teaching* relationship type

Figure 5.18 Prohibition of "team-teaching" modeled in the *Teaching* relationship type

Modeling Complex Relationships

Teaching relationship type must become a **1**. At this point, this design permits use of multiple classrooms for the same Section which is not a semantically practical option.

Suppose, at this point, we introduce a new business rule (V4R2): *A Section cannot span multiple classrooms*. This change is accomplished by removing CLASS_ROOM from the cluster entity type, SECTION, modeling a 1:m relationship type *Assigned_to* between CLASS_ROOM and SECTION, and restricting the classrooms used for a Section to **1** through the specification of maximum cardinality of **1** on the Taught_in edge of the *Assigned_to* relationship type (see Figure 5.19).

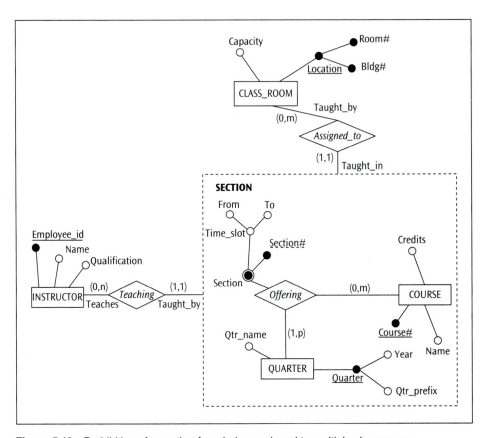

Figure 5.19 Prohibition of a section from being assigned to multiple classrooms

Interestingly, the design in Figure 5.19 does not prevent sections of two or more different courses from being taught in the same classroom at the same time during a quarter. One can argue that this is intended to permit cross-listing of multiple courses. Suppose cross-listing of courses is not permitted. How should the ERD in Figure 5.19 be altered to handle such a business rule? This is left as an exercise for the reader to pursue.

Now, let us embellish the Madeira College story further by stating a couple of additional business rules: *(1) While a course section need not use a textbook, it is also possible that a course section may sometimes use more than one textbook and that a textbook may be used in multiple course sections (V4R3). (2) Likewise, a student must enroll in one or more course sections and a course section must have more than one student (V4R4).*

Does the ERD shown in Figure 5.20 (a relationship type of degree five) accurately describe this scenario? Not unless it is okay for a course section in a quarter when held in one classroom(s) to use one textbook(s) for one student(s) and use a different textbook for another student(s); and a course section in a quarter to use a different textbook(s) for the same student(s) in a different classroom(s). Clearly, while the ERD is syntactically correct, it does not reflect the semantics conveyed by the business rules stated above. SECTION, to begin with, should be a cluster entity type akin to the design shown in Figure 5.18b. Then, it is possible to establish an m:n relationship type between SECTION and TEXTBOOK as well as between SECTION and STUDENT. The ERD in Figure 5.21 is a correct rendition of this scenario where a student enrolls in a course section and uses the same textbook(s) that all other students in that course section are using, and the classroom location of the course section does not change with a student(s) or a textbook(s).

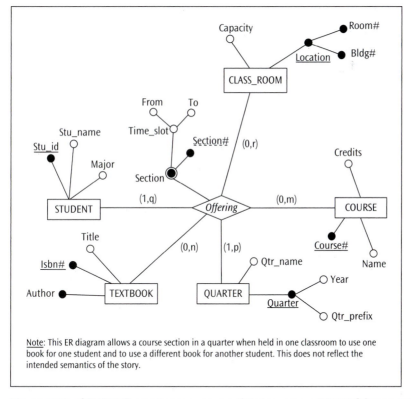

Note: This ER diagram allows a course section in a quarter when held in one classroom to use one book for one student and to use a different book for another student. This does not reflect the intended semantics of the story.

Figure 5.20 STUDENT enrolled in a course *Offering* using a TEXTBOOK modeled as a quintary relationship type

The final ERD reflecting all the business rules stated in the rest of the story for Madeira College appears in Figure 5.22.

5.2.6 Clustering a Recursive Relationship Type

ER modeling constructs lend themselves to use in innovative ways. Let's look at a couple of examples that illustrate interesting ways to use the ER modeling grammar.

Companies often use consultants on a contract basis to do some work and often have different types of boilerplate contracts that can be simply reviewed on a year-to-year basis. This can be modeled in the form of the binary relationship shown in Figure 5.23. As one can see in this figure, a consultant is also a company (the common attribute names indicate that the entity types are identical). In this case, the ER diagram reduces to the recursive relationship type shown in Figure 5.24.

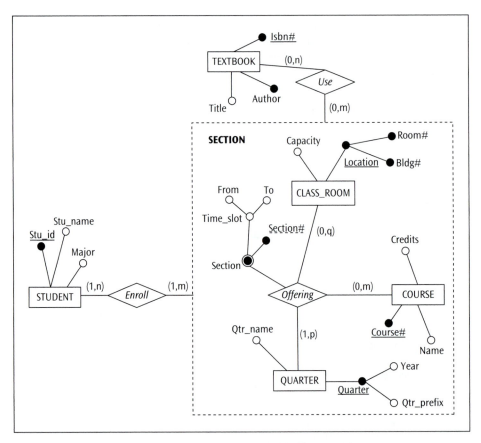

Figure 5.21   A correct rendition of the design intended in Figure 5.20

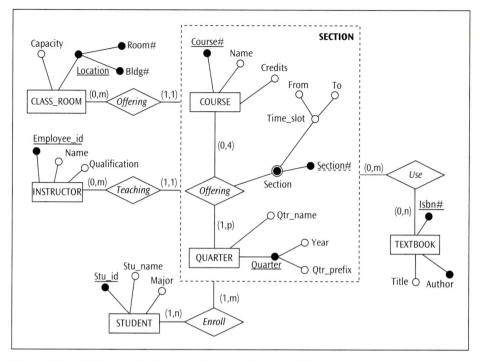

Figure 5.22 ER diagram for Madeira College - The rest of the story

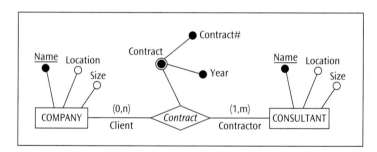

Figure 5.23 *Contract* as a binary relationship type

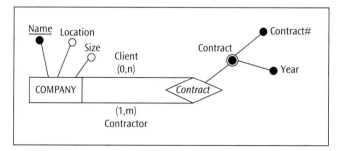

Figure 5.24 *Contract* as a recursive relationship type

Modeling Complex Relationships

Now, what if *each contract is associated with one project and a project may include several contracts?* Because the ER modeling grammar does not permit a relationship to be established between two relationship types (see Figure 5.25a), creating the cluster entity type CONTRACT permits the relationship type *Linked_with* to be specified (see Figure 5.25b).

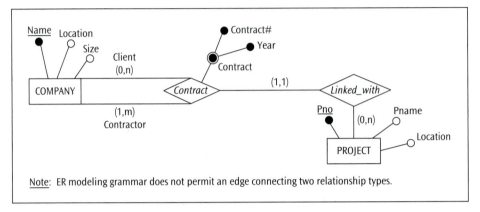

Figure 5.25a Incorrect representation of the *Linked_with* relationship type

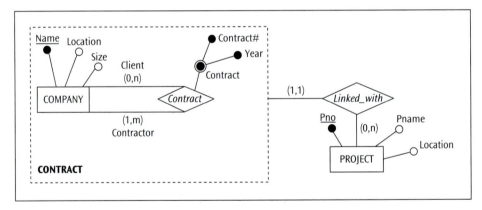

Figure 5.25b Involving the cluster entity type CONTRACT in the *Linked_with* relationship type

Here is another real-world scenario similar to the one described in the previous example. A flight often connects to several other flights at an airline's hub. Thus, a flight's arrival information and the connecting flight's departure information are crucial to the airline and its passengers. Figure 5.26 depicts this scenario.

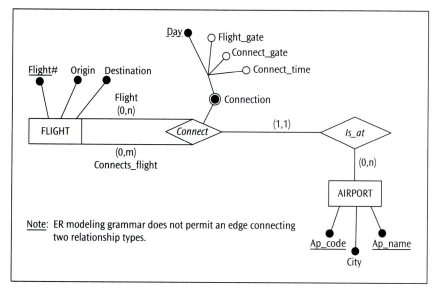

Figure 5.26a Syntactically incorrect representation of the *Is_at* relationship type

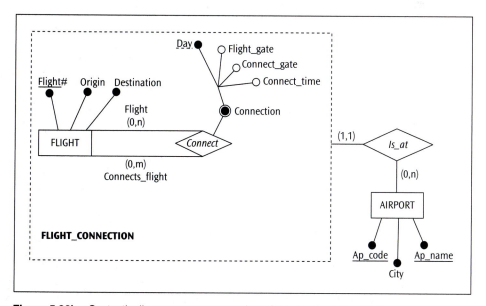

Figure 5.26b Syntactically correct representation of the *Is_at* relationship type

So far, we have seen several real-world scenarios that require relationship types beyond the conventional binary relationship types. The presentation is by no means exhaustive, because other innovative ways of combining the various modeling constructs of the ER modeling grammar are possible. An understanding of the domain of the business problem normally leads to the emergence of such uses. The objective of this section has been to sensitize data modelers and database designers to the rich modeling opportunities available at their disposal via the ER modeling grammar.

5.3 The Weak Relationship Type

Recall that the business rules for Bearcat Incorporated (Section 3.1) indicate that a plant has employees and every employee works in one and only one plant. Also, some plant employees hold the position of managers of these plants. Let us now change this business rule as follows: *All employees need not be working at the plants because some are in the corporate office and others are in the regional offices of Bearcat Incorporated. However, at present, we are modeling just the activities in the plants.* Also, let us add a new business rule: *A plant manager should be a plant employee.*

Figure 5.27 is an excerpt from Stage 1 of the Coarse-granular Design-Specific ER diagram for Bearcat Incorporated developed in Chapter 3 (Figure 3.5) that pertains to this scenario.

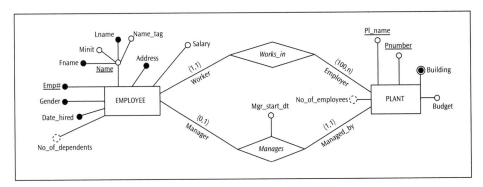

Figure 5.27   The *Works_in* and *Manages* relationship types from Figure 3.5

In order to incorporate the change in business rule, the participation constraint of EMPLOYEE in the *Works_in* relationship type must be made partial (min = 0) since not all Bearcat employees work at the plants. The new business rule essentially suggests a precedence relationship between *Works_in* and *Manages*. In other words, for a *Manages* relationship instance between an employee and a plant to exist, a corresponding *Works_in* instance between the same two entities must be present. Dey, Storey, and Barron (1999) introduced a new ER modeling construct called the **weak relationship type** to indicate an inter-relationship integrity constraint. The symbol used to denote a weak relationship type is the same as the identifying relationship type (a double-diamond symbol—refer back to Figure 3.2) as shown in Figure 5.28.[5] A solid arrow from a regular relationship type to the weak relationship type indicates an inter-relationship integrity constraint, implying that the latter relationship set is included in (i.e., a subset of) the former relationship set. That

---

[5] This does not cause any conflict in the ER modeling grammar because an identifying relationship type can exist only between a weak entity type and its identifying parent entity type(s), while a weak relationship type can relate only to a regular relationship type. Further, interpretation in context will clarify if a double diamond is an identifying relationship type or a weak relationship type or both.

is, in order for an instance of the *Manages* relationship type to occur, an instance of the *Works_in* relationship type between the same entity pair should be present; essentially, a manager of a plant must work in the same plant. This constraint specification is referred to as **inclusion dependency**.[6] Dey, Storey, and Barron define a weak relationship type as "... a relationship, the existence of whose instances depends on the [presence of] instances of (one or more) other relationships." (1999, p. 465). The inclusion dependency is shown as *Manages* $\subseteq$ *Works_in*.

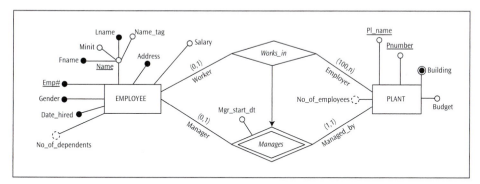

Figure 5.28 *Manages* as a (condition-precedent) weak relationship type

A weak relationship type arises in many real-world situations when two relationship types are linked by (1) a **condition precedence sequence** (based on meeting a condition), or (2) an **event precedence sequence** (based on the occurrence of an event). The weak relationship type *Manages* in Figure 5.28 is a condition-precedent weak relationship type because the condition that one has to be an employee of the plant in order to be a manager of that plant semantically precedes that employee being the manager of that plant. Another opportunity for a condition-precedent weak relationship can be seen in the scenario reflected in Figure 5.2. An appropriate excerpt from Figure 5.2 is presented in Figure 5.29. Consider a business rule: *In order for an instructor to teach a course, he or she must be capable of teaching that course.* This refinement of the story is incorporated in the ER diagram in Figure 5.29. *Teaches* here is a condition-precedent weak relationship type that is inclusion-dependent on *Can_teach*, as shown in (*Teaches* $\subseteq$ *Can_teach*). This is captured in the ERD by the solid arrow drawn from *Can_teach* to *Teaches*.

Figure 5.30 is a subset of the ERD for the story narrated in Vignette 2 about Get Well Pharmacists, Inc. (see Section 5.1.2 and Figure 5.6) where a new constraint is imposed via the business rule: *A medication must be stocked by a pharmacy before it can be dispensed to a patient.* The inclusion dependency *Dispenses* $\subseteq$ *Stocks* captures this business rule and is shown in the ERD by the solid arrow drawn from *Stocks* to *Dispenses*.

---

[6] Inclusion dependency is different from the inclusive arc construct of the ER modeling grammar in that it conveys directionality through a subset relationship, while the inclusive arc conveys the idea of mutuality in inclusiveness. Inclusion dependency is discussed further in Chapter 6 when mapping a conceptual schema to a logical schema.

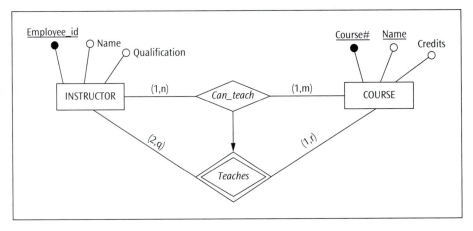

Figure 5.29 *Teaches* as a (condition-precedent) weak relationship type

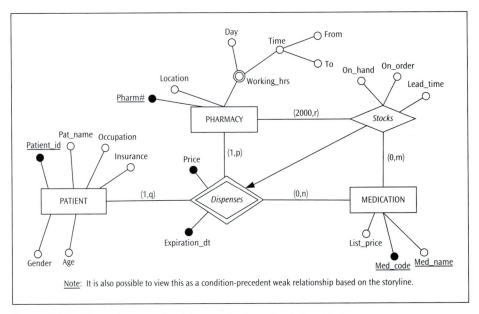

Figure 5.30 *Dispenses* as an (event-precedent) weak relationship type

Dispenses here is labeled as an event-precedent weak relationship type, but it may also be viewed as a condition-precedent weak relationship based on the story line.[7]

Suppose a rental agency rents an array of vehicles (e.g., cars, trucks, vans, boats, etc.). A plausible business rule in this context is: *Before the event "return of a vehicle by a customer" happens, the event "rental of that specific vehicle by that particular customer"*

---

[7] What is important is that this indicates a weak relationship type. Interpretation of whether it is event-precedent or condition-precedent has only amusement value.

should transpire. Figure 5.31 captures this requirement as an event-precedent weak relationship type.

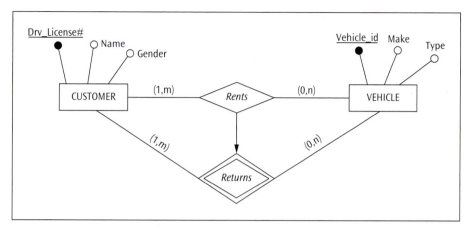

Figure 5.31 *Returns* as an (event-precedent) weak relationship type

While the weak relationship type can be used to indicate inclusion dependency in terms of a unidirectional subset relationship between two relationship types, it can also be used to model scenarios of bidirectional (mutual) exclusion between two relationship types. Referred to as **exclusion dependency**, this construct in the ER modeling grammar is conceptually equivalent to the ER modeling construct "exclusive arc" defined in Chapter 3 (see Figure 3.2 in Chapter 3). Thus, these two constructs are mutually substitutable. As an example, consider a scenario from the Bearcat Incorporated case (Chapter 3) where employees as well as dependents of employees have accounts with Bearcat Credit Union (Figure 5.32). The exclusive arc in this figure indicates that the relationship types *Held_by_E* and *Held_by_D* are mutually exclusive, conveying the business rule: *A joint-account between an employee and a dependent is prohibited*. The ERD in Figure 5.33 portrays the same business rule by making *Held_by_E* and *Held_by_D* mutually exclusive weak relationship types. The exclusion dependency is shown by a non-directional dotted edge connecting the two weak relationship types. Note that both *Held_by_E* and *Held_by_D* are rendered as weak relationship types only because the exclusiveness is bidirectional—i.e., it is not possible to identify one of them as weak relationship and the other a regular relationship, as can be done in the case of the inclusion dependency constraint.

It is interesting to observe that *Dependent_of* is an identifying relationship type (double diamond), as the weak entity type DEPENDENT is identification-dependent on the base entity type EMPLOYEE. However, neither *Held_by_E* nor *Held_by_D* (double diamond) is an identifying relationship type because the child entity type in these two relationship types, namely BCU_ACCOUNT, is not a weak entity type identification-dependent on either EMPLOYEE or DEPENDENT. *Held_by_E* and *Held_by_D* are weak relationship types mutually constrained by an exclusion dependency implementing the business rule: *A joint-account between an employee and a dependent is prohibited*.

Modeling Complex Relationships

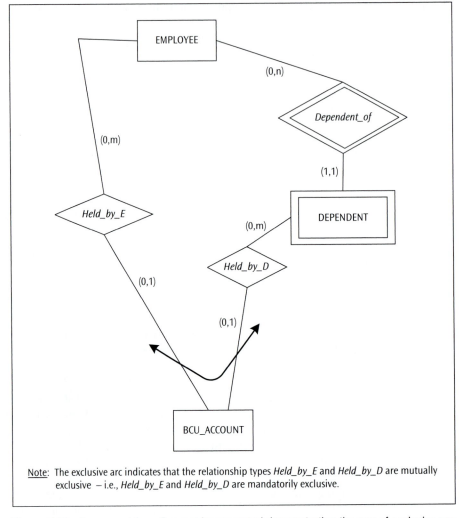

Note: The exclusive arc indicates that the relationship types *Held_by_E* and *Held_by_D* are mutually exclusive − i.e., *Held_by_E* and *Held_by_D* are mandatorily exclusive.

Figure 5.32 An excerpt from Bearcat Incorporated demonstrating the use of exclusive arc

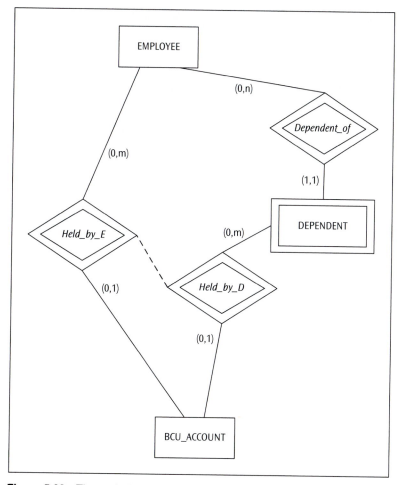

Figure 5.33 The exclusive arc of Figure 5.32 portrayed using exclusion dependency between weak relationship types

The diagram contains:

EMPLOYEE

(0,n)

Dependent_of

195

(0,m)

(1,1)

Held_by_E

DEPENDENT

(0,m)

(0,1)

Held_by_D

(0,1)

BCU_ACCOUNT

5.4 Composites of Weak Relationship Types

Weak relationship types also enable the data modeler to capture richer business situations in an ER diagram, as illustrated in the following examples.

5.4.1 Inclusion Dependency in Composite Relationship Types

Consider a scenario where a particular surgery type requires certain skills and only nurses possessing these skills can be assigned to this type of surgery. The set of skills required for assisting in all surgery types as well as the list of skills possessed by each nurse are also available.

Figure 5.34 depicts this scenario. The relevant entity types and their attributes are arbitrarily assigned. In this case, the composite relationship type ($Nurse\_skill \otimes Req\_skill$) captures which nurse(s) have the skills to assist in what surgery type(s). Therefore, the relationship type *Assigned_to* is transformed to a weak relationship type and the inclusion dependency $Assigned\_to \subseteq (Nurse\_skill \otimes Req\_skill)$[8] as denoted by the solid arrow drawn from the composite ($Nurse\_skill \otimes Req\_skill$) to *Assigned_to* in the ERD ensures incorporation of the business rule in the ER diagram.

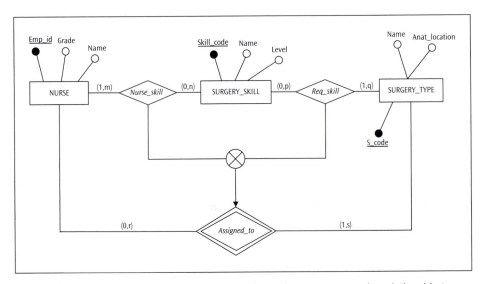

Figure 5.34 A weak relationship type inclusion-dependent on a composite relationship type

Next, let's examine another practical scenario. *Restaurants cater to banquets. A banquet contains a menu of food items and a restaurant caters various food items. Unless a restaurant is capable of preparing the set of food items contained in a banquet's menu, the restaurant cannot cater that particular banquet.*

---

[8] The symbol $\otimes$, referred to as a "composite" in $A \subseteq (B \otimes C)$, implies a projection from the natural join of B and C that is union-compatible with A.

The ERD in Figure 5.35 portrays the scenario for this story line. The relevant entity types, their respective attributes, and relationship types are arbitrarily assigned—e.g., a restaurant "can cater" to at least 50 banquets and a banquet can be catered by at least seven restaurants, etc. The specific business rule stated above is captured through the weak relationship *Can_cater*. Note that the ER diagram reflects the inclusion dependency (*Can_cater* ⊆ *Caters* ⊗ *Contains*) meaning that in order for a relationship to exist in the *Can_cater* set, it should be a subset of the composite set of *Caters* and *Contains*.

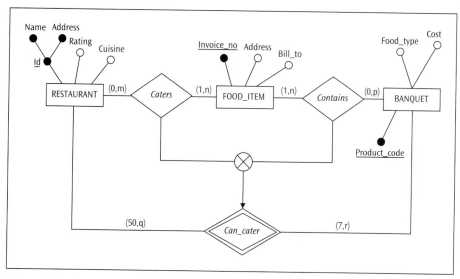

Figure 5.35 A weak relationship type inclusion-dependent on a composite relationship type: A second example

5.4.2 Exclusion Dependency in Composites of Weak Relationship Types

The expressive power of an exclusive arc is limited to portraying mutual exclusiveness between two relationship types. In other words, an exclusive arc is not capable of dealing with composites of relationship types similar to the ones illustrated in the previous section. The exclusion dependency construct, however, is capable of handling composites of relationship types.

As an example, consider the scenario where *professors can be authors of papers and/or reviewers of papers. That is, any professor can be an author as well as a reviewer of papers.* Figure 5.36 depicts this scenario as an ER diagram. The relevant entity types and their attributes are arbitrarily assigned.

Suppose we want to impose a business rule: *A refereeing assignment cannot be made if there is a conflict of interest between an author and a reviewer.* Clearly, the author reviewing his or her own paper or the paper written by his or her advisor, student, etc., amounts to conflict of interest. Can this conflict of interest be captured in this ER diagram? Following Dey, Storey, and Barron (1999), this story line can be expressed as an exclusion dependency between (*Writes* ⊗ *Referees*) and *Conflict_of_interest*, as shown in Figure 5.37. Here, the composite (*Writes* ⊗ *Referees*) represents the reviewer who referees an author's paper and *Conflict_of_interest* captures the reviewer who has a conflict of

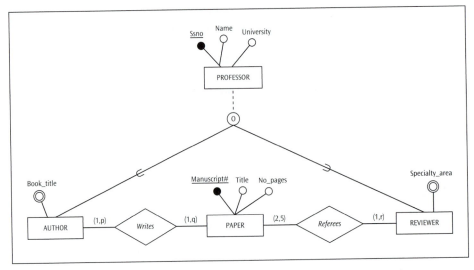

Figure 5.36 An ER diagram depicting a professor as an author and reviewer of papers

interest with the author. That is, *Conflict_of_interest* and (*Writes* ⊗ *Referees*) are mutually exclusive. In other words, if *Conflict_of_interest* exists, then the composite (*Writes* ⊗ *Referees*) cannot exist and thus the author cannot be assigned to review the paper. On the other hand, if the composite (*Writes* ⊗ *Referees*) exists, then the *Conflict_of_interest* relationship cannot exist. Thus, the exclusion is bidirectional. Notice that all three relationships (*Writes*, *Referees*, and *Conflict_of_interest*) are modeled as weak relationship types since, unlike in inclusion dependency, there is no directionality in expressing an exclusion dependency. In other words, the order of the relationship types is immaterial. However, a relationship instance cannot be part of both the composite (weak) relationship type and the other weak relationship type at the same time. The exclusion dependency itself is indicated by a dotted line with no directional pointer connecting the weak relationship type *Conflict_of_interest* and the composite (weak) relationship type (*Writes* ⊗ *Referees*).

5.5 Decomposition of Complex Relationship Constructs

Some of the relationship constructs presented so far in this chapter require further decomposition before they are ready for mapping to the logical schema. In this section, these constructs are identified and their mapping to the Fine-granular Design-Specific layer is demonstrated.

5.5.1 Decomposing Ternary and Higher-Order Relationship Types

In Section 3.2.4, we saw how a binary relationship type with an m:n cardinality ratio cannot be mapped to a logical schema "as is." Therefore, the relationship was decomposed to a gerund entity type. The concept is the same for any n-ary relationship type. A relationship beyond degree two cannot be expressed as is in a logical schema. Thus the decomposition of a ternary relationship type should transform the relationship such that the resulting transformation contains nothing beyond a set of binary relationship types and

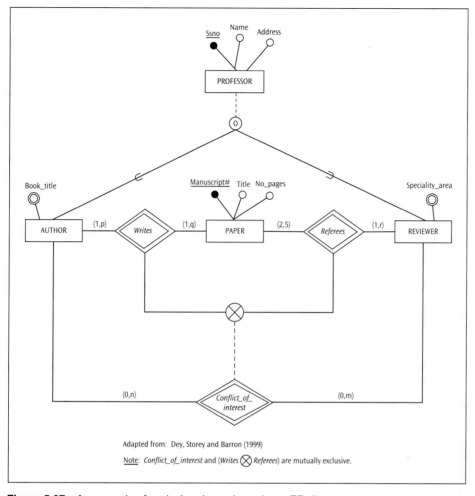

Figure 5.37   An example of exclusion dependency in an ER diagram

that too with cardinality ratios of the form 1:m. Converting the relationship type to a gerund entity type with all the participating base entity types as its identifying parents will accomplish this.

Consider the simple ternary relationship type *Uses* among INSTRUCTOR, COURSE, and BOOK depicted earlier in the Coarse-granular Design-Specific ER diagram shown in Figure 5.5. For the reader's convenience, this diagram is reproduced in Figure 5.38. Based on the rationale stated above, the decomposition to the Fine-granular Design-Specific ER diagram is shown in Figure 5.39. Observe that the gerund entity type USE is a weak entity type with no partial key and three identifying parents (since *Use* is a ternary relationship type). The participation of USE in each of the three identifying relationship types *Selects, Adopts,* and *Finds* is total because USE exhibits existence dependency (i.e., min = 1) in all three identifying relationships.

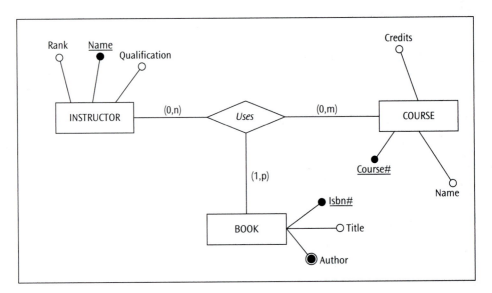

Figure 5.38   The ternary relationship type *Uses*

The transformation process is the same for quaternary (degree four), quintary (degree five), and higher order relationship types. The reader may, as an exercise, map the quaternary relationship *Performs* (Figure 5.13) to a gerund entity type PERFORMS.

5.5.2 Decomposing a Relationship Type with a Multi-valued Attribute

Relationship types with a multi-valued attribute present another interesting case. Consider the *Contract* relationship type shown earlier in Figure 5.23. For convenience, this figure is reproduced as Figure 5.40a. First of all, the cardinality ratio in this binary relationship type is of the form m:n. Thus the relationship needs decomposition preparatory to a logical schema mapping. In the absence of the multi-valued attribute **Contract** in *Contract*, the relationship type would have been decomposed to a gerund entity type CONTRACT with two identifying parents, COMPANY and CONSULTANT. With the composite multi-valued attribute, the decomposition is somewhat similar except that the multi-valued attribute becomes the partial key of the gerund entity type CONTRACT, as shown in Figure 5.40b. In other words, this example clarifies that a gerund entity type may, sometimes, have a partial key.

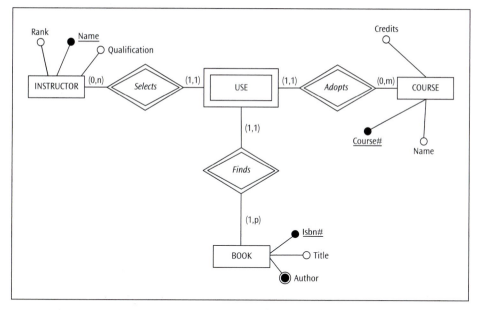

Figure 5.39 Decomposition of the ternary relationship type *Uses* to the gerund entity type USE

Let us build on the simple scenario of this contract between companies and consultants using additional business rules. While the contract may be a boilerplate as described in Section 5.2.6, let us just say that *as part of the contract there is a retainer fee that some companies pay their consultant(s), which can change as part of the contract provisions.* This entails only a minor change in the design ERD as shown in Figure 5.41a. Since the **Retainer** is specified as a part of the contract, it is included as part of the composite multi-valued attribute **Contract**. Observe that since only part of the attribute **Contract** is necessary to serve as the partial key, this part is explicitly identified here as the partial key (dotted underline). The decomposition to fine granularity is shown in Figure 5.41b. Although the decomposition here as well as in Figure 5.40b has a superficial appearance of a ternary relationship type, the semantics conveyed by them are the same as their coarse-granular counterparts (Figures 5.41a and 5.40a, respectively).[9] In fact, the degree of this relationship in the decomposition is not three. In order for this to qualify as a ternary relationship, *Execute* cannot be an identifying relationship type.

Let us now tweak the story line to refine the business rule about the retainer as: *The retainer is independent of the contract.* That is, a company pays a retainer (a fixed amount) to a consultant for availability at a short notice with or without a contract; this amount may change, but not as a part of the contract provisions. Moving the attribute **Retainer** out of the composite attribute **Contract** as in Figure 5.42a is the obvious solution. While this solution does indeed implement the desired business rule, there is a potential problem associated with this solution that may crop up when mapping this ERD to the fine

---

[9] In Chapter 6, we will see how the mapped logical schema conveys the semantics of the conceptual schema correctly.

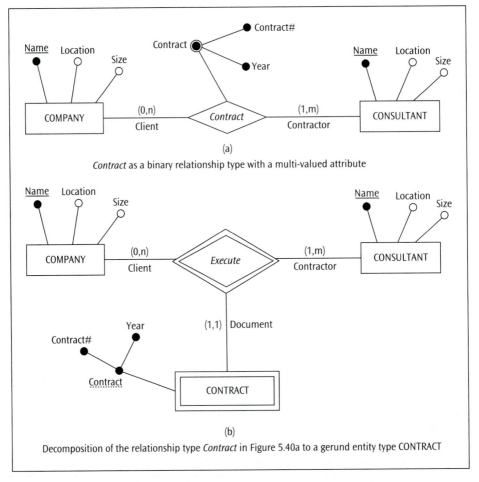

(a)

Contract as a binary relationship type with a multi-valued attribute

(b)

Decomposition of the relationship type *Contract* in Figure 5.40a to a gerund entity type CONTRACT

Figure 5.40   Decomposition of a relationship type with a multi-valued attribute

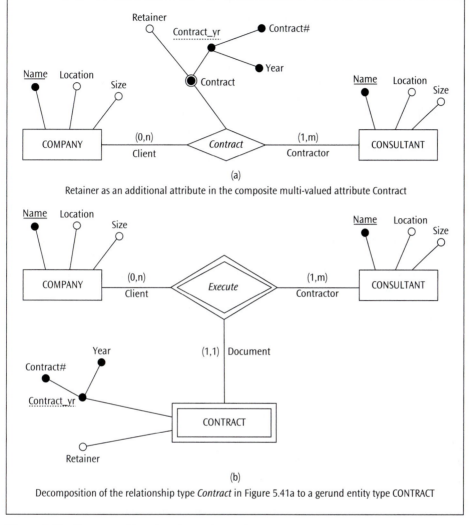

(a)

Retainer as an additional attribute in the composite multi-valued attribute Contract

(b)

Decomposition of the relationship type *Contract* in Figure 5.41a to a gerund entity type CONTRACT

Figure 5.41 Decomposition of a relationship type with a partial key in a multi-valued attribute

granularity. Decomposing the multi-valued attribute of the m:n relationship type *Contract* as was done in the previous scenario (see Figures 5.40 and 5.41), we end up with a gerund entity type CONTRACT with the two identifying parents COMPANY and CONSULTANT. Since **Retainer** is independent of CONTRACT, it cannot be an attribute of CONTRACT; instead, it is mapped as an attribute of the identifying relationship type *Execute*, as shown in Figure 5.42b. Is this mapping correct? As CONTRACT is the only child entity type in the relationship type *Execute*, **Retainer,** the attribute of this relationship type will end up in CONTRACT, the child entity type of this relationship type as in Figure 5.41b. But then, this design as in Figure 5.41b maps back to the ERD shown in Figure 5.41a at the coarse-granular level—not the ERD in Figure 5.42a. In short, the mapping shown in Figure 5.42b fails to honor the business rule: *The retainer is independent of the contract* expressed in Figure 5.42a.

An alternative two-step solution to the decomposition of the design depicted in the ERD of Figure 5.42a to fine granularity is presented below. As a first step, the m:n relationship type between COMPANY and CONSULTANT is preserved with **Retainer** as the attribute of the relationship type *Retains*. The multi-valued attribute **Contract** is decomposed as a weak entity child of the cluster entity type AGREEMENT resulting from the *Retains* relationship type between COMPANY and CONSULTANT. This ERD appears in Figure 5.43a. In this design, **Retainer** exists independent of the CONTRACT even if no contract is in place, thus preserving the stated business rule of independence between **Retainer** and the multi-valued attribute **Contract**. The second step of the decomposition, shown in Figure 5.43b, transforms this design to Fine-granular Design-Specific ERD.

Figure 5.44 is yet another alternative decomposition of the design presented in Figure 5.42a. Here, the modeling variation portrays CONTRACT as the weak entity child with two identifying parents, COMPANY and CONSULTANT, instead of it being the weak entity child of the relationship between COMPANY and CONSULTANT, that is, AGREEMENT. Please note that the semantics conveyed by the decomposition shown in Figures 5.43b and 5.44 are not the same. The former suggests that in order for a contract to exist between a company and a consultant an agreement with or without a retainer is required. However, the presence of an agreement does not mandate a contract between the same pair. The latter permits a company and a consultant to sign a contract without a formal agreement and also to have a retainer-based or retainer-free agreement without a contract.

Finally, formation of weak relationship types results from business rules that otherwise will have to be incorporated as semantic integrity constraints in the conceptual data model. None of them, however, entail any decomposition before being mapped to a logical schema.

5.5.3 Decomposing a Cluster Entity Type

A cluster entity type is essentially a virtual depiction sometimes requiring no further decomposition if the only purpose served is simply to express the cluster as a single collective semantic construct in a conceptual data model to enhance understanding. However, no matter how simple (Figure 5.10) or how involved (Figure 5.13 or 5.22) a cluster entity type may be, if it is related to an entity type outside the cluster, then the cluster entity type will have to be decomposed, because a cluster entity type cannot be expressed as is in the logical schema. The solution is to configure the cluster entity type either as a weak entity type or a gerund entity type. Invariably there is a nucleus relationship within a cluster

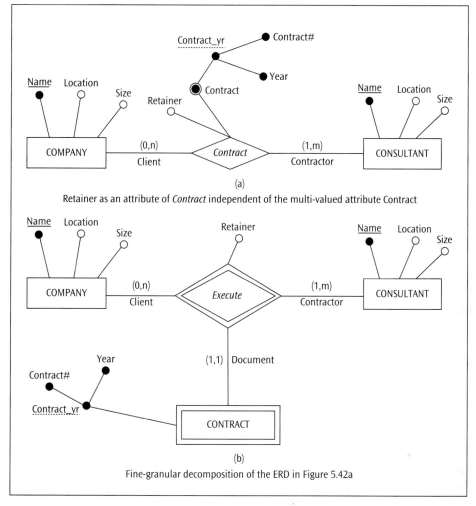

(a)

Retainer as an attribute of *Contract* independent of the multi-valued attribute Contract

(b)

Fine-granular decomposition of the ERD in Figure 5.42a

Figure 5.42 ERD with a single-valued as well as a multi-valued attribute of a relationship type

from which the cluster entity type emerges. Often this relationship type gets distilled to an entity type due to previous decompositions, such as a gerund entity type resulting from the decomposition of an m:n binary or recursive relationship type or any other relationship type of higher degree. Otherwise, at this time, this nucleus relationship from which the cluster entity type emerges can be condensed to a gerund or weak entity type. For example, notice in Figure 5.14 that the relationship type *Offering* can be expressed as a gerund entity type to represent a cluster entity type called SECTION. However, in Figure 5.17 SECTION will decompose to a weak entity child of a relationship among INSTRUCTOR, COURSE, and QUARTER. The reader may work these out as exercises.

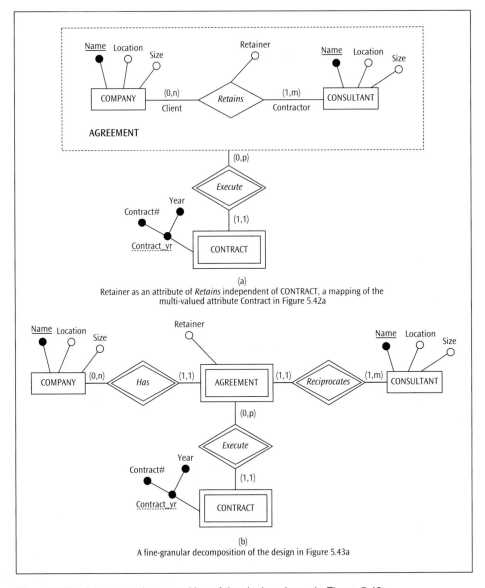

Figure 5.43 A two-step decomposition of the design shown in Figure 5.42a

5.5.4 Decomposing a Weak Relationship Type

Business rules expressed through weak relationship types sometimes do not require any decomposition before mapping to a logical schema, as seen in the earlier example in Figure 5.28. At other times, decomposition is necessary and possible. There are also times when some decomposition coupled with specification of semantic integrity constraints are required to fully capture a business rule. Occasionally, a business rule expressed through a weak relationship type may not be amenable to any decomposition and so will have to

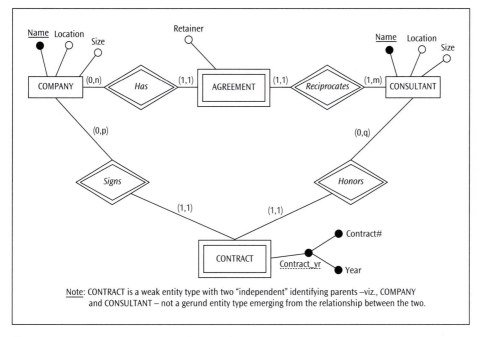

Note: CONTRACT is a weak entity type with two "independent" identifying parents –viz., COMPANY and CONSULTANT – not a gerund entity type emerging from the relationship between the two.

Figure 5.44   An alternative design for the ER diagram in Figure 5.43b

be carried forward as a semantic integrity constraint. Since the various cases in this chapter are highly context-sensitive, they are not covered with an exhaustive set of illustrations. However, in order to get a general idea about how to decompose a weak relationship type to render it ready for transformation to a logical schema, let us review a situation where a decomposition is necessary using the example that appears in Figure 5.29. For convenience, this figure is reproduced as Figure 5.45. The two binary relationships *Can_teach* and *Teaches* between INSTRUCTOR and COURSE require no further elaboration. However, how does one interpret the business rule: *In order to teach a course, an instructor should be capable of teaching that course*, implemented by the inclusion dependency *Teaches* ⊆ *Can_teach*, in the ER diagram? The answer is that the above rule translates to the gerund entity type TEACHING having existence dependency on the gerund entity type CAN_TEACH. This can be implemented in the design via a partial specialization of CAN_TEACH with TEACHING as its subclass, as shown in Figure 5.46. Figure 5.47 shows an alternative decomposition that is equivalent to the design appearing in Figure 5.46.

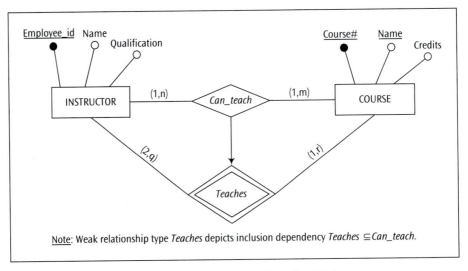

Note: Weak relationship type *Teaches* depicts inclusion dependency *Teaches ⊆ Can_teach*.

Figure 5.45 *Teaches* as a condition-precedent weak relationship type

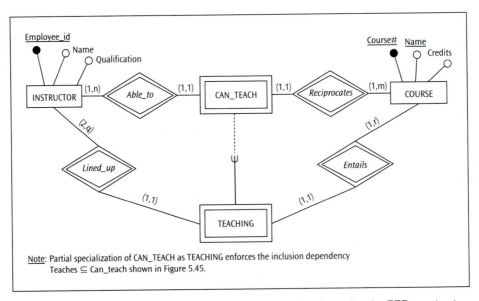

Note: Partial specialization of CAN_TEACH as TEACHING enforces the inclusion dependency Teaches ⊆ Can_teach shown in Figure 5.45.

Figure 5.46 Decomposition of the weak relationship type *Teaches* using the EER construct, Specialization

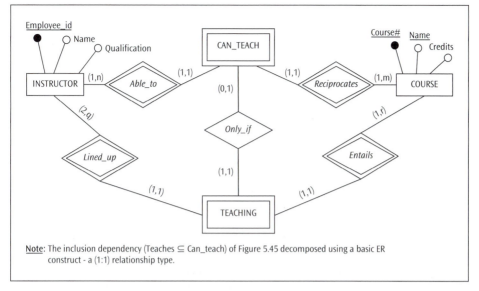

Note: The inclusion dependency (Teaches ⊆ Can_teach) of Figure 5.45 decomposed using a basic ER construct - a (1:1) relationship type.

Figure 5.47 Decomposition of the weak relationship type *Teaches*: An alternative design

5.6 Validation of the Conceptual Design

Up to this point in the book, we have gone through a detailed process of requirements-to-model using the ER modeling grammar and have implicitly assumed that this process takes care of or explains to the reader the outcome, namely, the conceptual model. A formal or informal investigation of how well the developed conceptual model answers the explicit and/or implicit questions present in the requirements specification has not been undertaken. In other words, the ER model has not undergone a critical inspection in light of the specified requirements and been validated accordingly. Having identified the entity types, their respective attributes, and the various types of relationships among the entity types, it is also necessary to verify that the conceptual model thus developed is a true representation of the "universe of interest" being modeled.

A religious adherence to the rules of the modeling grammar ensures that the conceptual model is syntactically correct. A clear understanding of the business rules implied in the requirements specification often leads to a semantically correct conceptual model. Occasionally, misinterpretations of the semantics conveyed by the requirements specification lead to a semantically incomplete conceptual model. This pitfall can be avoided, however, through a systematic validation of the developed conceptual model. Since this book uses the ER modeling grammar for conceptual modeling, this section is devoted to validation of ER models. The textual component of the ER model, namely, the semantic integrity constraints, lists the business rules not captured in the ER diagram. Thus, the focus of this section is on the validation of the semantics captured in the ER diagram—in particular, the semantics captured by the relationship types modeled.

Howe (1989) refers to errors caused in an ER diagram by the misinterpretation of relationships as **connection traps**. Two basic types of connection traps are common: the **fan trap** and the **chasm trap**. Since connection traps emanate from the structural aspects of the

ER modeling grammar, an ER diagram may contain several potential connection traps. Many of these, however, may be of no significance in the context of the requirements specification while others can be eliminated by restructuring the ER diagram appropriately. The idea is to evaluate each potential connection trap for possible obstruction to semantic completeness.

Misinterpretation of the meaning of any relationship implied by the requirements specification is fundamental to connection traps in general. In order to avoid errors of misinterpretation, it is imperative that the analyst carefully defines the relationship semantics and the designer thoroughly understands them.

5.6.1 Fan Trap

A **relationship fan** emerges when two or more relationship types "fan out" from (cardinality constraint of 1:n) or "fan in" to (cardinality constraint of n:1) a particular entity type. In other words, when an entity type serves as the focal point (parent or child) in more than one relationship type on which it is not identification-dependent a relationship fan occurs.[10] A fan trap results when the pathway between certain entities in the relationship fan becomes ambiguous. The following example illustrates a fan trap.

Suppose a library has a large membership and each patron is a member of exactly one library. Every library also stocks a lot of books, while a specific book can only be in one library. Every patron borrows at least one book and every book is borrowed by a patron. A patron may borrow books only from the library in which he or she is a member. Since we are considering the database environment at a given point in time, a book is borrowed by only one patron.

The ERD in Figure 5.48a models this scenario. The relevant entity types and their attributes are arbitrarily assigned. At the outset, it may appear that the connection of PATRON to BOOK via LIBRARY will facilitate deduction of which book(s) is/are available for borrowing to which patron. But a closer scrutiny of the ER diagram reveals this not to be the case. For instance, from the description of the scenario one may reasonably expect questions about the following to be answered by the design shown in Figure 5.48a:

1. Number of members in a given library
2. Library in which a particular patron has membership
3. Library in which a particular book is present
4. Number of books in a given library
5. Number of books borrowed by a patron
6. The patron who has borrowed a particular book

Observe that it is impossible to answer items 5 and 6 above from the current design. The design then, while semantically correct, is not semantically complete. The cause of this error could be a fan trap present in the design. Clearly, relationship types *Member_of* and *Available_in* are "fanning out" from LIBRARY. Thus, a relationship fan does exist. What

---

[10] Decomposition of an m:n relationship type results in a gerund entity type which, when viewed as the focal point (in this case, child) into which two relationship types fan in, should not be misconstrued as a relationship fan because the gerund entity type does not have an independent existence as an entity type— i.e., it is identification-dependent on the entity types constituting the original m:n relationship type. Thus, the apparent fan structure emanating from the gerund entity type is at best a trivial relationship fan incapable of a potential fan trap.

we need to investigate is whether the relationship fan, in this case, creates ambiguity in the pathway between patrons and books resulting in a fan trap.

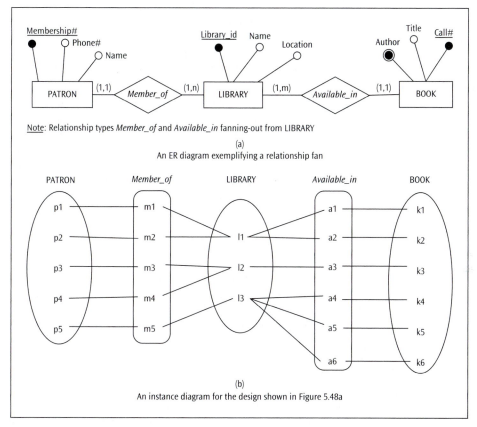

Figure 5.48 An example of a fan trap

The instance diagram (Figure 5.48b), reflecting a legal state of the relationships indicated in the ERD, facilitates the investigation. The pathway in the relationship fan clearly shows that p5 can borrow three books (k4, k5, and k6); no one else can borrow these three books. But can we know the number of books borrowed by patron p1 or p2? The answer is "No." Also, can we know who borrowed the books k1 and k2? The answer, again, is "No." We do know that p1 and p2 are members of the library, l1, and could have borrowed a book only from l1. We also know that the books k1 and k2 are available for borrowing only from l1. From these two facts it is impossible to infer who among p1 and p2 borrowed which of the two books, k1 and k2. Likewise, it is impossible to infer who among p3 and p4 could have borrowed the book k3. That is, the pathway connecting patrons and books through the relationships has ambiguity and it is not possible to answer items 5 and 6 in the above list of questions from this design. This is because the relationship fan here is causing this ambiguity and so is a fan trap. The reason this is a "trap" is because superficially the design appears to provide an unambiguous pathway between PATRON and BOOK while in reality it doesn't.

Figure 5.49 depicts an alternative design. This design is also syntactically and semantically correct. Further, the design answers items 5 and 6 in the above list of questions that are not answerable using the design presented in Figure 5.48. So, is this design the correct solution? Is it semantically complete in the context of the stated scenario and the list of anticipated questions? Let us investigate. The instance diagram in Figure 5.49b reflects a legal state of the relationships indicated in the ERD that appears in Figure 5.49a. As per the design shown in the ERD (Figure 5.49a), a patron may borrow several books. So, the patron p4 has borrowed the books k4 and k5. Likewise, a library may have many books. Observe that libraries l1, l2, and l3 have two books each. A book, however, can be in only one library. Accordingly the book k4 is in library l2 and k5 is in l3. In short, the instance diagram does not violate any relationship constraint defined in the ERD. If we navigate through the available pathway in the design from p4 to the library entities, it is seen that p4 is linked to both libraries l2 and l3. Incidentally, the only link available from the entity type PATRON to the entity type LIBRARY is through the entity type BOOK. So, the inevitable inference about membership from this design is that p4 is a member of two libraries. Thus the design violates a business rule of the stated scenario: *each patron is a member of exactly one library*. Consequently, the design also yields wrong answers to items 1 and 2 of the list of questions. The cause of this error could be a fan trap present in the design. Clearly, relationship types *Borrowed_by* and *Available_in* are "fanning in" to BOOK. So, a relationship fan does exist. Our investigation reveals that the relationship fan, in this case, does create ambiguity in the pathway between patrons and libraries resulting in a fan trap. The design then, while syntactically correct, is not semantically correct, but far less complete.

One way to prove that the relationship fans in the designs shown in Figures 5.48 and 5.49 are indeed fan traps is to demonstrate absence of ambiguous pathways among the entities in a design that is free of fan traps (proof by contradiction). Figure 5.50 is a design of the same scenario restructured to eliminate relationship fans. In fact, the ERD depicts a relationship hierarchy. The instance diagram of Figure 5.50b demonstrates that questions pertaining to all six items about the scenario listed above are answered using the design shown in the Figure 5.50a. The ERD that appears in Figure 5.50a is:

- syntactically correct and so are the ERDs in Figure 5.48a and 5.49a.
- semantically correct and so is the ERD in Figure 5.48a in that both portray the scenario specified equally accurate. Figure 5.49a has been shown to be semantically incorrect.
- semantically complete because it does not have a fan trap, while the designs in Figures 5.48a and 5.49a are plagued by fan traps and therefore are semantically incomplete.

That said, it is important to understand that all relationship fans are not necessarily fan traps. If structurally apparent fan traps are not of any semantic significance in the context of the requirements specification prevailing over its scenario, then those relationship fans are not fan traps. Thus, unconditional avoidance of relationship fans in ERDs limits the richness of the ER modeling grammar and is not recommended. For example, the scenario depicted by the ERD in Figure 5.51 expresses the following story line. *A music group has several musicians. Every musician owns one or more vehicles and a given vehicle is owned*

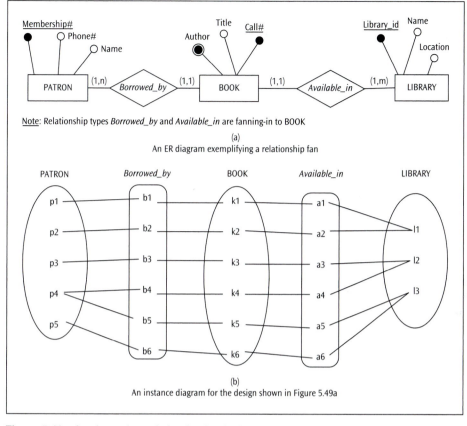

Note: Relationship types *Borrowed_by* and *Available_in* are fanning-in to BOOK

(a)
An ER diagram exemplifying a relationship fan

(b)
An instance diagram for the design shown in Figure 5.49a

Figure 5.49   An alternative solution for the design shown in Figure 5.48

by only one musician. Likewise, a musician can play several instruments, but in this group an instrument is played by only one musician.

To begin with, the relevant entity types and their attributes in the ERD (Figure 5.51) are arbitrarily assigned. Observe that a relationship fan exists in this design—two distinct relationship types, *Owned_by* and *Played_by*, fan out of the entity type MUSICIAN. Is there a fan trap inherent in the design? It is true that there can be ambiguities in the pathway between VEHICLE and INSTRUMENT. If questions like "what vehicle does a musician own while playing a guitar" or "how many instruments does a musician play while owning an SUV" are semantically relevant, then this relationship fan indeed constitutes a fan trap. In other words, since a pathway between VEHICLE and INSTRUMENT is not semantically relevant in the story line, any ambiguity in the pathway caused by the structural arrangement in the design is irrelevant. Therefore, the relationship fan, in this case, does not cause a fan trap.

5.6.2 Chasm Trap

A chasm trap occurs where a design models certain relationship types, but a pathway does not exist between certain entities through the defined relationships. A chasm trap may

Modeling Complex Relationships

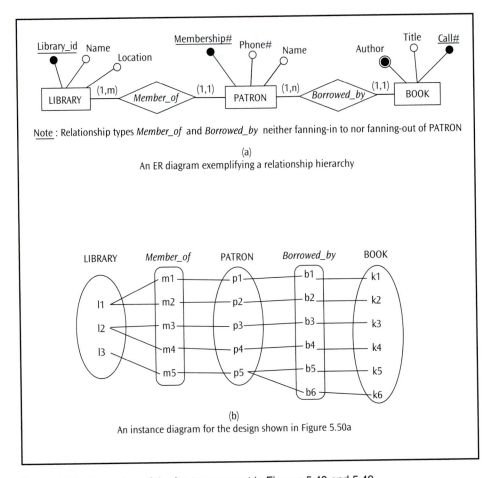

Note : Relationship types *Member_of* and *Borrowed_by* neither fanning-in to nor fanning-out of PATRON

(a)
An ER diagram exemplifying a relationship hierarchy

(b)
An instance diagram for the design shown in Figure 5.50a

Figure 5.50 Resolution of the fan trap present in Figures 5.48 and 5.49

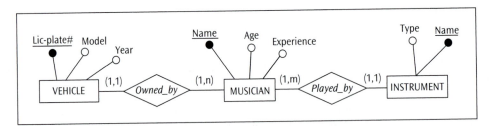

Figure 5.51 A relationship fan semantically irrelevant as a fan trap

occur when there is at least one relationship type in the pathway with optional participation of the referencing (child) entity type.

Suppose we impose a new business rule in the library scenario we discussed in the previous section: *Some books are not borrowed by any patron*. Since the design presented in Figure 5.50 is free of any fan traps, let us impose this new business rule on the ERD in

Figure 5.50a. In fact, this design also models a direct relationship between patrons and books. The above business rule is incorporated in the design by rendering the participation of BOOK in the relationship type *Borrowed_by* optional. The revised ERD and a corresponding instance diagram are presented in Figure 5.52. Observe that book k6 is not borrowed by any patron. This is reflected in the ERD by the min = 0 in the participation of BOOK in *Borrowed_by*. Suppose one asks: "Which library holds the book k6?" The current design fails to answer this question. In fact, the current design fails to answer items 3 and 4 in the list of questions stated in Section 5.6.1. The original designs shown in Figures 5.48 and 5.49 do indeed answer this particular question, but are unacceptable solutions because the presence of fan traps in these designs raise other semantic issues relevant to the scenario. So, what is the solution?

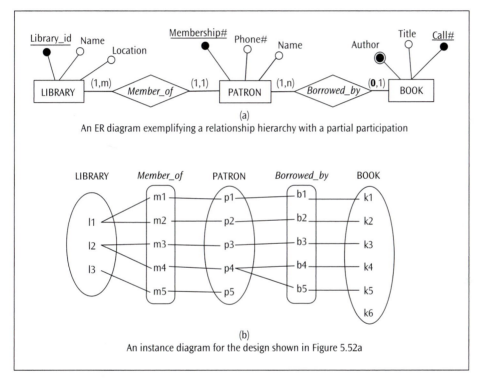

(a)
An ER diagram exemplifying a relationship hierarchy with a partial participation

(b)
An instance diagram for the design shown in Figure 5.52a

Figure 5.52   An example of a chasm trap

An ERD with an additional independent relationship type *Available_in* connecting LIBRARY and BOOK directly, as shown in Figure 5.53, results in a design free of connection traps in the context of the scenario and the associated business rules being modeled here. It is important to understand that the addition of this new relationship type

Available_in is intended to supplant the effect of the chasm trap induced by the optional participation (min = 0) of BOOK in the *Borrowed_by* relationship type. In other words, mandatory participation (min = 1) of BOOK in the *Borrowed_by* relationship type pre-empts the presence of chasm trap, rendering the addition of relationship type *Available_in* redundant. However, it is not the prerogative of the designer to make up business rules to resolve connection traps. The purpose of this discussion is to sensitize the designer to the nuances in the available design options so that he or she can carry on informed interaction with the user community during the development of the requirements specification for the application domain and avoid deceptive pitfalls in the design.

5.6.3 Miscellaneous Semantic Traps

While fan traps and chasm traps are common occurrences in a conceptual design and are also easily recognizable due to their defined structural patterns, there are other semantic traps that are more difficult to identify because they do not necessarily conform to predefined structural patterns. Presumably, such semantic traps are also less common occurrences in simple conceptual designs. In this section we review a couple of such complex connection traps.

5.6.3.1 Vignette 5

Suppose vendors supply products to projects. A critical thing to know is which vendor supplies what product to which project and the frequency of the supplies. In addition, say there is another business rule: A project can get a specific product only from one vendor. This does not preclude a project from getting other products from the same and other vendors. It only restricts a project from getting the same product from several vendors.

Based on the several examples provided at the beginning of this chapter (see Section 5.1), it appears that a ternary relationship among entity types VENDOR, PRODUCT, and PROJECT with **Frequency** as the attribute of the relationship type captures this scenario. This design is presented in the ERD shown in Figure 5.54a. Does this ERD capture the business rule: *A project can get a specific product only from one vendor?* The answer is 'No.' From a modeling perspective, what needs to be accomplished is that a {product, project} pair must be restricted to a relationship with just one vendor. The instinctive reaction to this constraint specification is to change the structural constraints of VENDOR in the *Supplies* relationship from (1, n) to (1, 1). This is a semantic trap in that the change does more than what the business rule specifies. That is, not only can a project get a specific product from just one vendor as required by the business rule, but also a vendor can supply no more than one product, and, that too, to no more than one project. This unexpected side effect amounts to a semantic trap.

The solution lies in restricting the relationship of a {product, project} pair to just one vendor while permitting a vendor to relate to multiple {product, project} pairs. This cannot be done by manipulating the structural constraints of a ternary relationship type among PRODUCT, PROJECT, and VENDOR. This is accomplished by restructuring the ternary relationship type to two binary relationships:

- Rendering an m:n relationship type connecting PRODUCT and PROJECT to a cluster entity type, and
- Specifying a 1:n relationship type between VENDOR and the cluster entity type.

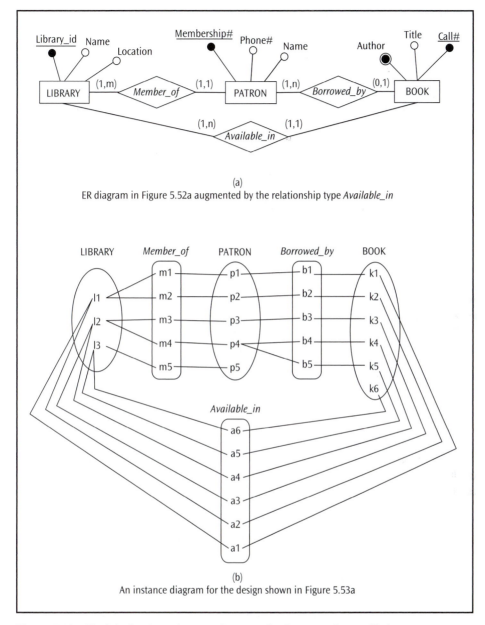

(a)

ER diagram in Figure 5.52a augmented by the relationship type *Available_in*

(b)

An instance diagram for the design shown in Figure 5.53a

Figure 5.53   Final design free of connection traps for the scenario specified

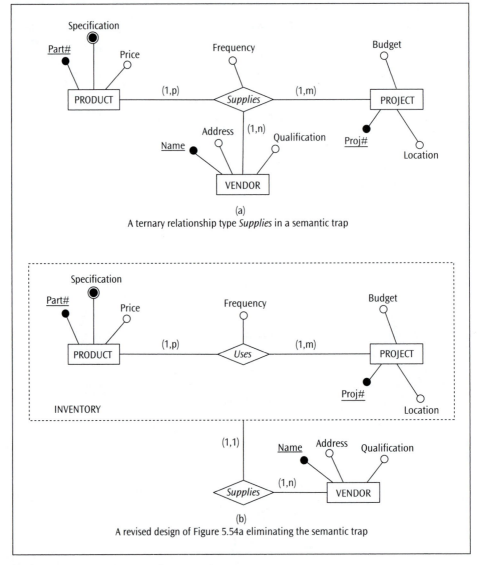

(a)
A ternary relationship type *Supplies* in a semantic trap

(b)
A revised design of Figure 5.54a eliminating the semantic trap

Figure 5.54 Demonstration of a semantic trap

The ERD for this revised design is depicted in Figure 5.54b. Observe that **Frequency** is now the attribute of the binary relationship *Uses*. Also, the cluster entity type has been arbitrarily named INVENTORY. In essence, what appeared to be a possible ternary relationship at first glance is not the correct model to capture the complete scenario portended in the story line. Does the revised model (Figure 5.54b) continue to preserve the requirement as to which vendor supplies what product to which project and the frequency of supplies? It certainly does capture the frequency of use of a certain product by a certain product in the *Uses* relationship in the cluster entity type INVENTORY. Since there is only one vendor related to this {product, project} pair, as indicated by the (1, 1) structural constraint of INVENTORY in the *Supplies* relationship, the requirement is intact in the revised design.

A couple of other examples are available in Section 5.2.2 where the cluster entity type as an ER modeling construct is first introduced.

5.6.3.2 Vignette 6

Let us revisit another earlier example to examine a concealed semantic trap. The second example in Section 5.4.1 describes a scenario of restaurants catering to banquets. For convenience, the exact description of the scenario is reproduced below along with a reproduction of the ER model developed then (Figure 5.35) as Figure 5.55a. *Restaurants cater to banquets. A banquet has a menu of food items and a restaurant caters various food items. Unless a restaurant is capable of preparing the set of food items contained in a banquet's menu, the restaurant cannot cater that particular banquet.*

A possible list of obvious questions pertaining to this scenario is:

1. Which banquets does a restaurant cater to?
2. How many restaurants does a particular banquet use?
3. What is the menu of a particular banquet?
4. What items in a banquet's menu are a given restaurant not capable of preparing?
5. What are the banquets whose menus no single restaurant is capable of preparing alone?

A quick scrutiny of the ERD in Figure 5.55a reveals that it is not possible to answer the first two items listed above using this ERD, even though the ERD is syntactically correct and can in fact be claimed as semantically correct too. From the ERD it is possible to list the banquets a particular restaurant is capable of catering to, but that does not indicate whether the restaurant actually catered to any or all of these banquets (Question 1). Likewise, it is possible to find out the number of restaurants a banquet can use given its menu. However, one cannot identify the number of restaurants that actually catered to a given banquet (Question 2). Thus, the ERD is certainly not semantically complete. In other words, a semantic trap is present in the design. It is a trap simply because its presence is concealed and the design was completed without recognizing the presence of a semantic trap, a case in point for the importance of data model validation.

An alternative design for the same scenario is portrayed in Figure 5.55b. Structurally, the entity types BANQUET and FOOD_ITEM are rearranged so that a direct relationship between RESTAURANT and BANQUET is enabled. This should facilitate answering questions 1 and 2. This rearrangement triggers another structural change in order to preserve the other business rules prevailing over the scenario. A composite of *Caters* and

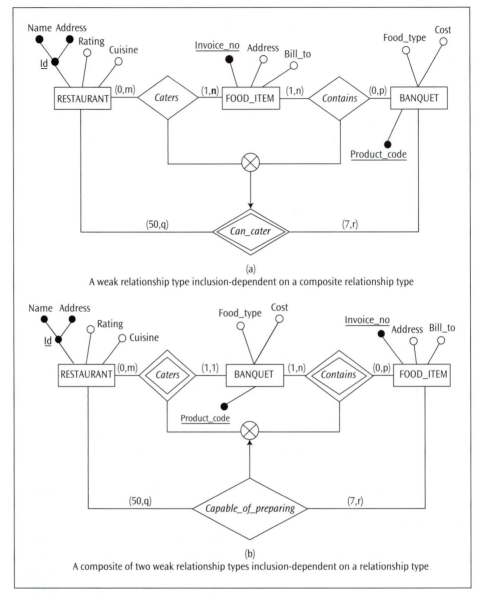

(a)
A weak relationship type inclusion-dependent on a composite relationship type

(b)
A composite of two weak relationship types inclusion-dependent on a relationship type

Figure 5.55   Inclusion dependency in composite relationship types: Two contrasting examples

Contains become inclusion-dependent on *Capable_of_preparing*. As a consequence, pursuant to ER modeling grammar rules, *Caters* and *Contains* become weak relationship types (double diamond) and the direction of the solid arrow in the ERD is accordingly reversed. Is this design superior to the one developed earlier (Figure 5.55a)? The answer to this question is context-dependent. Given the scenario and the list of probable questions pertaining to the scenario, the alternative design just developed (Figure 5.55b) is indeed superior since it fully captures all the specified semantics and specifically eliminates the semantic trap identified in the original ERD.

The more important question, though, is: How do we arrive at this solution? What is the approach used to solve the problem? Unfortunately, semantic traps often do not indicate a pattern like the connection traps where a structural configuration of the ERD capable of generating a specific connection trap is known as the approach(es) to resolve the particular connection trap. The two examples presented here bear no similarity to each other regarding the identification or resolution of the semantic trap. Trial and error and experimentation based on possible hints discerned in the scenario and its business rules seem to be the only available approach. The lesson to be learned here is simply that validation of the conceptual design is a crucial step in the data modeling process. Inadvertent misinterpretation of the semantics embedded in a requirements specification is an unavoidable aspect of conceptual modeling. The awareness of the possibility of connection traps and other semantic traps sensitizes a designer to pay close attention to the import of the requirements specification during the conceptual modeling process and inclusion of a formal step of model validation in the conceptual modeling process enhances the quality of the conceptual modeling script (e.g., ER model).

5.7 Cougar Medical Associates

The rest of the chapter shows how the advanced modeling techniques introduced thus far can be used to model complex real world scenarios using a comprehensive case titled Cougar Medical Associates.

Cougar Medical Associates (CMA) is a clinic located in Kemah, Texas, owned by a group of medical corporations and individual physicians. Clinic personnel include physicians, surgeons, nurses, and support staff. All clinic personnel except the surgeons are on an annual salary. Surgeons do not receive a salary but work for Cougar Medical Associates on a contract basis. It is possible for a physician to have an ownership position in the clinic.

Since surgeons perform surgery on patients as needed, it is required that a surgery schedule keep track of the operation theatre where a surgeon performs a certain surgery type on a particular patient and when that surgery type is performed. Some patients need surgeries and others don't. Surgeons perform surgeries in the clinic; some do a lot, and others just a few. Some surgery types are so rare that they may not yet have been performed in the clinic; but there are others that are performed numerous times. In addition, there is the need to keep track of nurses who can be assigned to a specific surgery type since all nurses cannot be assigned to assist in all types of surgeries. A nurse cannot be assigned to more than one surgery type. It is the policy of the clinic that all types of surgery have at least two nurses. The clinic maintains a list of surgery skills. A surgery type requires at least one but often many surgery skills. However, all surgery skills are not utilized in the clinic while some surgery skills are utilized in numerous surgery types. Nurses possess one or more of these surgery skills. There are certain surgery skills for which no nurse in the clinic qualifies; at the same time there are other surgery skills that have several qualified nurses. In order to assign a nurse to a surgery type, a nurse should possess one or more of the skills required for the surgery type.

Depending on the illness, some patients may stay in the clinic for a few days; but most require no hospitalization. In-patients are assigned a room and a bed. A nurse attends to several in-patients, but must have at least five. No more than one nurse attends to an in-patient;

but some in-patients may not have any nurse attending to them. If a nurse leaves the clinic, temporarily remove the association of all in-patients previously attended to by that nurse in order to allow these patients to be transferred to another nurse at a later time. Every physician serves as a primary care physician for at least seven patients; however, no more than 20 patients are allotted to a physician. If a physician leaves the clinic, temporarily assign the physician's patients to the clinic's chief of staff. Clinic personnel can also become ill and be treated in the clinic. A patient is assigned one physician for primary care.

Physicians prescribe medications to patients; thus it is necessary to capture which physician(s) prescribes what medication(s) to which patient(s) along with dosage and frequency. In addition, no two physicians can prescribe the same medication to the same patient. If a physician leaves the clinic, all prescriptions prescribed by that physician should also be removed because this information is also retained in the archives. A person affiliated with the clinic as a surgeon cannot be deleted as long as a record of all surgeries performed by the surgeon is retained.

A patient may be taking several medications, and a particular medication may be taken by several patients. However, in order for a patient to take a medicine the medicine must be prescribed to that patient. As a medicine may interact with several other medicines, the severity of such interaction must be recorded in the system. Possible interactions include S = Severe interaction, M = Moderate interaction, L = Little interaction, and N = No interaction.

A patient may have several illnesses and several patients may have the same illness. In order to qualify as a patient, a patient must have at least one illness. Also, a patient may have several allergies.

All clinic personnel have an employee number, name, gender (male or female), address, and telephone number; with the exception of surgeons, all clinic personnel also have a salary (which can range from $25,000 to $300,000), but salaries of some can be missing. Each person who works in the clinic can be identified by an employee number. For each physician, his or her specialty is captured whereas for each surgeon data pertaining to his or her specialty and contract are captured. Contract data for surgeons include the type of contract and the length of the contract (in years). Grade and years of experience represent the specific data requirements for nurses.

A surgery code is used to identify each specific type of surgery. In addition, the name, category, anatomical location, and special needs are also captured for each surgery type. There are two surgery categories: those that require hospitalization (category = H), and those that can be performed on an outpatient basis (category = O). A surgery skill is identified by its description and a unique skill code. Data for patients consists of personal data and medical data. Personal data includes patient number (the unique identifier of a patient), name, gender (male or female), date of birth, address, and telephone number. Medical data includes the patient's blood type, cholesterol (consisting of HDL, LDL, and triglyceride), blood sugar, and the code and name of all the patient's allergies.

For both clinic personnel and patients, a Social Security number is collected. For each illness, a code and description are recorded. Additional data for each in-patient consists of a required date of admission along with the patient's location (nursing unit, room number, and bed number). Nursing units are numbered 1 through 7, rooms are located in either the Blue or Green wing, and the bed numbers in a room are labeled A or B. Medications are identified by their unique medication code and also include name, quantity on hand, quantity on order, unit cost, and year to date usage. For medical corporations with ownership interest in the clinic, the corporation name and headquarters are obtained. Corporation name uniquely identifies a medical corporation. The percentage ownership of each clinic owner is also recorded.

The physicians who work in the clinic have recently embarked on a program to monitor the cholesterol level of its patients because cholesterol contributes to heart disease. Risk of heart disease is classified as N (None), L (Low), M (Moderate), and H (High). The ratio of a person's total cholesterol divided by HDL is used in the field of medicine as one indicator of heart risk. Total cholesterol is calculated as the sum of the HDL, LDL, and one-fifth of triglycerides. A Total cholesterol/HDL ratio less than 4 suggests no risk of heart disease due to cholesterol, a ratio between 4 and 5 reflects a low risk, and a ratio greater than 5 is a moderate risk. The high risk category is not coded as a function of cholesterol.

5.7.1 Conceptual Model for CMA: The Genesis

Recall that conceptual modeling is a heuristic as opposed to a scientific process. Therefore, the analyst must draw on intuition and expect to iteratively enhance the data model through several states of punctuated equilibriums before arriving at a final conceptual schema. Then, additional iterative enhancement of the conceptual schema may become necessary during logical and sometimes even physical design. With this in mind, let us begin modeling Cougar Medical Associates (CMA).

After several focused readings of the narrative, one may discern that nurses, surgeons, physicians, support staff, and patients emerge as the "human" entities in this scenario. One can also see that the first four in this list can be grouped under clinic personnel. Having learned about Superclass/subclass (SC/sc) relationships in Chapter 4, knowing from the CMA story that clinic personnel may also be patients from time to time, and noticing several attributes shared between clinic personnel and patients (Social Security number, name, gender, etc.), one would instinctively generalize these common attributes to form a base entity type to serve as the root of a specialization/generalization hierarchy. Sometimes, at this stage, it may be useful simply to list all the attributes involved in this group of entities and examine the list for obvious entity types and relationship types. Such a process in this case reveals a specialization that with an entity type labeled, say, PERSON at the helm as a superclass with two overlapping subclasses, say, CLINIC_PERSONNEL and PATIENT, seems feasible. All the clinic personnel except the surgeons happen to be salaried employees of the clinic. So, we can group all the salaried employees as the entity type SALARIED_EMPLOYEE, which can be a subclass along with SURGEON as the other subclass in a disjoint specialization of CLINIC_PERSONNEL. At first glance SUPPORT_STAFF appears to be a valid subclass along with NURSE and PHYSICIAN in a disjoint specialization of SALARIED_EMPLOYEE. A closer inspection reveals that there are no specific attributes for SUPPORT_STAFF beyond the only attribute of SALARIED_EMPLOYEE, nor do any specific relationships exist for SUPPORT_STAFF. Thus SUPPORT_STAFF can be absorbed in SALARIED_EMPLOYEE by simply making the specialization a partial one (i.e., completeness constraint is partial). Figure 5.56 shows this conceptualization as an ER diagram. Observe that this ERD is a multi-tier specialization hierarchy.

A slightly simplified alternative design can model NURSE, PHYSICIAN, SURGEON, and SUPPORT_STAFF as subclasses of a disjoint specialization of CLINIC_PERSONNEL. Observe that SUPPORT_STAFF is modeled as a subclass instead of subsumed in CLINIC_PERSONNEL because Salary cannot be included as an attribute of CLINIC_PERSONNEL—surgeons are not salaried members of clinic personnel. Therefore, in this

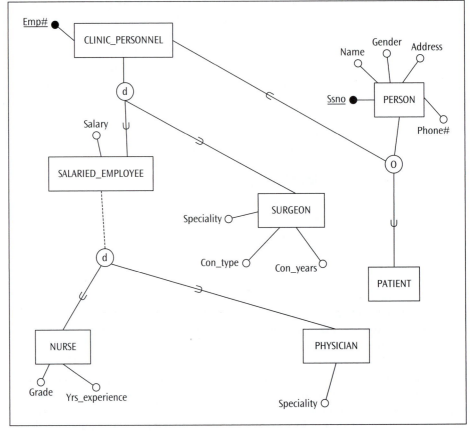

Figure 5.56   Presentation Layer ER diagram for Cougar Medical Associates – Stage 1

design, Salary is included as an attribute of NURSE, PHYSICIAN, and SUPPORT_STAFF. The ERD depicting this design appears in Figure 5.57. This design has the same number of entity types as the one in Figure 5.56 and actually has one less tier in the specialization hierarchy. Therefore, this design is used for the ER model of CMA.

The list of attributes for the entity type PATIENT is relatively large and appears clearly demarcated as personal and medical data about a patient. Although all of these attributes can certainly be recorded under PATIENT, it may be worthwhile modeling PERSONAL_ INFO and MEDICAL_INFO as separate entity types, especially if CMA expects to treat a large number of patients and the use of personal and medical information is clearly divided between administrative and medical personnel of the clinic. Since personal and medical information together make up patient information, the aggregation construct seems an appropriate way to depict this relationship. Figure 5.58 is an increment over Figure 5.57 incorporating the said aggregation construct.

Next, a patient having several illnesses and certain illnesses afflicting many patients can be captured as an m:n relationship between a new entity type ILLNESS and PATIENT. But then, observe that **Allergy** is modeled as a multi-valued attribute of a patient's MEDICAL_INFO instead of as an entity type ALLERGY. Why? The requirements specified

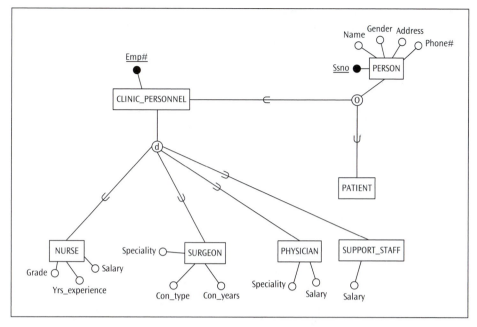

Figure 5.57 Presentation Layer ER diagram for Cougar Medical Associates – Stage 1 (An alternative design)

in the narrative simply state: "*a patient may have several allergies*" and nothing more. That **Allergy** can be a multi-valued attribute of PATIENT is quite clear from this statement. Attempting to specify an entity type ALLERGY amounts to speculating beyond the requirements specification and is incorrect in the context of the CMA scenario.

A relationship between PATIENT and PHYSICIAN also seems obvious from the story. MEDI-CATION appears to be another candidate for being modeled as an entity type. Once again, an m:n relationship between PATIENT and MEDICATION appears imminent. Which physician(s) prescribes what medication(s) to which patient(s) appears to fit the mold of a ternary relationship type with dosage and frequency as the attributes of the relationship type. Likewise, medicines interacting with other medicines convey an ideal recursive relationship type. Finally, the fact that a medicine must be prescribed in order for a patient to take that medicine can be handled by an inter-relationship constraint. These, as well as the relationships that follow, are incorporated in the Presentation Layer ER diagram shown in Figure 5.59.

Another ternary relationship looms in the story line, "Surgeons perform surgery on patients as needed." For this, the creation of an entity type called SURGERY_TYPE is necessary. Incidences of surgery performed by a surgeon on a patient indicating the time and operation theatre for these events are captured by this ternary relationship type. Note that the structural constraints of the ternary relationship types cannot be accurately reflected in the presentation layer because of the use of "look-across" notation in the grammar. It is important to note that nurses are assigned to surgery types (heart surgery, knee surgery, etc.), not to scheduled surgery events. Since nurses also attend to in-patients and other attributes are specific to in-patients that do not apply to patients in general, specializing IN_PATIENT as a subclass of PATIENT and relating it to NURSE makes sense.

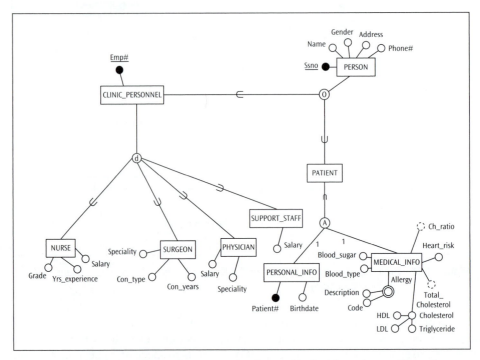

Figure 5.58 Presentation Layer ER diagram for Cougar Medical Associates – Stage 2

At first glance, it is conceivable to think of a nurse's surgery skills and skills required for a surgery type as multi-valued attribute(s). However, this will not be adequate to model the kinds of associations among nurses, surgery types, and surgery skills specified in the requirements. A closer perusal of the story will justify the creation of an entity type for SURGERY_SKILL and relate it to SURGERY_TYPE as well as to NURSE.

Finally, the fact that the clinic can have multiple owners and the owners can be medical corporations and individual physicians (belonging to two different entity classes) leads to the modeling of CLINIC_OWNER as a category arising from a subset of the union between PHYSICIAN and another new entity type MEDICAL_CORPORATION. Alternatively, when the system is being developed for just one clinic, the ownership information may be captured in the entity types PHYSICIAN and MEDICAL_CORPORATION. MEDICAL_CORPORATION, in this case, will be a stand-alone entity type in the ERD which, while technically acceptable, may not be considered an elegant design.

This gives an initial version of the ER diagram to serve as the input for further refinement and specification of other semantic integrity constraints that cannot be incorporated in the Presentation Layer ER diagram shown in Figure 5.59. Notice that the cardinality ratios and participation constraints as culled out from the narrative are shown in the ER diagram. The business rules that cannot be expressed in the ER diagram are included as semantic integrity constraints in Table 5.3.

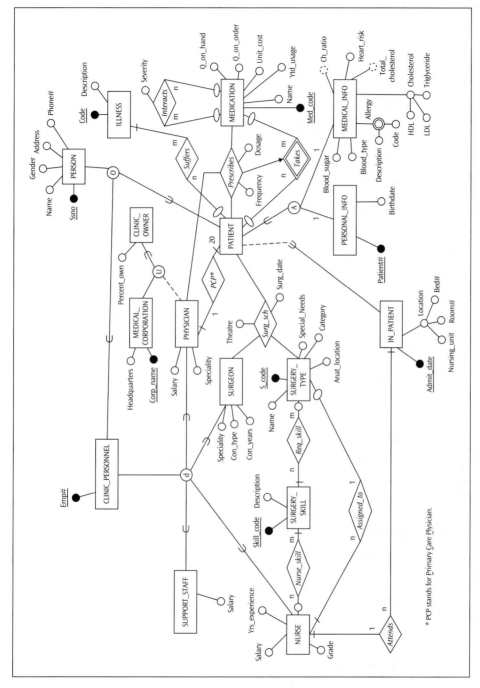

Figure 5.59 Presentation Layer ER diagram for Cougar Medical Associates – The genesis

Attribute-Level Business Rules

1. The gender of a person (i.e., a person affiliated with the clinic or a patient) is either male or female.
2. Nursing unit numbers range from 1 to 7.
3. Salaries of clinic personnel range from $25,000 to $300,000.
4. A surgery category can be either H – require hospitalization or O – be performed on an outpatient basis.
5. Rooms in the clinic are located in either the B = Blue wing or G = Green wing.
6. Bed numbers in a room are labeled either A or B.
7. Severity of medication interaction can be N = No interaction; L = Little interaction; M = Moderate interaction; and S = Severe interaction.
8. Heart risk can be N = No risk; L = Low risk; M = Moderate risk; and H = High risk.

Entity-Level Business Rules

1. A patient's heart risk is (a) "N" when the cholesterol ratio is below 4; (b) "L" when the cholesterol ratio is between 4 and 5; and (c) "M" when the cholesterol ratio is greater than 5.

Business Rules Governing Entity Deletion

1. If a nurse leaves the clinic, temporarily remove the association of all patients previously attended to by that nurse in order to allow these patients to be transferred to another nurse sooner or later.
2. If a physician leaves the clinic, temporarily assign the physician's patients to the clinic's chief of staff.
3. If a physician leaves the clinic, all prescriptions prescribed by that physician should be removed because this information will be retained in the archives.
4. A person affiliated with the clinic as a surgeon cannot be deleted as long as a record of surgeries performed by that surgeon is retained.

Miscellaneous Business Rules

1. A physician serves as a primary care physician for at least seven but no more than twenty patients.
2. Each nurse is assigned a minimum of five patients.
3. All types of surgery require at least two nurses.
4. A nurse is assigned to no more than one surgery type.

Table 5.3 Semantic integrity constraints for the Presentation Layer ER model for Cougar Medical Associates

5.7.2 Conceptual Model for CMA: The Next Generation

A reappraisal of the requirements specification reveals that the following business rules have not been incorporated in the initial Presentation Layer ER diagram shown in Figure 5.59: (1) *No two physicians can prescribe the same medication to the same patient*, and (2) *In order to assign a nurse to a surgery type, a nurse should possess one or more of the skills required for the surgery type.*

Either these rules should be reflected in the Presentation Layer ER diagram, or the rules should become part of the semantic integrity constraints that accompany the Presentation Layer ER diagram to uphold the principle of information preservation. Let us explore if and how these rules can be captured in the ER diagram.

As for business rule (1) above, a similar condition is discussed earlier in this chapter (see Section 5.1.2). The initial "gut reaction" to this business rule is to make the maximum cardinality of PHYSICIAN in the relationship type *Prescribes* a **1**. This will certainly accomplish the goal of no two physicians prescribing the same medication to the same patient; unfortunately, it will also prevent a physician from prescribing more than one medication to the same patient and also the same medication to another patient. That is, maximum cardinality of **1** for the PHYSICIAN in *Prescribes* will limit the relationship of a physician to one {patient, medication} pair. The chances are that this side effect is unintended and unacceptable. If so, then this is a case of a semantic trap discussed using Vignette 5 in Section 5.6.3.1. The solution lies in restricting the relationship of a {patient, medication} pair to just one physician while permitting a physician to relate to multiple {patient, medication} pairs. This cannot be done by manipulating the structural constraints of a ternary relationship type among PATIENT, PHYSICIAN, and MEDICATION. This is accomplished by restructuring the ternary relationship type to two binary relationships:

- Rendering an m:n relationship type connecting PATIENT and MEDICATION (*Prescribes*) as a cluster entity type (PRESCRIPTION), and
- Specifying a 1:n relationship type (*Writes*) between PHYSICIAN and the cluster entity type, PRESCRIPTION.

This solution implements the stated business rule without causing any unexpected side effects. This modification to the design is shown in Figure 5.60. An interesting question that may arise at this point is: Is there a need for a cluster entity type? How about specifying a base entity type called PRESCRIPTION which is more appealing to the conventional wisdom? Strict adherence to the story line of the CMA scenario reveals that a base entity type called PRESCRIPTION is not feasible because from the information available from the requirements specification, a unique identifier for such an entity type cannot be culled out. Further, a closer observation of the design reveals that the cluster entity type PRESCRIPTION will get decomposed to a gerund entity type PRESCRIPTION at the fine-granular level.

Business rule (2) above can be incorporated in the ER diagram by transforming the *Assigned_to* relationship type to a weak relationship type with an inclusion dependency on the composite of *Skill_set* and *Req_skill*. This particular design is discussed earlier in this chapter (see Section 5.4.1).

At this point the ER model consists of the ER diagram in Figure 5.60, the list of semantic integrity constraints in Table 5.3, and the attribute specifications in Table 5.4.

5.7.3 The Design-Specific ER Model for CMA: The Final Frontier

As we learned in Chapter 3, the user-oriented Presentation Layer ER diagram is at a high level of abstraction. It does not capture attribute characteristics and includes constructs that cannot be directly mapped to a logical schema. While it serves as an effective vehicle for the analyst to interact with the user community, it is not adequately equipped for migration to the next tier of data modeling—the logical tier. The Design-Specific ER model is aimed at preparing the Presentation-Layer ER model for direct mapping to the logical schema. While the recommended approach is to first develop the Coarse-granular Design-Specific ER model and then convert it to the Fine-granular model, for brevity,

Modeling Complex Relationships

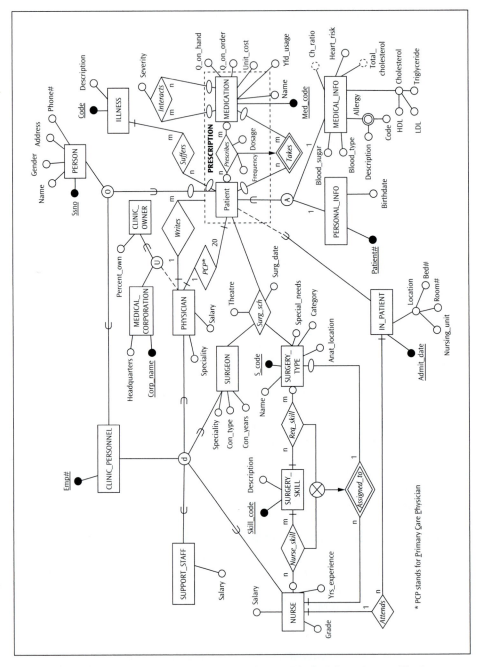

Figure 5.60 Presentation Layer ER diagram for Cougar Medical Associates – Final

Entity/Relationship Type Name	Attribute Name	Data Type	Size	Domain Contraint
PERSON	Ssno	Alphanumeric	9	
PERSON	Name	Alphabetic	30	
PERSON	Gender	Alphabetic	1	M or F
PERSON	Address	Alphanumeric	50	
PERSON	Phone#	Alphanumeric	10	
CLINIC_PERSONNEL	Emp#	Alphanumeric	6	
PHYSICIAN	Speciality	Alphabetic	20	
PHYSICIAN	Salary	Numeric	6	Salaries range from 25000 to 350000
SURGEON	Speciality	Alphabetic	20	
SURGEON	Con_type	Alphabetic	20	
SURGEON	Con_years	Numeric	(1.1)*	
NURSE	Grade	Alphabetic	20	
NURSE	Yrs_experience	Numeric	2	
NURSE	Salary	Numeric	6	Salaries range from 25000 to 350000
SUPPORT_STAFF	Salary	Numeric	6	Salaries range from 25000 to 350000
SURGERY_TYPE	S_code	Alphanumeric	3	
SURGERY_TYPE	Name	Alphabetic	30	
SURGERY_TYPE	Category	Alphabetic	1	O = Outpatient; H = Hospitalization
SURGERY_TYPE	Anat_location	Alphabetic	20	
SURGERY_TYPE	Special_needs	Alphanumeric	5000	
SURGERY_SKILL	Skill_code	Alphabetic	3	
SURGERY_SKILL	Description	Alphabetic	20	
PERSONAL_INFO	Patient#	Alphanumeric	10	
PERSONAL_INFO	Birthdate	Date	8	
MEDICAL_INFO	Blood_sugar	Numeric	3	
MEDICAL_INFO	Blood_type	Alphanumeric	2	
MEDICAL_INFO	Allergy_code	Alphabetic	3	
MEDICAL_INFO	Allergy_description	Alphabetic	20	
MEDICAL_INFO	HDL	Numeric	3	
MEDICAL_INFO	LDL	Numeric	3	
MEDICAL_INFO	Triglyceride	Numeric	3	
MEDICAL_INFO	Total_cholesterol**	Numeric	3	Computed as HDL + LDL + (0.20*Triglyceride)

* $(n_1.n_2)$ is used to indicate n_1 places to the left of the decimal point and n_2 places to the right of the decimal point
** Derived attribute (not intended to be stored in the database)

Table 5.4 Semantic integrity constraints for the Coarse-granular Design-Specific ER model for Cougar Medical Associates

Entity/Relationship Type Name	Attribute Name	Data Type	Size	Domain Contraint
MEDICAL_INFO	Ch_ratio**	Numeric	(1.2)*	Computed as Total_cholesterol /HDL
MEDICAL_INFO	Heart_risk	Alphabetic	1	N = No Risk; L = Low Risk; M = Moderate Risk; H = High Risk
IN_PATIENT	Admit_date	Date	8	
IN_PATIENT	Nursing_unit	Numeric	1	Integer values from 1 to 7
IN_PATIENT	Room#	Alphanumeric	3	ANN, where A = B or G (for the Blue or Green wing); NN is a two-digit room number
IN_PATIENT	Bed#	Alphabetic	1	A or B
MEDICATION	Med_code	Alphabetic	3	
MEDICATION	Name	Alphanumeric	30	
MEDICATION	Q_on_hand	Numeric	4	
MEDICATION	Q_on_order	Numeric	4	
MEDICATION	Unit_cost	Alphanumeric	30	
ILLNESS	Code	Alphabetic	3	
ILLNESS	Description	Alphabetic	20	
MEDICAL_CORPORATION	Corp_name	Alphabetic	30	
MEDICAL_CORPORATION	Headquarters	Alphabetic	15	
CLINIC_OWNER	Percent_own	Numeric	(2.1)*	
Interacts	Severity	Alphabetic	1	N = No; L = Little; M = Moderate; S = Severe
Prescribes	Frequency	Numeric	2	
Prescribes	Dosage	Numeric	4	
Schedule	Theatre	Alphabetic	15	
Schedule	Surg_date	Date	8	

* $(n_1.n_2)$ is used to indicate n_1 places to the left of the decimal point and n_2 places to the right of the decimal point
** Derived attribute (not intended to be stored in the database)

Entity Level Domain Constraints
1. A patient's heart risk is (a) N when the cholesterol ratio is below 4; (b) L when the cholesterol ratio is either between 4 and 5; and (c) M when the cholesterol ratio is greater than 5.

Table 5.4   Semantic integrity constraints for the Coarse-granular Design-Specific ERmodel for Cougar Medical Associates (continued)

only the Fine-granular Design-Specific ER model is shown here. The reader may work on developing the Coarse-granular Design-Specific ER diagram as an exercise.

Two major tasks in this transformation process are:

1. Decompose m:n relationship types to gerund entity types.
2. Transform multi-valued attributes to single-valued attributes.

Since each of the above two steps is discussed in detail in Chapters 3 and 4, suffice it to say here that there is only one multi-valued attribute (**Allergy**) requiring evaluation, and the m:n binary relationships requiring decomposition to gerund entity types are: *Nurse_skill, Req_skill, Prescribes, Takes, Interacts,* and *Suffers.* In addition, the ternary relationship type *Surg_sch* requires decomposition in preparation for mapping to the logical tier. How to decompose a ternary relationship has been illustrated earlier in Section 5.5.1. Following that procedure the decomposition of *Surg_sch* results in a gerund entity type SURG_SCH with three identifying parents, SURGEON, SURGERY_TYPE, and PATIENT as shown in Figure 5.61.

Next, the weak relationship type *Takes,* inclusion-dependent on the regular relationship type *Prescribes,* needs attention. Note that both *Prescribes* and *Takes,* being relationship types depicting the m:n cardinality ratio, will first be decomposed to the gerund entity types PRESCRIPTION and MED_TAKEN, respectively. Then, following the procedure prescribed in Section 5.5.4 for transforming a weak relationship type to a Fine-granular Design-Specific state, a partial specialization involving PRESCRIPTION and MED_TAKEN is modeled as superclass and subclass respectively (see Figure 5.61).

The situation with the weak relationship type *Assigned_to* in the Presentation Layer (Figure 5.60) is somewhat different because it has inclusion dependency on a composite of the two relationship types *Nurse_skill* and *Req_skill.* Remember, the story line states that a nurse must have some of the skills required for the surgery type in order to get assigned to that surgery type. Our first task is to recognize that *Nurse_skill* and *Req_skill* first get translated to the gerund entity types NURSE_SKILL and REQ_SKILL, respectively, because these two are relationship types bearing an m:n cardinality ratio. The task here is two-fold: (1) rendering the inclusion dependency *Assigned_to* $\subseteq$ (*Nurse_skill* $\otimes$ *Req_skill*) implementable, and (2) maintaining the structural constraints of the relationship type *Assigned_to* intact.

As shown in Figure 5.61, using a cross-referencing design for mapping *Assigned_to,* the weak entity type ASSIGNMENT is identification-dependent on NURSE. This meets the specification in item (2) above. The relationship type *Subject_to* between ASSIGNMENT and NURSE_SKILL and the relationship type *Depends_on* between ASSIGNMENT and REQ_SKILL followed by an inclusive arc across these two relationship types partially captures the specification in item (1) above. In addition, a constraint specifying that {Nurse, Surgery_type} projection from ASSIGNMENT should be a subset of the {Nurse, Surgery_type} set resulting from the intersection of NURSE_SKILL and REQ_SKILL must be included in the list of semantic integrity constraints.[11]

The final version of the Fine-granular Design-Specific ER diagram ready for conversion to a logical schema appears in Figure 5.61. Note that in addition to all the necessary decompositions, the deletion rules and all miscellaneous business rules from Table 5.3 and attribute characteristics from Table 5.4 have been incorporated in this ER diagram. Table 5.5 records the remaining semantic integrity constraints to be carried forward to the logical modeling phase, thus rendering the Design-Specific ER model fully information preserving.

---

[11] Projection is a vertical subset of attributes from an entity set. Intersection of two entity sets results in a third entity set containing entities common to the first two. A precise definition and detailed explanation of projection and intersection are present in Chapters 6 and 11.

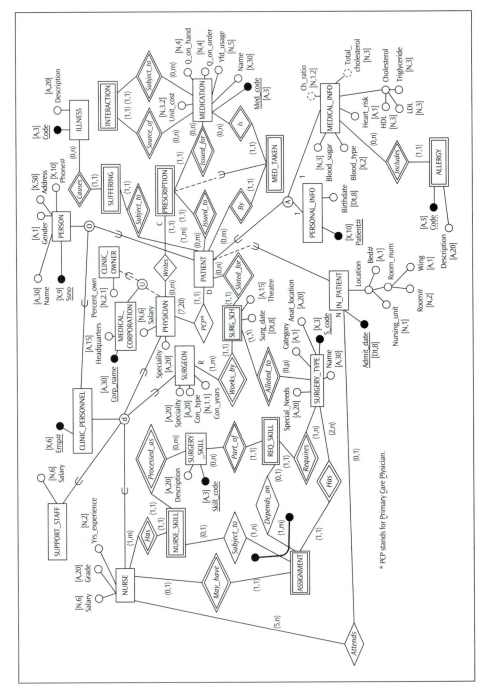

Figure 5.61 Fine-granular Design-Specific ER diagram for Cougar Medical Associates

> Constraint	Gender	IN ('M', 'F')
> Constraint	Salary	IN (25000 through 300000)
> Constraint	Category	IN ('O', 'H')
> Constraint	Heart_risk	IN ('N', 'L', 'M', 'H')
> Constraint	Nursing_unit	BETWEEN (1 and 7)
> Constraint	Wing	IN ('B', 'G')
> Constraint	Bed#	IN ('A', 'B')
> Constraint	Severity	IN ('N', 'L', 'M', 'S')
> Constraint	Ch_ratio	(< 4 and Heart_risk = 'N') OR
		((BETWEEN 4 and 5) and Heart_risk = 'L') OR
		(> 5 and Heart_risk = 'M') OR
		Heart_risk = 'H'

Constraints Carried Forward to Logical Design

1. (Nurse, Surgery_type) projection from ASSIGNMENT should be a subset of the
   (Nurse, Surgery_type) set resulting from the intersection of NURSE_SKILL and REQ_SKILL

Table 5.5 Semantic integrity constraints for the Fine-granular Design-Specific ER model for
Cougar Medical Associates

Chapter Summary

This chapter focuses on modeling relationships beyond binary relationships, characteristics of weak relationship types, the decomposition of complex relationships, and model validation.

How to handle ternary (degree 3), quaternary (degree 4), and quintary (degree 5) relationships is shown through a series of application scenarios and vignettes. The cluster entity type is a way to represent entity types that naturally emerge from a higher-order relationship type and/or a group of entity types and associations among them.

The relationship construct known as the weak relationship type was originally defined by Dey, Storey, and Barron (1999). A weak relationship type occurs when two relationship types are linked by either an event-precedent sequence or a condition-precedent sequence. An event-precedent weak relationship type occurs when an event associated with the occurrence of one relationship type must precede an occurrence of the weak relationship type. A condition-precedent weak relationship type occurs when a condition that triggers the occurrence of one relationship type must precede the occurrence of the weak relationship type.

The decomposition of ternary and higher-order relationship types is very similar to the decomposition of binary relationship types with an m:n cardinality ratio. It involves converting the relationship type to a gerund entity type such that the resulting transformation contains nothing but a set of binary relationships with cardinality ratios of 1:m. Binary relationship types with multi-valued attributes and design alternatives in their decomposition are also introduced. A brief discussion about the alteration of weak relationship types in preparation for logical model mapping follows.

It is important that a conceptual model be an accurate representation of the "universe of interest." Such an objective can only be achieved through the careful evaluation of how well the developed conceptual model addresses the explicit and/or implicit questions associated with the requirements specification. This is accomplished through a formal design validation step in the conceptual modeling process.

The Cougar Medical Associates case emulates a real-world scenario and provides an opportunity to employ several complex relationship constructs described in the chapter. This case is developed all the way to a Fine-granular Design-Specific ER model ready for transformation to the logical tier.

Exercises

1. What is a cluster entity type?

2. What is a weak relationship type? Contrast a condition-precedent weak relationship type with an event-precedent weak relationship type.

3. What is a composite relationship type?

4. What is required in order to decompose a ternary and higher-order relationship type in preparation for mapping to a logical schema? What is the cardinality ratio of each of the resulting set of binary relationships?

5. What is required to decompose a cluster entity type in preparation for mapping to a logical schema?

6. Figures 5.4, 5.7, and 5.39 each show the decomposition of a ternary relationship. Explain why the entity types SCHEDULE and PRESCRIPTION in Figures 5.4 and 5.7 are shown as base entity types while the entity type USE in Figure 5.39 is shown as a weak entity type.

7. Consider the ER diagram of Figure 5.5. State meaningful semantics for additional binary relationships among the entity types in the diagram and update the ER diagram accordingly with full specification of the structural constraints of the relationship types you have proposed. Can the three binary relationships among the three entity types present collectively capture the semantics conveyed by the ternary relationship type? State the condition under which the answer is "yes" and the condition under which the answer is "no."

8. Consider the Coarse-granular Design-Specific ER diagram of Figure 5.13. As we know, a relationship type of degree > 2 cannot be modeled *as is* in a logical schema. Specify the Fine-granular Design-Specific ER diagram that will render this design ready for mapping to a logical schema.

9. Decompose the Coarse-granular Design-Specific ER diagrams of Figures 5.17 and 5.18 to the Fine-granular Design-Specific stage and notice the differences.

10. As a more general exercise for transforming a Coarse-granular Design-Specific ER diagram to the Fine-granular Design-Specific stage for complex relationship types, convert the Coarse-granular Design-Specific ER diagram in Figure 5.22 to a Fine-granular Design-Specific ER diagram.

11. Transform the Coarse-granular Design-Specific ER diagram in Figure 5.28 to a Fine-granular Design-Specific ER diagram.

12. The weak relationship type shown in Figure 5.32 requires further decomposition preparatory to mapping to a logical schema. Develop the Fine-granular Design-Specific ER diagram.

13. Business Process, Inc. (BPI), a consulting company offering business process reengineering and application system development expertise, wants to develop a prototype of a simple University Registration System (UNIVREG) to handle student/faculty information, course/section schedule, and co-op and lab information. Many small universities are in need of such a system. BPI believes that the profit potential from economies of scale alone in custom-fitting such an IS application to small universities who primarily offer a small number of programs is an attractive business opportunity. You have recently been hired by BPI and assigned to make use of the following data specifications and develop the conceptual design for this application.

A university has several departments and these departments employ faculty (professors) for purposes of teaching, research and administration. A department may have many professors but has to employ at least five. A professor, however, belongs to only one department at any time. In addition to teaching, some of the professors may work as department heads. Each department has a department head, but no more than one. Every department should continue to exist as long as it has at least one professor associated with it or it offers at least one course. If a faculty member serving as a department head leaves the university, some other professor (often, the most senior faculty member of the department) assumes the role by default.

The departments may offer several courses as part of their academic mission. However, any particular course is offered by only one department. Not all courses are offered all the time—but, every course is offered sometime. When offered, multiple sections of some of the courses may be offered during a quarter in a year. If a particular course is no longer offered, all offerings (sections) of that course should be deleted unless there are students enrolled in the course sections. If, however, a student leaves, that student's enrollment in all associated course sections should be removed. A course may be a prerequisite for several other courses, but a course may have no more than one prerequisite. A course cannot be removed from the database as long as it is a prerequisite for other course(s); however, if it only has prerequisites, then its deletion should be accompanied by the removal of its links to all of its prerequisites. Some professors in the university also write textbooks. Sometimes, more than one professor may be co-authoring a book, but all textbooks used by the university need not have one of its professors as an author. Some professors are also authors of multiple textbooks. There is no plan to record authorship of professors that are not working in the university in this system. When a professor leaves the university, the university no longer keeps track of books written by that professor. Likewise, if a textbook is no longer in use, the authorship of the textbook is not preserved either. The system also needs to record which professor uses what book in which course. Removal of a textbook from the database is prohibited if it is used by a professor in a course. However, if a course is removed from the catalogue, its link to a professor using a particular textbook is also removed. Likewise, if a faculty member leaves the university, the link to the textbook used by the professor in a specific course is also deleted. All professors teach, and a professor may teach several course sections. A course section, however, is taught by just one professor and must have some professor assigned to teach it. Some of the course sections may have multiple lab sessions in a quarter. Each lab session caters to only one course section—that is, no joint lab sessions. If a course section has an associated lab session, cancellation of the course section is not permitted.

Students enroll in course sections. In fact, to remain a student, one has to take at least one course (section), but university rules forbid a student from taking more than six courses (sections) in a quarter. Each section has to have at least ten students enrolled; otherwise it will be cancelled. If a student has registered for a section, the section should continue to exist. Also, when a professor is assigned to teach a section, deletion of the professor record is prohibited. The university admits mostly graduate and undergraduate students, but a few non-matriculating students are also admitted. The undergraduate students may, as part of their academic program, enroll for professional practice (co-op) sessions with companies. Several students may be enrolled in the same co-op session and a co-op session has at least one undergraduate student enrolled in it. An undergraduate student can co-op more than once. When a student leaves the university for whatever reason, if it is a graduate student the associated graduate student record is deleted from the database; but, if it is an undergraduate student, the deletion of the student record is stopped so that the co-op status of the student can be properly verified. If, having verified the co-op status, the decision is made to drop the undergraduate student information from the system all co-op enrollments for that student should also be erased. Cancellation of a co-op session is prohibited if there is/are student(s) enrolled in it. As part of their academic experience some of the graduate students are assigned to conduct one or more lab sessions. A lab session can be conducted by at most one graduate student, but some lab sessions are not assigned

238

to any graduate student. When a graduate student graduates, the lab sessions assigned to him/her cannot be cancelled; instead, the capability should exist to indicate that, for the present, the lab session is not handled by a graduate student.

Students borrow books from a single (main) library on the campus. A student may borrow a lot of books and a book may be borrowed by several students when available. Book-returns by students are also recorded in this system. The return pattern is the same as that of borrowing. Deletion of a student record is not allowed if he or she has any borrowed books outstanding. If a book is removed from the library catalogue, all borrow and return links for that book are removed. When a student leaves, the book-return links for that student are also discarded. It is important to note that a book should have been borrowed in order for it to be returned.

The registration system should capture student information like the name [o], address, and a unique student ID for each student. In addition the status of the student should be recorded. For undergraduate students, data on concentration should be available—all undergraduate students have multiple (at least two) concentrations. Thesis option [o] and the undergraduate major of each graduate student should be captured by this system. A co-op session is identified by year and quarter and each co-op session has a session manager [o]. A particular student during a particular co-op session works in a company and the database should record the name of the company and co-op assessment [o] for the student for each co-op session. Every professor has a name, employee ID, office [o], and phone [o]. Both professor name and employee ID have unique values. Data gathered about a department are: the department name [o], department code [o], location, and phone# [o]. For a department, the name and code are both unique. The courses offered have data on course name, credit hours, college [o], and course#. The course# is used to distinguish between courses. Each course may have multiple course sections with data including the classroom [o], class time, class size [o], section number, quarter, and year. There is no unique identifier for course section, because the course section has existence dependency on course—section number, quarter, and year together in conjunction with course# can uniquely identify course sections. The grade a student makes in a particular course should be available through the system. The lab sessions have information about the topic [o], time, lab location, and the lab session number for a given course section. Attributes of textbooks include ISBN, the unique identifier, Year [o], Title, and Publisher [o]. The library books, on the other hand are identified by a call#. The ISBN# and Copy# together also identify a copy of the book. The name of the book [o] and author [o] are also recorded. *Note that optional attributes are marked by an [o]; so, the rest of the attributes are mandatory.*

a. Develop a Presentation Layer ER model for the University Registration System (UNIVREG). The ERD should be fully specified with the unique identifiers, other attributes for each entity type, and the relationship types that exist among the various entity types. All business rules that can be captured in the ERD must be present in the ERD. Any business rule that cannot be captured in the ERD should be specified as part of a list of semantic integrity constraints.

b. Incorporate the following business rule into the Presentation Layer ER model: no two courses can be taught by the same professor using the same textbook.

c. Develop a Coarse-granular Design-Specific ER model for UNIVREG.

d. Transform the above design to Fine-granular ER diagram. Note that attribute character-istics are not provided and thus need not appear in the diagram.

Selected Bibliography

Dey, D.; Storey, V. C.; and Barron, T. M. (1999) "Improving Database Design through the Analysis of Relationships," *ACM Transactions on Database Systems*, 24, 4 (December) 453–486.

Elmasri, R. and Navathe, S. B. (2003) *Fundamentals of Database Systems*, Fourth Edition, Addison-Wesley.

Hoffer, J. A.; Prescott, M. B.; and McFadden, F. R. (2002) *Modern Database Management*, Sixth Edition, Prentice-Hall.

Howe, D. R. (1989) *Data Analysis for Data Base Design*, Second edition, Edward Arnold.

Teorey, T. J. (1999) *Database Modeling and Design*, Third Edition, Morgan Kaufman.

240

PART II

LOGICAL DATA MODELING

INTRODUCTION

At the completion of the conceptual modeling phase, the systems analyst/application developer usually has a reasonably clear understanding of the data requirements for the application system at a high level of abstraction. It is important to note that a conceptual data model is technology-independent. During conceptual modeling, the analysis and design activities are not constrained by the technology that will be used for implementation. A conceptual schema may contain constructs not directly compatible with the technology intended for implementation. In addition, some of the design may require refinement to eliminate data redundancy. The next step after conceptual modeling, then, is to transform the conceptual schema into a logical schema that is more compatible with the implementation technology of choice.

Part II introduces logical data modeling, which serves as the transition between the conceptual schema and the physical design. Logical data modeling begins with the creation of a technology-independent logical schema, proceeds through the normalization process,[1] and concludes with a technology-dependent logical schema expressed using the relational modeling grammar. Figure II.1 points to where we are now in the database development process.

Part II discusses techniques and procedures for mapping a conceptual schema to a logical schema in the relational data model. Because database management systems are based on logical data models, and because the relational data model forms the basis for the data modeling and database design discussed in this textbook, Chapter 6 shows how logical data modeling is performed in the context of the entity-relationship (ER) and enhanced entity-relationship (EER) conceptual data models. As part of the discussion we will also see that the process of mapping a conceptual schema to its logical counterpart can be either information-reducing or information-preserving, depending upon the approach used.[2]

---

[1] Normalization is the subject of Chapters 7, 8, and 9 in Part III.

[2] Two basic logical data structures, inverted tree and network, underlie conventional database design; and three basic data model architectures, relational, hierarchical, and CODASYL, employ one or both of the logical data structures. A brief discussion of the inverted tree and network data structures and an overview of how the hierarchical and CODASYL data model architectures express the inverted tree and network data structures respectively are presented in Appendix A. Object-oriented concepts have drawn considerable attention among researchers and practitioners since the late 1980s, and have significantly influenced efforts to incorporate in the DBMS the ability to process complex data types beyond just storage and retrieval. Appendix B briefly introduces the reader to object-oriented concepts exclusively from a database, or to be more precise, from a data modeling perspective.

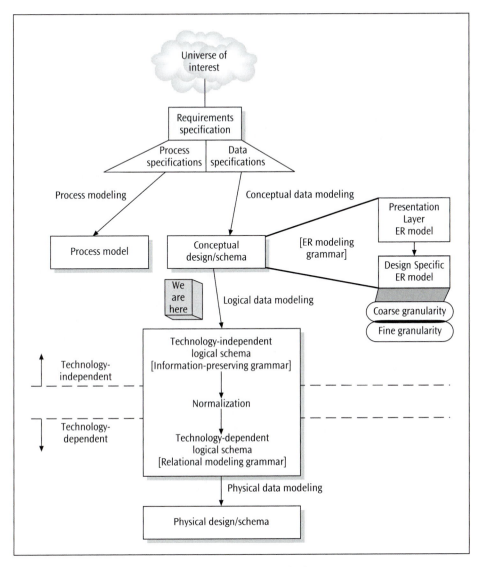

Figure II.1   Roadmap for data modeling and database design

CHAPTER **6**

The Relational Data Model

This chapter introduces the relational data model and the process of mapping a conceptual schema that is technology-independent to a logical schema (in this case, a relational data model) that provides the transition to a technology-dependent database design. Recall that in Part I of this book, the ER modeling grammar was used to develop a conceptual data model in the form of a Design-Specific ER model. It is this model that serves as the input for the logical data modeling activity.

The chapter flows as follows. Section 6.1 formally defines a relation, and Section 6.2 gives an informal description of a relation. The next section (Section 6.3) discusses the data integrity constraints pertaining to a relational data model. This is followed by a brief introduction in Section 6.4 to relational algebra as a means to specify the logic for data retrieval from a series of relations. The concepts of views and materialized views as different ways of looking at data stored in relations are introduced in Section 6.5. The rest of the chapter discusses mapping a conceptual schema to its logical counterpart using several examples. Section 6.6 introduces the idea of information preservation in data model mapping. Sections 6.7.1.1 and 6.7.1.2 present a detailed discussion of fundamental methods for transforming basic ER constructs (entity types, relationship types) to the logical tier. Section 6.7.1.3 demonstrates mapping techniques using the Design-Specific ER diagram of Bearcat Incorporated as the source (conceptual) schema. The solution highlights the information-reducing nature of the transformation process. Then, an information-preserving grammar for the logical schema is presented in Section 6.7.2. A discussion of the heuristics for mapping EER constructs to the logical schema and the metadata lost in the transformation process follows in Section 6.8.1. Finally, Section 6.8.2 presents the information-preserving grammar for modeling EER constructs at the logical tier.

6.1 Definition

C. J. Date (2004) asserts that "The foundation of modern database technology is without question the relational model; it is that foundation that makes the field a science. Thus, any book on the fundamentals of database technology that does not include a thorough coverage of the relational model is by definition shallow. Likewise, any claim to expertise in the database field can hardly be justified if the claimant does not understand the relational model in depth."

E. F. Codd proposed the relational data model in 1970 as a logically sound basis for describing the structure of data as well as data manipulation operations. The model uses the concept of mathematical relations as its foundation and is based on set theory. The relational data model includes a group of basic data manipulation operations called **relational algebra**. This chapter discusses the structural aspect of the relational data model and contains a brief introduction to relational algebra. Relational algebra is covered in more detail in Chapter 11.

The simplicity of the concept and its sound theoretical basis are two reasons why the relational data model has gained popularity as a logical data model for database design. As the name implies, the **relational data model** represents a database as a collection of relation values, or relations for short, where a relation resembles a two-dimensional table of values presented as rows and columns. A row in the table represents a set of *related* data values and is called a **tuple**. All values in a column are of the same data type. A column is formally referred to as an **attribute**. The set of all tuples in the table goes by the name **relation**. A relation consists of two parts: (a) an empty shell called the **heading**, which is a tuple of attribute names, and (b) a **body** of data which inhabits the shell; this body of data is a set of tuples all having the same heading. The heading of a relation is also referred to in the literature as a **relation schema, schema, scheme,** or **intension**. When the heading is called intension, then the body of the relation is referred to as **extension**.

Recall from Chapter 2 that the domain of an attribute is the set of possible values for the attribute. In a relational data model, a **domain** is defined as a set of atomic values for an attribute and an attribute is the name of a role played by a domain in the relation.

In formal terms, an attribute, A, is an ordered pair (N, D) where N is the name of the attribute and D is the domain that the named attribute represents. If r is a relation whose structure is defined by a set of attributes $A_1, A_2. ..., A_n$, then $R (A_1, A_2. ..., A_n)$ is called the relation schema[1] of the relation r. In other words, a relation schema, R, is a named collection of attributes (R, C) where R is the name of the relation schema and C is the set $\{ (N_1, D_1), (N_2, D_2), ... (N_n, D_n)\}$ where $N_1, N_2. ..., N_n$ are *distinct* names. **r** is the relation (or relation state) over the schema **R**. The domain of A_i (i = 1, 2, ..., n) is often denoted as **Dom** (A_i). The number of attributes (n) in R is called the **degree** (or **arity**) of R. A relation state r of the relation schema $R (A_1, A_2. ..., A_n)$, also denoted as r (R), is a set of n-tuples $\{t_1, t_2. ..., t_m\}$. Each n-tuple t_j (j = 1, 2, ..., m) in r (R) is an ordered list of n values $<v_{1j}, v_{2j}, ..., v_{nj}>$ where each v_{ij} (i = 1, 2, ..., n; j = 1, 2, ..., m) is an element of **Dom** (A_i) (i = 1, 2, ..., n) or when allowed a missing value represented by a special value called **null**. The number of tuples, **m**, in the relation state is called the **cardinality** of the relation. Figure 6.1 shows an example of a relation schema of degree 3 and two relation states of cardinality 5 and 4 respectively for that relation schema.

6.2 Characteristics of a Relation

The definitions in Section 6.1 ascribe certain characteristics to a relation schema and its relation states. These characteristics are as follows:

- A relation is a mathematical term approximately equivalent to a two-dimensional table.
- A relation has a heading, that is, a tuple of attribute names (also known as intension, relation schema, relvar (relation variable), or shell) and a body, which is a set of tuples all having the same heading (also known as extension, a relation state).

---

[1] A relation schema is sometimes loosely referred to as a relation. C. J. Date (2004) has coined the term *relvar* (for relation variable) to distinguish a relation schema from a relation. It is important to notice the difference between a relation and a relation schema as well as between a relation schema and a *relational* schema. A **relational schema** defines a set of relation schemas in a relational data model.

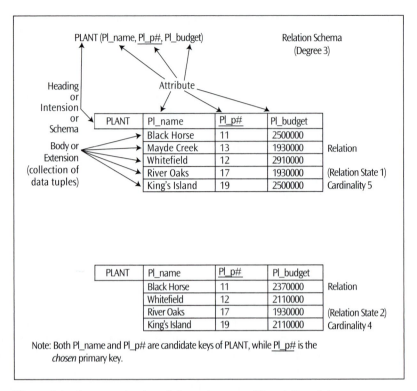

Figure 6.1 An example of a relation schema and its relation states

- Attributes of a relation schema have unique names.[2]
- Values of an attribute in a relation come from the same domain.
- The order of arrangement of tuples is immaterial.
- Each attribute value in a tuple is atomic; hence composite and multi-valued attributes are not allowed in a relation schema.[3]
- The order of arrangement of attributes in a relation schema is immaterial as long as the correspondence between attributes and their values in the relation is maintained.
- Derived attributes are not captured in a relation schema.
- All tuples in a relation must be distinct (that is, a relation schema must have a unique identifier).

Since the relational database theory stipulates that attribute names must be unique over the entire relational schema, the following guidelines represent the approach used in this chapter for developing attribute names:

[2] An implicit assumption of the relational database theory is that attributes have unique names over the entire *relational* schema. Additional discussion of this issue follows in Section 6.6.2.

[3] A relation maps to a flat file at the physical level. This is supported by the assumption of the theory behind the relational model called the First Normal Form assumption. A discussion of normal forms begins in Chapter 8.

- Each attribute name begins with a prefix of up to three letters that represents a meaningful abbreviation of the name of the relation schema to which the attribute belongs. This prefix is followed by an underscore character.
- Only the first letter of the prefix is capitalized.
- Following the underscore character is a suffix that corresponds to the attribute name itself. This suffix may contain only lowercase letters, pound sign (#) and underscore characters, and corresponds to the name of the attribute in the conceptual data model.

These guidelines were used in developing the attribute names for the PLANT relation schema in Figure 6.1.

6.3 Data Integrity Constraints

Data integrity constraints, generally referred to as just **integrity constraints**, are rules that govern the behavior of data at all times in a database. Data integrity constraints are technical expressions of the business rules that emerge from the user requirements specification for a database application system. To that extent, these integrity constraints prevail across all tiers of data modeling—conceptual, logical and physical.

In general, integrity constraints are considered to be part of the schema in that they are declared along with the structural design of the data model (conceptual, logical, and physical) and hold for all valid states of a database that correctly model an application (Ullman and Widom, 1997). Although it is possible to specify all the integrity constraints as a part of the modeling process, some cannot be expressed explicitly or implicitly in the schema of the data model. For instance, an ER diagram (conceptual schema) is not capable of expressing domain constraints of attributes. Likewise, constraints that are not declarative in nature and therefore require procedural intervention (for example, "no more than two projects can be in one location at given time") cannot be directly specified in a relational schema. As a consequence, such constraints are carried forward through the data modeling tiers in textual form and are often referred to as **semantic integrity constraints**,[4] while the constraints directly expressed in the schema of the data model are labeled as **schema-based** or **declarative constraints**.[5] Domain constraints, key constraints, entity integrity constraints, referential integrity constraints and functional dependency constraints are part of the declarative constraints of a relational data model. (Entity integrity constraints and referential integrity constraints are discussed in the following sections. Functional dependency constraints are discussed in Chapter 7.) Uniqueness constraints and structural constraints of a relationship type are declarative constraints specified at the conceptual level in an ER model. Semantic integrity constraints that require procedural intervention can always be specified and enforced through application programming code and are called **application-based constraints**. Procedural enforcement of integrity constraints is also often possible via mechanisms incorporated in the DBMS that use general purpose constraint specification languages, such as triggers and assertions.

---

[4] Note, however, that all integrity constraints pertain to the "semantics"of the database application.

[5] A third category of constraints are inherent to the data model and are called inherent model-based constraints and do not necessarily emerge from the semantics of the application. For example, the characteristics of a relation stated in Section 6.2 are inherent to a relational schema. Similarly, an inherent constraint of an ER diagram is that a relationship type can link only entity types.

Since every valid state of a database must satisfy the declarative and procedural forms of the integrity constraints noted above, these constraints are collectively referred to as **state constraints**. Sometimes, integrity constraints may have to be specified to define legal transitions of state. For example, the value of the attribute **Marital_status** can change from Married to Divorced or Widowed, but not to Single. These types of constraints are referred to as **transition constraints** and invariably require procedural language support either through application programs or general-purpose constraint specification languages of the DBMS.

6.3.1 The Concept of Unique Identifiers

An attribute or a collection of attributes can serve as a unique identifier of a relation schema. A set of one or more attributes, which taken collectively, uniquely identifies a tuple of a relation is referred to as the **superkey** of the relation schema. Accordingly, no two distinct tuples in any state r of a relation schema R can have the same value for the superkey. For example, consider the data in the PRESCRIPTION-A relation shown in Table 6.1. There are ten superkeys in PRESCRIPTION-A. They are:

> **Rx_rx#**
> **(Rx_rx#, Rx_pat#)**
> **(Rx_rx#, Rx_medcode)**
> **(Rx_rx#, Rx_dosage)**
> **(Rx_pat#, Rx_medcode)**
> **(Rx_rx#, Rx_pat#, Rx_medcode)**
> **(Rx_rx#, Rx_pat#, Rx_dosage)**
> **(Rx_rx#, Rx_medcode, Rx_dosage)**
> **(Rx_pat#, Rx_medcode, Rx_dosage)**
> **(Rx_rx#, Rx_pat#, Rx_medcode, Rx_dosage)**

In general, if K is a superkey, any **superset**[6] of K (i.e., any collection of attributes that includes K) is also a superkey.

Another type of unique identifier is called a **candidate key**. A candidate key is defined as a superkey with no **proper subsets**[7] that are superkeys. In short, a candidate key has two properties:

- **Uniqueness** – Two tuples of a relation schema cannot have identical values for the collection of attribute(s) that constitute the candidate key.
- **Irreducibility** – No proper subset of the candidate key has the uniqueness property.

The uniqueness property is common to both a superkey and a candidate key, while the irreducibility property is present only in a candidate key. This indicates that a superkey may contain superfluous attributes while a candidate key does not.

---

[6] A set S_1 is a *superset* of another set S_2 if every element in S_2 is in S_1. S_1 *may* have elements which are not in S_2.

[7] A set S_2 is a *subset* of another set S_1 if every element in S_2 is in S_1. S_1 *may* have exactly the same elements as S_2. If S_2 is a subset of S_1, S_1 is a superset of S_2. A set S_2 is a *proper subset* of another set S_1 if every element in S_2 is in S_1 and S_1 has some elements which are not in S_2.

PRESCRIPTION-A

Rx_rx#	Rx_pat#	Rx_medcode	Rx_dosage
A100	7642	PCN	3
A103	4678	TYL	2
A102	4772	CLR	2
A101	6742	ASP	2
A104	4772	ZAN	3
A105	7456	CLR	2
A107	2222	TYL	2
A106	4772	VAL	2
A108	7384	CLR	3
A109	7384	ZAN	2
A110	7642	VAL	2

PRESCRIPTION-B

Rx_rx#	Rx_pat#	Rx_medcode	Rx_dosage
B100	7642	PCN	3
B103	4678	TYL	2
B102	4772	CLR	2
B101	6742	ASP	2
B102	4772	ZAN	**2**
B105	7456	CLR	2
B107	2222	TYL	2
B106	4772	VAL	2
B108	7384	CLR	3
B109	7384	ZAN	2
B100	7642	VAL	2

Table 6.1 The PRESCRIPTION-A and PRESCRIPTION-B relations

As shown in Table 6.2, only **Rx_rx#** and **(Rx_pat#, Rx_medcode)** are candidate keys of PRESCRIPTION-A. For example, **(Rx_rx#, Rx_pat#)**, **(Rx_rx#, Rx_medcode)** and **(Rx_rx#, Rx_dosage)** are not candidate keys of PRESCRIPTION-A, because a proper subset of each, **(Rx_rx#)**, is also a superkey of PRESCRIPTION-A. **(Rx_pat#, Rx_medcode)** is a candidate key of PRESCRIPTION-A, because neither of its proper subsets **(Rx_pat#)** and **(Rx_medcode)** is a superkey of PRESCRIPTION-A. In essence, any superset of attributes that contains either **(Rx_rx#)** or **(Rx_pat#, Rx_medcode)**, while being a superkey, is not a candidate key of PRESCRIPTION-A.

As noted earlier, **Rx_rx#** is a candidate key of PRESCRIPTION-A. However, **Rx_rx#** does not uniquely identify tuples of PRESCRIPTION-B as seen in Table 6.1. That is, **Rx_rx#** is not a superkey of PRESCRIPTION-B. But, as can be seen in Table 6.2, six superkeys exist in PRESCRIPTION-B. They are:

(Rx_rx#, Rx_medcode)
(Rx_pat#, Rx_medcode)
(Rx_rx#, Rx_pat#, Rx_medcode)
(Rx_rx#, Rx_medcode, Rx_dosage)
(Rx_pat#, Rx_medcode, Rx_dosage)
(Rx_rx#, Rx_pat#, Rx_medcode, Rx_dosage)

The superkey **(Rx_rx#, Rx_medcode)** is a candidate key of PRESCRIPTION-B because neither of its proper subsets **(Rx_rx#)**, **(Rx_medcode)** is a superkey of PRESCRIPTION-B. Similarly, **(Rx_pat#, Rx_medcode)** is also a candidate key of PRESCRIPTION-B because neither of its proper subsets is a superkey of PRESCRIPTION-B. Next, let's evaluate whether the superkey **(Rx_rx#, Rx_pat#, Rx_medcode)** is a candidate key of PRESCRIPTION-B. This can be done by investigating if any of its proper subsets is a superkey of PRESCRIPTION-B. The proper subsets of **(Rx_rx#, Rx_pat#, Rx_medcode)** are:

(Rx_rx#)
(Rx_pat#)
(Rx_medcode)
(Rx_rx#, Rx_pat#)
(Rx_rx#, Rx_medcode)
(Rx_pat#, Rx_medcode)

While **(Rx_rx#)**, **(Rx_pat#)**, **(Rx_medcode)**, and **(Rx_rx#, Rx_pat#)** are not superkeys of PRESCRIPTION-B, **(Rx_rx#, Rx_medcode)** and **(Rx_pat#, Rx_medcode)** are indeed superkeys of PRESCRIPTION-B. Therefore, **(Rx_rx#, Rx_pat#, Rx_medcode)** is not a candidate key of PRESCRIPTION-B because two of its proper subsets are superkeys of PRESCRIPTION-B.

The same procedure can be used to evaluate if any other superkey of PRESCRIPTION-B is also a candidate key of PRESCRIPTION-B. A review of the data in Table 6.1 indicates that none of the other superkeys is also a candidate key of PRESCRIPTION-B.

Every attribute, atomic or composite, plays one of three roles in a relation schema. It is a key attribute, a non-key attribute, or a candidate key. Any attribute that is a proper subset of a candidate key is a **key attribute**. An attribute that is not a candidate key is a **non-key attribute**. For example, in PRESCRIPTION-A, since **(Rx_pat#, Rx_medcode)** is a candidate key, **Rx_pat#** and **Rx_medcode** are key attributes. Since **Rx_rx#**, an atomic attribute, is a candidate key of PRESCRIPTION-A, it is, by definition, *not* a key attribute in PRESCRIPTION-A because **Rx_rx#** is not a proper subset of **Rx_rx#**. However, **Rx_rx#** is not a non-key attribute either because it is a candidate key of PRESCRIPTION-A.

PRESCRIPTION (Rx_rx#, Rx_pat#, Rx_medcode, Rx_dosage)

	PRESCRIPTION-A		PRESCRIPTION-B	
	Superkey	Candidate Key	Superkey	Candidate Key
Rx_rx#	Yes	Yes	No	No
Rx_pat#	No	No	No	No
Rx_medcode	No	No	No	No
Rx_dosage	No	No	No	No
Rx_rx#, Rx_pat#	Yes	No	No	No
Rx_rx#, Rx_medcode	Yes	No	Yes	Yes
Rx_rx#, Rx_dosage	Yes	No	No	No
Rx_pat#,Rx_medcode	Yes	Yes	Yes	Yes
Rx_pat#, Rx_dosage	No	No	No	No
Rx_medcode, Rx_dosage	No	No	No	No
Rx_rx#, Rx_pat#, Rx_medcode	Yes	No	Yes	No
Rx_rx#, Rx_pat#, Rx_dosage	Yes	No	No	No
Rx_rx#, Rx_medcode, Rx_dosage	Yes	No	Yes	No
Rx_pat#, Rx_medcode, Rx_dosage	Yes	No	Yes	No
Rx_rx#, Rx_pat#, Rx_medcode, Rx_dosage	Yes	No	Yes	No

Table 6.2 Superkeys and candidate keys in the PRESCRIPTION-A and PRESCRIPTION-B relations

In short, *a candidate key in itself is neither a key attribute nor a non-key attribute in R.* Further discussion of this appears in Chapter 7.

A **primary key** serves the role of uniquely identifying tuples of a relation. A primary key is a candidate key (an irreducible unique identifier) with one additional property. This additional property results from what is known as the **entity integrity constraint** which specifies that the primary key of a relation schema cannot have a "missing" value (i.e., a null value), essentially assuring identification of every tuple in a relation. Given a set of candidate keys for a relation schema, exactly one is chosen as the primary key. Since the entity integrity constraint applies exclusively to a primary key, by implication the rest of the candidate keys[8] not chosen as the primary key apparently tolerate "missing" values; otherwise, the entity integrity rule will apply to all candidate keys (Date, 2004). In short, when a candidate key of a relation schema is chosen to be the primary key of that relation schema, it is bound by the entity integrity constraint and from that point forward, the alternate keys (other candidate keys) may entertain "missing" values.[9] What happens when an alternate key is chosen as the primary key at a later time? The answer is inherent in the definition of the primary key – i.e, the entity integrity constraint must be enforced in order for the alternate key to become the primary key of the relation schema.

[8] Once the primary key is chosen from among the set of candidate keys, the remaining candidate keys are referred to as alternate keys. The choice of a primary key of a relation schema from among its candidate keys is essentially arbitrary.

[9] This indeed is in contradiction with the strict definition of a relation – i.e., a relation cannot have missing value for any of its attributes. Nonetheless, relaxation of this constraint is very common in practice.

Since the primary key value is used to identify individual tuples in a relation, if null values are allowed for the primary key, some tuples cannot be identified; hence the entity integrity constraint, disallowing null values for a primary key. A primary key of a relation schema is denoted by underlining the atomic attribute(s) that constitute a primary key. In Table 6.3, **Pl_p#** is the primary key of PLANT. Observe that both **Pl_name** and **Pl_p#** are candidate keys of PLANT as modeled in the conceptual schema (refer to Figure 3.11) and **Pl_p#** has been chosen by the systems analyst/database designer as the primary key. This implies that the entity integrity constraint is imposed only on **Pl_p#** and not on **Pl_name**.

6.3.2 Referential Integrity Constraint in the Relational Data Model

While the key constraints (superkey and candidate key) and entity integrity constraint (primary key) pertain to individual relation schemas, a **referential integrity constraint** is specified between two relation schemas, R1 and R2. It is common for tuples in one relation to reference tuples in the same or other relations. A referential integrity constraint stipulates that a tuple in a relation, **r2** (i.e., **r(R2)**), that refers to another relation, **r1** (i.e., **r(R1)**), refers to *a tuple that exists* in **r1**. R2 is called the *referencing* relation schema and R1 is known as the *referenced* relation schema. An important type of referential integrity specification is known as the **foreign key constraint**. A foreign key constraint (a) establishes an explicit association between the two relation schemas, and (b) maintains the integrity of such an association (i.e., maintains consistency among tuples of the two relations). In order to establish a foreign key constraint between two relation schemas, R1 and R2, it is necessary that every value of an attribute A_2, atomic or composite, in relation **r2** must either occur as the value of a candidate key A_1 in some tuple of the relation **r1** that **r2** refers to, or, if allowed, be null.[10] The attribute in R2 that meets this condition is called a **foreign key** of R2.[11] In other words, the definition of a foreign key requires that the set of referenced attributes be a candidate key in the referenced relation schema. Incidentally, **r1** and **r2** need not be distinct relations. A referential integrity constraint, on the other hand, is defined more generally in that it *does not* require the referenced set of attributes to be any sort of a key in the referenced relation. In relational database theory, this constraint is referred to as an **inclusion dependency** and is algebraically expressed as follows:

$$R2.\{A_2\} \subseteq R1.\{A_1\}$$

A foreign key constraint is just a special kind of inclusion dependency. In terms of the foreign key constraint, the foreign key value in **r2** represents a *reference* to the tuple containing the matching candidate key value in the referenced tuple in **r1**. When a relation schema includes a foreign key that references some candidate key in the same relation schema, then the relation schema is said to be *self-referencing* (equivalent to a recursive relationship type in the ER diagram).

---

[10] If the participation of R2 in this relationship with R1 is partial, the foreign key attribute, A_2, can have null values.

[11] The relational model originally required that foreign keys reference, very specifically, the primary key, not just candidate keys. This limitation is unnecessary and undesirable in general, though it might often constitute good discipline in practice (Date, 2004, p. 274).

PLANT (Pl_name, Pl_p#, Pl_budget)

PROJECT (Prj_name, Prj_p#, Prj_location, *Prj_pl_p#)*   Version 1
        OR
PROJECT (Prj_name, Prj_p#, Prj_location, *Prj_pl_name*)   Version 2

Note: The Foreign keys are italicized.

PLANT	Pl_name	Pl_p#	Pl_budget
	Black Horse	11	1230000
	Mayde Creek	13	1930000
	Whitefield	12	2910000
	River Oaks	17	1930000
	King's Island	19	2500000
	Ashton	15	2500000

PROJECT	Prj_name	Prj_n#	Prj_location	Prj_pl_p#	
	Solar Heating	41	Sealy	11	
	Lunar Cooling	17	Yoakum	17	
	Synthetic Fuel	29	Salem	17	Version 1
	Nitro-Cooling	23	Parthi	12	
	Robot Sweeping	31	Ponca City	11	
	Robot Painting	37	Yoakum	19	
	Ozone Control	13	Parthi	19	

Note: PROJECT.**Prj_pl_p#** is the foreign key referencing PLANT.**Pl_p#**, the primary key of PLANT.

PROJECT	Prj_name	Prj_n#	Prj_location	Prj_pl_name	
	Solar Heating	05 41	Sealy	Black Horse	
	Lunar Cooling	17	Yoakum	River Oaks	
	Synthetic Fuel	29	Salem	River Oaks	Version 2
	Nitro-Cooling	23	Parthi	Whitefield	
	Robot Sweeping	31	Ponca City	Black Horse	
	Robot Painting	37	Yoakum	King's Island	
	Ozone Control	13	Parthi	King's Island	

Note: PROJECT.**Prj_pl_name** is the foreign key referencing PLANT.**Pl_name**, a candidate key of PLANT.

Table 6.3 Enforcement of referential integrity constraint

As an example, consider a scenario where plants undertake projects. All plants need not undertake projects, but any plant may undertake several projects. Likewise, a project is controlled by only one plant and not all projects are controlled by plants. Version 1 of the PROJECT relation in Table 6.3 uses **Prj_pl_p#** as the foreign key referencing the primary key of PLANT (**Pl_p#**). Observe that the name of the foreign key attribute begins with the prefix **Prj** that represents an abbreviation of the referencing relation schema name; the suffix is the name of the referenced attribute **Pl_p#** from the referenced relation schema. The constraint is expressed as:

$$\text{PROJECT.}\{\text{Prj\_pl\_p\#}\} \subseteq \text{PLANT.}\{\text{Pl\_p\#}\} \text{ or } \varnothing$$

where Ø indicates "null," meaning that a project need not be controlled by a plant.

In Version 2 of the PROJECT relation, **Prj_pl_name** in PROJECT references **Pl_name** in PLANT, a candidate key of PLANT, not the primary key of PLANT.

The constraint is expressed as:

$$\text{PROJECT.}\{\text{Prj\_pl\_name}\} \subseteq \text{PLANT.}\{\text{Pl\_name}\} \text{ or } \varnothing$$

The naming convention applied here to name a foreign key attribute in the referencing relation schema consists of (a) the prefix used in conjunction with the attribute names in the referencing relation schema, (b) an underscore, and (c) the referenced attribute name. The example in Table 6.3 demonstrates enforcement of a referential integrity constraint along with this naming convention. Observe that in its current state, all projects in the PROJECT relation are controlled by some plant.

6.4 A Brief Introduction to Relational Algebra

The relational data model includes a group of basic data manipulation operations. On the basis of its theoretical foundation in set theory, these data manipulation operations include Union, Difference, and Intersection. Five other operations also exist: Selection, Projection, Cartesian product, Join, and Division. Collectively, these eight operations comprise what is known as relational algebra.

This section contains a brief introduction to six of these operations and provides examples of their use in the context of the AW_PLANT, TX_PLANT, and PROJECT relations that appear in Table 6.4.[12] The AW_PLANT and PROJECT relations are the same as those that appear in Table 6.3. The AW_PLANT relation contains data on award-winning plants while the TX_PLANT relation contains data on plants located in the state of Texas.

6.4.1 Unary Operations: Selection (σ) and Projection (π)

The Selection operation is used to create a second relation by extracting a horizontal subset of tuples from a relation that matches specified search criteria. On the other hand, the Projection operation creates a second relation by extracting a vertical subset of columns from a relation. Selection and Projection are referred to as unary operations because they produce a new relation by manipulating only a single relation.

---

[12] A formal discussion of relational algebra, including the Cartesian product and Division operations, appears in Chapter 11.

AW_PLANT (Aw_pl_name, Aw_pl_p#, Aw_pl_budget)

TX_PLANT (Tx_pl_name, Tx_pl_p#, Tx_pl_budget)

PROJECT (Prj_name, Prj_p#, Prj_location, *Prj_aw_pl_p#*)

Note: The foreign keys are italicized.

AW_PLANT	Aw_pl_name	Aw_pl_p#	Aw_pl_budget
	Black Horse	11	1230000
	Mayde Creek	13	1930000
	Whitefield	12	2910000
	River Oaks	17	1930000
	King's Island	19	2500000
	Ashton	15	2500000

TX_PLANT	Tx_pl_name	Tx_pl_p#	Tx_pl_budget
	Southern Oaks	16	1230000
	River Oaks	17	1930000
	Kingwood	18	2910000

PROJECT	Prj_name	Prj_n#	Prj_location	*Prj_aw_pl_p#*
	Solar Heating	41	Sealy	11
	Lunar Cooling	17	Yoakum	17
	Synthetic Fuel	29	Salem	17
	Nitro-Cooling	23	Parthi	12
	Robot Sweeping	31	Ponca City	11
	Robot Painting	37	Yoakum	19
	Ozone Control	13	Parthi	19

Table 6.4 Sample relations for relational algebra examples

Example of a Selection operation: *Which award-winning plants have a budget that exceeds $2,000,000?*

Result:[13]

R_aw_pl_name	R_aw_pl_p#	R_aw_pl_budget
Whitefield	12	2910000
King's Island	19	2500000
Ashton	15	2500000

Example of a Projection operation: *What is the plant number and budget of each award-winning plant?*

Result:

R_aw_pl_p#	R_aw_pl_budget
11	1230000
13	1930000
12	2910000
17	1930000
19	2500000
15	2500000

Should each attribute involved in a Projection operation not be unique, it is possible for the new relation produced to have duplicate tuples. If this occurs these duplicate tuples are deleted.

6.4.2 Binary Operations: Union (U) , Difference (-), and Intersection (∩)

Each of the three binary operations involves the manipulation of two union compatible relations in order to produce a third relation. **Union compatibility** requires that each relation be of the same degree and that corresponding attributes in the two relations come from (or share) the same domain. The union of two relations is formed by adding the tuples from one relation to those from a second relation to produce a third relation consisting of tuples that are in *either* the first relation *or* the second relation. The difference of two relations is a third relation that contains the tuples from the first relation that are not in the second relation. The intersection of two relations is a third relation containing tuples common to the two relations.

---

[13] The result obtained in this and all other examples in this section produces the new relation RESULTS, which contains its own unique attribute names.

Example of the Union operation: *What plants are either located in Texas or are award-winning plants?*
Result:

R_aw_pl_name	R_aw_pl_p#	R_aw_pl_budget
Black Horse	11	1230000
Mayde Creek	13	1930000
Whitefield	12	2910000
River Oaks	17	1930000
King's Island	19	2500000
Ashton	15	2500000
Southern Oaks	16	1930000
Kingwood	18	1930000

Observe that duplicate tuples are omitted (i.e., the River Oaks plant, an award-winning plant located in Texas appears only once in the result). In addition, note that AW_PLANT and TX_PLANT are union compatible.

Example of the Difference operation: *Which Texas plants are not award-winning plants?*
Result:

R_aw_pl_name	R_aw_pl_p#	R_aw_pl_budget
Southern Oaks	16	1930000
Kingwood	18	1930000

Note the difference between this result and the result obtained when the question involves the award-winning plants not located in Texas:
Result:

R_aw_pl_name	R_aw_pl_p#	R_aw_pl_budget
Black Horse	11	1230000
Mayde Creek	13	1930000
Whitefield	12	2910000
King's Island	19	2500000
Ashton	15	2500000

Example of the Intersection operation: *Which award-winning plants are located in Texas?*
Result:

R_aw_pl_name	R_aw_pl_p#	R_aw_pl_budget
River Oaks	17	1930000

6.4.3 The Natural Join (*) Operation

The Join operation combines two relations into a third relation by matching values for attributes in the two relations that come from the same domain. The tuples in the new relation consist of the tuples extracted from the first relation concatenated with each tuple in the second relation where there is a match on the joining attributes. When the new relation contains all the attributes from the first relation plus all the attributes from the second relation, but does not redundantly carry the joining attributes, the result is called a Natural Join.[14]

Example of a Natural Join operation: *Perform a natural join of the award winning plant and project relations.*
Result:

R_prj_name	R_prj_n#	R_prj_location	R_aw_pl_p#	R_aw_pl_p#
Solar Heating	41	Sealy	11	Black Horse
Lunar Cooling	17	Yoakum	17	River Oaks
Synthetic Fuel	29	Salem	17	River Oaks
Nitro-Cooling	23	Parthi	12	Whitefield
Robot Sweeping	31	Ponca City	11	Black Horse
Robot Painting	37	Yoakum	19	King's Island
Ozone Control	13	Parthi	19	King's Island

Note that the joining attributes here are: **Aw_pl_p#** and **Prj_aw_pl_p#**. Relational algebra operations can be combined to form more complex expressions. Use of the Natural Join operation above followed by a Selection operation on plant number 11 and a Projection operation on the project name yields a relation that contains the names of the projects controlled by plant number 11 (i.e., Solar Heating and Robot Sweeping).

---

[14] Three other types of joins exist: Equijoin, Theta Join, and Outer Join. These are discussed in Chapter 11.

6.5 Views and Materialized Views in the Relational Data Model

A **view** is defined as a named "virtual" relation schema constructed from one or more relation schemas through the use of one or more relational algebra operations. Unlike a relation schema, a view does not store actual data but is just a logical window to view selected data (attributes and tuples) from one or a set of relations. The value of a view at any given time is a derived relation *state* and results from the evaluation of a specified relational expression at that time. In a database environment views play a number of roles:

- Views allow the same data to be seen by different users in different ways at the same time. This makes it possible for different users with different requirements to "view" only the portions of the database that are relevant to them.
- Views provide security by restricting user access to a predetermined set of tuples and attributes from predetermined relations.
- Views hide data complexity from the user because data selected from several relations can be made to appear to the user as a single object.

Implicit in this description is the concept of logical data independence. Recall that logical data independence is the *immunity of external schema* (the top tier of the three-schema architecture outlined in Chapter 1) *to changes in the logical structure of the conceptual schema* (the middle tier of the three-schema architecture). Views are the means by which logical data independence is implemented in a relational database system.

Changes to the logical structure of the conceptual schema consist of two aspects: **growth** and **restructuring** (Date, 2004).

- Growth is expansion of existing relation schema(s) in the relational data model by adding new attributes. Addition of a new relation schema to the existing relational data model is another dimension of growth. Both imply incorporation of new information in the conceptual schema.
- Restructuring is about changes to the logical structure of the conceptual schema other than growth. The idea is about restructuring the schema that is *information-equivalent* to the current conceptual schema—i.e., the restructuring should be reversible. For instance, replacement of a relation schema with multiple relation schemas for some reason (e.g., normalization)[15] amounts to restructuring that is information-equivalent.[16]

Observe that growth does not have any impact on the existing external schema – that is, the existing user views and programs. The impact of restructuring can be compensated for by recreating the original structure to serve as the source schema(s) for the construction of the external schema (e.g., a denormalized view of the normalized relation schemas that simulates the original logical structure). Views are an effective means to achieve logical data independence as long as any restructuring of the conceptual schema creates a version that is information-equivalent to the original conceptual schema.

A **materialized view** (also known as a **snapshot**), despite the similarity in name, is not a view. Like a view, it is constructed from one or more relation schemas; but unlike a view, a materialized view is stored in the database and refreshed when updates occur to the

---

[15] The topic of normalization is covered in Part III.

[16] Deletion of an attribute or a relation schema usually does not yield an information-equivalent logical structure – if it does, then it implies that the original conceptual schema had information redundancy.

relation schemas from which the materialized view is generated. Materialized views are often used to freeze data as of a certain moment without preventing updates to continue on the data in the relation schemas on which they are based. A materialized view is often deleted when it is not used for a period of time and then reconstructed from scratch as future needs dictate.

6.6 The Issue of Information Preservation

In Part I, Conceptual Modeling, it was argued that a database design is primarily driven by the business rules specified by the user community, and it is imperative that a comprehensive set of business rules be captured during the requirements analysis. While the resulting conceptual schema (ER model) adds significant value to the database design activity, it also carries an attendant burden of preserving the information content of the business rules through the remainder of the design steps. The goal of any schema transformation method ought to be the preservation of the information capacity of the source data model such as an ER model in the target data model (in this case the relational data model).

Many data modeling scholars have noted that representations of logical schema (e.g., relational schema) continue to suffer from the limitations of information-reducing transformations from the conceptual model (Batini, Ceri, and Navathe, 1992; Navathe, 1992; Fahrner and Vossen, 1995). To quote Fahrner and Vossen (1995, p. 220), "the major difficulty of logical database design, i.e., of transforming an ER schema into a schema in the language of some logical model, is the *information preservation issue*. Indeed, assuring a complete mapping of all modeling constructs and constraints which are inherent, implicit, or explicit in the source schema is problematic since constraints of the source model often cannot be represented directly in terms of the structures and constraints of the target model. In such cases they must be realized through application programs; or, alternatively, an information-reducing transformation must be accepted."[17] Thus, it is crucial that a logical schema captures as much of the inherent, implicit, and explicit constructs and constraints conveyed by the conceptual schema as possible through the logical modeling grammar, and carry forward the remainder of the semantic integrity constraints to the next step in the design process, physical data modeling.

However, popular mapping techniques presently in vogue that map directly to a relational schema are information-reducing in nature. Section 6.7 discusses these techniques and points out their information-reducing aspects. Then, Section 6.8 presents a new, information-preserving logical modeling grammar[18] for transforming ER and EER models to their logical counterparts.

---

[17] Formally, a transformation where all possible database instances that can be represented in a source schema can all be represented in a target schema implies information preservation (Fahrner and Vossen, 1995). To that extent the spirit of the discussion here pertains to *design* information preservation.

[18] An early version of this grammar was presented in the Workshop on Information Technology and Systems (WITS) in December 2000 at Brisbane, Australia (Umanath and Chiang, 2000).

6.7 Mapping an ER Model to a Logical Schema

The major task in this step is to convert the Fine-granular Design-Specific ER diagram to a target logical schema. Constructs and constraints that cannot be captured in the target logical schema cannot be ignored but must be carried forward to the next step in the design process. In addition, the list of semantic integrity constraints that supplements the Fine-granular Design-Specific ER diagram needs to be evaluated for mapping to the target logical schema. The product of this activity is a logical data model comprising: (a) a logical schema expressed using a logical modeling grammar, and (b) a list of semantic integrity constraints not incorporated into the logical schema to be carried forward to the next step in the database design activity. The logical modeling grammar presented here is a variation of a relational schema that captures additional constructs and constraints typically not present in a relational schema. The reason for this choice stems from the fact that the technology-dependent version is intended for deployment in a relational database management system (RDBMS).

6.7.1 Information-Reducing Mapping of ER Constructs

A popular technique for representing the logical counterpart of an ER diagram is described in this section, and evaluated for how well it preserves in the logical schema the information available from the ER model. To illustrate the transformation process, different variations of subsets from the Fine-granular Design-Specific ER diagram previously developed in Chapter 3 (Figure 3.11) for Bearcat Incorporated will be used.

6.7.1.1 Mapping Entity Types

Review the ER diagram in Figure 6.2. The first step in logical data modeling is to create a relation schema for each base entity type in the ER diagram. Only the stored attributes are translated to the logical level. In the case of composite attributes, only their constituent atomic components are recorded. A primary key is chosen from among the candidate keys for each relation schema. The atomic attribute(s) that make(s) up the primary key of the relation schema is/are underlined. In the ER diagram in Figure 6.2, either the composite attribute **[Emp_e#a, Emp_e#n]** or the composite attribute **[Emp_fname, Emp_minit, Emp_lname, Emp_nametag]** can serve as the primary key of EMPLOYEE because both are defined as candidate keys of EMPLOYEE in the ER diagram. Likewise, either **Pl_name** or **Pl_p#** becomes the primary key of PLANT. Observe in Figure 6.3 that **[Emp_e#a, Emp_e#n]** and **Pl_p#** have been chosen as the primary keys of the relation schema, EMPLOYEE and PLANT respectively as indicated by the underlining of these attributes.

A weak entity type in the ER diagram is also mapped in a like manner except that the primary key of each identifying parent of the weak entity type is added to the relation schema. The attributes thus added plus the partial key of the weak entity type together form the primary key of the relation schema representing the weak entity type. In other words, there is no such thing called a "weak" relation schema. All relation schemas are "strong"—meaning they have a primary key. Observe the mapping of the weak entity type BUILDING in Figure 6.3. The primary key of PLANT, the only identifying parent of BUILDING, has been concatenated to the partial key, **Bld_building**, to form the primary key of the relation schema BUILDING.

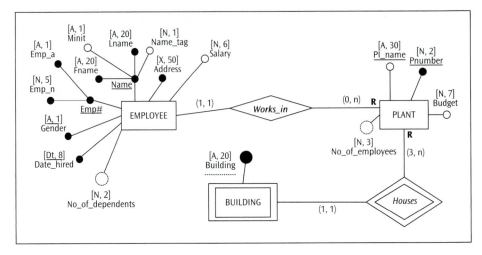

Figure 6.2 Excerpt from Fine-granular Design-Specific ER diagram of Figure 3.12

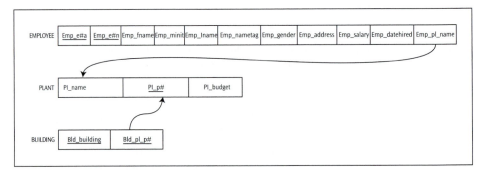

Figure 6.3 Logical schema for the ER diagram in Figure 6.2: Foreign key design

6.7.1.2 Mapping Relationship Types

As was the case with the entity types, each relationship type (diamond symbol) in the ER diagram is individually accounted for in the logical model. Remember that considerable effort has been expended to prepare the Fine-granular Design-Specific ER model to be suitable for direct mapping to the logical tier. As a result, notice that all relationship types in the Fine-granular Design-Specific ER diagram are binary (or recursive) in nature and exhibit a cardinality ratio of **1:1** or **1:n** –there are no complex relationship types and no **m:n** cardinality ratios. In addition, there are no multi-valued attributes. The mapping of a relationship type is accomplished by simply imposing a referential integrity constraint between the relation schemas over the relationship type being mapped (see Section 6.3.2).

Mapping the Cardinality Ratio of 1:n: Foreign Key Design   When the cardinality ratio of the relationship type under consideration is **1:n**, the entity type on the *n-side* of the relationship type is the child in the PCR (parent-child relationship) and hence it becomes the **referencing relation schema** in the logical tier. Because each tuple in the child relation is related to at most one tuple in the referenced (parent) relation, the placement of foreign key attribute(s) in the *referencing* relation schema maps the relationship type specified in the ER diagram. Of course, the foreign key placed in the referencing schema shares the same domain with a candidate key (invariably and preferably, but not necessarily the primary key) of the **referenced relation schema**. This is expressed diagrammatically by drawing a *directed arc* originating from the foreign key attribute(s) to the relation schema it references with the arrow head terminating at the referenced candidate (or primary) key. In addition, a rule requiring that all foreign key values match some value of the referenced candidate key except, of course, when the foreign key value is null (i.e., referential integrity constraint) is implied. This technique is often labeled as the **foreign key technique/design**.

In the example in Figure 6.2, the *Works_in* relationship type represents a PCR where PLANT is the parent and EMPLOYEE is the child. Therefore, the mapping of *Works_in* is implemented in the logical schema (Figure 6.3) by adding an attribute **Emp_pl_name** to EMPLOYEE, which fulfills the role of a foreign key by referencing **Pl_name**, an alternate key of PLANT.[19] Also, in Figure 6.3, notice the directed arc originating from **EMPLOYEE.Emp_pl_name** and pointing at **PLANT.Pl_name**. The attributes of a relationship type in the ER diagram, if any, are also added to the referencing (child) relation schema along with the foreign key attribute(s). Next, while the primary key of BUILDING is **[Bld_building, Bld_pl_p#]**, **BUILDING.Bld_pl_p#** also serves as the foreign key referencing **PLANT.Pl_p#**, the primary key of the (identifying) parent, PLANT, thus accurately portraying the presence of the identifying relationship type, *Houses*.

The clarity of the foreign key design deteriorates very quickly as the number of relation schemas in the relational data model increases because the spaghetti of directed arcs becomes difficult to trace. Alternatively, instead of the directed arcs, it is possible to express the relationship types via the specification of *inclusion dependencies*. This method of expressing a referential integrity constraint is shown in Figure 6.4. Both methods fail to map the participation constraints available in the ER diagram to the logical tier – a case in point for information reduction in mapping.

When using the foreign key design, it is important to note what happens should the foreign key attribute inadvertently be placed in the parent relation (what should be the referenced relation schema) instead of in the child relation (what should be the referencing relation schema). For example, mapping the *Works_in* relationship type by placing the primary key of EMPLOYEE **[Pl_emp_e#a, Pl_emp_e#n]** as a foreign key attribute in PLANT amounts to a reversal of the cardinality constraint and results in a serious semantic error.

Mapping the Cardinality Ratio of 1:n: Cross-Referencing Design   Suppose an employee does not have to necessarily work in a plant. This is modeled by changing the participation of EMPLOYEE in *Works_in* from a **1** to a **0**. This condition is handled at the logical tier by allowing the foreign key, **EMPLOYEE.Emp_pl_name** to have null values (see Figures 6.3 and 6.4). There is an alternative modeling approach that guarantees the absence of

---

[19] Instead, the foreign key added to EMPLOYEE could have been an attribute that references **Pl_p#**, the primary key of PLANT.

EMPLOYEE (<u>Emp_e#a</u>, <u>Emp_e#n</u>, Emp_fname, Emp_minit, Emp_lname, Emp_nametag, Emp_gender, Emp_address, Emp.salary, Emp_datehired, Emp_pl_name)
EMPLOYEE.{Emp_pl_name} ⊆ PLANT.{Pl_name}
PLANT (Pl_plname, Pl_p#, Pl_budget)
BUILDING (<u>Bld_building</u>, Bld_pl_p#)
BUILDING.{Bld_ pl_p#} ⊆ PLANT.{Pl_p#}

Figure 6.4 Referential integrity constraints in Figure 6.3 expressed as inclusion dependencies: Foreign key design

null values in the foreign key.[20] This **cross-referencing design** entails the creation of a relation schema to represent the relationship type, and is illustrated in Figures 6.5 through 6.7. Figure 6.6 portrays this design where WORKS_IN is a relation schema. No relation state in this design need have null values for the foreign key(s). Any relation state of the relation schema, WORKS_IN, will contain tuples for only those employees who work in one of the plants. In other words, employees who do not work in any of the plants may still be employees of the corporation, and tuples in the EMPLOYEE relation will capture this information. But then, since these particular employees do not work in any of the plants, WORKS_IN will not have tuples for any of these employees. In the previous approach (the foreign key design illustrated in Figures 6.2 through 6.4), if there is a need to add an employee who does not work for any plant, a tuple is added to the EMPLOYEE relation with a null value for the foreign key **EMPLOYEE.Emp_pl_name**. In the current cross-referencing design, however, an employee can be added without any concern about whether he or she works in a plant or not, and yet not use a null value in EMPLOYEE to portray this. Figure 6.7 is a representation of the same cross-referencing design using inclusion dependencies.

The cross-referencing design is not as compact as the foreign key design and can proliferate the logical data model with numerous relation schemas in a hurry. Even when the participation of EMPLOYEE in *Works_in* is optional, it is probably practical to use the foreign key design and allow null values for the foreign key **EMPLOYEE.Emp_pl_name** in the interest of a somewhat more efficient design. Nonetheless, as explained in the aboveexample (total participation of PLANT in *Works_in*), the cross-referencing design has definite utility in specific situations. We will encounter another such condition in the following discussion about mapping **1:1** relationship types to the logical tier.

Mapping the Cardinality Ratio of 1:1   When the cardinality ratio of a relationship type is **1:1**, mapping such a relationship type to the logical tier becomes somewhat complicated because either one of the entity types engaged in this relationship type can be the parent or the child. Three solutions are possible and each is conducive to specific situations – situations occasioned by particular participation constraints in the relationship.

Case 1: The participation constraint of *one* of the entity types participating in the relationship type is *total*, as shown in Figure 6.8.

---

[20] There is a sense of tentativeness associated with the status of "null" values in data. Therefore, a database design that avoids presence of null values is expected to be relatively more robust than the ones that freely allow null values in the data.

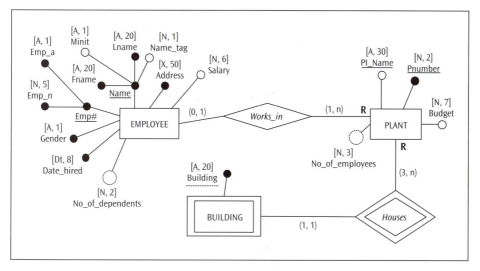

Figure 6.5 Reproduction of Figure 6.3 with a change in participation constraints of *Works_in*

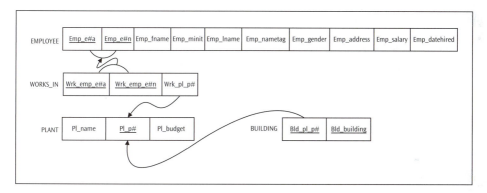

Figure 6.6 Logical schema for the ER diagram in Figure 6.5: Cross-referencing design

EMPLOYEE (Emp_e#a, Emp_e#n, Emp_fname, Emp_minit, Emp_lname, Emp_nametag, Emp_gender, Emp_address, Emp.salary, Emp_datehired)

WORKS_IN (Wrk_emp_e#a, Wrk_emp_e#n, Wrk_pl_p#)
\# WORKS_IN.{Wrk_emp_e#a, Wrk_emp_e#n} ⊆ EMPLOYEE.{Emp_e#a, Emp_e#n}
\# WORKS_IN.{Wrk_pl_p#} ⊆ PLANT.{Pl_p#}

PLANT (Pl_plname, Pl_p#, Pl_budget)

BUILDING (Bld_building,Bld_pl_p#)
\# BUILDING.{Bld_pl_p#} ⊆ PLANT.{Pl_p#}

Figure 6.7 Referential integrity constraints in Figure 6.6 expressed as inclusion dependencies: Cross-referencing design

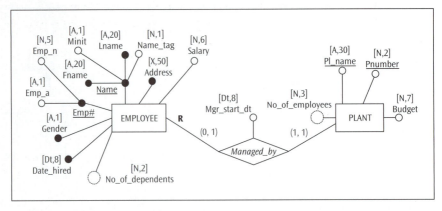

Figure 6.8 A 1:1 relationship type with total participation of PLANT in *Managed_by*

The best way to handle this case is to choose the entity type with total participation in the relationship type to assume the role of child in the PCR. Then the foreign key design described in the Section 6.7.1.2.1 for mapping a **1:n** cardinality ratio can be directly applied here as well, as shown in Figures 6.9 and 6.10. As the foreign key design amounts to an implicit specification of **1:n** cardinality ratio, an additional constraint explicitly specifying that the foreign key value must be unique is necessary to convey the **1:1** cardinality ratio which incidentally renders the foreign key an alternate key (candidate key) of the child entity type. Equally important, the total participation of the child in the relationship type is incorporated in the design via an explicit specification of "no missing value" for the foreign key. For instance, consider the *Managed_by* relationship type in Figure 6.8. The participation of PLANT in this relationship type is total, as indicated by the (min) value of **1**. Therefore, if attribute(s) representing the foreign key is/are added to the relation schema PLANT and this foreign key shares the same domain with the primary key, **[Emp_e#a, Emp_e#n]** or alternatively, any other candidate key, **[Emp_fname, Emp_minit, Emp_lname, Emp_nametag]** of the relation schema EMPLOYEE (see Figure 6.9), then with additional constraint specifications of uniqueness and "not null" on the foreign key in PLANT, the *Managed_by* relationship type will be fully implemented in the relational data model. *These two constraints can be specified declaratively.* On the other hand, suppose EMPLOYEE is chosen as the child in this **1:1** PCR. Then, under the foreign key design, either **Emp_pl_name** or **Emp_pl_p#** will be added to the relation schema EMPLOYEE as the foreign key to depict the *Managed_by* relationship type. However, since the participation of EMPLOYEE in the *Managed_by* relationship is only partial, the corresponding foreign key values can legitimately have null values in some of the EMPLOYEE tuples. As a consequence, addition of a tuple in PLANT will require a procedural intervention in EMPLOYEE that ensures at the least a concurrent assignment of an employee to manage a plant, because every plant must have a manager (total participation of PLANT in the *Managed_by* relationship type). In other words, a plant added to the PLANT relation must reference some employee in the relation, EMPLOYEE. Thus, it is fairly obvious that including the foreign key in the relation schema that has mandatory participation in the relationship type is the better solution.

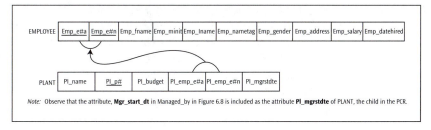

Note: Observe that the attribute, **Mgr_start_dt** in Managed_by in Figure 6.8 is included as the attribute **Pl_mgrstdte** of PLANT, the child in the PCR.

Figure 6.9 Logical schema for the ER diagram in Figure 6.8: Foreign key design

EMPLOYEE (Emp_e#a, Emp_e#n, Emp_fname, Emp_minit, Emp_lname, Emp_nametag, Emp_gender, Emp_address, Emp_salary, Emp_datehired)

PLANT (Pl_name, Pl_p#, Pl_budget, Pl_emp_e#a, Pl_emp_e#n, Pl_mgrstdte)

\# PLANT.{Pl_emp_e#a, Pl_emp_e#n} ⊆ EMPLOYEE. {Emp_e#a, Emp_e#n}

Figure 6.10 Referential integrity constraint in Figure 6.9 expressed as an inclusion dependency: Foreign key design

Case 2: The participation constraints of *both* entity types in the relationship type are *partial*, as shown in Figure 6.11.

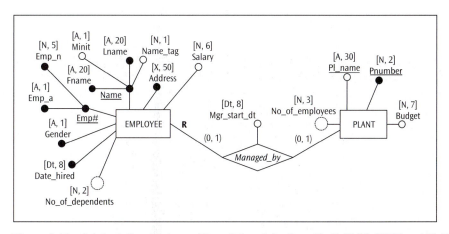

Figure 6.11 A 1:1 relationship type with partial participation of both EMPLOYEE and PLANT in Managed_by

In this case, from a strictly design perspective, addition of a foreign key in either one of the relation schemas involved in the **1:1** relationship type is sufficient. Figures 6.11, 6.12, and 6.13 display an example for Case 2. The ER diagram in Figure 6.11 is a simple variation of the earlier example (Figure 6.8) in that the participation of PLANT in *Managed_by* is also partial, as reflected by the (min) value of **0**. Since the participation of both EMPLOYEE and PLANT in the *Managed_by* relationship type is partial, either one of PLANT and EMPLOYEE can assume the role of the child in this relationship. Figure 6.12

The Relational Data Model

shows the foreign key design using a directed arc where EMPLOYEE, as the child, carries the foreign key. The same design using inclusion dependency appears in Figure 6.13.

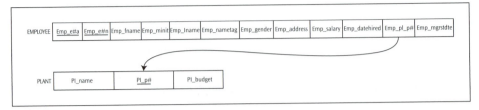

Figure 6.12   Logical schema for the ER diagram in Figure 6.11: Foreign key design

EMPLOYEE (Emp_e#a, Emp_e#n , Emp_fname, Emp_minit, Emp_lname, Emp_nametag, Emp_gender, Emp_address, Emp.salary, Emp_datehired, Emp_pl_p#, Emp_mgrstdte)

EMPLOYEE. {Emp_pl_p#} ⊆ PLANT. {Pl_p#} or ∅

PLANT (Pl_name, Pl_p#, Pl_budget)

Figure 6.13   Referential integrity constraint in Figure 6.12 expressed as an inclusion dependency: Foreign key design

Other semantic or operational considerations may sometimes suggest inclusion of the foreign key in a specific relation schema. For instance, the user may have a predisposition towards the semantics of the relationship. Sometimes, one of the entity types may have a small entity set relative to the other in which case it is operationally efficient to designate it as the child in the PCR. In certain cases, optimal data access is facilitated by *mutual-referencing*—i.e., when the two relation schemas directly reference each other by placing foreign keys in both. In this situation, cross-referencing ought to be considered instead of mutual-referencing, because mutual-referencing between two relation schemas entails specification of *additional* constraints to ensure consistency maintenance (i.e., reference to the correct tuple). Such constraints can only be implemented procedurally, and the ramifications are further clarified in the discussion of Case 3 below. The cross-referencing design eliminates imposition of such a constraint, however, at the expense of adding a relation schema to portray the relationship type. Sometimes such an alternative may be worth considering, such as if the expected size of the intervening relation is small relative to the two base relations in the relationship type. The cross-referencing designs (using directed arcs and inclusion dependencies) for the ERD shown earlier in Figure 6.11 and reproduced in Figure 6.14 are portrayed in Figures 6.15 and 6.16 respectively.

Case 3: The participation constraints of *both* entity types in the relationship type are total.

Here, it is first necessary to add a foreign key in both relation schemas engaged in the relationship – in other words, mutual-referencing. Only by constraining both foreign keys to be unique can it be ascertained that the cardinality ratio is 1:1. By virtue of this constraint, the defined foreign keys also become alternate keys of the respective relation

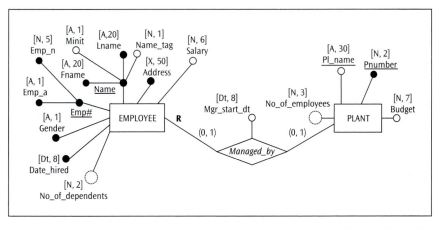

Figure 6.14 A 1:1 relationship type with partial participation of both EMPLOYEE and PLANT in *Managed_by*

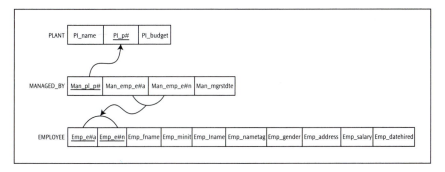

Figure 6.15 Logical schema for the ER diagram in Figure 6.14: Cross-referencing design

PLANT (Pl_name, Pl_p#, Pl_budget)

MANAGED_BY (Man_pl_p#, Man_emp_e#a, Man_emp_e#n, Man_mgrstdte)

\#   MANAGED_BY {Man_pl_p#} ⊆ PLANT. {Pl_p#}
\#   MANAGED_BY {Man_emp_e#a, Man_emp_e#n} ⊆ EMPLOYEE. {Emp_e#a, Emp_e#n}

EMPLOYEE (Emp_e#a, Emp_e#n , Emp_fname, Emp_minit, Emp_lname, Emp_nametag, Emp_gender, Emp_address, Emp.salary, Emp_datehired)

Figure 6.16 Referential integrity constraints in Figure 6.15 expressed as inclusion dependencies: Cross-referencing design

schemas. Total participation of both entity types in the relationship type is incorporated in the design by not allowing null values for the two foreign keys. The presence of foreign keys in both relation schemas referencing each other creates two problems: (1) It becomes necessary to make sure that the [primary/candidate key, foreign key] pairs in the two relations match. This cannot be done using declarative constraints; procedural intervention is necessary to accomplish this. For instance, if employee A12357 manages plant 19, since mutually referencing foreign keys are present in both relations, plant 19 must be managed by employee A12357 and nobody else. This is an additional constraint and can only be implemented via procedural intervention. Notice that in a cross-referencing design (as shown in Figure 6.15), the fact that A12357 manages plant 19 and *vice versa* is captured in the MANAGED_BY relation; this eliminates the need for any procedural intervention. (2) The two relation schemas referencing each other create a *cycle*. Therefore, enforcement of at least one of the two referential integrity constraints must be deferred to run time. An alternative solution to preempt these problems is to merge the two relation schemas into a *single-schema design*. However, it is not always possible to adopt a single-schema design, especially if the relationship involves two distinct entity types and/or the entity types also participate independently in other relationship types.

A variation of the *Managed_by* relationship type (the variation is intended only for lending better semantic sense) appears in Figure 6.17. MANAGER is a partial specialization of EMPLOYEE. A given plant is managed by exactly 1 manager and every manager manages exactly 1 plant each. The total participation of both MANAGER and PLANT in *Managed_by* implies mutual-referencing between PLANT and MANAGER.

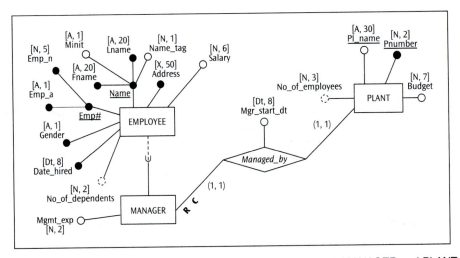

Figure 6.17   A 1:1 relationship type with total participation of both MANAGER and PLANT in *Managed_by*

The mutual-referencing designs in Figures 6.18 and 6.19 reflect this relationship. Clearly, a procedural constraint is required to verify that the pairs of (**[Mgr_emp_e#a, Mgr_emp_e#n, Mgr_pl_p#], [Pl_emp_e#a, Pl_emp_e#n, Pl_p#]**) values from MANAGER and PLANT match. In addition, deferred enforcement of at least one of the two referential integrity constraints (Figure 6.19) is also necessary. Additional procedures may be required to manage other constraints, for example, the deletion rule that restricts deletion of a tuple in MANAGER if a matching tuple in PLANT exists. Incidentally, the example here includes a partial specialization of EMPLOYEE as MANAGER. Mapping of enhanced ER model constructs is discussed later in this chapter (see Section 6.8).

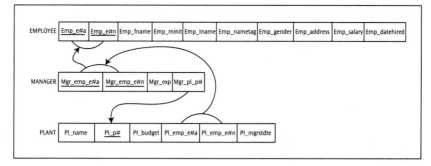

Figure 6.18 Logical schema for the ER diagram in Figure 6.17: Mutual-referencing design

EMPLOYEE (Emp_e#a, Emp_e#n, Emp_fname, Emp_minit, Emp_lname, Emp_nametag, Emp_gender, Emp_address, Emp_salary, Emp_datehired)

MANAGER (Mgr_emp_e#a, Mgr_emp_e#n, Mgr_exp, Mgr_pl_p#)

\#   MANAGER.{Mgr_emp_e#a, Mgr_emp_e#n} $\subseteq$ EMPLOYEE. {Emp_e#a, Emp_e#n}
\#   MANAGER. {Mgr_pl_p#} $\subseteq$ PLANT. {Pl_p#}

PLANT (Pl_name, Pl_p#, Pl_budget, Pl_emp_e#a, Pl_emp_e#n, Pl_mgrstdte)

\#   PLANT.{Pl_emp_e#a, Pl_emp_e#n} $\subseteq$ MANAGER.{Mgr_emp_e#a, Mgr_emp_e#n}

Figure 6.19 Referential integrity constraints in Figure 6.18 expressed as inclusion dependencies: Mutual-referencing design

Note that the redundant inclusion of foreign keys in both relation schemas referencing each other (mutual-referencing) can be done in all three cases above as long as one is willing to incur the penalty of maintaining consistency. Similarly, in any relationship type that has a **1:1** cardinality ratio, combining the two entity types into a single relation schema requires evaluation because a single-schema design, when feasible, minimizes complex integrity constraints and attribute redundancies.

Also, the self-referencing property inherent to a recursive relationship type does not pose any special problems beyond what has been considered in the discussions in this section.

6.7.1.3 A Comprehensive Example

At this point, let us go through the process of mapping the Fine-granular Design-Specific conceptual model to a relational schema using a more comprehensive example. This process is summarized in Table 6.5.

- Create a relation schema for each base entity type in the ER diagram.

- Only the stored attributes are translated to the logical level – derived attributes are not mapped.

- In the case of composite attributes, only their constituent atomic components are recorded.

- A primary key is chosen from among the candidate keys for each relation schema. The atomic attribute(s) that make(s) up the primary key of the relation schema is/are underlined.

- When a weak entity type in the ER diagram is mapped, the primary key of each identifying parent of the weak entity type is added to the relation schema. The attributes thus added plus the partial key of the weak entity type together form the primary key of the relation schema representing the weak entity type. *There is no such thing as a "weak" relation schema.*

- Based on the cardinality ratio and participation constraints associated with a relationship type, choose either the foreign key design, the cross-referencing design, or the mutual-referencing design.

- The placement of foreign key attribute(s) in the *referencing* relation schema (child in the PCR) maps the relationship type specified (1:n or 1:1) and facilitates enforcement of the *referential integrity constraint.*
  Note: *A candidate key (invariably and preferably, but not necessarily the primary key) of the parent (referenced relation schema) is the referenced attribute(s).*
- Attribute(s) of a 1:n or a 1:1 relationship type is (are) added to the referencing relation schema (child in the PCR).

Table 6.5   A stepwise guide for mapping a Design-Specific ER diagram (Fine granularity) to a relational schema using the foreign key design

The Fine-granular Design-Specific ER diagram for Bearcat Incorporated, the domain constraints on the attributes, and a few other semantic integrity constraints not recorded in the ER diagram were given in Chapter 3, but are reproduced here in Figure 6.20 and Table 6.6 for convenience.[21] The objective of this section is to develop the corresponding logical data model. To this end, the section first focuses on mapping the ER diagram to a logical schema.

Using the mapping guide in Table 6.5 and applying the foreign key design discussed thus far in this chapter, the logical schema shown in Figure 6.21 and Table 6.7 can be obtained (the reader may wish to do this as an exercise and verify the result against Figure 6.21). The corresponding relational schema using inclusion dependencies is shown in Figure 6.22.

---

[21] Recall that an ER model constitutes the ER diagram and the semantic integrity constraints not specified in the ER diagram.

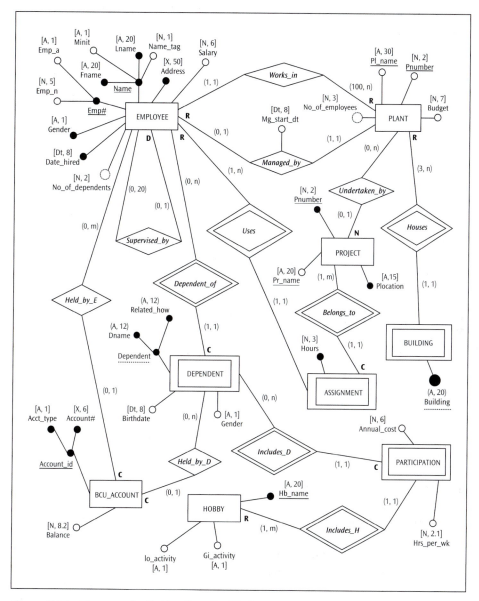

Figure 6.20 Fine-granular Design-Specific ER diagram for Bearcat Incorporated

273

> Constraint   PLANT.Pnumber   IN (10 through 20)

> Constraint   Nametag   IN (1 through 9)

> Constraint   Gender   IN ('M', 'F')

> Constraint   Salary   IN (35000 through 90000)

> Constraint   PROJECT.Pnumber   IN (1 through 40)

> Constraint   Plocation   IN ('Bellaire', 'Blue Ash', 'Mason', 'Stafford', 'Sugarland')

> Constraint   Acct_type   IN ('C', 'S', 'I')

> Constraint   Io_activity   IN ('I', 'O')

> Constraint   Gi_activity   IN ('G', 'I')

> Constraint   Related_how   IN ('Spouse') OR
                       Related_how   IN (('Mother', 'Daughter') AND Gender IN ('F')) OR
                       Related_how   IN (('Father', 'Son') AND Gender IN ('M'))

> Constraint   Building   COUNT (not < 3)

> Constraint   No_of_employees   NOT < 100

Constraints Carried Forward to Logical Design

1. An employee cannot be his or her own supervisor.
2. A dependent can have a joint account only with an employee of Bearcat Incorporated with whom he or she is related.
3. The salary of an employee cannot exceed the salary of the employee's supervisor.
4. Either PLANT.Pnumber or Pl_name must have a value.
5. Every plant is managed by an employee who works in the same plant.

Table 6.6   Semantic integrity constraints for the Fine-granular Design-Specific ER model (A reproduction of Table 3.3)

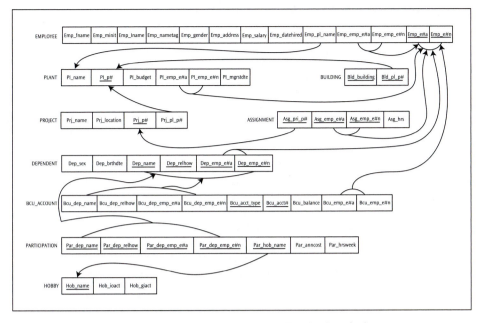

Figure 6.21   Logical schema for Bearcat Incorporated: Foreign key design using directed arcs

L1: EMPLOYEE (Emp_fname, Emp_minit, Emp_lname, Emp_nametag, Emp_emp_e#a, Emp_emp_e#n, Emp_address, Emp_salary, Emp_pl_name, Emp_gender, Emp_datehired, <u>Emp_e#a, Emp_e#n</u>)

\#   EMPLOYEE.{Emp_emp_e#a , Emp_emp_e# n} ⊆ EMPLOYEE.{Emp_ e#a, Emp_e#n}   or   Ø
    EMPLOYEE.{Emp_pl_name} ⊆ PLANT.{Pl_name}

L2: PLANT (<u>Pl_p#</u>, Pl_budget, Pl_name, Pl_emp_e#a, Pl_emp_e#n, Pl_mgrstdte)

\#   PLANT.{Pl_emp_e#a, Pl_emp_e#n} ⊆ EMPLOYEE.{Emp_ e#a, Emp_ e#n}

L3: BUILDING (<u>Bld_building, Bld_pl_p#</u>)

\#   BUILDING.{Bld_pl_p#} ⊆ PLANT.{Pl_p#}

L4: PROJECT (Prj_name, Prj_location, <u>Prj_p#</u>, Prj_pl_p#)

\#   PROJECT.{Prj_pl_p#} ⊆ PLANT.{Pl_p#}   or   Ø

L5: ASSIGNMENT (<u>Asg_prj_p#, Asg_e mp_e#a, Asg_emp_e#n</u>, Asg_hrs)

\#   ASSIGNMENT.{Asg_prj_p# } ⊆ PROJECT.{Prj_p#}
    ASSIGNMENT.{Asg_emp_e#a, Asg_emp_e#n} ⊆ EMPLOYEE.{Emp_ e#a, Emp_e#n}

L6: DEPENDENT (Dep_sex , Dep_brthdte , <u>Dep_name, Dep_relhow , Dep_e mp_e#a, Dep_emp_e#n</u>)

\#   DEPENDENT.{Dep_emp_e#a, Dep_emp_e#n} ⊆ EMPLOYEE.{Emp_ e#a, Emp_e#n}

L7: BCU_ACCOUNT (Bcu_dep_name, Bcu_dep_relhow, Bcu_dep_emp_e#a, Bcu_dep_emp_e#n, <u>Bcu_acct_type</u>, <u>Bcu_acct#</u>, Bcu-balance, Bk_emp_e#a, Bcu_emp_e#n)

\#   BCU_ACCOUNT.{Bcu_emp_e#a, Bcu_emp_e#n} ⊆ EMPLOYEE.{Emp_e#a, Emp_e#n}   or   Ø
    BCU_ACCOUNT.{Bcu_dep_name, Bcu_dep_relhow, Bcu_dep_emp_e#a, Bcu_dep_emp_e#n} ⊆ DEPENDENT.{ Dep_name, Dep_relhow, Dep_emp_e#a, Dep_emp_e#n}   or   Ø

L8: PARTICIPATION (<u>Par_dep_name, Par_dep_relhow, Par_dep_emp_e#a, Par_dep_emp_e#n, Par_hob_name</u>, Par_anncost, Par_hrsweek)

\#   PARTICIPATION.{Par_hob_name}⊆HOBBY.{Hob_name}
    PARTICIPATION.{Par_dep_name, Par_dep_relhow, Par_dep_emp_e#a, Par_dep_emp_e#n}⊆DEPENDENT.{Dep_name, Dep_relhow, Dep_emp_e#a, Dep_emp_e#n}

L9: HOBBY (<u>Hob_name</u>, Hob_loact, Hob_giact)

Figure 6.22   Logical schema for Bearcat Incorporated: Foreign key design using inclusion dependencies

While the foreign key, cross-referencing, and mutual-referencing designs using directed arcs are good visual tools to understand mapping of relationship constructs from an ER diagram to a logical schema in isolated examples, and replacing directed arcs with inclusion dependencies may add more precision to the expression of the relationships *per se*, both methods give up a significant amount of information (metadata) in the process of transforming a Fine-granular Design-Specific ER diagram to a relational schema. The information lost is listed below:

- Both methods are incapable of distinguishing between 1:1 and 1:n cardinality ratios.
- Both methods do not map the participation constraints of a relationship type.
- The optional/mandatory property of an attribute is not retained in the transformation.
- Alternate keys (candidate keys not chosen as the primary key) can no longer be identified.
- The composite nature of some collection of atomic attributes is ignored in the mapping process.
- Derived attributes specified in the ER diagram are not carried forward.
- Deletion rules are not mapped to the logical schema.
- Attribute type and size specified in the ER diagram are not carried forward.

None of the information lost is trivial. These items convey the business rules specified by the user community and design characteristics explicitly expressed in the Design-Specific ER diagram that serves as the source schema for the mapping process. Most of this

> Constraint	Pl_p#	IN (10 through 20)
> Constraint	Nametag	IN (1 through 9)
> Constraint	Gender	IN ('M', 'F')
> Constraint	Salary	IN (35000 through 90000)
> Constraint	Prj_p#	IN (1 through 40)
> Constraint	Plocation	IN ('Bellaire', 'Blue Ash', 'Mason', 'Stafford', Sugarland')
> Constraint	Acct_type	IN ('C', 'S', 'I')
> Constraint	Io_activity	IN ('I', 'O')
> Constraint	Gi_activity	IN ('G', 'I')
> Constraint	Related_how Related_how Related_how	IN ('Spouse') OR IN (('Mother', 'Daughter') AND Gender IN ('F')) OR IN (('Father', 'Son') AND Gender IN ('M'))
> Constraint	Building	COUNT (not < 3)
> Constraint	No_of_employees	NOT < 100

Constraints Carried Forward to Physical Design

1. An employee cannot be his or her own supervisor.
2. A dependent can have a joint account only with an employee of Bearcat Incorporated with whom he or she is related.
3. The salary of an employee cannot exceed the salary of the employee's supervisor.
4. Every plant is managed by an employee who works in the same plant.

Table 6.7 Semantic integrity constraints for the logical schema

metadata is needed to implement the physical data model correctly. One alternative is to include all the lost information in a list of semantic integrity constraints at the logical tier. Alternatively, the conceptual data model (the Fine-granular Design-Specific ER diagram and the semantic integrity constraints) can be used to supplement the underdeveloped logical schema. In that case, the very utility of a logical schema in the systematic development of a database design becomes questionable. The next section presents a logical modeling grammar that can produce an information-preserving script (logical schema).

6.7.2 An Information-Preserving Mapping

The steps involved in the information-preserving mapping are the same as discussed in Section 6.7.1, except that the grammar used for expressing the logical schema is different. Constraints that define a relation schema limit expression of several conceptual modeling constructs in the logical tier such as alternate keys, participation constraints, optional/ mandatory attributes, composite attributes, derived attributes, and deletion rules. In this book, we use the term **logical scheme** instead of relation schema for an entity type transformed to the logical tier in order to relax these constraints. A set of logical schemes becomes a logical schema. The logical schema thus developed can certainly be implemented in a relational database environment. The systematic steps to map a Fine-granular Design-Specific ER diagram to a logical schema are as follows:

```
Lx: SCHEME                                       min Ly max min Lz max
         ┌───────┬─────────┬──────┬──────┬──────┬──────┬──────┬────────┬────────┬─────┬──────┐
         │ Att 1 │ Q[Att2] │ Att3•│ Att4 │[Att5 │ Att6]│ Att7 │ FKAtt1 │ FKAtt2 │ ... │ AttZ │
         │ (t, s)│ (t, s)  │ (t, s)│(t, s)│(t, s)│(t, s)│(t, s)│ (t, s) │ (t, s) │     │(t, s)│
         └───────┴─────────┴──────┴──────┴──────┴──────┴──────┴────────┴────────┴─────┴──────┘
                                                        min D max  min D max

                                    or

                                         min Ly max    min Lz max
Lx: SCHEME (Att 1, Q[Att2] , Att3•, Att4, [Att5, Att6], Att7, ...   FKAtt1,    FKAtt2,..., AttZ)
            (t, s)  (t, s)   (t, s) (t, s) (t, s) (t, s) (t, s)       (t, s)     (t, s)    (t, s)
                                                                   min D max   min D max
```

Step 1: Specify a logical scheme for each base and weak entity type in the ER diagram following the grammar described below.

Where x = (1, 2, 3, 4, ..., N)

- SCHEME is the name of the entity type being mapped. (Use all capital letters.)
- Lx is a label for the SCHEME. (Or it could be an abbreviated short name of the SCHEME.)
- N is the number of SCHEMEs in the logical schema.
- Att1, . . . AttZ, are the names of the atomic attributes from the entity type. (Capitalize first letter only.)[22] Note that the attribute list enclosed in parentheses above lists atomic attributes separated by commas.
- The primary key is underlined—the constituent attributes *need not* be recorded successively.
- Attributes specified as mandatory in the ER diagram (•) are marked by an • following the attribute.
- The data type (**t**) and size (**s**) of an attribute mapped from the ER diagram are recorded immediately below the attribute.

---

[22] The *Universal Relation Schema (URS) assumption* dictates that every attribute name must be unique because attributes have a global meaning in a database schema. Therefore, if an attribute name appears in several relation schemas, all of these denote the same meaning—i.e., the attributes are semantically join-compatible. In an ER model, however, the same attribute name is allowed to appear in different entity types since they imply different roles for the attribute name. We have adopted the URS assumption for the logical schema presented here. Thus, mapping of attributes from an ER model to a logical schema requires careful attention in order to ensure unique attributes names in the logical schema. Note that a referencing foreign key and corresponding referenced primary (or alternate) key having the same attribute name in a logical schema does not violate the URS assumption.

- Composite (molecular) attributes are enclosed by square brackets [this reason, the constituent atomic attributes are recorded next to each
- Alternate keys are enclosed in square brackets and marked by a Q (me unique) preceding the alternate key attribute(s).
- Derived attributes not stored in the database are denoted by a dotted underline (- - - -).

Step 2: Map each relationship type in the ER diagram using either the foreign key design, cross-referencing design, or mutual-referencing design as discussed in Section 6.7.1.2.

- Each relationship type (diamond in the ER diagram) is accounted for by adding to the child scheme in the PCR a foreign key, say, *FKAtt1* - attribute(s) that share the same domain with the primary key of the parent scheme (the highlighting with italics is just intended to draw attention). *While the reference implied in general is to the primary key of the parent scheme, if the design intention is to refer, in some cases, to an alternate key of the parent scheme, it is accomplished by coding the foreign key name to exactly match the name of the alternate key that is the target of the reference. This does not violate the universal relation schema assumption (see footnote 22).*
- The label (e.g., Ly, Lz) of the parent relation is coded on top of the foreign key box that depicts the relationship. The explicit reference connoted by the foreign key is supplemented via this notation for ease of locating the referenced scheme. Alternatively, the label can be prefixed to the foreign key attribute's name.

Step 3: Incorporate the structural constraints of the relationships (i.e., cardinality ratio and participation constraints) using (min, max) as described below.

- The structural constraints of the relationship (i.e., cardinality ratio and participation constraint) are expressed using the (min, max) notation as follows: *The (min, max) expression coded on the parent edge of the PCR in the ER diagram is shown on the top of the foreign key and the (min, max) expression coded on the child edge of the PCR is shown on the bottom of the foreign key.*
- min: participation constraint ($min = 0$) indicates partial participation; ($min \geq 0$) indicates total participation.
- max: cardinality ratio ($max \geq min$).

Step 4: Indicate the deletion rule parameter (**C, N, D,** or **R**) as prescribed above.

- A **D** between min and max recorded below the foreign key specifies the deletion rules—i.e., action to be taken when a tuple from the parent scheme in the relationship type (PCR) is deleted. Four options are possible: C = Cascade; N = Set null; D = Set default value provided; R = Restrict.

Despite its significant capacity to preserve design information, the grammar presented here is not capable of preserving all metadata (that is, expressed in the ER diagram and semantic integrity constraints) declaratively. For instance, the domain

ributes often listed as semantic integrity constraints are not cap-
, and so will have to be carried forward via a list of semantic integ-
ed at the logical tier. Also, the specific names and the roles of the
not preserved in the mapping process even though the relationship
ully captured.

ve grammar, an example for an information-preserving mapping of a
Specific ER diagram to a logical schema is explicated in Figure 6.23.
t the top of the figure) is a combination of the ER diagrams shown in
long with a recursive relationship type *Supervised_by*. First the two
LOYEE and PLANT, and the weak entity type BUILDING are con-
nes with corresponding names; the three logical schemes are labeled
tively, shown at the bottom of Figure 6.23.

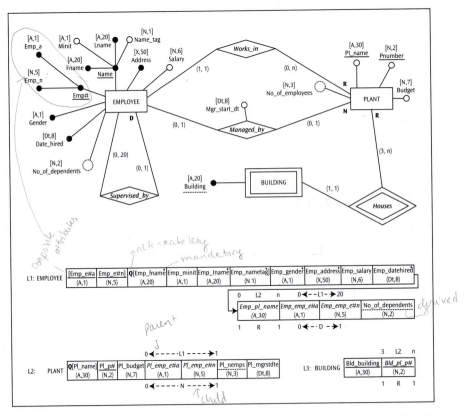

Figure 6.23 A Fine-granular Design-Specific ER diagram and its associated information-preserving logical schema

The attribute type and size for each attribute are shown right below the attribute name. The mandatory property of an attribute is marked by • above the attribute. The primary key for each logical scheme is denoted by the underline. Observe that when a primary key is a composite attribute, then all atomic attributes constituting the primary key are underlined (for example, **[Bld_building, Bld_pl_p#]**)[23]. Alternate key(s) of a logical scheme is/are identified by the letter Q preceding the bracketed attribute(s), for example **Q[Pl_name]**. **Q[Emp_fname, Emp_minit, Emp_lname, Emp_nametag]** means that this composite attribute is unique – not the constituent atomic attributes. Foreign keys are highlighted by italics (**Emp_pl_name** in EMPLOYEE, **Bld_pl_p#** in BUILDING). The derived attribute, **Pl_nemps** of PLANT is also mapped to the logical tier, the dotted underline indicating the derived nature of the attribute. The label of the parent in a PCR and the (min, max) on the parent edge of the relationship type are stated on top of the foreign key representing the relationship type. For example, **0 L2 n** above **Emp_pl_name** in EMPLOYEE signifies that L2 (i.e., PLANT) is the parent in the PCR *Works_in*, and *Works_in* is represented in the logical scheme by the foreign key **Emp_pl_name** in EMPLOYEE. The (min, max) comes from the edge connecting PLANT to *Works_in*. Likewise, the (min, max) on the edge connecting the child in the PCR (i.e., EMPLOYEE) to *Works_in*, and the deletion rule for *Works_in* are stated right below the foreign key representing that relationship type (i.e., **1 R 1** below **Emp_pl_name** in EMPLOYEE). Notice that the attribute pair **[Emp_emp_e#a, Emp_emp_e#n]** in EMPLOYEE references L1 which is EMPLOYEE itself, thus capturing the recursive relationship type *Supervised_by*. In the same fashion, *Managed_by* and *Houses* are captured by **PLANT.[Pl_emp_e#a, Pl_emp_e#n]** and **BUILDING.Bld_pl_p#** respectively. In essence, the logical schema in Figure 6.23 preserves all the design information portrayed in the ER diagram above it.

The information-preserving logical schema for the Fine-granular Design-Specific ER diagram of Bearcat Incorporated shown in Figure 6.20 is given in Figure 6.24. In addition to incorporating all metadata conveyed by the ER diagram in the logical schema, item 4 of the semantic integrity constraints for the Fine-granular Design-Specific ER model (Table 6.6) is also implicitly captured in the logical schema. The rest of the semantic integrity constraints in Table 6.7 along with the logical schema (Figure 6.24) complete the logical data model which then becomes fully information-preserving.

6.8 Mapping Enhanced ER Model Constructs to a Logical Schema

Chapter 4 discussed four different constructs of Superclass/subclass (SC/sc) relationships: specialization/generalization hierarchy, specialization/generalization lattice, categorization, and aggregation. In all these constructs, the cardinality ratio of an SC/sc relationship is **1:1**. In addition, the participation of a subclass in the relationship is always total. Thus, the mapping of these constructs to the logical tier can follow the same strategy as Case 1 and Case 3 discussed in Section 6.7.1.2.3, depending on the type of participation of the superclass (partial or total) in the relationship.

6.8.1 Information-Reducing Mapping of EER Constructs

This section presents the foreign key design using directed arcs to depict the SC/sc relationships of the various EER modeling constructs. EER diagrams from earlier chapters are used to illustrate the mapping process.

---

[23] It is not mandatory that atomic attributes comprising the primary key of a logical scheme be listed contiguously, though it is a good practice to do so.

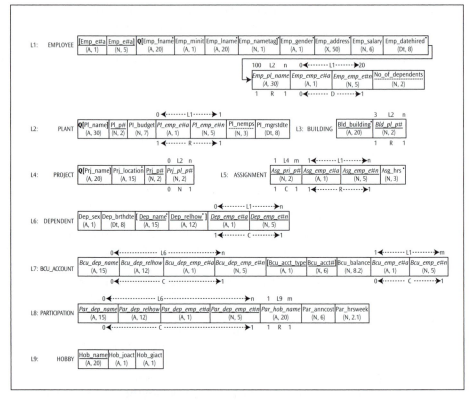

Figure 6.24 Information-preserving logical schema for the Fine-granular Design-Specific ER diagram for Bearcat Incorporated in Figure 6.20

6.8.1.1 Mapping a Specialization

Figure 6.25 is a variation of the EER diagram that appears in Figure 4.17 depicting a simple specialization. Three different logical schema solutions are possible (Figure 6.26). The most general form of solution that supports all four combinations of disjointness and completeness constraints (disjoint/partial, disjoint/total, overlapping/partial, overlapping/total) is shown in Figure 6.26, Solution 1. Here the inheritance property of the specialization is implicitly defined. The second solution merges all subclasses into the superclass yielding just one relation schema, shown in Figure 6.26, Solution 2. Thus, what are otherwise subclasses are inherent in the single schema. Figure 6.26, Solution 3, implements the inheritance property of a specialization explicitly.

The first solution yields one logical scheme each for the subclasses and the superclass in the specialization. Observe that the inheritance property of the specialization yields a candidate key for every subclass (sc) in the specialization which also serves as the foreign key referencing the superclass (SC) in the specialization. Sometimes, when the number of attributes in the subclasses is very small, the efficiency of this design with all the referential integrity constraints becomes questionable. The single-schema design (Solution 2) also supports all four combinations of disjointness and completeness constraints by essentially eliminating the need for these constraints. The absence of referential integrity constraints and a single-schema

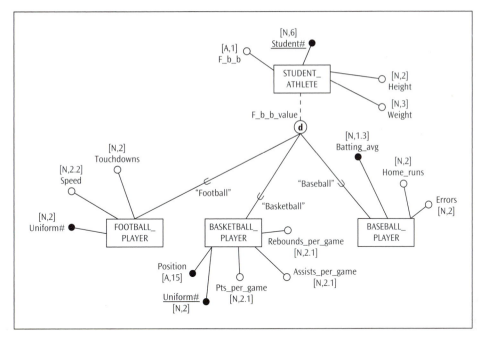

Figure 6.25 An example of a specialization

design certainly enhances operational efficiency. Unless the specialization is overlapping (as opposed to disjoint), the database will have null values for several attributes in each tuple. If subclasses contain lots of attributes, this may be an issue. In addition, any specific relationship types in which one or more of the subclasses independently or collectively participate cannot always be optimally implemented. The third solution will have to be rejected if the completeness constraint is partial, as in the example under illustration. The information that there are student-athletes that are neither football players, nor basketball players, nor baseball players will be lost in this design. Assuming that the completeness constraint is total, this solution can be an optimal middle ground among the three, especially when the number of attributes in the superclass is minimal. However, if the specialization is overlapping, some data redundancy is to be expected. The alternative foreign key design using inclusion dependencies in place of the directed arcs is left as an exercise for the reader.

The remainder of the discussion in this section demonstrates only the foreign key design using the directed arc notation shown in Figure 6.26, Solution 1.

6.8.1.2 Mapping a Specialization Hierarchy

The ER model and the corresponding logical schema (foreign key design using directed arcs) displayed in Figures 6.27 and 6.28 represent a specialization hierarchy. Notice that STUDENT_ATHLETE has been specialized in three different ways. In all three cases, it is a disjoint specialization and all but one are partial specializations. The general solution in Figure 6.28 allows for a team to have more than one captain but for a student-athlete to serve as captain of at most one team. Once again, the reader may translate the solution to depict the directed arcs by inclusion dependencies.

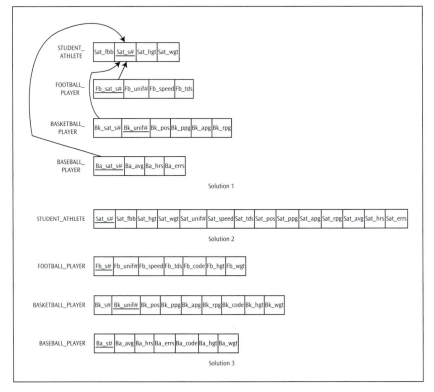

Figure 6.26 Three possible logical schemas for the specialization in Figure 6.25

6.8.1.3 Mapping a Specialization Lattice and a Categorization

The example shown in Figure 6.29 is an excerpt from the Fine-granular Design-Specific ER diagram of Bearcat Incorporated (see Figure 4.22 in Chapter 4). The example includes a total category SPONSOR as a union of CHURCH, SCHOOL, and INDIVIDUAL; a partial specialization of SPONSOR as NOT_FOR_PROFIT_ORGANIZATION; and a specialization lattice where PUBLIC_SCHOOL is a shared subclass participating in two partial specializations, one with NOT_FOR_PROFIT_ORGANIZATION as the superclass and the other with SCHOOL as the superclass. Before embarking on the logical schema mapping task, note that SPONSOR does not have a candidate key. Therefore, a surrogate key called, say, **Sp_s#** (N,5) needs to be created. Further, while the same attribute name reflecting different roles in different entity types is permissible in the ER diagram, pursuant to the universal relation schema assumption, attribute names in the logical schema must be unique. Therefore, let us rename **CHURCH.Name** as **Ch_name** and **SCHOOL.Name** as **Sch_name** in the corresponding logical schemes. Figure 6.30 portrays the logical schema for the ER diagram in Figure 6.29 using the foreign key design with directed arcs. It must be noted that in a categorization a candidate key of the subclass (i.e., SPONSOR) is added to each superclass (i.e., CHURCH, SCHOOL, and INDIVIDUAL) as the foreign key instead of the other way around as is done in a specialization.

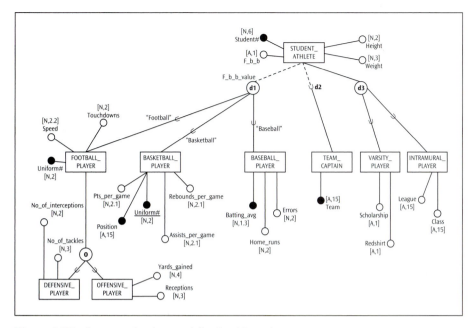

Figure 6.27 An example of a specialization hierarchy

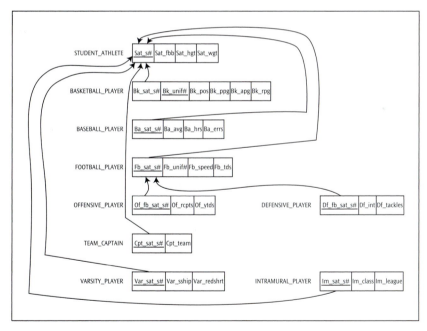

Figure 6.28 Logical schema for the specialization hierarchy in Figure 6.27: Foreign key design using directed arcs

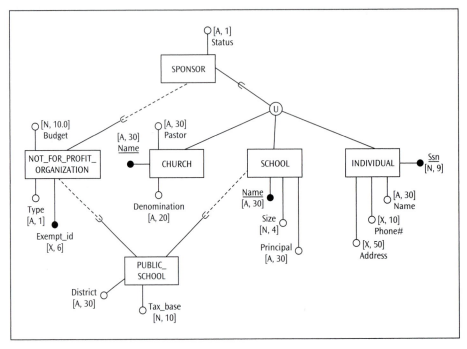

Figure 6.29   An example of a specialization lattice and a categorization

6.8.1.4 Mapping an Aggregation

Aggregation along with a categorization is demonstrated in Chapter 4, Figure 4.19. For the convenience of the reader, the ER diagram is reproduced in Figure 6.31. The basic difference between PROPERTY and TAXABLE_PROPERTY is that a lot and a building together constituting a "property" is identified by a single address, while every individual lot and building that is a "taxable property" is identified by a unique **Txp_taxid**. The logical schema reflecting the foreign key design with directed arcs is shown in Figure 6.32. Incidentally, **Txp_taxid** is a surrogate key of TAXABLE_PROPERTY that is "manufactured" since TAXABLE_PROPERTY does not have a candidate key. Also, **Lot_txp_taxid** and **Bld_txp_taxid** are alternate keys for LOT and BUILDING respectively. Likewise, **Lot_pr_address** and **Bld_pr_address** are also alternate keys for LOT and BUILDING, respectively. As in a categorization, a candidate key of an aggregate (subclass) is mapped as a foreign key in every superclass that participates in that aggregation. The reader may wish to develop the logical schema using inclusion dependencies as an exercise.

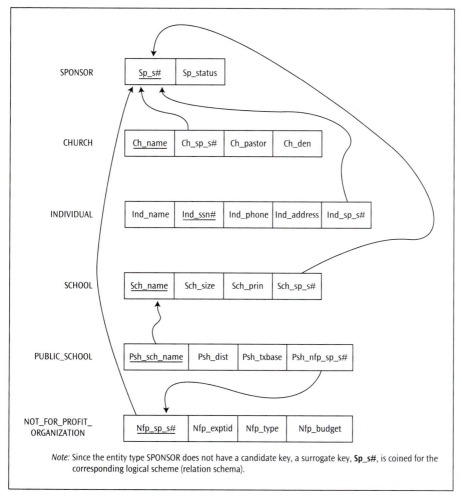

Note: Since the entity type SPONSOR does not have a candidate key, a surrogate key, **Sp_s#**, is coined for the corresponding logical scheme (relation schema).

Figure 6.30 Logical schema for the specialization lattice and categorization in Figure 6.29: Foreign key design using directed arcs

6.8.1.5 Information Lost While Mapping EER Constructs to the Logical Tier

Once again, the techniques discussed in Section 6.8.1 are information-reducing. Two kinds of metadata information are lost: user-specified business rules and design features. In addition to all but the first two information loss items listed in Section 6.7.1.3, the following information is lost as well:

- The type of relationship (e.g., specialization/generalization, categorization, aggregation) is not carried forward to the logical schema.
- SC/sc (i.e., intra-entity class) relationships become indistinguishable from the regular (inter-entity class) relationships.
- The disjointness constraint of a specialization/generalization is lost during the conversion process.
- Multiple specializations of the same superclass are not captured.

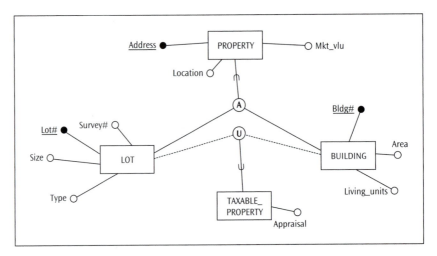

Figure 6.31 A category and an aggregate contrasted

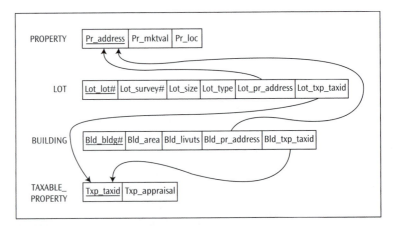

Figure 6.32 Logical schema for the category and aggregate in Figure 6.31: Foreign key design using directed arcs

- Specialization lattices are not discernible.
- The number of subclasses participating in a specialization and the number of superclasses participating in a categorization and/or aggregation are lost in the mapping.
- The completeness constraint of an SC/sc relationship is not present in the logical schema.

Are these losses of information trivial? If so, why are they collected in the conceptual data model to begin with? The argument that the ER diagram and the list of semantic integrity constraints can be used to supplement the logical schema during physical design defeats the purpose of even attempting to develop a logical schema. Furthermore, there is nothing wrong in attempting to develop a self-sufficient logical data model. *Simplicity of design as the rationale for an underdeveloped logical schema is not a worthy compromise.* The next section describes an extension to the information-preserving logical modeling grammar presented in Section 6.7.2, as applied to EER modeling.

6.8.2 Information-Preserving Grammar for Enhanced ER Modeling Constructs

Once again, the steps involved in the mapping process for EER constructs are the same as in Section 6.7.2, except that the grammar used for expressing the logical schema is different and is shown below.

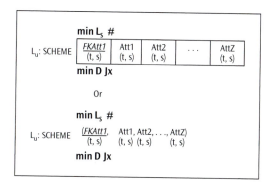

Where $u = (1, 2, ...N)$

- min: completeness constraint

  (min = 0) indicates partial completeness; (min = 1) indicates total completeness.

  Note: Since the cardinality ratio in an SC/sc relationship is always (**1:1**) the max part of (min,max) is always a 1. Therefore, the max is inherently preserved.

- L_S is the label of the parent scheme coded on top of the foreign key that depicts the SC/sc relationship.

- \# denotes the number of scs in the specialization/generalization or the number of SCs in the categorization, specialization lattice, and aggregation.

- Jx is the SC/sc label where

  J is the disjointness constraint value: d = disjoint; o = overlapping; u = union; a = aggregate; L = specialization lattice; - = none.

  * x is the marker for the specialization/categorization/aggregation/ specialization lattice (e.g., 1, 2, 3, . . .)

- The **D** between min and Jx under the foreign key specifies action to be taken when a tuple from the parent entity in a relationship is deleted. C = Cascade; N = Set null; D = Set default value provided; R = Restrict.

While the grammar notation is almost the same as that of the one employed for the ER constructs, two syntactical markers are specific to intra-entity class (SC/sc) relationships: **Jx** and **#**. **J** in the **Jx** reflects the value of the disjointness constraint in the cases of specialization/generalization (**o** for overlapping and **d** for disjoint), the union property (**u**) in the case of categorization and the aggregate property (**a**) in the case of aggregation. The **x** in the **Jx** marks the specific specialization type occurrence in which the scheme participates. The same is true for categorization and aggregation as well. The **#** marker above the foreign key is an integer that indicates the number of subclasses in the specialization flagged by **Jx**. In the cases of categorization, aggregation, and specialization lattice, this will be the number of superclasses in the particular SC/sc construct.

The general form of the specialization lattice follows. Note that each specialization in which the shared subclass is a member is captured by an independent foreign key. L_L is the shared subclass in the lattice and L_{S1}, L_{S2}, ... point to parents in the SC/sc relationships.

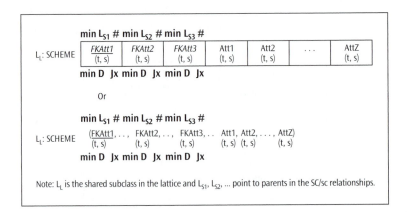

The following examples illustrate the extension to the information-preserving grammar prescribed above for mapping EER model constructs to the logical schema. To begin with, let us revisit the specialization hierarchy depicted in Figure 6.33. This is a reproduction of Figure 6.27. As pointed out in Section 6.8.1.5, both the foreign key design using directed arcs approach for logical model mapping (Figure 6.28) and its inclusion dependency alternative are information-reducing. Following the information-preserving grammar rules described in this section produces the logical schema shown in Figure 6.34. The primary key, data type and size, optional properties of attributes (mandatory or optional value), alternate key, and foreign key are mapped following the information-preserving grammar described in Section 6.7.2. As for the structural constraints of the specialization constructs, the **min** component of the (min, max)—i.e., the participation constraint—is captured as is, and the **max** component is *not* mapped because the two max values depicting the cardinality ratio are always **1** in a specialization/generalization relationship. Instead, in the *max slot* on top of the foreign key (**#**) is utilized to record the number of subclasses in the specialization and the *max slot* below the foreign key (**Jx**) is used to show the specialization label. For example, the value **3** for the # in FOOTBALL_PLAYER, BASKETBALL_PLAYER, and BASEBALL_PLAYER each denote that there are three subclasses in the specialization **d1** (**Jx** value in all three logical schemes) of STUDENT_ATHLETE marked by **L1**. The value **d1** for **Jx** in all three indicate that the three logical schemes are subclasses of the *same* specialization of STUDENT_ATHLETE marked **d1**. Specialization **d3**, on the other hand, has only two subclasses. This is reflected by the value for the # in both VARSITY_PLAYER and INTRAMURAL_PLAYER. The specialization label **d2** may appear superfluous at first, because with just one subclass in this specialization the disjoint property itself has no meaning. Nonetheless, the label is needed to record the fact that the relationship type between TEAM_CAPTAIN and STUDENT_ATHLETE is indeed a specialization/generalization. Thus, the fact that STUDENT_ATHLETE participates as the superclass (SC) in three distinct specializations, and the related metadata are fully captured in the logical schema with no loss of information in the transformation process. The overlapping specialization of the superclass FOOTBALL_PLAYER as {OFFENSIVE_PLAYER, DEFENSIVE_PLAYER} is coded by the value **o** for **Jx** (instead of **o1** or **o2**) because there is only one specialization of FOOTBALL_PLAYER.

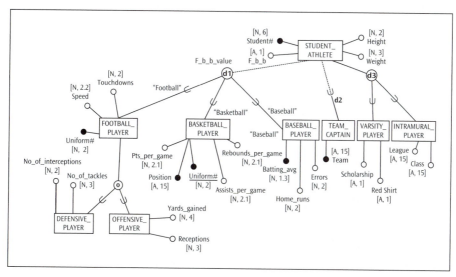

Figure 6.33 An example of a specialization hierarchy

Let us turn our attention now back to Figure 6.29 where a categorization and a specialization lattice are portrayed. In fact, while the relational schema in Figure 6.30 indicates the presence of these relationships, the fact that these are SC/sc relationships, and represent a categorization and a specialization lattice, are completely lost in Figure 6.30. It is true that the logical schema in Figure 6.30 is correct in that the physical implementation of the design will work, but the mapping is not complete, thus the implemented database system will not be robust. Indeed, some of the metadata lost in the transformation may not matter in implementations in certain relational DBMSs. The lost information, however, is integrity constraints arising from the business rules of the application, that is, metadata of the data model. While declarative implementation of some of these constraints may not be possible, procedural implementation methods can be used to make up for it. In sum, during database design, losing metadata through the data modeling tiers cannot be casually accepted and should be strictly avoided.

In this spirit, let us examine an information-preserving logical schema for this ER model (reproduced as Figure 6.35) that appears in Figure 6.36. Notice once again that the logical mapping of a categorization construct is somewhat opposite to that of a specialization. Here, despite the fact that a subclass inherits attributes and relationship types from the superclasses (selective type inheritance in the case of a categorization; see Chapter 4), the primary key of the category (subclass) is carried to the superclasses as foreign keys in order

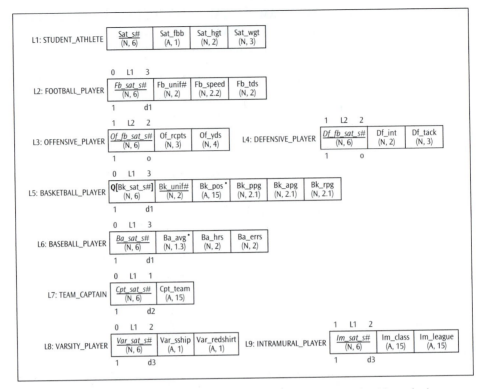

Figure 6.34 Information-preserving logical schema for the specialization hierarchy in Figure 6.33

to establish the relationships. This is so because if the primary key of the superclasses are mapped as foreign keys in the subclass as in a specialization, the category will not only include several foreign keys (one corresponding to each superclass in the categorization), but also in each tuple of the category, all but one foreign key will have null values. This obviously is not a desirable modeling practice.

Next, in Figure 6.35, observe that the category, SPONSOR, does not have a candidate key. While this is okay in the conceptual tier, absence of a candidate key in a logical scheme is, by definition, unacceptable. Therefore, a surrogate key, **Sp_s#**, has been coined for the logical scheme of SPONSOR in Figure 6.36 and mapped as the foreign key in each of the three superclasses in the categorization: CHURCH, SCHOOL, and INDIVIDUAL. Also, conforming to the universal relation schema assumption, some of the attributes in the logical schema have been renamed so that the attribute names are globally unique in the logical schema, for example **Ch_name** in CHURCH and **Sch_name** in SCHOOL. Note that

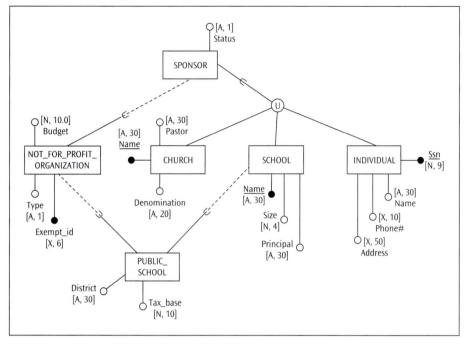

Figure 6.35 An example of a specialization lattice and a categorization

these foreign keys also end up as candidate keys of the respective logical schemes due to the cardinality ratio of 1:1 implicitly present in the design.[24]

In the specialization lattice, however, PUBLIC_SCHOOL is a shared subclass in two specializations. It inherits attributes and relationship types from both SCHOOL and NOT_FOR_PROFIT_ORGANIZATION. Since this is a lattice of "specialization," the shared subclass, PUBLIC_SCHOOL assumes the role of child in the two PCRs of specialization. Therefore, there are two foreign keys, **Psh_sch_name** and **Psh_nfp_sp_s#**, in PUBLIC_SCHOOL which capture the relationships with SCHOOL and NOT_FOR_PROFIT_ORGANIZATION respectively. Observe that the two foreign keys are also candidate keys of PUBLIC_SCHOOL because of the 1:1 cardinality ratios inherent in specialization relationships. Since **Psh_sch_name** is chosen as the primary key, **Psh_nfp_sp_s#** automatically becomes an alternate key of the logical scheme.

How do you convert an aggregation construct to the logical tier? This was discussed earlier in this chapter and illustrated in Figures 6.31 and 6.32 (see Section 6.8.1.4). The information-preserving mapping of aggregation is similar to that of categorization except that the aggregate is denoted by **a** for the **Jx** value instead of a **u**. Recall that the # above the foreign key indicates the value of the number of superclasses participating in the

---

[24] Selective type inheritance as a distinguishing property of categorization is inherent in the relationship construct. This property is not explicitly seen in the logical schema either. The constraint means that the same value of **Sponsor#** cannot occur in more than one of the relations, CHURCH, SCHOOL, and INDIVIDUAL. The subtle nature of this constraint deems it necessary to be explicitly specified in the list of semantic integrity constraints that accompanies the local schema.

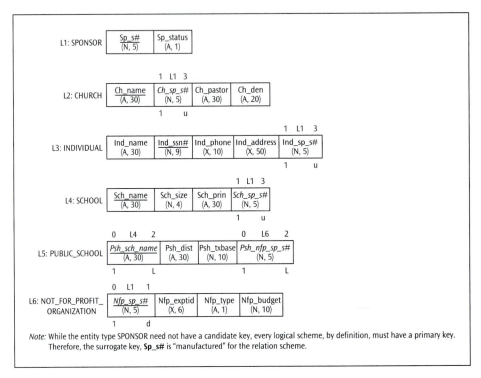

Figure 6.36 Information-preserving logical schema for the specialization lattice and categorization in Figure 6.35

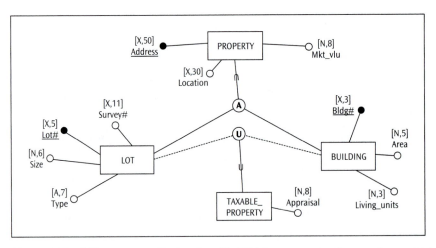

Figure 6.37 A Fine-granular Design-Specific ER Diagram contrasting a category and an aggregate

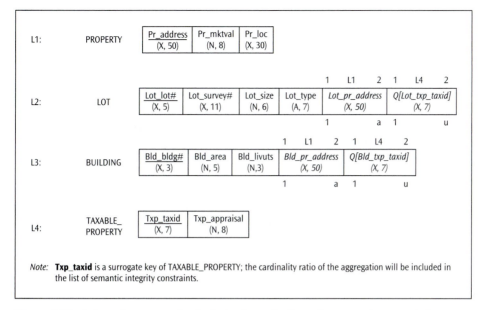

Figure 6.38 Information-preserving logical schema for the category and aggregate in Figure 6.37

aggregation. Therefore, when the cardinality ratio of an SC/sc relationship in an aggregation is not a 1:1, the associated metadata is not captured in the logical schema; instead, it is recorded in the list of semantic integrity constraints that accompanies the logical schema. Figure 6.37 shows the design-specific rendition of the ERD (a repeat of Figure 6.31, for the reader's convenience). The information-preserving logical schema for the aggregation and categorization shown in Figures 6.31 and 6.37 appears in Figure 6.38.

In Chapter 4, the Bearcat Incorporated story was further enriched to incorporate elements that would readily translate to EER constructs. The associated Fine-granular Design-Specific EER model (the EER diagram and the semantic integrity constraints) are collectively depicted in Figure 4.22 and Table 4.5. The reader is encouraged to develop the corresponding logical data model, first using the foreign key design with both the directed arcs and inclusion dependencies alternatives, and then using the information-preserving grammar. Such an exercise will not only reinforce understanding of the information-preserving grammar, but also, clarify the information-preserving nature of this new grammar. The logical schemas in Figures 6.21, 6.22, and 6.24 may be used as springboards to embark on this adventure.

Chapter Summary

The relational data model was first outlined in a paper published by E. F. Codd in 1970. The model uses mathematical relations as its foundation and is based on set theory. The simplicity of the concept and the sound theoretical basis are two reasons why the relational data model has become the model on which most commercial database systems are based. The relational data model represents a database as a collection of relations, where a relation resembles a two-dimensional table of values presented as rows and columns. A row in the table represents a set of related data values and is called a tuple. All values in a column are of the same data type. A column is formally referred to as an attribute. The set of all tuples in the table goes by the name relation. A relationship (association) between two relations in the relational data model takes the form of a referential integrity constraint.

An attribute or collection of attributes can serve as a unique identifier of a relation. A super-key is a set of one or more attributes, which taken collectively, uniquely identifies a tuple. A second type of unique identifier is called a candidate key. A candidate key is defined as a superkey with no proper subsets that are superkeys. A candidate key has two properties: uniqueness and irreducibility. The uniqueness property is common to both a superkey and a candidate key, while the irreducibility property is present only in a candidate key. Every attribute plays only one of three roles in a relation schema: it is a candidate key, a key attribute, or a non-key attribute of the relation schema. Any attribute that is a constituent part (proper subset) of a candidate key of the relation schema is a key attribute. An attribute that is not a candidate key is a non-key attribute. A primary key serves the role of uniquely identifying tuples of a relation. In addition to possessing the uniqueness and irreducibility properties, a primary key is not allowed to have a missing (i.e., null) value (this property is known as the entity integrity constraint).

The relational data model includes a group of basic manipulation operations that involves relations. Collectively these operations comprise what is known as relational algebra. Section 6.4 discusses six (selection, projection, union, minus, intersection, and natural join) of the eight basic relational algebra operations. A more in-depth treatment of relational algebra appears in Chapter 11. The relational data model also includes views and materialized views. A view is defined as a named "virtual" relation schema constructed from one or more relation schemas through the use of one or more relational algebra operations. In a database environment, views (a) allow the same data to be seen by different users in different ways at the same time, (b) provide security by restricting user access to predetermined data, and (c) hide complexity from the user. Unlike a view, a materialized view is real and contains its own separate data. Materialized views are used to freeze data as of a certain point in time without preventing updates to continue on the data in the relation schemas on which they are based.

Sections 6.6 through 6.8 describe ways to map conceptual schema (both ER and EER models) to logical schema. Approaches for mapping ER constructs to a logical schema begin by creating a relation schema for each base and weak entity type present in the Fine-granular Design-Specific ER diagram. Only the stored attributes of the entity type become attributes of the relation schema. In the case of composite attributes, only their constituent atomic components are recorded. For each relation schema based on a base entity type, the atomic attribute(s) serving as the primary key is/are underlined. The primary key of a relation schema for a weak entity type includes the partial key of the weak entity type plus the primary key of each identifying parent of the weak entity type.

A Fine-granular Design-Specific ER diagram contains binary and recursive relationship types that exhibit a cardinality ratio of **1:n** and **1:1**. In cases where the cardinality ratio is **1:n**, the entity type on the *n-side* of the relationship type is the child in the parent-child relationship and the child (or referencing relation schema) contains the foreign key attribute(s). This approach, where the foreign key in the child shares the same domain with a candidate key (most of the time, the primary key) of the parent, is known as the foreign key design. The foreign key design can be expressed diagrammatically via the use of directed arcs or by the specification of inclusion dependencies. An alternative to the foreign key design is the cross-referencing design (which can also be expressed using directed arcs or inclusion dependencies) which entails the creation of a relation schema to represent the relationship type. This approach can be used if the absence of null values in the foreign key is an important consideration.

In situations where the cardinality ratio of the relationship type is **1:1**, either one of the entity types can be the parent or child. Approaches for handling a **1:1** cardinality ratio depend on the nature of the participation constraint that characterizes the relationship. In cases where the participation constraint of only one of the entity types participating in the relationship type is total, the entity type with total participation in the relationship type assumes the role of the child in the parent-child relationship and the foreign key design is applied. When the participation constraints of both entity types in the relationship type are partial, a variety of approaches can be considered: (a) a foreign key design with the addition of a foreign key in either one of the relation schemas involved in the **1:1** relationship type, (b) mutual-referencing where the two relation schemas directly reference each other via foreign keys included in both, and (c) a cross-referencing design. A third case is where the participation constraints of both entity types in the relationship type are total. Situations of this type are handled by using mutual-referencing. Mutual-referencing here must be accompanied by the imposition of several constraints, some of which can be established declaratively and some of which require procedural intervention. Merging the entity types involved in this or any other type of **1:1** relationship into a single relation schema is always a possibility but often not employed if the distinct nature of the entity types is lost or the entity types also participate independently in other relationship types.

The information-reducing nature of design approaches that make use of directed arcs and inclusion dependencies for mapping ER constructs (i.e., entity types and relationship types) to a logical schema is illustrated via their application to the mapping of the Fine granular Design-Specific ER Model for Bearcat Incorporated to a logical schema. Information lost (i.e., ignored) in the transformation process includes: the nature of the cardinality ratios, the participation constraints of each relationship type, the optional/mandatory property of an attribute, the identification of alternate keys, the composite nature of certain atomic attributes, the existence of derived attributes, deletion rules, and attribute type and size. A logical modeling grammar capable of producing a logical schema that is information-preserving is described and then applied to the mapping of the Fine-granular Design-Specific ER Model for Bearcat Incorporated

The chapter concludes with a discussion of the application of the foreign key design approach using directed arcs to map the EER constructs of (a) the specialization/generalization hierarchy, (b) the specialization/generalization lattice, (c) categorization, and (d) aggregation. With the exception of aggregation, given the presence of only **1:1** cardinality ratios, the mapping of these constructs can make use of strategies that are similar to those used to map relationship types with **1:1** cardinality ratios, depending on the type of participation of the superclass (partial or total) in the relationship. Since aggregation permits the relaxation of the inherent property of the **1:1** cardinality ratio in an SC/sc relationship (an aggregate is a subclass that is a subset of the aggregation of the superclasses in the relationship), the logical mapping of an aggregation is similar to that of the foreign key design in a **1:n** relationship type. As was the case when discussing the application of the foreign key design using directed arcs and inclusion dependencies in the context of ER constructs, use of these techniques in the context of EER constructs to develop a logical schema can also be shown as information-reducing and thus amenable to an extension of the information-preserving logical modeling grammar described previously.

Exercises

1. Define the terms tuple, attribute, and relation.

2. What is a relation schema? What is the difference between a relation, a relation schema, and a relational schema?

3. What is the difference between a derived attribute and a stored attribute in terms of their representation in a relation schema?

4. What is a null value? What gives rise to null values in a relation?

5. Distinguish between a subset and a proper subset?

6. What is a candidate key? How does a candidate key differ from a superkey?

7. What is a primary key? How do the properties of a primary key differ from those of a candidate key?

8. Identify the superkeys, candidate key(s), and the primary key for the following relation instance of the STU-CLASS relation schema.

StudentNumber	StudentName	StudentMajor	ClassName	ClassTime
0110	KHUMAWALA	ACCOUNTING	BA482	MW3
0110	KHUMAWALA	ACCOUNTING	BD445	TR2
0110	KHUMAWALA	ACCOUNTING	BA491	TR3
1000	STEDRY	ANTHROPOLOGY	AP150	MWF9
1000	STEDRY	ANTHROPOLOGY	BD445	TR2
2000	KHUMAWALA	STATISTICS	BA491	TR3
2000	KHUMAWALA	STATISTICS	BD445	TR2
3000	GAMBLE	ACCOUNTING	BA482	MW3
3000	GAMBLE	ACCOUNTING	BP490	MW4

9. Define the term, referential integrity constraint. Why is referential integrity important? How is the term foreign key used in the context of referential integrity?

10. This exercise refers to the relations R1 and R2 given on the following page. Show the relations created as a result of the following relational algebra operations.

Relation R1		
R1.a	R1.b	R1.c
30	a	20
45	b	32
75	a	24

Relation R2		
R2.x	R2.y	R2.z
30	b	24
75	c	12
30	b	20

a. The union of R1 and R2

b. The difference of R1 and R2

c. The difference of R2 and R1

d. The intersection of R1 and R2

e. The natural join of R1 and R2. Assume that **R1.a** and **R2.x** are the joining attributes.

11. Exercise 11 refers to the DRIVER, TICKET_TYPE, and TICKET relations given below.

DRIVER (Dr_license_no, Dr_name, Dr_city, Dr_state)

TICKET_TYPE (Ttp_offense, Ttp_fine)

TICKET (Tic_ticket_no, Tic_ticket_date, Tic_dr_license, Tic_ttp_offense)

An instance of each of these relations follows. Use the data shown below to (a) write the answer to each question, and (b) list the relational algebra operation(s) required to obtain the answer.

Dr_license_no	Dr_name	Dr_city	Dr_state
MVX 322	E. Mills	Waller	TX
RVX 287	R. Brooks	Bellaire	TX
TGY 832	L. Silva	Sugarland	TX
KEC 654	R. Lence	Houston	TX
MQA 823	E. Blair	Houston	TX
GRE 720	H. Newman	Pearland	TX

Ttp_offense	Ttp_fine
Parking	15
Red Light	50
Speeding	65
Failure To Stop	30

The Relational Data Model

Tic_Ticket_no	Tic_ticket_date	Tic_dr_license_no	Tic_ttp_offense
1023	20-Dec-2007	MVX 322	Parking
1025	21-Dec-2007	RVX 287	Red Light
1397	03-Dec-2007	MVX 322	Parking
1027	22-Dec-2007	TGY 832	Parking
1225	22-Dec-2007	KEC 654	Speeding
1212	06-Dec-2007	MVX 322	Speeding
1024	21-Dec-2007	RVX 287	Speeding
1037	23-Dec-2007	MVX 322	Red Light
1051	23-Dec-2007	MVX 322	Failure To Stop

    a.   What are the names of all drivers?

    b.   What are the license numbers of all drivers who have been issued a ticket?

    c.   What are the license numbers of those drivers who have never been issued a ticket? Hint: Consider the use of the minus operator along with one other relational algebra operator.

    d.   What are the names of all drivers who have been issued a ticket?

12. What would cause a relational schema for a database to contain more relation schemas than there are entity types?

13. Discuss the concept of information preservation in data model mapping.

14. What is required to map a base entity type to a relation schema? Describe how this approach differs for a weak entity type.

15. What is required to map a relationship type that exhibits a 1:n cardinality ratio?

16. What is the difference between the referencing relation schema and the referenced relation schema? How are these terms incorporated into the foreign key design?

17. What is the purpose of the cross-referencing design?

18. What complicates the mapping of 1:1 cardinality ratios?

19. Describe mutual-referencing and the complexities that it introduces.

20. What information is lost by the use of the information-reducing grammar?

21. Describe how to map a specialization hierarchy, a specialization lattice, and a categorization.

22. What do you think are the ultimate consequences of failure to "preserve information" in the data model mapping process?

23. For the excerpt from an ER diagram given on the next page, specify the logical (relational) schema as per the foreign key design:

    a.   using directed arcs

    b.   in terms of inclusion dependencies

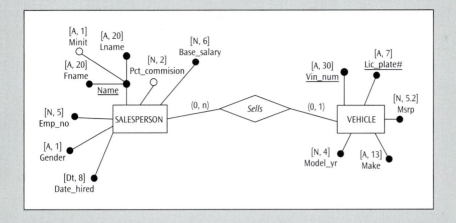

24. List (tabulate) the metadata available in the ER diagram for Exercise 23 and indicate the ones captured in the logical schema of design 23(a) and design 23(b).

25. For the ER diagram for Exercise 23, specify the logical (relational) schema as per the cross-referencing design:

    a. using directed arcs

    b. in terms of inclusion dependencies

    Explain the merits and demerits of this cross-referencing design over the foreign key design solution.

26. For the excerpt from an ER diagram that follows, specify the logical (relational) schema using either directed arcs or in terms of inclusion dependencies according to the foreign key design, cross-referencing design, and mutual-referencing design for the following three cases:

    a. when x = 0 and y = 1

    b. when x = 0 and y = 0

    c. when x = 1 and y = 1

    In each case, offer a comparative discussion of the merits and demerits of the three design options.

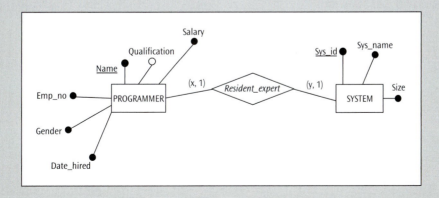

27. For the excerpt from an ER diagram that follows, specify the logical (relational) schema. Choose a design such that there is no need to use null values to indicate partial participation. Specify the schema:

a. using directed arcs

b. in terms of inclusion dependencies

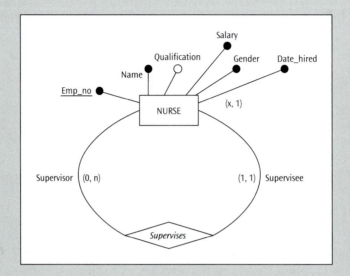

28. How can the design requirement in Exercise 27 be satisfied if the "Supervisee" part of the relationship has the structural constraints (0, 1)? Again, show a design using both directed arcs and in terms of inclusion dependencies.

29. Specify the logical schema for the ER diagram for Exercise 23 using the information-preserving grammar and indicate the metadata present in the ER diagram (Exercise 24) captured by this logical schema.

30. Specify the logical schema for the ER diagram for Exercise 27 using the information-preserving grammar.

31. Specify the logical schema for the ER diagram that follows using the foreign key design:

a. with a directed arc

b. in terms of an inclusion dependency

c. using the information-preserving grammar

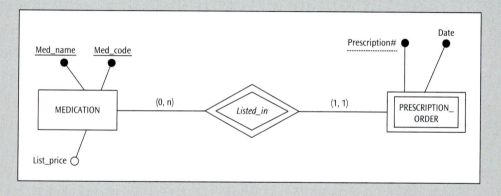

32. Specify the logical schema for the Presentation Layer ER diagram that follows using the foreign key design:

    a. with directed arcs

    b. in terms of inclusion dependencies

    c. using the information-preserving grammar

    Also, list the metadata present in the ER diagram and indicate the ones captured by each of the three designs of the logical schema.

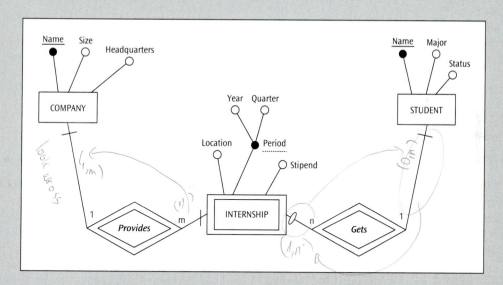

33. Specify the logical (relational) schema for the ER diagram shown in Figure 5-61 in Chapter 5 for the Cougar Medical Associates as per the foreign key design:

    a. using directed arcs

    b. in terms of inclusion dependencies

Th

Selected Bibliography

Batini, C., Ceri, S. and Navathe, S. B. (1992) *Conceptual Database Design: An Entity Relationship Approach*, Benjamin Cummins.

Batini, C., Lenzerini, M. and Navathe, S. B. (1986) "A Comparative Analysis of Methodologies for Database Schema Integration," *ACM Computing Surveys*, 8, 4 (December) 323–364.

Codd, E. F. (1970) "A Relational Model for Large Shared Data Banks," *Communications of the ACM*, 13, 6 (June) 377–387.

Connolly, T. M. and Begg, C. E. (2002) *Database Systems*, Third Edition, Addison-Wesley.

Date, C. J. (2004) *An Introduction to Database Systems*, Eighth Edition, Addison-Wesley.

Date, C. J. and Darwen, H. (1998) *Foundation for Object/Relational Databases*, Addison-Wesley.

Date, C. J. "Back to the Relational Future," *http://www.dbsummit.com/or/date.html*

Elmasri, R. and Navathe, S. B. (2003) *Fundamentals of Database Systems*, Fourth Edition, Addison-Wesley.

Fahrner, C. and Vossen, G. (1995) "A Survey of Database Design Transformations Based on the Entity-Relationship Model," *Data & Knowledge Engineering*, 15, 3 213–250.

Johnson, J. L. (1997) *Database: Models, Languages, Design*, Oxford University Press.

Kifer, M., Bernstein, A., and Lewis, P. M. (2005) *Database Systems: An Application-Oriented Approach*, Second Edition, Addison-Wesley.

Kroenke, D. M. (1977) *Database Processing*, Science Research Associates, Inc.

Mannila, H. and Raiha, K. (1992) *The Design of Relational Databases*, Addison-Wesley.

Markowitz, V. M. and Shoshani, A. (1992) "Representing Extended Entity-Relationship Structures in Relational Databases," *ACM Transactions on Database Systems*, 17, 423-464.

Navathe, S. B. (1992) "Evolution of Data Modeling for Databases," *Communications of the ACM*, 35, 9 (September) 112-123.

Ram, S. (1995) "Deriving Functional Dependencies from the Entity Relationship Model," *Communications of the ACM*, 38, 9 (September). 95–107

Ramakrishnan, R. and Gehrke, J. (2000) *Database Management Systems*, McGraw-Hill.

Shepherd, J. C. (1990) *Database Management: Theory and Application*, Richard D. Irwin, Inc.

Silberschatz, A., Korth, H. F. , and Sudarshan, S. (2002) *Database System Concepts*, Fourth Edition, McGraw-Hill.

Stonebraker, M., Rowe, L., Lindsay, B., Gray, P., Carie, M., Brodie, M.L., Bernstein, P., and Beech D. (1990) "Third-Generation Database System Manifesto," *ACM SIGMOD Record*, 19, 31-44.

Storey, V. C. (1991) "Relational Database Design Based On the Entity-Relationship Model," *Data & Knowledge Engineering*, 7, 47-83.

Teorey, T. J., Yang, D., and Fry, J. P. (1986) "A Logical Design Methodology for Relational Databases Using the Extended Entity-Relationship Model," *Computing Surveys*, 18, 2 (June) 197-222.

Umanath, N. S. and Chiang, R. (2000) "An Information-Preserving Representation of Relational Schema," *Workshop on Information Technology & Systems (WITS)*, (December), 91-96.

PART  III

NORMALIZATION

INTRODUCTION

At this point in the database design life cycle, we are in the logical tier and a logical data model comprising a logical schema and a list of semantic integrity constraints has been developed. The modeling process to this point has been more heuristic and intuitive than scientific, and in fact the source schema (ER diagram) for the logical modeling process was conceived intuitively.

Now that we are at the doorstep of implementing a database system using this design, a valid question to consider concerns the "goodness" of the design. What do we know about the quality of the data model we have in our hands? How do we vouch for the goodness of the initial conceptual model and the quality of the process of transforming the conceptual data model to its logical counterpart? The data models on hand at this point are probably "good" for user-analyst interaction purposes. But how can we make sure that the database design, if implemented, will work without causing any problems? No matter what approach is taken,[1] grouping of attributes is an intuitive process and so requires validation for design quality. How do we go about doing this? The answer is normalization.

A major problem that often escapes attention during semantic considerations in data modeling is data redundancy.[2] Data redundancy creates the potential for inconsistencies in the stored data. Normalization is a technique that systematically eliminates data redundancies in a relational database. The principles of normalization have been developed as a part of relational database theory. While the dependency preserving logical data model developed in Part II accommodates constructs beyond what is permissible in a relational data model, the issues and answers addressed by normalization principles apply equally to all data models in the logical tier. Since contemporary database systems are dominated by relational data models, we confine our attention to the relational data model and relational database systems. Figure III.1 points out our current location in the data modeling journey.

Chapter 7 looks at data redundancy in a relation schema and why it is a problem. The problem is then traced to its source, that is, undesirable functional dependencies. Functional dependencies are examined through inference rules called Armstrong's axioms. Next, we study techniques to derive the candidate keys of a universal relation schema for a given set of functional dependencies. Chapter 8 is dedicated to developing a solution to data redundancy problems triggered by undesirable functional dependencies; in other words, normalization. After discussing normal forms associated with functional dependencies in isolation, we examine the side effects of normalization—namely, the lossless-join property and dependency preservation property. Chapter 8 presents a comprehensive approach to resolving various normal form violations triggered by a set of functional dependencies in a universal relation schema. This is followed by a brief discussion of how to "reverse engineer" a normalized relational schema to the conceptual tier, which often forges a better understanding of the database design. Chapter 9 completes the discussion of normalization by examining the impact of multi-valued dependency and join-dependency on a relation schema.

---

[1] This book uses a top-down approach to database design (also known as design by analysis) as shown in Figure III.1 and the other Part-introductory figures. A bottom-up approach to database design, based on the early binary modeling work by Abrial (1974), is also possible. Somewhat less popular than the top-down approach, this design by synthesis approach is the basis for the NIAM model.

[2] Redundancy means "superfluous repetition" that does not add any new meaning.

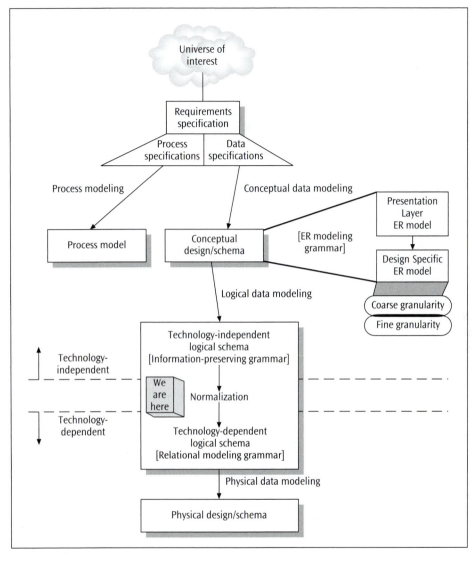

Figure III.1

## CHAPTER 7

Functional Dependencies

In developing ER models, entity types and relationships among them are intuitively distilled from the requirement specifications; then attributes are assigned to each entity type and sometimes to relationship types. Alternatively, all discernible data elements in the requirement specifications are treated as attributes and these attributes are grouped based on apparent commonalities. The clusters of attributes are then labeled as entity types and related to each other as semantically obvious. Unfortunately, there is no objective means to validate the attribute allocation process during conceptual modeling.[1]

Normalization, the topic of the next chapter, is a technique that facilitates systematic validation of participation of attributes in a relation schema from a perspective of data redundancy. The building block that enables a scientific analysis of data redundancy and the elimination of anomalies caused by data redundancy through the process of normalization is called functional dependency. This chapter introduces the concept of functional dependency and its role in the normalization process.

This chapter begins with a simple example in Section 7.1 that highlights the issues pertaining to "goodness" of design of a conceptual/logical data model. Section 7.2 introduces functional dependency and how this concept can be used to scientifically evaluate the "goodness" of a conceptual/logical design from the perspective of data redundancy. This section includes a definition of functional dependency, a discussion of inference rules that govern functional dependencies called Armstrong's axioms, and the idea of a minimal cover for a set of functional dependencies. Application of Armstrong's axioms to systematically derive the candidate keys of a relation schema, given a set of functional dependencies that hold on the relation schema, is presented in Section 7.3.

7.1 A Motivating Exemplar

Figure 7.1a is an excerpt from a larger ER diagram. This example focuses on the entity type STOCK that appears in Figure 7.1a. The relation schema for STOCK is shown in Figure 7.1b, and an instance of STOCK (i.e., a representative state of the Relation[2] hereafter referred to as a **relation instance**) appears in the data set in Figure 7.1c. Since we have chosen the relational database architecture for the implementation of the logical design, this data set is indeed a table.

---

[1] In fact, some people question the efficacy of the conceptual modeling step in a database design. Hypothetically speaking, it is possible to develop a relational data model directly from user requirement specifications by transforming all business rules to domain constraints and functional dependencies. However, we subscribe to the school that advocates conceptual modeling as a necessary and useful step in the database development process.

[2] A representative state means that all characteristics of the real, complete relation can be inferred from the instance shown. That is, the tuples in the *relation instance* have been hand-picked to fully represent all the characteristics of the source relation. For instance, one can infer that each Product has exactly one Price from Figure 7.1c. It is incorrect to argue about the possibility of the **Price** of a **Product** varying from store to store on common sense grounds. After all, common sense varies from person to person! In other words, any inference about the properties of this relation must be made from the instance of the relation presented—that is why the instance is made available.

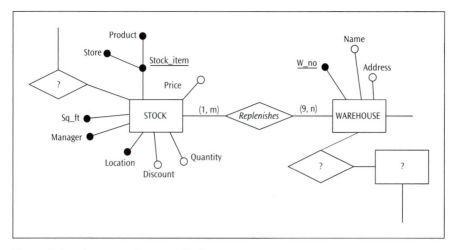

Figure 7.1a An excerpt from an ER diagram

STOCK (Store, Product, Price, Quantity, Location, Discount, Sq_ft, Manager)

Figure 7.1b Relation schema for the entity type, STOCK

STOCK

Store	Product	Price	Quantity	Location	Discount	Sq_ft	Manager
15	Refrigerator	1850	120	Houston	5%	2300	Metzger
15	Dishwasher	600	150	Houston	5%	2300	Metzger
13	Dishwasher	600	180	Tulsa	10%	1700	Metzger
14	Refrigerator	1850	150	Tulsa	5%	1900	Schott
14	Television	1400	280	Tulsa	10%	1900	Schott
14	Humidifier	55	30	Tulsa		1900	Schott
17	Television	1400	10	Memphis		2300	Creech
17	Vacuum Cleaner	300	150	Memphis	5%	2300	Creech
17	Dishwasher	600	150	Memphis	5%	2300	Creech
11	Computer		180	Houston	10%	2300	Creech
11	Refrigerator	1850	120	Houston	5%	2300	Creech
11	Lawn Mower	300		Houston		2300	Creech

Figure 7.1c An instance of the relation schema, STOCK

A quick review of the contents of Figure 7.1c confirms **{Store, Product}** to be a candidate key of STOCK as denoted in the ER diagram. As a consequence, it can be the primary key of the relation schema STOCK as indicated by the underlining here:

STOCK (Store, Product, Price, Quantity, Location, Discount, Sq_ft, Manager)

A cursory look at the table in Figure 7.1c indicates all sorts of redundancy in its content—literally every attribute value appears to be duplicated. A closer inspection reveals that there is some data redundancy in the table, but not all data that appear on the surface to be redundant are actually redundant. For instance, there are lots of duplicate values of **Quantity**. Does this mean there is data redundancy in the attribute **Quantity**? No, because there is no "superfluous repetition" of data values of **Quantity** in STOCK. It is true that a given **Product** has the same **Quantity** in more than one row of the table. This would be redundant only if this is the case irrespective of the store in which it is stocked. Since that is not the case, presence of duplicate values of **Quantity** in STOCK does not signify redundancy. On the other hand, the **Price** of a **Product** in STOCK is the same irrespective of any other fact in the table (e.g., any store). Therefore, duplication of the **Price** of a **Product** in multiple rows in the table amounts to redundant data. Based on similar reasoning, notice that there is redundancy in the data for **Location** as well as **Discount** in STOCK. It is a good exercise for the reader to reason this out.

The next issue to investigate is the "so what?" question—that is, why does the data redundancy matter? While the wasted storage space need not be a serious issue, there are more significant problems. Suppose we want to add Washing Machine to the stock with a **Price**. We cannot do this without knowing a **Store** where washing machines are stocked. This is because **{Store, Product}** is the primary key of this table STOCK, and the entity integrity constraint stipulates that neither **Store** nor **Product** can have "null" values in this table. This is what is called an **insertion anomaly**.[3] This is a serious problem because it may be an unreasonable imposition on the user community. Now say that store 17 is closed. In order to remove store 17, not only do we need to remove several rows from the STOCK table, but we inadvertently lose the information that the vacuum cleaner is priced at $300 since no other store presently stocks vacuum cleaners. This is a **deletion anomaly**. If, for instance, we want to change the **Location** of store 11 from Houston to Cincinnati, we need to update all rows in the STOCK table that are store 11. Failure to do so will result in store 11 being located in both Houston and Cincinnati. This is referred to as an **update anomaly**. In this and other chapters of the book, we use the umbrella term **modification anomalies** to refer to insertion, deletion, and update anomalies collectively. One way of addressing modification anomalies is to decompose the STOCK table into other relations, as shown in Figure 7.2. The three relations (tables) STORE, PRODUCT, and INVENTORY are decompositions that collectively replace the table STOCK shown in Figure 7.1.

Now, if we want to add Washing Machine and its price, we can add the information to the PRODUCT table; whenever a store begins to stock washing machines we can add the necessary data to the INVENTORY table, which eliminates the insertion anomaly in STOCK. If store 17 is closed, a *single* row in the STORE table is deleted and there is no other loss of information (we still know that a vacuum cleaner costs $300)—an example of removing the deletion anomaly. Changing the location for store 11 requires the update of a single row in the STORE table, as opposed to modifying several rows in the original STOCK table, removing the update anomaly. It is true that the decomposed design in Figure 7.2 may be less efficient for data retrieval; but then, that is the price for eliminating modification anomalies caused by redundant data.

---

[3] An anomaly, according to the *Random House Dictionary*, means a deviation from the rule, type, or form; an irregularity or abnormality.

STORE

Store	Location	Sq_ft	Manager
15	Houston	2300	Metzger
13	Tulsa	1700	Metzger
14	Tulsa	1900	Schott
17	Memphis	2300	Creech
11	Houston	2300	Creech

PRODUCT

Product	Price
Refrigerator	1850
Dishwasher	600
Television	1400
Humidifier	55
Vacuum Cleaner	300
Computer	
Lawn Mower	300
Washing Machine	750

INVENTORY

Store	Product	Quantity	Discount
15	Refrigerator	120	5%
15	Dishwasher	150	5%
13	Dishwasher	180	10%
14	Refrigerator	150	5%
14	Television	280	10%
14	Humidifier	30	
17	Television	10	
17	Vacuum Cleaner	150	5%
17	Dishwasher	150	5%
11	Computer	180	10%
11	Refrigerator	120	5%
11	Lawn Mower		

Figure 7.2 A decomposition of the STOCK instance in Figure 7.1c

STORE and PRODUCT now do not have any data redundancy. How about INVENTORY? Are **Store** and **Product** in this relation redundant because these attributes are already present in the other two relations? The answer is "No," simply because the repetition of the attributes in INVENTORY is *not* superfluous. These attributes in INVENTORY convey more semantics than what is present in STORE and PRODUCT. However, it is obvious from the table instance INVENTORY as well as from the original table instance of STOCK that **Discount** values are redundantly stored. After all, for a given **Quantity** there is only a single, specific **Discount** value. The solution is a simple further decomposition shown in Figure 7.3.

However, there are times when some data redundancy is willfully tolerated as a tradeoff for efficiency of querying (data retrieval). Suppose the discount structure, according to the user, is relatively stable (i.e., minimal changes). Then, the design in Figure 7.2 may be a more optimal design than that in Figure 7.3 despite the data redundancy in INVENTORY. Such a redundancy is sometimes referred to as **controlled redundancy**.

The table STOCK in Figure 7.1c is said to be "unnormalized," that is, it has many data redundancies; the set of tables in Figure 7.3 is said to be fully "normalized," having no data redundancies. The questions at this point ought to be:

- How do we systematically identify data redundancies?

STORE

Store	Location	Sq_ft	Manager
15	Houston	2300	Metzger
13	Tulsa	1700	Metzger
14	Tulsa	1900	Schott
17	Memphis	2300	Creech
11	Houston	2300	Creech

PRODUCT

Product	Price
Refrigerator	1850
Dishwasher	600
Television	1400
Humidifier	55
Vacuum Cleaner	300
Computer	
Lawn Mower	300
Washing Machine	750

INVENTORY

Store	Product	Quantity
15	Refrigerator	120
15	Dishwasher	150
13	Dishwasher	180
14	Refrigerator	150
14	Television	280
14	Humidifier	30
17	Television	10
17	Vacuum Cleaner	150
17	Dishwasher	150
11	Computer	180
11	Refrigerator	120
11	Lawn Mower	

DISC_STRUCTURE

Quantity	Discount
120	5%
150	5%
180	10%
280	10%
30	
10	

Figure 7.3   A redundancy-free decomposition of the STOCK instance in Figure 7.1c

- How do we know how to decompose the base relation schema under investigation?
- How do we know that the decomposition is correct and complete without looking at sample data?

The remainder of this chapter is dedicated to answering these questions, and the process that emerges as a result is called *normalization*. But before embarking on this journey, let us explore the "normalized" design represented by the set of tables in Figure 7.3.

What is the relational schema that will yield this set of tables? Obviously it will contain four relation schemas, one each for STORE, PRODUCT, INVENTORY, and DISC_STRUCTURE. Figure 7.4a displays this relational schema.

Note that the structural constraints of relationships emerge from the data in the tables. Since deletion rules are not discernible from the tables, the default value of restrict (R) is adopted. At this point, we are able to further "reverse engineer" the relational schema to a Design-Specific ER diagram, as shown in Figure 7.4b.

Observe that INVENTORY is a gerund entity type with two identifying parents, STORE and PRODUCT. Finally, we can also abstract the Design-Specific ER diagram to the Presentation layer by unraveling the m:n relationship type implicit in the gerund entity type. Since ER modeling grammar does not allow a relationship type to be related to another

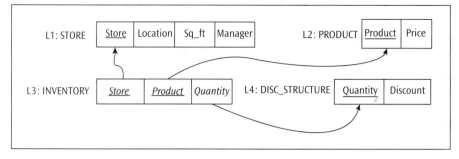

Figure 7.4a A reverse-engineered logical schema for the set of tables in Figure 7.3

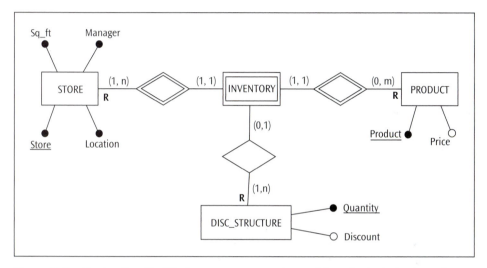

Figure 7.4b Design-Specific ER diagram reverse-engineered from the logical schema in Figure 7.4a

relationship type, INVENTORY necessarily becomes a cluster entity type. The Presentation Layer ER diagram that will yield the Design-Specific ER diagram in Figure 7.4b is presented in Figure 7.4c, and we have just completed what is known in data modeling circles as "reverse engineering." By comparing the ER diagrams in Figures 7.4c and 7.1, we ought to be able to appreciate the design problems due to data redundancy hidden in the original entity type STOCK.

Similar problems may be present in WAREHOUSE (see Figure 7.1a) and every other relation schema in a relational data model. At this point in the database design life cycle, each relation schema in the relational data model must be independently scrutinized and "normalized" where needed. While conceptual modeling is not a scientific process, organized application of intuition in the development of the conceptual data model and careful mapping of the conceptual schema to the logical tier often yields a logical (relational) schema where most of the constituent relation schemas are fully normalized, that is, free from modification anomalies. Nonetheless, as shown in the example, a systematic verification of the "goodness" of the design at this stage of data modeling is imperative lest the implemented database

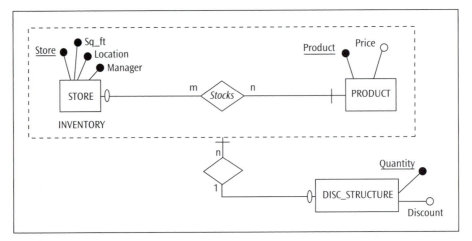

Figure 7.4c Presentation Layer ER diagram reverse-engineered from Figure 7.4b

system should fail to meet user expectations. To that end, let us proceed to investigate the "How to" questions:

- How do we systematically identify data redundancies?
- How do we know how to decompose the base relation schema under investigation?
- How do we know that the decomposition is correct and complete without looking at sample data?

In order to engage in this enquiry, it is necessary to understand the concept of functional dependency.

7.2 Functional Dependencies

A crucially important aspect of relational database theory is the concept of **functional dependency (FD)**. Functional dependency is the building block of normalization principles. *Undesirable* functional dependencies in a relation schema are the seeds of data redundancies that lead to modification anomalies. Functional dependencies, desirable or otherwise, are essentially technical translations of user-specified business rules expressed as constraints in a relation schema and so cannot be ignored or discarded when undesirable.

7.2.1 Definition of Functional Dependency

In simple terms, an attribute A (atomic or composite) in a relation schema R functionally determines another attribute B (atomic or composite) in R if for a given value a_1 of A there is a *single, specific* value b_1 of B in relation r of R.[4] Note that r represents all possible states (instances) of R. A symbolic expression of this FD is:

$$A \rightarrow B$$

---

[4] As a matter of technicality, functional dependencies exist only when attributes or other elements involved have unique and singular identifiers (Kent, 1983, p. 121). Since a relation schema by definition has a candidate key, the functional dependency concept is applicable to a relation schema.

where A (the left-hand side of the FD) is known as the **determinant** and B (the right-hand side of the FD) is known as the **dependent**.

In other words, if A functionally determines B (i.e., B is functionally dependent on A) in R, then it is invalid to have two (or more) tuples that have the same A value, but different B values in R. That is, in order for the functional dependency A → B to hold in R, if two tuples of r(R) agree on their A values, they must necessarily agree on their B values. A corollary of this definition is:

- If A is a candidate key of R (i.e., no two tuples in r(R) can have the same values for A), then A → B for any subset of attributes, B of R.

Equally important, A → B in R *does not* ever imply B → A in R.

For example, from the instance of the relation schema STOCK (see Figure 7.1c), we can infer that **Product** → **Price** because all tuples of STOCK with a given **Product** value also have the same **Price** value. Likewise, it can also be inferred that **Store** → **Location**.

Notice that **Location** does not determine **Store** in STOCK, because for a given value of **Location** in STOCK, there is more than one value of **Store**. This can be shown as:

Location ↛ Store

When the determinant and/or dependent in an FD is a composite attribute, the constituent atomic attributes are enclosed by braces as shown in the following example:

{Store, Product} → **Quantity**

Once again, an FD is a property of the semantics (meaning) of the relationship among attributes emerging from the business rules. The sample data shown in Figure 7.1c is intended to clarify these semantics. Any extrapolation of meaning beyond the facts conveyed in the relation instance is incorrect. Note that an FD is a property of the relation schema R, not of particular relation state r of R. Therefore, an FD cannot be automatically inferred from *any* relation state r of R. It must be explicitly specified as a constraint, and the source for this specification is the business rules of the application domain. However, a *relation instance* is, by our definition, a specially prepared relation state r of R conforming to the set of FDs specified on R and can be used to infer the FDs in R.

7.2.2 Inference Rules for Functional Dependencies

Given a set of *semantically obvious* FDs for a relation schema R[5] emerging from the business rules of an application domain, it is possible to deduce all the other FDs that hold in R, but are not explicitly stated. The set of FDs that are explicitly specified on a relation schema is usually referred to as **F**. As an example, in the relation schema STOCK, if

fd1:   **{Store, Product}** → **Quantity** and
fd2:   **Quantity** → **Discount**

are specified in F, one can infer an FD

fd3:   **{Store, Product}** → **Discount**

---

[5] Here it is useful to assume the entire database for the application domain is a single universal relation schema.

Therefore, it is not necessary to explicitly state fd3. The notation

F ⊨ fd3 or F ⊨ {Store, Product} → Discount

is used to indicate that fd3 is inferred from the set of functional dependencies, F.

If a set of functional dependencies F prevails over the relation schema R, F ⊨ X → Y also holds in R. That is, whenever a relation state r of R satisfies all FDs in F, then the FD X → Y, if inferred from F, is also satisfied in that r. It is useful to define the idea of all possible FDs that hold in R. Referred to as *closure*, this includes all possible FDs that can be inferred from the set F. Denoted as **F⁺**, the closure of F includes the set of FDs stated in F plus all other FDs that can be inferred from F. While in practice there is little need to compute **F⁺**, it is often necessary to know if a particular FD exists in **F⁺**.

Armstrong (1974) proposed a systematic approach to derive FDs that can be inferred from F. Usually referred to as **Armstrong's axioms**, this set of inference rules can also be used to derive precisely the closure F⁺.

Given R (X, Y, Z, W) where X, Y, Z, and W are arbitrary subsets (atomic or composite) of the set of attributes of a universal relation schema R, three fundamental inference rules that can be directly proven from the definition of functional dependency are:

- **Reflexivity rule**: If Y is a subset of X (e.g., if X is {A,B,C,D} and Y is {A,C}), then X → Y.

Note: The reflexivity rule defines what is called a **trivial dependency**. A dependency is trivial if it is impossible to *not* satisfy it.[6]

- **Augmentation rule**: If X → Y, then {X,Z} → {Y,Z}; also, {X,Z} → Y.

- **Transitivity rule**: If X → Y, and Y → Z, then X → Z.

Using the illustration in Figure 7.1,
{Store, Product} → Store exemplifies the *reflexivity* rule.
The *augmentation* rule works as follows:
If **Store → Location**, then **{Store, Product} → {Location, Product}**
and
{Store, Product} → Location
The *transitivity* rule is demonstrated as follows:
If **{Store, Product} → Quantity** and **Quantity → Discount**, then transitively
{Store, Product} → Discount.

Four more inference rules can be constructed from the above three. Some of these additional rules help simplify the practical task of generating F⁺. They are:

- **Decomposition rule**: If X → {Y,Z}, then X → Y and X → Z.
- **Union (or additive) rule**: If X → Y, and X → Z, then X → {Y,Z}
- **Composition rule**: If X → Y, and Z → W, then {X,Z} → {Y,W}
- **Pseudotransitivity rule**: If X → Y, and {Y,W} → Z, then {X,W} → Z.

---

[6] An FD in R is trivial if and only if the dependent is a subset of the determinant. Since trivial dependencies do not provide any additional information (i.e., do not add any new constraints on R), they are usually removed from F⁺.

Again, using the relation instance of STOCK in Figure 7.1,

If **Store → {Location, Sq_ft, Manager}**, then per the *decomposition* rule
Store → Location and **Store → Sq_ft** and **Store → Manager**.

Given
Store → Location and **Store → Sq_ft**,
Store → {Location, Sq_ft} exemplifies the *union* rule.

This example demonstrates the *composition* rule:
If **Store → Location** and **Product → Price**, then **{Store, Product} → {Location, Price}**

The *pseudotransitivity* rule is a handy corollary of the transitivity rule and works the following way:
If **Manager → Store** and **{Store, Product} → Quantity**, then **{Manager, Product} → Quantity**.
The inference rules for functional dependencies known as Armstrong's axioms are summarized in Table 7.1. Several of the inference rules discussed above can be derived from Darwen's *General Unification Theorem* (1992):
If $X \rightarrow Y$, and $Z \rightarrow W$, then $X \cup \{Z-Y\} \rightarrow \{Y,W\}$
In principle, the closure F^+ of a given set of FDs, F, can be computed by the use of a rather inefficient algorithm: *Repeatedly apply Armstrong's axioms on F until it stops producing new FDs* (Date, 2004).

Table 7.1 Inference rules for functional dependencies: Armstrong's axioms

Rule	Definition
Reflexivity	If Y is a subset of X [i.e., if X is (A,B,C,D) and Y is (A,C)], then $X \rightarrow Y$. (The reflexivity rule defines trivial dependency as a dependency that is impossible to *not* satisfy.)
Augmentation	If $X \rightarrow Y$, then $\{X,Z\} \rightarrow \{Y,Z\}$; also, $\{X,Z\} \rightarrow Y$.
Transitivity	If $X \rightarrow Y$, and $Y \rightarrow Z$, then $X \rightarrow Z$.
Decomposition	If $X \rightarrow \{Y,Z\}$, then $X \rightarrow Y$ and $X \rightarrow Z$.
Union (or additive)	If $X \rightarrow Y$, and $X \rightarrow Z$, then $X \rightarrow \{Y,Z\}$.
Composition	If $X \rightarrow Y$, and $Z \rightarrow W$, then $\{X,Z\} \rightarrow \{Y,W\}$.
Pseudotransitivity	If $X \rightarrow Y$, and $\{Y,W\} \rightarrow Z$, then $\{X,W\} \rightarrow Z$.

References: Armstrong, W. W. "Dependence Structures of Data Base Relationships" *Proc. IFIP Congress*, Stockholm, Sweden (1974); Darwen, H. "The Role of Functional Dependencies in Query Decomposition," In C.J. Date and H. Darwen, *Relational Database Writings 1989 – 1991*, Addison-Wesley (1992).

7.2.3 Minimal Cover for a Set of Functional Dependencies

If every FD in a set of FDs, F, can be inferred from another set of FDs G, then G is said to *cover* F. In this case, every FD in F is also present in G^+—that is, $F \subset G^+$. Two sets of FDs, F and G, are considered *equivalent* if $F^+ = G^+$. This equivalence of the two sets of FDs is expressed as $F \equiv G$, implying that $F \subset G^+$ and $G \subset F^+$; in other words, G covers F and F covers G. From a practical perspective, when $F \equiv G$ one can choose to enforce either F or G and the valid database states will remain the same.

It is likely that a set of FDs, F, translated from user-specified business rules has redundant (extraneous) attributes and sometimes redundant (extraneous) FDs[7] because as narratives requirement specifications are prone to be somewhat repetitive. For example, consider the attribute set **{Store, Product, Price}** and an associated set of FDs, F culled from the requirement specification where fd1: **{Store, Product}** → **Price** and fd2: **Product** → **Price**. Here, fd1 is redundant because (F – fd1) is equivalent to F, meaning removal of fd1 will not change F^+. Alternatively, **Store** in fd1 is a redundant attribute, and removal of **Store** from fd1 will not change F^+. Likewise, given an attribute set **{Store, Product, Price, Quantity}** and an associated set of FDs, G, where fd1: **{Store, Product}** → **{Quantity, Price}** and fd2: **Product** → **Price**, the attribute **Price** is redundant in fd1—that is, removal of **Price** from fd1 will not change G^+.

Suppose that a set of FDs, F, prevails over a relational schema. This means that whenever a user performs an update that entails changes to one or more relations in the database, the DBMS must ensure that the update does not violate any of the FDs in F. All FDs in F must hold in the updated database state; otherwise, the DBMS must roll back the updates and restore the database to the original state that prevailed before the update. It is always useful to identify a simplified set of FDs, G_c, equivalent to F, that is, having the same closure (F^+) as F and not further reducible. This G_c is not only equivalent to F, but further reduction of G_c destroys the equivalence. The practical value of such a simplified set of FDs, G_c is that the effort required to check for violations in the database is minimized because a database that satisfies G_c will also satisfy F and vice versa. G_c in this case is called the **canonical cover** or **minimal cover** of F. Formally, a set of FDs, G_c is a minimal cover for another set of FDs F if the following conditions are satisfied:

- $G_c \equiv F$ (G_c and F are equivalent.)
- The *dependent* (right-hand side) in every FD in G_c is a singleton attribute. This is known as the *standard or canonical form* of an FD and is intended to simplify the conditions and algorithms that ensure absence of redundancies in F.
- No FD in G_c is redundant—i.e., no FD from G_c can be discarded without converting G_c into some set not equivalent to G_c and therefore not equivalent to F.
- The *determinant* (left-hand side) of every FD in G_c is irreducible—i.e., no attribute can be discarded from the determinant of any FD in G without converting G_c into some set not equivalent to G_c.

The process of deducing minimal covers is illustrated in the following examples.

---

[7] An FD, f, in F is redundant (extraneous) if removing f from F does not change F^+ - i.e., (F – f) is equivalent to F. An attribute, A, in f is considered redundant (extraneous) if it can be removed without changing the closure F^+ of F.

7.2.3.1 Example 1

Here is a simple example. Consider the set of attributes **{Tenant, Apartment, Rent}** and the set of FDs, F:

fd1: **Tenant → {Apartment, Rent}**;	fd2: **Apartment → Rent**;
fd3: **{Tenant, Apartment} → Rent**;	fd4: **{Tenant, Rent} → Apartment**

Using Armstrong's axioms, we can deduce from F that **Tenant → Apartment** is in F^+. Note that the set of FDs, G:

fd1: **Tenant → {Apartment, Rent}**;	fd2: **Apartment → Rent**;
fd3: **{Tenant, Apartment} → Rent**;	fd4: **Tenant → Apartment**

is a cover for F. However, is G a minimal cover for F? No. It will be a minimal cover only if there are no redundant attributes and redundant FDs in G. An examination of G reveals that, given fd2, the attribute **Tenant** is redundant in fd3. Removal of the redundant attribute from fd3 renders fd2 and fd3 identical; so one of these two FDs (say, fd3) is redundant and can be deleted. Next, given fd1, fd4 is redundant and can be removed without any consequence. Thus, we are left with G': {fd1, fd2} where

fd1: **Tenant → {Apartment, Rent}**; and fd2: **Apartment → Rent**

Is G' a minimal cover for F? The answer is still no, because **Tenant → Rent** in fd1 is still redundant. Removing this redundancy from fd1, we have G_c: {fd1, fd2} where

fd1: **Tenant → Apartment** and fd2: **Apartment → Rent**

which now is a minimal cover for F, G, and G'.

Is G_x: {fd1, fd2} where

fd1: **Tenant → Apartment** and fd2: **Tenant → Rent**

a minimal cover for F?

The answer is no because G_x is not equivalent to F in that G_x^+ is *not* = F^+. For instance, the FD **Apartment → Rent** present in F^+ is not present in G_x^+.

In short, a minimal cover G_c of a set of FDs F is not only equivalent to F (that is, $F \equiv G_c$, meaning $F^+ = G_c^+$), but also contains neither redundant FDs nor redundant attributes. Every set of FDs, F possesses a minimal cover. F can be its own minimal cover too. In fact, careful construction of F from the business rules often yields F itself as a minimal cover of the set of FDs in it. Furthermore, there can be several minimal covers for F. When several sets of FDs qualify as minimal covers of F, additional criteria are used to choose a minimal cover, such as the minimal cover with the least number of FDs or the minimal cover that most closely resembles F.

7.2.3.2 Example 2

Consider a set of attributes {A, B, C} and an associated set of FDs F: {fd1, fd2, fd3, fd4} where:

fd1: **A → C**;	fd2: **(A, C) → B**;	fd3: **B → A**;	fd4: **C → (A, B)**

fd4 can be rewritten in standard form as:

fd4a: $C \rightarrow A$ and fd4b: $C \rightarrow B$

Based on fd4b, using Armstrong's axioms, we infer that A in fd2 is a redundant attribute. Removal of A from fd2 renders fd2 and fd4b identical. Therefore, one of these two FDs can be dropped with no consequence. fd4a is also a redundant FD since it is derived by the rule of transitivity applied on fd4b and fd3. The resulting set of FDs, G_c: {fd1, fd2, fd3} where

fd1: $A \rightarrow C$;	fd2: $C \rightarrow B$;	fd3: $B \rightarrow A$

is a minimal cover of F.

Targeting attributes and FDs for evaluation in a different sequence, it can be shown that G_m: {fd1, fd2, fd3} where

fd1: $A \rightarrow B$;	fd2: $B \rightarrow C$;	fd3: $C \rightarrow A$

is also a minimal cover of F. The reader is encouraged to verify this.

An algorithm to compute the minimal cover for a given set of functional dependencies F is as follows:

    i.   Set G to F.
    ii.   Convert all FDs in G to *standard (canonical) form*—i.e., the right-hand side (dependent attribute) of every FD in G should be a singleton attribute.
    iii.   Remove all redundant attributes from the left-hand side (determinant) of the FDs in G.
    iv.   Remove all redundant FDs from G.

Two conditions in the algorithm are noteworthy:

- The execution of this algorithm may yield different results depending on the order in which the candidates for removal (both attributes and FDs) are evaluated, thus confirming the fact that it is possible to have multiple minimal covers for F.
- Steps iii and iv are not interchangeable—that is, executing step iv before step iii will not always return a minimal cover.

The next example applies the above algorithm to derive the minimal cover of a set of FDs.

7.2.3.3 Example 3

Consider the set of attributes {Student, Advisor, Subject, Grade} and an associated set of FDs, F:

    fd1:   {Student, Advisor} → {Grade, Subject};
    fd2:   Advisor → Subject;
    fd3:   {Student, Subject} → {Grade, Advisor}

G, the expression of F in standard form, can be written as:

fd1a: {Student, Advisor} → Grade;	fd1b: {Student, Advisor} → Subject;
fd2: Advisor → Subject;	
fd3a: {Student, Subject} → Grade;	fd3b: {Student, Subject} → Advisor

Given fd2, **Student** in fd1b is a redundant attribute. Elimination of this redundant attribute from fd1b renders fd1b entailed by fd2. Thus, fd1b becomes a redundant FD and can be removed. There are no other redundant attributes in the left-hand side (determinant) of any other FD in G. Next, fd1a is a redundant FD since it is entailed by the set of FDs {fd2, fd3a}. There are no other redundant FDs in G. Thus, we have G_c: {fd2, fd3a, fd3b} where:

fd2: Advisor → Subject;	
fd3a: {Student, Subject} → Grade;	fd3b: {Student, Subject} → Advisor;

G_c is a minimal cover of F and G.

Another example follows.

7.2.3.4 Example 4

Consider the attribute set

{Product, Store, Vendor, Date, Quantity, Unit_price, Discount, Size, Color}

and the set of FDs, F:

fd1:   **Product → {Size, Color}**
fd2:   **{Vendor, Quantity} → {Unit_price, Discount};**
fd3:   **{Product, Store, Date, Quantity} → {Vendor, Discount};**
fd4:   **{Product, Size, Color, Store, Date} → Vendor**

What is the minimal cover of F?

The first step is to express F in standard form, say, G {fd1a, fd1b, fd2a, fd2b, fd3a, fd3b, fd4} where:

fd1a: Product → Size;	fd1b: Product → Color;
fd2a: {Vendor, Quantity} → Unit_price;	fd2b: {Vendor, Quantity} → Discount;
fd3a: {Product, Store, Date, Quantity} → Vendor;	fd3b: {Product, Store, Date, Quantity} → Discount;
fd4: {Product, Size, Color, Store, Date} → Vendor	

Now that G is in the standard form, the next step in the algorithm to deduce the minimal cover is to identify and remove redundant attributes from the left-hand side (determinant) of the FDs constituting G. Accordingly, based on fd1a and fd1b, **Size** and **Color** in fd4 are redundant. Thus, fd4 reduces to fd4a: **{Product, Store, Date} → Vendor**. Based on fd4a, **Quantity** in fd3a becomes a redundant attribute and elimination of this attribute from fd3a renders fd4a and fd3a identical. Thus one of them (fd4a) entails the other (fd3a) which then becomes a redundant FD and so can be removed with no consequence. Next, fd3b, implied by {fd2b, fd4a}, becomes a redundant FD and so can be deleted. Thus we have, G_c {fd1a, fd1b, fd2a, fd2b, fd4a} where:

fd1a: **Product** → **Size**;	fd1b: **Product** → **Color**;
fd2a: {**Vendor, Quantity**} → **Unit_price**;	fd2b: {**Vendor, Quantity**} → **Discount**;
fd4a: {**Product, Store, Date**} → **Vendor**	

A close examination of F and G_c reveals that G_c is a cover for F since $F \equiv G_c$ meaning $F^+ = G_c{}^+$. Since G_c does not contain any redundant attributes or redundant FDs, G_c is then, by definition, a *minimal* cover for F. How do we know that G_c does not contain any redundant attributes or redundant FDs? This is tested by finding an attribute or an FD in G_c, the removal of which from G_c does not disturb the equivalence of G_c to F. The reader is encouraged to try this as an exercise.

Since {**Product, Store, Date**} → **Vendor** (fd4a), could we have retained fd3b in G instead of fd2b? In other words, is G_x {fd1a, fd1b, fd2a, fd3b, fd4a}

fd1a: **Product** → **Size**;	fd1b: **Product** → **Color**;
fd2a: {**Vendor, Quantity**} → **Unit_price**;	
fd3b: {**Product, Store, Date, Quantity**} → **Discount**;	
fd4: {**Product, Store, Date**} → **Vendor**	

a minimal cover for F? The answer is no. In fact, G_x is not even a cover for F, let alone a minimal cover, because F is not $\equiv G_x$ meaning F^+ is not $= G_x{}^+$. Note that {**Vendor, Quantity**} → **Discount** does not exist in $G_x{}^+$.

Once again, the practical value of a minimal cover, G_c, of a set of FDs, F is that the effort required to check for violations in the database is minimized because a database that satisfies G_c will also satisfy F and vice versa.

7.2.4 Closure of a Set of Attributes

A concept akin to the closure of a set of FDs is the idea of the closure of a set of attributes. Given a relation schema R, a set of FDs, F that hold in R, and a subset Z of attributes of R, the closure Z^+ of Z under F is the set of all attributes of R that are functionally dependent on Z. Observe that a closure of a set of attributes is always subject to a set of specified FDs and is expressed as: $Z^+ = $ Closure [Z | F]. As an illustration, suppose R (A, B, C, D, E, G, H) is a relation schema over which the following set of FDs, F prevails:

fd1: **B** → {**G, H**};	fd2: **A** → **B**;	fd3: **C** → **D**;

What is the closure [A | F]? What is being sought here is the set of attributes in R that are functionally dependent on A under the set of FDs, F. A quick perusal of F—that is, fd1, fd2, and fd3—indicates that $A^+ = $ {A, B, G, H}.

Here is an algorithm to compute the closure of an attribute set, Z, under a set of FDs, F in a relation schema R:

1. Set Closure [Z | F] to Z;
2. For each FD of the form $X \rightarrow Y$ in F,
    If X $\subseteq$ Closure [Z | F],
    Set Closure [Z | F] to (Closure [Z | F] U Y);[8]
3. Iterate step 2 through F until no further change in the Closure [Z | F].

Suppose we want to compute $\{A,C\}^+$, the Closure [{A,C} | F] in R using the above algorithm.

Start: $\{A,C\}^+ = \{A,C\}$

First iteration through F:

*   In fd1, the determinant B is not a subset of $\{A,C\}^+$—so, no change in $\{A,C\}^+$;
*   In fd2, the determinant A is a subset of $\{A,C\}^+$—so, $\{A,C\}^+ = \{A,C\}^+$ U B = $\{A,C,B\}$;
*   In fd3, the determinant C is a subset of $\{A,C\}^+$—so, $\{A,C\}^+ = \{A,C\}^+$ U D = $\{A,C,B,D\}$.

Second iteration through F:

*   In fd1, the determinant B is a subset of $\{A,C\}^+$—so, $\{A,C\}^+ = \{A,C\}^+$ U $\{G,H\}$ =
* $\{A,C,B,D,G,H\}$;
*   In fd2, the determinant A is a subset of $\{A,C\}^+$—so, $\{A,C\}^+ = \{A,C\}^+$ U B = no change;
*   In fd3, the determinant C is a subset of $\{A,C\}^+$—so, $\{A,C\}^+ = \{A,C\}^+$ U D = no change.

Third iteration through F, as can be seen, is not necessary; so the algorithm terminates.
End: $\{A,C\}^+ = \{A,C,B,D,G,H\}$.

Two useful corollaries are worthy of attention:

*   Given F, it is possible to know if a specific FD $X \rightarrow Y$ follows from F by computing the attribute closure X^+. If and only if Y is a subset of X^+, can we infer that $X \rightarrow Y$ follows from F. Note that we are able to determine whether the FD $X \rightarrow Y$ follows from F without actually having to compute F^+.
*   Given F, it is possible to know if a certain subset K of the attributes of R is a superkey of R by computing K^+, the closure [K | F]. K is a superkey of R if and only if the closure [K | F] is precisely the set of all attributes of R. If K happens to be an irreducible superkey of R under F, then K is a candidate key of R.

Note that {A,C} is not a superkey of R in the above example because $\{A,C\}^+$ is not precisely the set of all attributes of R—that is, $\{A,C\}^+ = \{A,C,B,D,G,H\}$ does not contain the attribute E of R.

7.2.5 Whence Do FDs Arise?

Where do FDs come from? Why were they not considered during conceptual modeling? The answer is that FDs are technical expressions of user-specified business rules. They were considered implicitly during ER modeling. That is how attributes were collected as

---

[8] This is based on the application of the transitivity rule of Armstrong's axioms.

different entity types—that is, grouping a set of attributes that are independent of each other, but functionally dependent on a specific attribute(s) (e.g., the candidate key). While an ER model is certainly useful in defining the scope of the database design, the ER modeling process itself is still more of an intuitive process. Now that the ER model has been mapped to the logical tier, it is necessary to *formally* examine each relation schema and make sure that the initial assignment of attributes to the entity types at the conceptual tier is correct. The ratification at this point is no longer an intuitive process. It is a scientific analysis and is important because this is the final opportunity to correct errors and establish a robust design before implementing the database system.

The logical analysis of attribute assignment to a relation schema may require further consultations with the user community to sharpen the business rules and lead to an explicit specification of inter-attribute constraints (i.e., functional dependencies) that were intuitively considered during the development of the original entity types and their respective attribute sets. Note that the purpose here is the verification of the "goodness" of design. Although conceptual modeling may not be a scientific process, "organized" application of intuition in the development of the conceptual data model and careful mapping of the conceptual schema to the logical tier often yields a logical (relational) schema where most of the constituent relation schemas are fully normalized—i.e., free from modification anomalies arising from undesirable functional dependencies. Nonetheless, normalization is necessary to confirm that there are no modification anomalies, and to fine-tune the database design before embarking on implementation.

7.3 Candidate Keys Revisited

The term *candidate key* is introduced in Chapter 6 along with the terms *superkey* and *primary key* as a part of the foundation concepts of a relational data model and is defined as an irreducible unique identifier of a relation schema. Now it is time to develop a more formal understanding of the idea of a candidate key and explore methods to derive the candidate keys of a relation schema. Remember that the identification of unique identifiers of an entity type during ER modeling is an intuitive process in spite of the fact that the identification is guided by the business rules of the application domain being modeled.

Also remember that a relation schema can have multiple candidate keys and it is useful (sometimes, necessary) to know all the candidate keys of a relation schema. Two approaches are generally used to derive a candidate key of a relation schema. Given a set of FDs, F, it is possible to derive a candidate key for the universal relation schema (URS) constructed from the set of attributes present in F through a process of synthesis. This method is based on the principle of the closure of an attribute set. In short, this method seeks to derive an irreducible set of attributes whose closure is precisely all attributes of the URS. Alternatively, given a URS and the set of FDs, F that prevails over the URS, a candidate key can be derived through a process of decomposition of a superkey of the URS. After one candidate key is derived using either of these approaches, the method used to derive the rest of the candidate keys of the relation schema is the same. The synthesis approach is somewhat more heuristic in nature, while the decomposition approach is more algorithmic. Each of these approaches is illustrated in the following sections.

7.3.1 Deriving Candidate Key(s) by Synthesis

The steps involved in the heuristic procedure to progressively synthesize a candidate key from a set of FDs are listed in Table 7.2 (see steps 1 – 6). The following example demonstrates the use of this procedure.

Table 7.2 A heuristic for the derivation of candidate key(s) by synthesis, given a Universal Relation Schema (URS) and a set of functional dependencies, F, that prevails over it

Derivation of first candidate key of URS	
Step 1	Derive the minimal cover F_c for the set of functional dependencies (FDs) F that prevails over the URS.
Step 2	Select the FD with maximum number of attributes constituting the determinant as the starting point. Let us call this FD as the target FD, and the determinant of this first FD as the target determinant (TD1). *Note*: If more than one such target FD exists, select one of them for now. Compute the attribute closure $[TD1 \mid F_c]$.
Step 3	If $TD1^+$ is precisely the set of all attributes of URS, then TD1 is a candidate key of URS. If so, skip to step 7.
Step 4	If $TD1^+$ is not precisely the set of all attributes of URS, select a functional dependency whose determinant is not a subset of $TD1^+$ as the next target FD—the determinant of this FD will be TD2. Compute the attribute closure $[TD2 \mid F_c]$.
Step 5	If $TD2^+$ is precisely the set of all attributes of URS, then TD2 is a candidate key of URS. Otherwise, if $\{TD1^+ \cup TD2^+\}$ is precisely the set of all attributes of URS, then $\{TD1 \cup TD2\}$ is a candidate key of URS. If either one is true, skip to step 7.
Step 6	Otherwise, repeat steps 4 and 5 using the next target FD. Repeat steps 4 and 5 until an attribute set K is derived such that K^+ is precisely the set of all attributes of URS. Then, K is a candidate key of URS.
Derivation of other candidate key(s) of URS	
Step 7	If F_c contains an FD, fdx, where a candidate key of URS is a dependent, then the determinant of fdx is also a candidate key of URS.
Step 8	When a candidate key of URS is a composite attribute, for each key attribute (atomic or composite), evaluate if the key attribute is a dependent in an FD, fdy, in F_c. If so, then the determinant of fdy, by the rule of pseudotransitivity, can replace the key attribute under consideration, thus yielding additional candidate key(s) of URS.
Step 9	Repetition of steps 7 and 8 for every candidate key of URS will systematically reveal all the other candidate key(s), if any, of URS.

Suppose there is a set of eight functional dependencies, F, as follows:

fd1: {Store, Branch} → Location;	fd5: Product → Price;
fd2: Customer → Address;	fd6: {Store, Branch} → Manager;
fd3: Vendor → Product;	fd7: Manager → {Store, Branch};
fd4: {Store, Branch} → Sq_ft;	fd8: Store → Type;

A universal relation schema that includes all these functional dependencies is:
URS1 (Store, Branch, Location, Customer, Address, Vendor, Product, Sq_ft, Price, Manager, Type)

The set of FDs in F may also be expressed via a dependency diagram, as shown in Figure 7.5.

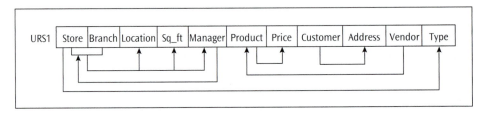

Figure 7.5 Dependency diagram for the relation schema, URS1

7.3.1.1 Deriving the First Candidate Key by Synthesis

The following steps demonstrate the application of the synthesis approach to derive the first candidate key of URS1 under the set of constraints in F. Section 7.3.1.2 hows the steps to derive the remaining candidate keys of URS1.

Step 1. Derive the minimal cover F_c for the set of functional dependencies (FDs) F that prevails over the URS.

There are no redundant attributes or redundant FDs in F. The reader can verify this using the algorithm given at the end of Section 7.2.3.2. Thus, in this example, F_c = F.

Step 2. Select the FD with maximum number of attributes constituting the determinant as the starting point. Let us call this FD as the target FD, and the determinant of this first FD as the target determinant (TD1). *Note:* If more than one such target FD exists, select one of them for now.

Compute the attribute closure [TD1 | F_c].

Target FD fd1: {Store, Branch} → Location;	*Target determinant* (TD1): {Store, Branch}

Start: **{Store, Branch}⁺ = {Store, Branch}**
First iteration through F:

- In fd1, the determinant **{Store, Branch}** is a subset of **{Store, Branch}⁺**—so, **{Store, Branch}⁺ = {Store, Branch}⁺ ∪ Location = {Store, Branch, Location}**;
- In fd2 and fd3, the respective determinants, **Customer** and **Vendor** are not subsets of **{Store, Branch}⁺**—so, no change in **{Store, Branch}⁺**;

- In fd4, the determinant **{Store, Branch}** is a subset of **{Store, Branch}**$^+$—so, **{Store, Branch}**$^+$ = **{Store, Branch}**$^+$ ∪ SQ_ft = **{Store, Branch, Location, Sq_ft}**;
- In fd5, the determinant, **Product** is not a subset of **{Store, Branch}**$^+$—so, no change in **{Store, Branch}**$^+$;
- In fd6, the determinant **{Store, Branch}** is a subset of **{Store, Branch}**$^+$—so, **{Store, Branch}**$^+$ = **{Store, Branch}**$^+$ ∪ Manager = **{Store, Branch, Location, Sq_ft, Manager}**;
- In fd7, the determinant **{Manager}** is a subset of **{Store, Branch}**$^+$—so, **{Store, Branch}**$^+$ = **{Store, Branch}**$^+$ ∪ **{Store, Branch}** = **{Store, Branch, Location, Sq_ft, Manager}**;
- In fd8, the determinant **{Store}** is a subset of **{Store, Branch}**$^+$—so, **{Store, Branch}**$^+$ = **{Store, Branch}**$^+$ ∪ **{Type}** = **{Store, Branch, Location, Sq_ft, Manager, Type}**;
- First iteration through F completed.

Second iteration through F:

- No further change in the Closure [**{Store, Branch}** | F] due to fd1 through fd8; so step 2 terminates.

End:

{Store, Branch}$^+$ = **{Store, Branch, Location, Sq_ft, Manager, Type}**

Step 3. If TD1$^+$ is precisely the set of all attributes of URS, then TD1 is a candidate key of URS. If so, skip to step 7.

{Store, Branch}$^+$ is not precisely the set of all attributes of URS1; therefore, **{Store, Branch}** is not a candidate key of URS1.

Step 4. If TD1$^+$ is not precisely the set of all attributes of URS, select a functional dependency whose determinant is not a subset of TD1$^+$ as the next target FD. The determinant of this FD will be TD2.

Compute the attribute closure [TD2 | F$_c$].

Target FD – fd2: **Customer → Address;**	*Target determinant* (TD2): **Customer**

Start: **Customer+ = Customer**

First iteration through F:

- In all FDs in F but fd2, the respective determinants, are not subsets of **Customer**$^+$—so, no change in **Customer**$^+$; due to fd1, fd3, fd4, fd5, fd6, fd7, and fd8.
- In fd2, the determinant **Customer** is a subset of **Customer**$^+$—so, **{Customer}**$^+$ = **Customer**$^+$ ∪ **Address** = **{Customer, Address}**;

Second iteration through F:

- No further change in the Closure [**Customer** | F]; so step 4 terminates.

End:

Customer$^+$ = **{Customer, Address}**

Step 5. If TD2$^+$ is precisely the set of all attributes of URS, then TD2 is a candidate key of URS. Otherwise, if {TD1$^+$ ∪ TD2$^+$} is precisely the set of all attributes of URS, then {TD1 ∪ TD2} is a candidate key of URS.

If either one is true, skip to step 7.

Customer$^+$ is not precisely the set of all attributes of URS1; therefore, **Customer** is not a candidate key of URS1.

{**Store, Branch**}$^+$ ∪ **Customer**$^+$ = {**Store, Branch, Location, Sq_ft, Manager, Type, Customer, Address**}.

{**Store, Branch**}$^+$ ∪ **Customer**$^+$ is not precisely the set of all attributes of URS1; therefore, {**Store, Branch, Customer**} is not a candidate key of URS1 either.

Step 6. Otherwise, repeat steps 4 and 5 using the next target FD. Repeat steps 4 and 5 until an attribute set K is derived such that K$^+$ is precisely the set of all attributes of URS.

Then, K is a candidate key of URS.

Target FD – fd3: **Vendor → Product;**	*Target determinant* (TD3): **Vendor**

Start: **Vendor**$^+$ = **Vendor**

First iteration through F_c:

- In all FDs in F except fd3 and fd5, the respective determinants, are not subsets of **Vendor**$^+$—so, no change in **Vendor**$^+$ due to fd1, fd2, fd4, fd6, fd7, and fd8.
- In fd3, the determinant **Vendor** is a subset of **Vendor**$^+$—so, **Vendor**$^+$ = **Vendor**$^+$ ∪ **Product** = {**Vendor, Product**};
- In fd5, the determinant **Product** is a subset of **Vendor**$^+$—so, **Vendor**$^+$ = **Vendor**$^+$ ∪ **Price** = {**Vendor, Product, Price**};
- First iteration through F_c completed.

Second iteration through F_c:

- No further change in the Closure [**Vendor** | F]; so step 6 terminates.

End:

Vendor$^+$ = {**Vendor, Product, Price**}

Vendor$^+$ is not precisely the set of all attributes of URS1; therefore, **Vendor** is not a candidate key of URS1.

{**Store, Branch**}$^+$ ∪ **Customer**$^+$ ∪ **Vendor**$^+$ = {**Store, Branch, Location, Sq_ft, Manager, Type, Customer, Address, Vendor, Product, Price**}

{**Store, Branch**}$^+$ ∪ **Customer**$^+$ ∪ **Vendor**$^+$ is indeed precisely the set of all attributes of URS1; therefore, {**Store, Branch, Customer, Vendor**} is a candidate key of URS1.

Observe that {**Store, Branch, Customer, Vendor**} is a superkey of URS1, and none of its proper subsets is a superkey of URS1 confirming that the irreducible superkey {**Store, Branch, Customer, Vendor**} is a candidate key of URS1.

7.3.1.2 Deducing the Other Candidate Key(s) of URS1 by Synthesis

Step 7. If F contains an FD, fdx, where a candidate key of URS is a dependent, then the determinant of fdx is also a candidate key of URS.

There are no FDs in F where {**Store, Branch, Customer, Vendor**} is a dependent. Therefore, proceed to step 8.

Step 8. When a candidate key of URS is a composite attribute, for each key attribute (atomic or composite): Evaluate if the key attribute is a dependent in an FD, fdy, in F. If so, then the determinant of fdy, by the rule of pseudotransitivity, can replace the key attribute under consideration, thus yielding additional candidate key(s) of URS.

Since in **Manager → {Store, Branch}** (see fd7), the dependent, **{Store, Branch}**, is a subset of the candidate key **{Store, Branch, Customer, Vendor}**, using the *rule of pseudotransitivity*, **{Manager, Customer, Vendor}** is extracted as another candidate key of URS1.

Since there is no other key attribute, atomic or composite, of the candidate key, **{Store, Branch, Customer, Vendor}**, that is a dependent in any other FD in F, continue to step 9.

Step 9. Repetition of steps 7 and 8 above for every candidate key of URS will systematically reveal all the other candidate key(s), if any, of URS.

The only other candidate key of URS1 so far is **{Manager, Customer, Vendor}**. Application of steps 7 and 8 above in this case do not yield any more candidate keys.

In summary, URS1 where the set of FDs denoted by F prevail has two candidate keys. They are:

> **{Store, Branch, Customer, Vendor}**
> **{Manager, Customer, Vendor}**

7.3.2 Deriving Candidate Keys by Decomposition

Since, by definition, a relation schema must have a superkey, every relation schema has at least one default superkey—the set of all its attributes. In the decomposition method for deriving a candidate key, for a given URS and the set of FDs that prevail over it, we start by setting the collection of all attributes of URS as its superkey, K. We then arbitrarily remove one attribute at a time from K and check if the collection of the remaining attributes, K', continues to satisfy the uniqueness condition of a superkey in URS. The checking is done using the FDs in F that prevail over URS. If the test fails, the attribute removed is restored to K'. The process continues until K' is reduced to a superkey that is not further reducible, thus becoming a candidate key of URS.

When URS has multiple candidate keys, the candidate key returned through this process of decomposition of a superkey depends on the order in which attributes are removed from URS. The method is algorithmically stated in Table 7.3. The following example demonstrates the application of this algorithm to generate the first candidate key of URS.

Table 7.3 An algorithm for the derivation of candidate key(s) by decomposition of the superkey given a universal relation schema: URS = $\{A_1, A_2, A_3, \ldots\ldots, A_n\}$ and the set of functional dependencies over URS: F = $\{fd_1, fd_2, fd_3, \ldots\ldots, fd_m\}$

Derivation of first candidate key of URS	
Step 1	Set superkey, K, of URS = $\{A_1, A_2, A_3, \ldots\ldots, A_n\}$
Step 2	Remove an attribute A_i, (i = 1, 2, 3, $\ldots\ldots$, n) from URS such that $\{K - A_i\}$ is still a superkey, K', of URS.
	Note: In order for K' to be a superkey of URS, the FD: $(K' \rightarrow A_i)$ should persist in F^+.
Step 3	Repeat step 2 above recursively until K' is further irreducible.
	The irreducible K' is a candidate key of URS under the set of FDs, F.

Table 7.3 An algorithm for the derivation of candidate key(s) by decomposition of the superkey given a universal relation schema: URS = {A₁, A₂, A₃, ... , Aₙ} and the set of functional dependencies over URS: F = {fd₁, fd₂ , fd₃, ... , fdₘ} (continued)

Derivation of other candidate key(s) of URS	
Step 4	If F contains an FD, fdx, where a candidate key of URS is a dependent, then the determinant of fdx is also a candidate key of URS.
Step 5	When a candidate key of URS is a molecular attribute, for each key attribute (atomic or molecular): • Evaluate if the key attribute is a dependent in an FD, fdy, in F. • If so, then the determinant of fdy, by the rule of pseudotransitivity, can replace the key attribute under consideration, thus yielding additional candidate key(s) of URS.
Step 6	Repetition of steps 4 and 5 above for every candidate key of URS will systematically reveal all the other candidate key(s), if any, of URS.

7.3.2.1 Deriving the First Candidate Key by Decomposition

Suppose there is a set of nine functional dependencies, F, as follows:

fd1: **Company → Location;**	fd2: **Company → Size;**	fd3: **Company → President;**
fd4: **{Product, Company} → Price;**	fd5: **Sales → Production;**	fd6: **{Company, Product} → Sales;**
fd7: **{Product, Company} → Supplier;**	fd8: **Supplier → Product;**	fd9: **President → Company**

A universal relation schema that includes all these functional dependencies is:
URS2 (Company, Location, Size, President, Product, Price, Sales, Production, Supplier)

The set of FDs in F may also be expressed via a dependency diagram, as shown in Figure 7.6.

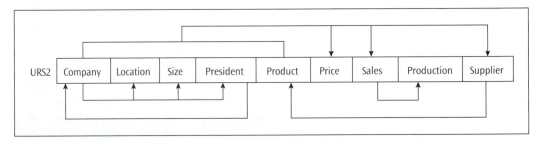

Figure 7.6 Dependency diagram for the relation schema, URS2

Here are the steps to determine the candidate key(s) of URS2 by decomposing the superkey.

Step 1. Set superkey, K, of URS = {A₁, A₂, A₃, ... , Aₙ}

K = {Company, Location, Size, President, Product, Price, Sales, Production, Supplier}

Step 2. Remove an attribute Aᵢ, (i = 1, 2, 3, ..., n) from URS such that {K − Aᵢ} is still a superkey, K', of URS.

Note: In order for K' to be a superkey of URS, the FD: $(K' \rightarrow A_i)$ must persist in F^+.

Let us arbitrarily remove the attribute from the left end of K—i.e., remove **Company** from K.

Then,

K' = (K – Company) = {Location, Size, President, Product, Price, Sales, Production, Supplier}

In order to know if **K'** is a superkey of URS2, we need to check if **(K' → Company)** is in F^+.

Since **(President → Company)** is in F (see fd9) and **(President ⊂ K')**, using the augmentation rule of Armstrong's axioms, we infer that **(K' → Company)** is in F^+. Therefore, **K'** is a superkey of URS2.

Step 3. Repeat step 2 above recursively until K' is further irreducible.

Continue removing attributes (arbitrarily) from the left of **K'**. Accordingly, next remove the attribute **Location**.

Then,

K' = (K' – Location) = {Size, President, Product, Price, Sales, Production, Supplier}

The pared down **K'** still remains a superkey of URS2 because, since fd1: **(Company → Location)** and **(K' → Company)** are in F^+, then so is **(K' → Location)**.

Likewise, using fd2, we can conclude that after the removal of **Size** from K',

K' = {President, Product, Price, Sales, Production, Supplier}

continues to remain a superkey of URS2.

Next, when **President** is removed from the current K', the pared down **K' = {Product, Price, Sales, Production, Supplier}** fails to persist as a superkey of URS2 anymore because **(K' → Company, Location, Size, President)** is no longer present in F^+.

So **President** is added back to K' restoring K' as a superkey of URS2:

K' = {President, Product, Price, Sales, Production, Supplier}

Even when **Product** is removed from K', the reduced **K' = {President, Price, Sales, Production, Supplier}** continues to remain as a superkey of URS2 because since **Supplier → Product** (fd8), we can infer that **K' → Product**. In addition, K' will also functionally determine all other dependent attributes in FDs where **Product** is the determinant. It just happens that in this example there are no non-trivial FDs in F^+ where **Product** is the determinant.

What happens when **Price** is removed from the current K'? Does the trimmed down **K' = {President, Sales, Production, Supplier}** persist as a superkey of URS2?

Since **Price** is not a part of a determinant in any FD in F, it is sufficient if **K' → Price** is in F^+ in order for K' to remain a superkey of URS2. From fd8, fd9, and fd4 we can conclude that **{President, Supplier} → Price**. Since **{President, Supplier} K'**, we further infer that **K' → Price** is in F^+. Therefore, K' remains a superkey of URS2.

Along similar lines of recursive refinement by dropping attributes from K' from left to right, it can be shown that **Sales** and **Production** can be removed from K' while the successively reduced K' persists as the superkey of URS2. However, removal of **Supplier** from K' at the end results in K' not persisting as a superkey of URS2 any longer. As a consequence, **Supplier** is added back to K', restoring K' as a superkey of URS2. At this point, it can be seen that

K' = {President, Supplier}

is a superkey of URS2 and does not remain a superkey of URS2 if further reduced. Consequently, **{President, Supplier}** becomes a candidate key (CK1) of URS2.

7.3.2.2 Deducing the Other Candidate Key(s) of URS2

Step 4. If F contains an FD, fdx, where a candidate of URS is a dependent, then the determinant of fdx is also a candidate key of URS.

There are no FDs in F where **{President, Supplier}** is a dependent. Therefore, proceed to step 5.

Step 5. When a candidate key of URS is a composite attribute, for each key attribute (atomic or composite): Evaluate if the key attribute is a dependent in an FD, fdy in F. If so, then the determinant of fdy, by the rule of pseudotransitivity, can replace the key attribute under consideration, thus yielding additional candidate key(s) of URS.

Given that **{President, Supplier}** is a candidate key of URS2, **President** is a key attribute and **Supplier** is a key attribute.

President is a dependent in fd3: **Company → President**. Therefore, **{Company, Supplier}** is a candidate key (CK2) of URS2.

Supplier is a dependent in fd7: **{Product, Company} → Supplier**. Therefore, **{President, Product, Company}** and **{Company, Product, Company}** result from the application of the *pseudotransitivity* rule on CK1 and CK2 respectively. However, since **President → Company** and **Company → Company**, **Company** is redundant in both cases. Therefore, **{President, Product}** and **{Company, Product}** become candidate keys (CK3 and CK4) of URS2.

To sum up, in this example we have discovered the following four candidate keys for URS2 under the set of functional dependencies, F:

{Company, Product};
{Company, Supplier};
{President, Product};
{President, Supplier}

Observe that any one of these could have been the first candidate key of URS2 derived through the process of superkey decomposition depending on the order of removal of attributes in the derivation process.

7.3.3 Deriving a Candidate Key—Another Example

A third example demonstrating the derivation of candidate key(s) from a set of functional dependencies is presented in this section. Given the universal relation schema:

URS3 (Proj_nm, Emp#, Proj#, Job_type, Chg_rate, Emp_nm, Budget, Fund#, Hours, Division)
and a set of functional dependencies, F prevailing over URS3, as follows:

fd1: **Proj# → Proj_nm**;	fd2: **Job_type → Chg_rate**;	fd3: **Emp# → Emp_nm**;
fd4: **Proj_nm → Budget**;	fd5: **Fund# → Proj_nm**;	fd6: **{Proj_nm, Emp#} → Hours**;
fd7: **Proj_nm → Proj#**;	fd8: **Emp# → Job_type**;	fd9: **{Proj#, Emp#} → Fund#**

What is/are the candidate key(s) of URS3?

The set of FDs in F may also be expressed via a dependency diagram, as shown in Figure 7.7.

For this example, the solution using both the decomposition technique and the synthesis technique are demonstrated.

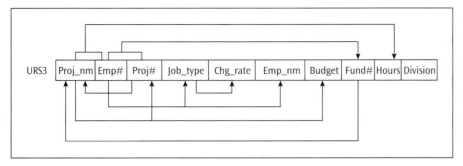

Figure 7.7 Dependency diagram for the relation schema, URS3

7.3.3.1 Deriving a Candidate Key by Decomposition

Step 1. Set superkey, K, of URS = $\{A_1, A_2, A_3, \ldots, A_n\}$

K = {Proj_nm, Emp#, Proj#, Job_type, Chg_rate, Emp_nm, Budget, Fund#, Hours, Division}

Step 2. Remove an attribute A_i, (i = 1, 2, 3, ..., n) from URS such that **{K – A_i}** is still a superkey, **K'**, of URS. *Note*: In order for **K'** to be a superkey of URS, the FD: **(K' → A_i)** must persist in F+.

Let us arbitrarily remove the attribute from the right end of K—i.e., remove **Division** from K.

Then,

K' = (K – Division) = {Proj_nm, Emp#, Proj#, Job_type, Chg_rate, Emp_nm, Budget, Fund#, Hours}

In order to know if **K'** is a superkey of URS3, we need to check if the FD **(K' → Division)** is in F^+.

A perusal of F indicates that none of the subsets of **K'** functionally determine **Division**. Therefore, **(K' → Division)** cannot be present in F^+ either. Accordingly, **K'** is not a superkey of URS3. In fact, **Division** cannot be removed from **K** in order to construct a superkey of URS3. More precisely, any superkey of URS3 must include **Division**.

Removal of **Hours** from **K** to create **K'** results in

K' = (K – Hours) = {Proj_nm, Emp#, Proj#, Job_type, Chg_rate, Emp_nm, Budget, Fund#, Division}

and **K'** now is a superkey of URS3 since the FD **(K' → Hours)** exists in F^+.

Working along similar lines using fd9, fd4, fd3, fd2, fd8, and fd7 respectively, we can see that removal of the attributes **Fund#**, **Budget**, **Emp_nm**, **Chg_rate**, **Job_type** and **Proj#** pares down **K'** to **{Proj_nm, Emp#, Division}** without affecting its role as a superkey of URS3. Furthermore, as a superkey, **K'** is further irreducible because none of its proper subsets **{Proj_nm}**, **{Emp#}**, **{Division}**, **{Proj_nm, Emp#}**, **{Proj_nm, Division}**, or **{Emp#, Division}** qualifies as a superkey of URS3. Thus, we infer that **{Proj_nm, Emp#, Division}** is a candidate key (CK1) of URS3.

7.3.3.2 Deducing the Other Candidate Key(s) of URS3

Step 3. If F contains an FD, fdx, where a candidate of URS is a dependent, then the determinant of fdx is also a candidate key of URS.

There are no FDs in F where **{Proj_nm, Emp#, Division}** is a dependent. Therefore, proceed to step 4.

Step 4. When a candidate key of URS is a composite attribute, for each key attribute (atomic or composite): Evaluate if the key attribute is a dependent in an FD, fdy, in F. If so, then the determinant of fdy, by the rule of pseudotransitivity, can replace the key attribute under consideration, thus yielding additional candidate key(s) of URS.

Given that **{Proj_nm, Emp#, Division}** is a candidate key of URS3, its proper subsets **{Proj_nm}, {Emp#}, {Division}, {Proj_nm, Emp#}, {Proj_nm, Division}, {Emp#, Division}** are all key attributes. All except **{Proj_nm}** do not participate in any FD in F as a dependent, and **{Proj_nm}** is a dependent in fd1 as well as fd5. Therefore, by applying the rule of *pseudotransitivity* we deduce from fd1 and fd5 that **{Proj#, Emp#, Division}** and **{Fund#, Emp#, Division}** are also candidate keys of URS3.

7.3.3.3 Deriving Candidate Key(s) by Synthesis

The following steps demonstrate the application of the synthesis approach to derive the candidate key(s) of URS3 under the set of constraints set F_c.

Step 1. Derive the minimal cover F_c for the set of functional dependencies (FDs) F that prevails over the URS.

There are no redundant attributes or redundant FDs in F. The reader can verify this using the algorithm given at the end of Section 7.3.2.2. Thus, in this example, F_c = F.

Step 2. Select the FD with maximum number of attributes constituting the determinant as the starting point. Let us call this FD as the target FD, and the determinant of this first FD as the target determinant (TD1). *Note*: If more than one such target FD exists, select one of them for now.

Compute the attribute closure [TD1 | F_c].

Target FD – fd6: **{Proj_nm, Emp#}** → **Hours;**	*Target determinant* (TD1): **{ Proj_nm, Emp#}**

Start: **{Proj_nm, Emp#}**$^+$ = **{Proj_nm, Emp#, Hours}**
First iteration through F_c:

- In fd1 and fd2, the respective determinants, **Proj#** and **Job_type** are not subsets of **{Proj_nm, Emp#}**$^+$—so, no change in **{Proj_nm, Emp#}**$^+$;
- In fd3, the determinant **{Emp#}** is a subset of **{Proj_nm, Emp#}**$^+$—so, **{Proj_nm, Emp#}**$^+$ = **{Proj_nm, Emp#}}**$^+$ U **Emp_nm** = **{Proj_nm, Emp#, Hours, Emp_nm}**;
- In fd4, the determinant **{Proj_nm}** is a subset of **{Proj_nm, Emp#}**$^+$—so, **{Proj_nm, Emp#}**$^+$ = **{Proj_nm, Emp#}}**$^+$ U **Budget** = **{Proj_nm, Emp#, Hours, Emp_nm, Budget}**;
- In fd5, the determinant, **Fund#** is not a subset of **{Proj_nm, Emp#}**$^+$—so, no change in **{Proj_nm, Emp#}**$^+$;
- fd6 has already been considered at the start of the first iteration through F—so, no change in **{Proj_nm, Emp#}**$^+$ at this time;
- In fd7, the determinant **{Proj_nm}** is a subset of **{Proj_nm, Emp#}**$^+$—so, **{Proj_nm, Emp#}**$^+$ = **{Proj_nm, Emp#}}**$^+$ ∪ **Proj#** = **{Proj_nm, Emp#, Hours, Emp_nm, Budget, Proj#}**;

- In fd8, the determinant **{Emp#}** is a subset of **{Proj_nm, Emp#}**$^+$—so, **{Proj_nm, Emp#}**$^+$ = **{Proj_nm, Emp#}}**$^+$ $\cup$ **Job_type** = **{Proj_nm, Emp#, Hours, Emp_nm, Budget, Proj#, Job_type}**;
- In fd9, the determinant **{Proj#, Emp#}** is a subset of **{Proj_nm, Emp#}**$^+$—so, **{Proj_nm, Emp#}**$^+$ = **{Proj_nm, Emp#}**$^+$ $\cup$ **Fund#** = **{Proj_nm, Emp#, Hours, Emp_nm, Budget, Proj#, Job_type, Fund#}**;
- First iteration through F_c completed.

Second iteration through F_c:

- fd1 does not lead to any changes to **{Proj_nm, Emp#}**$^+$ since both the determinant and dependent in fd1 (**Proj#** and **Proj_nm** respectively) are already present in **{Proj_nm, Emp#}**$^+$;
- In fd2, the determinant **{Job_type}** is a subset of **{Proj_nm, Emp#}**$^+$—so, **{Proj_nm, Emp#}**$^+$ = **{Proj_nm, Emp#}**$^+$ $\cup$ **Chg_rate** = **{Proj_nm, Emp#, Hours, Emp_nm, Budget, Proj#, Job_type, Fund#, Chg_rate}**;
- No further change in the Closure [**{Proj_nm, Emp#}** | F] due to fd3 through fd9; so step 2 terminates.

{Proj_nm, Emp#}$^+$ = **{Proj_nm, Emp#, Hours, Emp_nm, Budget, Proj#, Job_type, Fund#, Chg_rate}**;

Step 3. If TD1$^+$ is precisely the set of all attributes of URS, then TD1 is a candidate key of URS. If so, skip to step 7.

{Proj_nm, Emp#}$^+$ is not precisely the set of all attributes of URS3; therefore, **{Proj_nm, Emp#}** is not a candidate key of URS3.

Step 4. If TD1$^+$ is not precisely the set of all attributes of URS, select a functional dependency whose determinant is not a subset of TD1$^+$ as the next target FD - the determinant of this FD will be TD2.

Compute the attribute closure [TD2 | F_c].

There are no FDs in F_c whose determinant is not a subset of **{Proj_nm, Emp#}**$^+$.

Observe that the attribute **Division** does not participate in any FD in F while it is present in URS3. This indicates the independent state of this attribute in URS3. One way to formalize the presence of the attribute **Division** in URS3 is to portray it through an implicit trivial FD in F$^+$, viz., **Division → Division**. Therefore, let fd10 be **Division → Division**.

Target FD – fd10: **Division → Division;**	*Target determinant* (TD2): **Division**

Start: **Division**$^+$ = **Division;**
First iteration through F:

- In all FDs in F but fd10, the respective determinants, are not subsets of **Division**$^+$—so, no Change in **Division**$^+$ due to fd1, fd2, fd3, fd4, fd5, fd6, fd7, fd8 and fd9.
- Since fd10 is the target FD, it is already accounted for in **Division**$^+$.
- No further change in the Closure [**Division** | F]; so step 4 terminates.

End:
Division$^+$ = **Division**

Step 5. If TD2$^+$ is precisely the set of all attributes of URS, then TD2 is a candidate key of URS. Otherwise, if {TD1$^+$ U TD2$^+$} is precisely the set of all attributes of URS, then {TD1 U TD2} is a candidate key of URS.

If either one is true, skip to step 7.

Division$^+$ is not precisely the set of all attributes of URS3; therefore, **Division** is not a candidate key of URS3.

{Proj_nm, Customer#}$^+$ ∪ **Division**+ = **{Proj_nm, Emp#, Hours, Emp_nm, Budget, Proj#, Job_type, Fund#, Chg_rate, Division}**;

{Proj_nm, Customer#}$^+$ ∪ **Division**$^+$ is precisely the set of all attributes of URS3; therefore, **{Proj_nm, Customer#, Division}** is a candidate key of URS3 under F.

Observe that **{Proj_nm, Customer#, Division}** is a superkey of URS3, and none of its proper subsets is a superkey of URS3 confirming that the irreducible superkey **{Proj_nm, Customer#, Division}** is a candidate key of URS3.[9] Had we chosen fd9: **{Proj#, Emp#}** → **Fund#** as the target FD and therefore **{Proj#, Emp#}** as the target determinant (TD1) in the first step, we would have ended up with **{Proj#, Customer#, Division}** as the candidate key of URS3 under F.

7.3.3.4 Deducing the Other Candidate Key(s) of URS3 by Synthesis

The steps involved in discovering other candidate keys of URS3 are the same as employed in the decomposition method discussed in Section 7.3.3.2 and so are not reiterated here.

In summary, URS3 where the set of FDs denoted by F are preserved has three candidate keys:

{Proj#, Emp#, Division}
{Proj_nm, Emp#, Division}
{Fund#, Emp#, Division}

7.3.4 Prime and Non-prime Attributes

Chapter 6 introduced the concepts of key and non-key attributes (see Section 6.3.1). The ideas are reaffirmed here. An attribute, atomic or composite, in a relation schema, R, is called a *key attribute* if it is a proper subset of any candidate key of R. Likewise, attributes that are not subsets of a candidate key of R become non-key attributes. In other words, a *non-key attribute* is not a member of any candidate key of R.

In the example in Section 7.2.3.2, **Company, President, Product,** and **Supplier,** individually, are all key attributes of URS2 since each one is a proper subset of some candidate key of URS2. On the other hand, **Location, Size, Price, Sales,** and **Production** in URS2 are, individually or in groups, non-key attributes. The composite attribute, **{Company, Product}** is not a key attribute because it is not in the *proper* subset of the candidate key **{Company, Product}** of URS2. The same is true for every one of the other candidate keys of URS2: **{Company, Supplier}; {President, Product};** and **{President, Supplier}.**

Is **{Company, Location}** or **{Product, Sales}** a key attribute of URS2? The answer is "No," even though one of the attributes in each of these two composite attributes is a key attribute. Any attribute, atomic or composite, in a relation schema R that fails the test for a key attribute (being a proper subset of a candidate key) is a non-key attribute, with the

---

[9] For a review of the properties of superkey and candidate key, see Section 6.3.1 in Chapter 6.

exception of the candidate key(s) of R—which are, in fact, neither key attributes nor non-key attributes in R. This is because a candidate key is a subset of itself and thus fails the test for non-key attribute. It is, however, not a proper subset of itself and thus fails the test for a key attribute.

Based on this discussion, we have an alternative definition for a candidate key from this point forward:

A candidate key of a relation schema, R, fully functionally determines all attributes of R.

While the choice of primary key from among the candidate keys is essentially arbitrary, some rules of thumb are often helpful:[10]

- A candidate key with the least number of attributes may be a good choice.
- A candidate key whose attributes are numeric and/or of small sizes may be easy to work with from a developer's perspective.
- A candidate key that is a determinant in a functional dependency in F rather than F^+ may be a good choice because it is probably semantically obvious from the user's perspective.
- Surrogate keys (especially DBMS developed sequence numbers) should be used only as a last resort because they don't offer semantic reference points to the user community.

Primary keys were discussed in Chapter 6 (see Section 6.3.1). As a refresher, a primary key is an irreducible unique identifier like any other candidate key of a relation schema; but, a primary key is in addition bound by the entity integrity constraint—i.e., none of the attributes constituting a primary key is allowed to have "null" values. Suppose **{Company, Product}** is the primary key of URS2 in our example. Then, neither **Company** nor **Product** can have null values in any tuple (in any state) of the relation, URS2. Similar to a key attribute, any attribute, atomic or composite, in a relation schema, R, that is a proper subset of the primary key of R is called a **prime attribute**. An attribute of R that is not a member of the primary key is a **non-prime attribute** *except when it is a candidate key of R*. Any candidate key of R not chosen as the primary key is referred to as an *alternate key* of R and like the primary key, is neither a prime nor a non-prime attribute of R. Here are some examples that provide additional clarification.

Consider URS2, the example from Section 7.3.2.1,

URS2 (Company, Location, Size, President, Product, Price, Sales, Production, Supplier),
where, under the FDs specified in F with reference to URS2, the candidate keys are **{Company, Product}; {Company, Supplier}; {President, Product};** and **{President, Supplier}** and the chosen primary key is **{Company, Product}**.

Table 7. 4 shows the key and non-key attributes and the prime and non-prime attributes for URS2.

---

[10] No matter which candidate key has been chosen to be the primary key of a relation of schema, the normalization process may spontaneously force some changes. We will have an opportunity to observe this phenomenon in Chapter 9.

Table 7.4   Attribute roles in URS2

	Role of the Attribute	
Attribute	**Key/Non-key Attribute**	**Prime/Non-prime Attribute**
Company	Key attribute	Prime attribute
Location	Non-key attribute	Non-prime attribute
Size	Non-key attribute	Non-prime attribute
President	Key attribute	Non-prime attribute
Product	Key Attribute	Prime attribute
{Company, Product}	*Candidate key*	*Primary key*
{President, Product}	*Candidate key*	*Alternate key*
Price	Non-key attribute	Non-prime attribute
Sales	Non-key attribute	Non-prime attribute
Production	Non-key attribute	Non-prime attribute
Supplier	Key attribute	Non-prime attribute
{Company, Supplier}	*Candidate key*	*Alternate key*
{President, Supplier}	*Candidate key*	*Alternate key*

Note: Any composite attribute that includes one or more non-key attribute(s) is a non-key attribute.

As another example, consider URS3 from Section 7.3.3:

URS3 (Proj_nm, Emp#, Proj#, Job_type, Chg_rate, Emp_nm, Budget, Fund#, Hours, Division) where, under the FDs specified in F with reference to URS3, the candidate keys are **{Proj#, Emp#, Division}; {Proj_nm, Emp#, Division}; and {Fund#, Emp#, Division}** and the chosen primary key is **{Proj#, Emp#, Division}**.

Table 7.5 shows the key and non-key attributes and the prime and non-prime attributes for URS3.

Table 7.5 Attribute roles in URS3

Attribute	Role of the Attribute	
	Key/Non-key Attribute	Prime/Non-prime Attribute
Proj_nm	Key attribute	Non-prime attribute
Emp#	Key attribute	Prime attribute
{Proj_nm, Emp#}	Key attribute	Non-prime attribute
Proj#	Key attribute	Prime attribute
{Proj#, Emp#}	Key attribute	Prime attribute
Job_type	Non-key attribute	Non-prime attribute
Chg_rate	Non-key attribute	Non-prime attribute
Emp_nm	Non-key attribute	Non-prime attribute
Budget	Non-key attribute	Non-prime attribute
Fund#	Key attribute	Non-prime attribute
{Fund#, Emp#}	Key attribute	Non-prime attribute
Hours	Non-key attribute	Non-prime attribute
Division	Key attribute	Prime attribute
{Proj_nm, Division}	Key attribute	Non-prime attribute
{Emp#, Division}	Key attribute	Prime attribute
{Proj#, Division}	Key attribute	Prime attribute
{Fund#, Division}	Key attribute	Non-prime attribute
{Proj_nm, Emp#, Division}	*Candidate key*	*Alternate key*
{Proj#, Emp#, Division}	*Candidate key*	*Primary key*
{Fund#, Emp#, Division}	*Candidate key*	*Alternate key*

Note: Any composite attribute that includes one or more non-key attribute(s) is a non-key attribute.

Chapter Summary

Normalization is a technique that facilitates systematic validation of the participation of attributes in a relation schema from the perspective of data redundancy. One of the main concepts associated with normalization is functional dependency. A functional dependency (FD) in a relation schema, R, is a constraint of the form A → B where an attribute B (atomic or composite) is dependent on attribute A (atomic or composite), if each value, a, of A is associated with exactly one value, b, of B.

Examination of functional dependencies in a relation schema is important because certain functional dependencies (i.e., those which are undesirable) can lead to insertion, deletion, and update anomalies via data redundancies in the associated relation instances, collectively known as modification anomalies. Since functional dependencies, desirable or undesirable, arise from the business rules embedded in the user requirements specification, they cannot be conveniently discarded if undesirable. Therefore, the data redundancies and modification anomalies are removed by decomposing the relation schema such that the undesirable functional dependencies are rendered desirable. The example in Section 7.1 illustrates this.

The functional dependencies in a relation schema which are semantically obvious from the business rules are often explicitly specified and are collectively referred to as **F.** All possible FDs that can be inferred from the set F plus the set F itself constitute the closure of F. The closure of F denoted as F^+. Armstrong's axioms are a set of seven inference rules pertaining to functional dependencies that are used to derive F^+. Table 7.1 in Section 7.2.2 summarizes the inference rules for functional dependencies.

It is possible to progressively synthesize a candidate key from a set of functional dependencies through the systematic application of the principle of closure of an attribute set. A second method for identifying the candidate key(s) of a relation schema uses a top-down approach of decomposition. In this method, given a set of functional dependencies that prevail over a universal relation schema (URS), the superkey, K, consisting of all attributes of URS is progressively decomposed by arbitrarily removing one attribute at a time from K until K', a superkey that is not further reducible (i.e., no proper subset of K has the uniqueness property) results yielding a candidate key. The other candidate keys of URS are derived using F and the initial candidate key by the application of the pseudotransitivity rule of Armstrong's axioms. Three examples are used to illustrate the use of Armstrong's axioms to derive candidate keys of a relation schema using the method of synthesis and the method of decomposition. Finally, a definition of prime/non-prime and key/non-key attributes is presented along with a handful of rules of thumb to choose a primary key from among the candidate keys.

Exercises

1. What is the purpose of the normalization technique in the data modeling process?
2. Explain why data redundancy exists for the attributes **Discount** and **Location** in the STOCK table in Figure 7.1c.
3. Explain functional dependency between two attributes.
4. Why can functional dependency not be inferred from a particular relation state?

5. Identify the set of functional dependencies in the relation instance CAR shown below. Does this constitute the minimal cover for the set of functional dependencies present in CAR? If it is not a minimal cover, derive a minimal cover.

CAR

Model	#Cylinders	Origin	Tax	Fee
Camry	4	Japan	15	30
Mustang	6	USA	0	45
Fiat	4	Italy	18	30
Accord	4	Japan	15	30
Century	8	USA	0	60
Mustang	4	Canada	0	30
Monte Carlo	6	Canada	0	45
Civic	4	Japan	15	30
Mustang	4	Mexico	15	30
Mustang	6	Mexico	15	45
Civic	4	Korea	15	30

6. What is the difference between F, F^+, and F_c?

7. What is the purpose of Armstrong's axioms?

8. Suppose F {fd1, fd2} consists of the following functional dependencies:

   fd1: **Ssn → {Ename, Bdate, Address, Dnumber}**

   fd2: **Dnumber → {Dname, Dmgrssn}**

   Which of Armstrong's axioms allows the following additional functional dependencies to be inferred?

   a. **Ssn → {Dname, Dmgrssn}**

   b. **Ssn → Ssn**

   c. **Dnumber → Dname**

   d. **Ssn → Dname**

9. Why is it useful to know all the candidate keys of a relation schema?

10. Describe the two approaches used in this book to derive candidate keys.

11. What is the difference between (a) a prime attribute and a non-prime attribute and (b) a key and non-key attribute?

12. Given R (X, A, Z, B) and A → {B, Z}, what is the candidate key(s) of R?

13. Consider the universal relation schema INVENTORY (Store#, Item, Vendor, Date, Cost, Units, Manager, Price, Sale, Size, Color, Location) and the constraint set F {fd1, fd2, fd3, fd4, fd5, fd6, fd7} where:

fd1: {Item, Vendor} → Cost	fd2: {Store#, Date} → {Manager, Sale}
fd3: {Store#, Item, Date} → Units	fd4: Manager → Store#
fd5: Cost → Price	fd6: Item → {Size, Color}
fd7: Vendor → Location	

  a. Do the functional dependencies shown constitute a minimal cover of F? If not, derive a minimal cover?

  b. Derive the candidate key(s) of URS using the synthesis approach.

  c. Derive the candidate key(s) of URS by using the decomposition approach.

14. Given the set of functional dependencies F {fd1, fd2, fd3, fd4, fd5, fd6, fd7, fd8, fd9, fd10} where:

fd1: Tenant# → {Name, Job, Phone#, Address}	fd2: Job → Salary
fd3: Name → Gender	fd4: Phone# → Address
fd5: {Name, Phone#} → {Tenant, Deposit}	fd6: County → Tax_rate
fd7: Area → {Rent, County}	fd8: Survey# → Lot
fd9: {Lot, County} → {Survey#, Area}	fd10: {Survey#, Area} → County

  a. Construct the universal relation schema that includes (i.e., preserves) the set of functional dependencies in F.

  b. Do the functional dependencies shown constitute a minimal cover of F? If not, derive a minimal cover.

  c. Derive the candidate key(s) of F.

  d. Select the primary key and justify your choice.

  e. Considering your primary key and candidate key(s), distinguish between (1) key versus non-key attributes and (2) prime versus non-prime attributes.

15. Given the set of functional dependencies F {fd1, fd2, fd3, fd4, fd5, fd6, fd7, fd8, fd9, f10, f11} where:

fd1: Client → Office	fd2: Stock → {Exchange, Dividend}
fd3: Broker → Profile	fd4: Company → Stock
fd5: Client → {Risk_profile, Analyst}	fd6: Analyst → Broker

fd7: {Stock, Broker} → {Investment, Volume}   fd8: Stock → Company

fd9: Investment → {Commission, Return}   fd10: {Stock, Broker} → Client

fd11: Account → Assets

a. Construct the universal relation schema that includes (i.e., preserves) the set of functional dependencies in F.

b. Do the functional dependencies shown constitute a minimal cover of F? If not, derive a minimal cover.

c. Derive the candidate key(s) of F.

d. Select the primary key and justify your choice.

e. Considering your primary key and candidate key(s), distinguish between (1) key versus non-key attributes and (2) prime versus non-prime attributes.

Selected Bibliography

Abrial, J. (1974) "Data Semantics," In Klimbie and Koffeman (eds.): Data Base Management, North-Holland.

Armstrong, W. W. "Dependence Structures of Data Base Relationships" *Proc. IFIP Congress*, Stockholm, Sweden (1974).

Codd, E. F. (1970) "A Relational Model for Large Shared Data Banks," *Communications of the ACM*, 13, 6 (June).

Darwen, H. "The Role of Functional Dependencies in Query Decomposition," In C.J. Date and H. Darwen, *Relational Database Writings 1989 – 1991*, Addison-Wesley (1992).

Date, C. J. (2004) *An Introduction to Database Systems*, Eighth Edition, Addison-Wesley.

Elmasri, R.; and Navathe, S. B. (2003) *Fundamentals of Database Systems*, Fourth Edition, Addison-Wesley.

Kent, W. (1983) "A Simple Guide to the Five Normal Forms in Relational Database Theory," *Communications of the ACM*, 26, 2 (February), 120 – 125.

Kifer, M.; Bernstein, A.; and Lewis, P. M. (2005) *Database Systems: An Application-Oriented Approach*, Second Edition, Addison-Wesley.

Ramakrishnan, R. and Gehrke, J. (2000) *Database Management Systems*, Second Edition, McGraw-Hill.

Silberschatz, A.; Korth, H. F.; Sudarshan, S. (2002) *Database System Concepts*, Fourth Edition, McGraw-Hill.

Teorey, T. J.; Yang, D.; and Fry, J. P. (1986) "A Logical Design Methodology for Relational Databases Using the Extended Entity-Relationship Model," *Computing Surveys*, 18, 2 (June) 197–222.

CHAPTER **8**

Normal Forms Based on Functional Dependencies

In Chapter 7 we saw how certain functional dependencies (FDs) can create data redundancy problems in a relation schema. This chapter shows how the process of normalization can be used to resolve these problems. Recall that *normalization* is a technique that facilitates systematic validation of the participation of attributes in a relation schema from a perspective of data redundancy.

This chapter flows as follows. Section 8.1 introduces normalization as a technique to facilitate systematic validation of the goodness of design of a relation schema. The first, second, and third normal forms (1NF, 2NF, and 3NF) are explained with appropriate examples in subsections 8.1.1, 8.1.2, and 8.1.3, respectively. Boyce-Codd Normal Form (BCNF) is presented as a stronger version of the 3NF in Section 8.1.4. The lossless-join property and dependency preservation are then presented in Section 8.1.5 as two critical side effects that require attention in the normalization process. The motivating exemplar originally introduced in Section 7.1 of the previous chapter is revisited in Section 8.2 to provide an explanation for and illustration of the logic behind the normalization process. Section 8.3 gives a comprehensive example of normalizing a universal relation schema subject to a defined set of FDs. Section 8.4 briefly discusses denormalization, and Section 8.5 presents the use of reverse engineering in data modeling.

8.1 Normalization

Data redundancy and the consequent modification (insertion, deletion, and update) anomalies can be traced to "undesirable" functional dependencies in a relation schema. What is an undesirable functional dependency? Any FD in a relation schema, R, where the determinant is a candidate key of R is a **desirable FD** because it will not cause data redundancy and the consequent modification anomalies. Where the determinant of an FD in R is *not* a candidate key of R, the FD will cause data redundancy and the consequent modification anomalies and so is an **undesirable FD**.

So, what can we do with the undesirable FDs? The source of all FDs, desirable and undesirable, is the set of user-specified business rules and so must be incorporated in the database system. Therefore, FDs cannot be selectively ignored or discarded because they are undesirable. The only solution is to somehow render the undesirable FDs desirable, and the process of doing this is called **normalization**.

Normal forms (NFs) provide a stepwise progression toward the goal of a fully normalized relation schema that is guaranteed to be free of data redundancies that cause modification anomalies from a functional dependency perspective.[1] A relation schema is said to be in a particular normal form if it satisfies certain prescribed criteria; otherwise, the relation schema is said to violate that normal form. First normal form (1NF) reflects one of the properties of a relation schema—i.e., by definition a relation schema is in 1NF. The normal forms associated with functional dependencies are second normal form (2NF), third normal form (3NF), and Boyce-Codd normal form (BCNF).[2]

The violations of each of these normal forms signal the presence of a specific type of "undesirable" FD. When a relation schema violates a certain normal form, it can be interpreted as equivalent to an inadvertent mixing of entity types belonging to two different entity classes in a single entity type. Therefore, by appropriately decomposing the relation schema, the undesirable FD causing the violation of a specific normal form can be rendered desirable in the resulting set of relation schemas, that is, the relational schema.

It is important to note that the normalization process is anchored to the candidate key of a relation schema, R. The assessment of normal form can be based on the primary key or any candidate key of R. This is not an issue when R has only one candidate key. Even when R has multiple candidate keys, normalization based on any and every candidate key (including the primary key) will yield the same set of normalized relation schemas. Therefore, we will use the primary key as the basis for evaluating and normalizing a relation schema. This does not by any means contradict the assertion that an FD in R is undesirable only when the determinant of that FD is not a *candidate key* of R.

The following sections delineate each of the normal forms using meaningful examples. Later in the chapter, we will address the situation of generating a fully normalized relational schema from a given set of FDs.

8.1.1 First Normal Form (1NF)

First normal form (1NF) imposes conditions so that a base relation which is physically stored as a file does not contain records with a variable number of fields. This is accomplished by prohibiting multi-valued attributes, composite attributes, and combinations thereof in a relation schema. As a consequence, the value of an attribute in a tuple of a relation can be neither a set of values, nor another tuple. Such a constraint in effect prevents relations from containing other relations.[3] In essence, 1NF, by definition, requires that the domain of an attribute must include only atomic values and that the value of an attribute in a relation's tuple must be a single value from the domain of that attribute.

---

[1] A fully normalized relation schema from the perspective of functional dependencies need not be completely free of data redundancies and the consequent modification anomalies if multi-valued dependencies are present in the relation schema. This will be addressed in Chapter 9.

[2] E.F. Codd first proposed the 1NF, 2NF, and 3NF in 1972. Later it was discovered that under certain conditions (i.e., FDs) a relation schema in 3NF continues to have data redundancies causing modification anomalies. A revised, stronger definition of the 3NF was then proposed by Boyce and Codd in 1974 which came to be known as Boyce-Codd normal form (BCNF).

[3] This constraint is relaxed in object-relational database systems which allow non-1NF relations.

Consider the schema ALBUM and the corresponding instance of ALBUM shown in Figure 8.1. Here for a given **Album_no**, there is a single specific value of **Price** (i.e., **Album_no** → **Price**) and a single specific value of **Stock** (i.e., **Album_no** → **Stock**). On the other hand, either there are multiple **Artist_nms** associated with an **Album_no** or the domain of **Artist_nm** does not have atomic values. In either case, 1NF is violated. In fact, by definition, ALBUM is not even a relation. The solution to render ALBUM in 1NF is to simply expand the relation so that there is a tuple for each (atomic) **Artist_nm** for a given **Album_no**. This is shown in NEW_ALBUM which is in 1NF with **{Album_no, Artist_nm}** as its primary key.

8.1.2 Second Normal Form (2NF)

The **second normal form (2NF)** is based on a concept known as full functional dependency.

A functional dependency of the form $Z \rightarrow A$ is a full functional dependency if, and only if, no proper subset of Z functionally determines A. In other words, if $Z \rightarrow A$ and $X \rightarrow A$, and X is a proper subset of Z, then Z does not *fully* functionally determine A—i.e., $Z \rightarrow A$ is not a full functional dependency; it is a **partial dependency**.

Partial dependency of a non-prime attribute on the primary key of a relation schema, R, is one form of an undesirable FD, because this amounts to the presence of an FD in R where the determinant is not a candidate key of R.

Consider the 1NF relation instance (table) NEW_ALBUM in Figure 8.1. The relation schema for NEW_ALBUM and a copy of the relation instance are shown in Figure 8.2. A review of the data in NEW_ALBUM reveals the presence of the following minimal (canonical) cover of FDs in it:

F: fd1: **Album_no** → **Price**; fd2: **Album_no** → **Stock**

Note that:

Album_no ↛ **Artist_nm**

What is the primary key of NEW_ALBUM? Using Armstrong's axioms, we can infer that fd12: **Album_no** → **{Price, Stock}** and fd12x: **{Album_no, Artist_nm}** → **{Price, Stock}** exist in F^+. Therefore, **{Album_no, Artist_nm}** is a candidate key of NEW_ALBUM; being the only candidate key, it becomes the primary key of NEW_ALBUM. Are there any "undesirable" FDs in NEW_ALBUM? The answer is "Yes." Given that **{Album_no, Artist_nm}** is the primary key of NEW_ALBUM, fd1 and fd2 (or fd12) reflects a partial dependency of a non-prime attribute on the primary key of NEW_ALBUM. Therefore, 2NF is violated in NEW_ALBUM.

ALBUM

Album_no	Artist_nm	Price	Stock
BS123	Britney Spears	17.95	1000
JT111	Justin Timberlake	17.95	1200
BTL007	{John Lennon, Paul McCartney, George Harrison, Ringo Star}	23.95	
MJ100	Michael Jackson	17.95	
JM456	John Mayer	16.95	1000
JM151	John Mayer	16.95	1000
MX789	Madonna	11.95	500
DJM237	{John Denver, Michael Jackson, Madonna}	11.95	2000
DR711	Diana Ross	12.95	1000
PM137	Paul McCartney	19.95	

NEW_ALBUM

Album_no	Artist_nm	Price	Stock
BS123	Britney Spears	17.95	1000
JT111	Justin Timberlake	17.95	1200
BTL007	John Lennon	23.95	
BTL007	Paul McCartney	23.95	
BTL007	George Harrison	23.95	
BTL007	Ringo Star	23.95	
MJ100	Michael Jackson	17.95	
JM456	John Mayer	16.95	1000
JM151	John Mayer	16.95	1000
MX789	Madonna	11.95	500
DJM237	John Denver	11.95	2000
DJM237	Michael Jackson	11.95	2000
DJM237	Madonna	11.95	2000
DR711	Diana Ross	12.95	1000
PM137	Paul McCartney	19.95	

Figure 8.1   An example of 1NF violation and resolution

R: NEW_ALBUM (Album_no, Artist_nm, Price, Stock)

NEW_ALBUM

Album_no	Artist_nm	Price	Stock
BS123	Britney Spears	17.95	1000
JT111	Justin Timberlake	17.95	1200
BTL007	John Lennon	23.95	
BTL007	Paul McCartney	23.95	
BTL007	George Harrison	23.95	
BTL007	Ringo Star	23.95	
MJ100	Michael Jackson	17.95	
JM456	John Mayer	16.95	1000
JM151	John Mayer	16.95	1000
MX789	Madonna	11.95	500
DJM237	John Denver	11.95	2000
DJM237	Michael Jackson	11.95	2000
DJM237	Madonna	11.95	2000
DR711	Diana Ross	12.95	1000
PM137	Paul McCartney	19.95	

F: fd1: Album_no → Price;   fd2: Album_no → Stock

F^+:   fd12: Album_no → (Price, Stock);   fd12x: (Album_no, Artist_nm) → (Price, Stock);

Candidate key of NEW_ALBUM: (Album_no, Artist_nm);   Primary key: (Album_no, Artist_nm)

fd1 and fd2 (fd12) violate 2NF in NEW_ALBUM

Solution:

D: R1: ALBUM_INFO (Album_no, Price, Stock);   R2: ALBUM_ARTIST ($\overset{1\ \ R1\ \ n}{\underset{1\ \ R\ \ 1}{\underline{Album\_no}, Artist\_nm}}$)

ALBUM_INFO

Album_no	Price	Stock
BS123	17.95	1000
JT111	17.95	1200
BTL007	23.95	
MJ100	17.95	
JM456	16.95	1000
JM151	16.95	1000
MX789	11.95	500
DJM237	11.95	2000
DR711	12.95	1000
PM137	19.95	

ALBUM_ARTIST

Album_no	Artist_nm
BS123	Britney Spears
JT111	Justin Timberlake
BTL007	John Lennon
BTL007	Paul McCartney
BTL007	George Harrison
BTL007	Ringo Star
MJ100	Michael Jackson
JM456	John Mayer
JM151	John Mayer
MX789	Madonna
DJM237	John Denver
DJM237	Michael Jackson
DJM237	Madonna
DR711	Diana Ross
PM137	Paul McCartney

Figure 8.2   An example of 2NF violation and resolution

Normal Forms Based on Functional Dependencies

Now let us examine the relation instance of NEW_ALBUM to see if there are any data redundancies in it that lead to modification anomalies. It is obvious that both **Price** and **Stock** are repeated for every **Artist_nm** for a given **Album_no** (e.g., see **Album_no** BTL007). Are there any modification anomalies in NEW_ALBUM?

If we want to change the value of **Price** or **Stock** of **Album_no** BTL007, four tuples in the relation require update—a clear case of *update anomaly*. The anomaly is due to the fact that it is possible to erroneously post different values for **Price** and **Stock** for the same value of **Album_no** in the four different tuples.

Also, if we want to add a new tuple, (**Album_no**: XY111, **Price**: 17.95, and **Stock**: 100) to NEW_ALBUM, it is not possible to do so without knowing a value for **Artist_nm** because the primary key of NEW_ALBUM is **{Album_no, Artist_nm}** and a prime attribute, **Artist_nm**, cannot have a null value. This is an *insertion anomaly* simply because of the inability to add a genuine tuple to the database.

In order to delete **Album_no** BTL007, four tuples in the relation NEW_ALBUM must be deleted. This is an example of a *deletion anomaly*. If all four tuples are not deleted, the information conveyed by the data in the relation is distorted; hence the anomaly.

The next question is, how do we know that the undesirable FDs identified earlier (i.e., fd1 and fd2) indeed cause the data redundancies and the associated modification anomalies? The way this can be verified is to eliminate the undesirable FDs from NEW_ALBUM (that is, rendering them desirable) and see if the data redundancies and the associated anomalies persist. The resolution of 2NF violation is a two-step process that decomposes the target relation schema with undesirable FDs into multiple relation schemas that are free from undesirable FDs.

1. Pull out the undesirable FD(s) from the target relation schema as a separate relation schema.
2. Retain the determinant of the pulled-out relation schema as an attribute(s) (foreign key) of the leftover target relation schema to facilitate reconstruction of the original target relation schema.

Applying this two-step process to NEW_ALBUM, we have the decomposition, D:

D: ALBUM_INFO **(Album_no, Price, Stock)**; and ALBUM_ARTIST **(Album_no, Artist_nm)**

Note: ALBUM_INFO and ALBUM_ARTIST are arbitrary (meaningful) names assigned to the decomposed relation schemas.

All non-prime attributes in ALBUM_INFO are fully functionally dependent on the primary key of ALBUM_INFO, and in ALBUM_ARTIST there are no non-prime attributes. So, both are in 2NF. Reviewing the corresponding decomposed relations (see Figure 8.2), it is clear that there are no data redundancies causing modification anomalies in either relation. It is now possible to insert the tuple (**Album_no**: XY111, **Price**: 17.95, and **Stock**: 100) in the database (i.e., in ALBUM_INFO) without having a value for **Artist_nm**. If the **Price** or **Stock** of **Album_no** BTL007 changes, the corresponding update requires change of just one tuple in spite of the fact that the album involves multiple artists. So, we can infer that the resolution of undesirable FDs causing the 2NF violation eliminated the data redundancies and the associated modification anomalies.

How about the presence of **Album_no** as an attribute in both ALBUM_INFO and ALBUM_ARTIST? Is it not data redundancy? It is; but it is known as **controlled redundancy** (also see Section 1.4.3 in Chapter 1) as long as the referential integrity constraint between the two relation schemas is enforced by specifying the inclusion dependency:

ALBUM_ARTIST.{Album_no} ⊆ ALBUM_INFO.{Album_no}

which can also be used to reconstruct the original target schema, NEW_ALBUM. Observe that the violation of 2NF is a concern only when the primary key of a relation schema is a composite attribute. If the primary key is an atomic attribute, a partial dependency is impossible.

8.1.3 Third Normal Form (3NF)

The **third normal form (3NF)** is based on the concept of **transitive dependency**. Consider a relation schema R (X, A, B) where X, A, and B are pair-wise disjoint atomic or composite attributes, X is the primary key of R, and A and B are non-prime attributes. If $A \rightarrow B$ (or $B \rightarrow A$) in R, then B (or A) is also said to be transitively dependent on X, the primary key of R. This is another form by which an undesirable FD appears in a relation schema. Clearly, this form of FD is by definition undesirable, because the determinant in the FD is, again, not a candidate key of R. Fundamentally, the source of the problem is not the principle of transitivity itself, because had A (or B) been another candidate key (alternate key) of R, the transitive nature of $X \rightarrow B$ (or $X \rightarrow A$) does not yield an undesirable FD. In essence, the problem boils down to a non-prime attribute functionally determining another non-prime attribute. Nonetheless, this is not a violation of 2NF, because $A \rightarrow B$ (or $B \rightarrow A$) in R is not a partial dependency. In fact, R is in 2NF.

D E F I N I T I O N

3NF definition: A relation schema R is in 3NF if no non-prime attribute is functionally dependent on another non-prime attribute in R.

While violation of 2NF and 3NF are independent effects, since these two normal forms are labeled as such (i.e., 2NF and 3NF), it is customary to specify that in order for a relation schema to be in 3NF, it should also be in 2NF.

As an example, consider the relation schema:

FLIGHT **(Flight#, Origin, Destination, Mileage)**

and the set of FDs, F, where:

fd1: **Flight# → Origin**; fd2: **Flight# → Destination**; fd3: **{Origin, Destination} → Mileage**;

In order to assess the normal form of FLIGHT, we first need to identify the candidate keys of FLIGHT. To that end, using Armstrong's axioms, we can infer that F⁺ includes the following FDs:

fd12: **Flight# → {Origin, Destination}**; - Union rule

fd3x: **Flight# → Mileage**; - Transitivity rule

fd123: **Flight# → {Origin, Destination, Mileage}** - Union rule

Thus, we see that **Flight#** is a candidate key of FLIGHT. Since there are no other candidate keys of FLIGHT, **Flight#** becomes the primary key of FLIGHT. Since the primary key of FLIGHT is an atomic attribute, a 2NF violation is impossible in FLIGHT. So, FLIGHT is in 2NF. Is it also in 3NF? No, because fd3 (i.e., **{Origin, Destination} → Mileage**) causes a transitive dependency in FLIGHT, because a composite non-prime attribute, **{Origin,**

Normal Forms Based on Functional Dependencies

Destination}, functionally determines another non-prime attribute, **Mileage**. Thus 3NF is violated in FLIGHT.

Let us now explore if there are data redundancies in FLIGHT. Figure 8.3 displays an instance of a relation for FLIGHT. Note that the relation instance precisely reflects the FDs specified. Data redundancy is exemplified in FLIGHT by the repetition of distance 1058 from Chicago to Dallas in more than one tuple.

Are there any modification anomalies in FLIGHT?

If the tuple identified by **Flight#** DL507 is deleted, the information that Seattle to Denver is 1537 miles is inadvertently lost from the database, an example of a *deletion anomaly*.

Addition of a tuple to FLIGHT to indicate that the mileage for Cincinnati to Houston is 1100 (**Origin**: 'Cincinnati', **Destination**: 'Houston', **Mileage**: 1100) is not possible without a **Flight#** identifying this route. Since **Flight#** is the primary key of FLIGHT, it cannot have null values. This is an *insertion anomaly*.

Once again, if the normalization of FLIGHT to 3NF by the removal of the undesirable FD:

{Origin, Destination} → Mileage

eliminates the modification anomalies, we can infer that the data redundancy and the consequent modification anomalies are due to the presence of the undesirable FD in FLIGHT.

The resolution of 3NF violation is accomplished by applying the same two-step process used earlier to resolve the 2NF violation (see Section 8.1.2). To review, the two-step process is:

1. Pull out the undesirable FD(s) from the target relation schema as a separate relation schema.
2. Retain the determinant of the pulled-out relation schema as an attribute(s) (foreign key) of the leftover target relation schema to facilitate reconstruction of the original target relation schema.

Accordingly, we have the decomposition, D:[4]

D: DISTANCE **(Origin, Destination, Mileage)**; FLIGHT **(Flight#, Origin, Destination)**

Note that DISTANCE is an arbitrary (meaningful) name assigned to the decomposed relation schema. The leftover target relation schema retains the same name as FLIGHT.

There are no 3NF violations in the decomposed set of relation schemas, DISTANCE and FLIGHT. So, the solution has yielded a relational schema that is in 3NF. A review of the corresponding decomposed relations (see Figure 8.3) reveals that there are no data redundancies causing modification anomalies in either relation of the 3NF design. It is now possible to insert the tuple (**Origin**: 'Cincinnati', **Destination**: 'Houston', **Mileage**: 1100) in the database (i.e., in DISTANCE) without having a value for **Flight#**. Deletion of the tuple identified by **Flight#** 507 in FLIGHT no longer gets rid of the information that Seattle to Denver is 1537 miles from the database (i.e., from DISTANCE). So, we can infer that the resolution of undesirable FDs causing the 3NF violation eliminated the data redundancies and the associated anomalies. The controlled redundancy between DISTANCE and FLIGHT in the 3NF design via the referencing attributes **{Origin, Destination}** establishes referential integrity constraint between the two relation schemas as reflected in the inclusion dependency:

FLIGHT.{Origin, Destination} ⊆ DISTANCE.{Origin, Destination}

which can also be used to reconstruct the original target relation schema.

---

[4] There are two other ways to decompose FLIGHT. The merits/demerits of those solutions are discussed later in this chapter (see Section 8.1.5).

R: FLIGHT (Flight#, Origin, Destination, Mileage)

FLIGHT

Flight#	Origin	Destination	Mileage
DL507	Seattle	Denver	1537
DL123	Chicago	Dallas	1058
DL723	Boston	St. Louis	1214
DL577	Denver	Los Angeles	1100
DL5219	Minneapolis	St. Louis	580
DL357	Chicago	Dallas	1058
DL555	Denver	Houston	1100
DL5237	Cleveland	St. Louis	580
DL5271	Chicago	Cleveland	300

F: fd1: Flight# → Origin;   fd2: Flight# → Destination;
          fd3: (Origin, Destination) → Mileage

FLIGHT is in 1NF (No composite or multi-valued attributes in FLIGHT)

F^{+}:   F;
          fd12: Flight# → (Origin, Destination);   fd3x: Flight# → Mileage;
          fd123: Flight# → (Origin, Destination, Mileage)

          Candidate key of FLIGHT: Flight#   Primary key of FLIGHT: Flight#

fd1 & fd2 are desirable FDs – why?
                        Determinant in both cases is a candidate key (primary key) of FLIGHT

fd3 does not violate 2NF (not a partial dependency) – FLIGHT, therefore, is in 2NF
but, fd3 violates 3NF – why?
                        Non-prime determines another non-prime in FLIGHT

Solution:

                        1< - - - R2 - - ->n
D: R1: FLIGHT (Flight#, Origin, Destination);   R2: DISTANCE (Origin, Destination, Mileage)
                        1   R   1

FLIGHT

Flight#	Origin	Destination
DL507	Seattle	Denver
DL123	Chicago	Dallas
DL723	Boston	St. Louis
DL577	Denver	Los Angeles
DL5219	Minneapolis	St. Louis
DL357	Chicago	Dallas
DL555	Denver	Houston
DL5237	Cleveland	St. Louis
DL5271	Chicago	Cleveland

DISTANCE

Origin	Destination	Mileage
Seattle	Denver	1537
Chicago	Dallas	1058
Boston	St. Louis	1214
Denver	Los Angeles	1100
Minneapolis	St. Louis	580
Denver	Houston	1100
Cleveland	St. Louis	580
Chicago	Cleveland	300

Figure 8.3 An example of 3NF violation and resolution

8.1.4 Boyce-Codd Normal Form (BCNF)

After Codd (1972) proposed the first three normal forms, it was discovered that the 3NF did not satisfactorily handle a more general case of undesirable functional dependencies. In other words, data redundancies and the consequent modification anomalies due to functional dependencies can persist even after a relation schema is normalized to 3NF. In particular, modification anomalies persist if the following pertain:

- A relation schema has at least two candidate keys,
- Both candidate keys are composite attributes, and
- There is an attribute overlap between the two candidate keys.

In order to rectify this inadequacy, Boyce and Codd (1974) proposed an improved version of 3NF definition. Since the definition, while simpler, is strictly stronger than the original definition of 3NF, it is given a different name, **Boyce-Codd Normal Form (BCNF)**.

Consider a relation schema R (X, A, B, C) where X, A, B, and C are pair-wise disjoint atomic or composite attributes. Suppose the set of FDs (a minimal cover over R), F, given below prevail over R.

F: fd1: **{X, A} → B**; fd2: **{X, A} → C**; and fd3: **B → A**

Using Armstrong's axioms, we first infer from fd1 and fd2 that **{X, A}** is a candidate key of R. Then, based on fd3, we infer that **{X, B}** is another candidate key of R (using the pseudotransitivity rule of Armstrong's axioms). Choosing **{X, A}** as the primary key of R, R is in 2NF because there are no partial dependencies in R.

R is also in 3NF because there is no transitive dependency of a non-prime attribute on the primary key.

Note: **B → A** (fd3) does not violate 3NF because **A** is a *prime* attribute. In fact, in fd3 a *non-prime* attribute determines a *prime* attribute.

Therefore, R is, indeed, in 3NF when evaluated on the basis of **{X, A}** as the primary key of R. Observe that **B → A**, by definition, is an undesirable FD in R simply because **B** is not a candidate key of R.

DEFINITION

BCNF definition: A relation schema R is in BCNF if for every non-trivial functional dependency in R, the determinant is a superkey of R.

The immediate questions, then, ought to be: "Is there any data redundancy in R? If so, does it cause any modification anomalies?" To explore this condition, let us review an example.

STU_SUB (Stu#, Subject, Teacher, Ap_score)

F: fd1: **{Stu#, Subject} → Teacher**; fd2: **{Stu#, Subject} → Ap_score**; fd3: **Teacher → Subject**

It is obvious that **{Stu#, Subject}** is a candidate key of STU_SUB. Choosing this as the primary key of STU_SUB, fd1, fd2, and fd3 do not violate either 2NF or 3NF. In fact, it can be shown that no FD in F⁺ violates 2NF or 3NF.

A relation instance for STU_SUB appears in Figure 8.4 where one can observe that **{Subject, Teacher}** pairs are redundantly recorded. Does this cause any anomalies? Since **Teacher → Subject**, if we want to add a new **Teacher** for a **Subject** (e.g., Teacher: 'Salter', Subject: 'English'), it is not possible to do so unless a corresponding **Stu#** is also provided.

Likewise, if Campbell is no longer advising and is being replaced by, say, Smith, multiple tuples require modification. These are cases of *insertion anomaly* and *update anomaly*, respectively, in STU_SUB. In short, STU_SUB is in 3NF and yet modification anomalies are present in it. An examination of F (or F$^+$) reveals that fd3 **(Teacher → Subject)** in STU_SUB is an undesirable FD because the determinant in fd3, **Teacher**, is not a candidate key of STU_SUB. Since fd3 is not a trivial dependency (i.e., the dependent in fd3 is not a subset of the determinant of fd3), the fact that **Teacher** is not a superkey of STU_SUB causes a BCNF violation in STU_SUB per the definition of BCNF.

Again, if the removal of the undesirable FD, fd3 **(Teacher → Subject)** from STU_SUB eliminates the modification anomalies, we can infer that the data redundancy and the consequent anomalies are due to the presence of the undesirable FD in STU_SUB. Let us see if the removal of fd3, the undesirable FD, eliminates the BCNF violation also from STU_SUB. The resolution of BCNF violation is accomplished by applying the same two-step process used earlier to resolve the 2NF and 3NF violations (see Sections 8.1.2 and 8.1.3):

1. Pull out the undesirable FD(s) from the target relation schema as a separate relation schema.
2. Retain the determinant of the pulled-out relation schema as an attribute(s) (foreign key) of the leftover target relation schema to facilitate reconstruction of the original target relation schema.

Accordingly, we have the decomposition, D:

D: TEACH_SUB **(Teacher, Subject)**; STU_AP **(Stu#, Teacher, Ap_score)**

Note: TEACH_SUB and STU_AP are arbitrary (meaningful) names assigned to the decomposed relation schemas.[5]

There are no BCNF violations in the decomposed set of relation schemas TEACH_SUB and STU_AP, because in both relation schemas the determinant of the only FD present in each is a superkey of the respective relation schemas. So, the solution has yielded a relational schema that is in BCNF. Reviewing the corresponding decomposed relations (see Figure 8.4), it is seen that there are no data redundancies causing modification anomalies in either relation of the BCNF design. It is now possible to insert the tuple (**Teacher**: 'Salter', **Subject**: 'English') in the database (i.e., in TEACH_SUB) without having a value for **Stu#**. Replacement of Campbell as a teacher requires change of attribute value in just one tuple (i.e., in TEACH_SUB). So, we can infer that the resolution of the undesirable FD causing the BCNF violation eliminated the data redundancies and the associated modification anomalies. The **controlled redundancy** between TEACH_SUB and STU_AP in the BCNF design via the referencing attributes (**Teacher**) establishes referential integrity constraint between the two relation schemas as reflected in the inclusion dependency:

STU_AP.{Teacher} ⊆ TEACH_SUB.{TEACHER}

which can also be used to reconstruct the original target relation schema.

---

[5] Note that (Stu#, Teacher) → Ap_score is in F$^+$ and is also a candidate key of STU_SUB. By choosing (Stu#, Teacher) as the candidate key, the BCNF violation can be viewed as a 2NF violation and resolved accordingly to produce the same answer. There is another BCNF decomposition of STU_SUB. The merits/demerits of this solution are discussed later in this chapter (see Section 8.1.5.3).

R: STU_SUB (Stu#, Subject, Teacher, Ap_score)

STU_SUB

Stu#	Subject	Teacher	Ap_score
IH123	Chemistry	Raturi	4
IH123	English	Stephan	4
IH235	History	Walker	5
IH357	English	Campbell	4
IH571	Chemistry	Raturi	3
IH235	English	Campbell	4

F   fd1: (Stu#, Subject) ‡ Teacher;   fd2: (Stu#, Subject) ‡ Ap_score;
       fd3: Teacher ‡ Subject

STU_SUB is in 1NF (No composite or multi-valued attributes in STU_SUB)

       Candidate keys of STU_SUB are: (Stu#, Subject);   (Stu#, Teacher)

       Primary key of STU_SUB: (Stu#, Subject)   [Chosen for this example]

fd1 & fd2 are desirable FDs – why?
                        Determinant in both cases is a candidate key (primary key) of STU_SUB

fd3 does not violate 2NF (not a partial dependency)
STU_SUB, therefore, is in 2NF for the chosen primary key

fd3 does not violate 3NF (non-prime attribute not a determinant of another non-prime attribute)
 STU_SUB, therefore, is in 3NF

But, fd3 violates BCNF – why?
                        Determinant is not a candidate key of STU_SUB
                        (Non-prime determines a prime in STU_SUB)

Solution:

                                                              1 R1 n
D: R1:   TEACH_SUB (Teacher, Subject);   R2:   STU_AP (Stu#, Teacher, Ap_score)
                                                              1 R 1

TEACH_SUB

Teacher	Subject
Raturi	Chemistry
Stephan	English
Walker	History
Campbell	English

STU_AP

Stu#	Teacher	Ap_score
IH123	Raturi	4
IH123	Stephan	4
IH235	Walker	5
IH357	Campbell	4
IH571	Raturi	3
IH235	Campbell	4

Figure 8.4   An example of BCNF violation and resolution

8.1.5 Side Effects of Normalization

So far we have approached the normalization process strictly from the perspective of eliminating data redundancies that lead to modification anomalies. Normalization entails decomposition of the target relation schema, R, into multiple relation schemas. D: {R1, R2, ... Rn} such that the join of {R1, R2, ... Rn} is strictly equal to R. First, we must make sure that each attribute in the target relation schema, R, is present in some relation schema, Ri, in the decomposition, D—i.e., all attributes of R should collectively appear in D (no attribute can be lost in the decomposition). This basic condition of decomposition is called **attribute preservation**. In addition, two other critical aspects of design require attention during the decomposition that results from normalization. They are *dependency preservation* and *lossless-join*[6] *decomposition*. These two independent properties are an expected requirement of an "ideal" design and both are always tied to a set of functional dependencies, F that holds over the relation schema, R being normalized.

8.1.5.1 Dependency Preservation

A relation schema, R, under scrutiny for normalization preserves all FDs, F specified over R in a single relation schema. It is logical to expect that any decomposition of R continue to preserve the minimal cover of all FDs (F is always one of the covers, but not necessarily a minimal cover) that hold in R. Since the *specified* functional dependencies, F, arise from business rules, they will have to be accounted for in the database implementation. In other words, each FD in F represents a constraint on the database. Therefore, a robust logical database design should fully preserve the set of FDs, F specified on a relation schema, R, across the decomposition, D of R. More importantly, it is necessary that each FD in F should either directly appear in *single* decomposed relation schemas in D, or be inferable from the FDs that appear in *single* decomposed relation schemas. Only then is the decomposition considered **dependency-preserving**. If there is a need to join two or more decomposed relation schemas to ascertain an FD in F, then the decomposition is *not* dependency-preserving. At the same time, it is not necessary that the exact dependencies present in F appear in one of the relation schemas in the decomposition D. *It is sufficient if the union of the FDs that hold on individual relation schemas of D is a cover for F*—i.e., can generate F^+. When a modification (insertion, deletion, or update) is made to the database, the DBMS should be able to check that the semantics of the relations in the database are not corrupted by the modification—i.e., one or more FDs specified in F do not hold in the database any longer because of the modification. Efficient validation requires that there is no need to join two or more relations to verify FDs.

Let us review the examples of normalization from the previous sections to see if the normalized decompositions are dependency-preserving. In the 2NF example (see Section 8.1.2):

    F: fd1: **Album_no → Price**; fd2: **Album_no → Stock**
is specified over the relation schema:

    R: NEW_ALBUM **(Album_no, Artist_nm, Price, Stock)**
    The 2NF decomposition of NEW_ALBUM is:

D: R1: ALBUM_INFO **(Album_no, Price, Stock)**; R2: ALBUM_ARTIST **(Album_no, Artist_nm)**

---

[6] The lossless-join property is also referred to as non-additive join property or sometimes nonloss property.

First of all, the decomposition is attribute preserving since the union of all attributes in D is exactly the same as the attributes in R. The union of the FDs that hold on individual relation schemas of D are:

Album_no → Price; and Album_no → Stock

Since this set of FDs is exactly equivalent to F specified on R, the 2NF decomposition, D, is dependency-preserving. Note that it is impossible to decompose NEW_ALBUM any other way to achieve a 2NF design.

Also notice that F^+ contains other non-trivial FDs:

F^+: fd12: **Album_no → {Price, Stock};** fd12x: **{Album_no, Artist_nm} → {Price, Stock}**

fd12 and fd12x can be generated from the FDs preserved in D, specifically in this case, from ALBUM_INFO using Armstrong's axioms.

Next, consider the 3NF example in Section 8.1.3. Here,

F: fd1: **Flight# → Origin;** fd2: **Flight# → Destination;** fd3: **{Origin, Destination} → Mileage**

is specified over the relation schema:

R: FLIGHT **(Flight#, Origin, Destination, Mileage)**

The 3NF decomposition of FLIGHT is:

D: R1: FLIGHT **(Flight#, Origin, Destination);** R2: DISTANCE **(Origin, Destination, Mileage)**

Once again, the decomposition is attribute preserving since the union of all attributes in D is exactly the same as the attributes in R. The union of the FDs that hold on individual relation schemas of D are:

Flight# → Origin; and **Flight# → Destination;** (in R1) and **{Origin, Destination} → Mileage (in R2)**

Since this set of FDs is exactly equivalent to F specified on R, the 3NF decomposition, D, is dependency-preserving.

F^+ contains other non-trivial FDs listed here:

F^+: fd12: **Flight# → {Origin, Destination};** fd3x: **Flight# → Mileage;** fd123: **Flight# → {Origin, Destination, Mileage}**

All the FDs in F^+ can be generated from the FDs preserved in D using Armstrong's axioms.

Unlike the 2NF example, here, two other decompositions of FLIGHT are possible. They are:

D: R1a: FLIGHT_A **(Flight#, Origin, Destination);** R2a: DISTANCE_A **(Flight, Mileage)**

D: R1b: FLIGHT_B **(Flight#, Mileage);** R2b: DISTANCE_B **(Origin, Destination, Mileage)**

Interestingly, both decompositions yield a design that is in 3NF. Both decompositions are attribute preserving since the union of all attributes in D in each case is exactly the same as the attributes in R.

Let us explore if these two designs are dependency-preserving. The union of the FDs that hold on individual relation schemas of D: {R1a, R2a} are:

Flight# → Origin; and **Flight# → Destination;** (in R1a) and **Flight# → Mileage (in R2a)**

Observe that this set of FDs is not a cover for F because it is impossible to deduce fd3: **{Origin, Destination} → Mileage** of F from this set of FDs. Therefore, D: {R1a, R2a} is not a dependency-preserving design.

Likewise, for the second design, D: {R1b, R2b}, the union of the FDs that hold on individual relation schemas of D is:

Flight# → Mileage; {Origin, Destination} → Mileage

Observe that from this set of FDs it is impossible to derive fd1: **Flight# → Origin;** and fd2: **Flight# → Destination** of F. Therefore, D: {R1b, R2b} is not a dependency-preserving design either.

Let us now compare an example of a dependency-preserving decomposition (D: {R1, R2}) with another decomposition where dependency is not preserved (D: {R1a, R2a}) to get a better grip on what dependency preservation actually means. The table at the top of Figure 8.5 displays the relation instance for FLIGHT that is not in 3NF. Nonetheless, the FD **{Origin, Destination} → Mileage** is preserved in FLIGHT. The dependency-preserving decomposition D: {FLIGHT, DISTANCE} and a decomposition, D{FLIGHT_A, DISTANCE_A} that does not preserve the FD, **{Origin, Destination} → Mileage** are also included in Figure 8.5.

The clarification sought here is about what it means to not preserve the FD **{Origin, Destination} → Mileage**. Suppose we want to add a new flight (**Flight#**: DL111, **Origin**: Seattle, **Destination**: Denver, **Mileage**: 1300). In the 3NF decomposition that is dependency-preserving (the set of tables in the middle of Figure 8.5), it is legal to add the tuple (**Flight#**: DL111, **Origin**: Seattle, **Destination**: Denver) to FLIGHT, while it is not possible to add the tuple (**Origin**: Seattle, **Destination**: Denver, **Mileage**: 1300) to DISTANCE because **{Origin, Destination}** is the primary key of DISTANCE. Since a tuple with values (**Origin**: Seattle, **Destination**: Denver, **Mileage**: 1537) already exists in DISTANCE, another tuple that contains a duplicate value for the primary key cannot be added. Thus, **{Origin, Destination} → Mileage** continues to mean that for a given value of **{Origin, Destination}**, say ('Seattle', 'Denver'), there is a single, specific value of **Mileage**, say 1537—i.e., the FD is preserved. Also, when the two relations (R1 and R2) are joined, the FD **{Origin, Destination} → Mileage** will continue to be preserved as in the original FLIGHT table from which the decomposition occurred.

On the other hand, in the 3NF decomposition that is *not* dependency-preserving (the set of tables in the bottom of Figure 8.5), it certainly is legal to add the tuple (**Flight#**: DL111, **Origin**: Seattle, **Destination**: Denver) to FLIGHT, and it is equally legal to add the tuple (**Flight#**: DL111, **Mileage**: 1300) to DISTANCE. Clearly it is not possible to verify the FD **{Origin, Destination} → Mileage** from any *single* relation. If there is a need to combine (join) multiple relations (in this case, two) to check for an FD, then, by definition, that dependency is not preserved in the relational schema.

What does this entail? This can be seen by joining the two relations, R1a and R2a. When the two relations are joined, observe that the FD **{Origin, Destination} → Mileage** is not preserved because there is a tuple (**Flight#**: DL111, **Origin**: Seattle, **Destination**: Denver, **Mileage**: 1537) and another tuple (**Flight#**: DL111, **Origin**: Seattle, **Destination**: Denver, **Mileage**: 1300) in the joined relation; this means that for ('Seattle', 'Denver') we do not have a single, specific value of **Mileage**—there are two values: 1537 and 1300. This demonstrates the seriousness of failure to preserve the FD, **{Origin, Destination} → Mileage**. Essentially, the database has been contaminated with incorrect data. In short, failure to preserve the specified functional dependencies renders the resulting database vulnerable to contamination in the context of the business rules conveyed by the specified functional dependencies.

A similar analysis can be conducted about the other 3NF decomposition, D: {R1b, R2b}. It is important to note that the simple algorithm reflected by the two-step process prescribed will always yield a dependency-preserving decomposition for a relation schema that violates 2NF or 3NF (e.g., D: {R1, R2}).

R: FLIGHT (Flight#, Origin, Destination, Mileage)

FLIGHT

Flight#	Origin	Destination	Mileage
DL507	Seattle	Denver	1537
DL123	Chicago	Dallas	1058
DL723	Boston	St. Louis	1214
DL577	Denver	Los Angeles	1100
DL5219	Minneapolis	St. Louis	580
DL357	Chicago	Dallas	1058
DL555	Denver	Houston	1100
DL5237	Cleveland	St. Louis	580
DL5271	Chicago	Cleveland	300

1< - - - R2 - - -> n

D: R1: FLIGHT (Flight#, Origin, Destination);   R2: DISTANCE (Origin, Destination, Mileage)

1   R   1

FLIGHT

Flight#	Origin	Destination
DL507	Seattle	Denver
DL123	Chicago	Dallas
DL723	Boston	St. Louis
DL577	Denver	Los Angeles
DL5219	Minneapolis	St. Louis
DL357	Chicago	Dallas
DL555	Denver	Houston
DL5237	Cleveland	St. Louis
DL5271	Chicago	Cleveland
DL111	Seattle	Denver	Legal

DISTANCE

Origin	Destination	Mileage
Seattle	Denver	1537
Chicago	Dallas	1058
Boston	St. Louis	1214
Denver	Los Angeles	1100
Minneapolis	St. Louis	580
Denver	Houston	1100
Cleveland	St. Louis	580
Chicago	Cleveland	300
~~Seattle~~	~~Denver~~	~~1300~~	Illegal

1 R2 n

D: R1a: FLIGHT_A (Flight#, Origin, Destination);   R2a: DISTANCE_A (Flight, Mileage)

1 R 1

FLIGHT_A

Flight#	Origin	Destination
DL507	Seattle	Denver
DL123	Chicago	Dallas
DL723	Boston	St. Louis
DL577	Denver	Los Angeles
DL5219	Minneapolis	St. Louis
DL357	Chicago	Dallas
DL555	Denver	Houston
DL5237	Cleveland	St. Louis
DL5271	Chicago	Cleveland
DL111	Seattle	Denver	Legal

DISTANCE_A

Flight	Mileage
DL507	1537
DL123	1058
DL723	1214
DL577	1100
DL5219	580
DL357	1058
DL555	1100
DL5237	580
DL5271	300
DL111	1300	Legal

Note: Shading indicates that addition of the tuple is "legal." Strike-out indicates that addition of the tuple is "illegal."

Figure 8.5   A demonstration of the dependency preservation property

8.1.5.2 Lossless-Join (Non-Additive Join) Property

The basic principle behind a **lossless-join decomposition** is that the decomposition of a relation schema, R, should be strictly reversible—i.e., losslessly reversible in that the reversal should yield the original target relation intact with no loss of tuple or no additional spurious tuples. The condition stated holds on all legal states of a relation schema—legal state in the sense that the relation state conforms to the set of FDs, F, specified on the relation schema, R. It is important to note that the lossless-join property:

- Is always stated with respect to a set of FDs, and
- Is predicated on the premise that the join attributes in the decomposition are non-null values.

Even though the decomposition of a relation schema during normalization may ultimately yield a set of multiple relation schemas, in the stepwise process of normalization, each step typically involves a decomposition that produces two relation schemas—i.e., a **binary decomposition**. Therefore, we will address the lossless-join property in binary decompositions.

Formally, a decomposition D: {R1, R2} of a relation schema, R, is lossless (non-additive) with respect to a set of FDs, F specified on R, if for every relation state r of R that satisfies F, the natural join of r(R1) and r(R2) strictly yields r(R) from which the projections r(R1) and r(R2) emerged.

Note that the term "loss" in lossless-join implies loss of information—not loss of tuples. In fact, *loss join* occurs when the natural join of r(R1) and r(R2) yields r'(R) which includes additional spurious tuples beyond r(R) from which r(R1) and r(R2) are projected. The additional spurious tuples amount to loss of information because their presence corrupts the semantics of the source relation schema, R—i.e., the FDs specified on R no longer hold good in r'(R). Then, the decomposition is a loss-join decomposition, not a lossless-join decomposition.

At this point, let us revisit the FLIGHT example. The three 3NF solutions for this example are reproduced here:

F: fd1: **Flight# → Origin**; fd2: **Flight# → Destination**; fd3: **{Origin, Destination} → Mileage**

is specified over the relation schema:

R: FLIGHT **(Flight#, Origin, Destination, Mileage)**

The three 3NF solutions of FLIGHT are the decompositions:

D:	R1: FLIGHT (**Flight#**, Origin, Destination);	R2: DISTANCE (**Origin, Destination**, Mileage)
D:	R1a: FLIGHT_A (**Flight#**, Origin, Destination);	R2a: DISTANCE_A (**Flight#**, Mileage)
D:	R1b: FLIGHT_B (**Flight#**, Mileage);	R2b: DISTANCE_B (**Origin, Destination**, Mileage)

Figure 8.6a shows an instance of FLIGHT and the decompositions corresponding to the three 3NF solutions, each with its associated reversals. In case 1, D: {FLIGHT, DISTANCE} is strictly reversible because {FLIGHT * DISTANCE} yields exactly the source relation instance, {FLIGHT}. Thus, the decomposition is, by definition, a lossless-join

decomposition. In the second case also, the natural join {FLIGHT_A * DISTANCE_A} of the projections {FLIGHT_A} and {DISTANCE_A} produces the source relation instance {FLIGHT} intact. So, this solution is also a lossless-join decomposition. In case 3, however, the natural join {FLIGHT_B * DISTANCE_B} of the projections {FLIGHT_B} and {DISTANCE_B} produces a relation state of {FLIGHT} that contains four additional tuples beyond what is present in the source relation instance {FLIGHT}. A close examination of this relation state reveals that the presence of these new tuples changes the semantics incorporated in the original design of the relation schema, FLIGHT. For instance, the source relation instance {FLIGHT} precisely indicates that Flight# 577 flies from Denver to Los Angeles. However, from the case 3 solution we infer that Flight# 577 flies from Denver to Los Angeles as well as from Denver to Houston, which is incorrect according to the FDs specified over the relation schema, FLIGHT. Therefore, even though we have a 3NF solution, the result is not a lossless-join decomposition.

A review of this example in Section 8.1.5.1 combined with the above analysis indicates that while all three solutions yield 3NF decompositions,

- {FLIGHT, DISTANCE} delivers a lossless-join decomposition that is also dependency-preserving with respect to the FDs, F specified on the relation schema FLIGHT.
- {FLIGHT_A, DISTANCE_A} delivers a lossless-join decomposition. However, the decomposition is *not* dependency-preserving with respect to the FDs, F specified on the relation schema FLIGHT.
- {FLIGHT_B, DISTANCE_B} delivers a *loss-join* decomposition that also fails to preserve the FDs, F specified on the relation schema FLIGHT.

Clearly, the decomposition {FLIGHT, DISTANCE} is the only acceptable design. Once again, note that the simple algorithm incorporated in the two-step decomposition process always yields 2NF and 3NF solutions that are dependency-preserving and also possess the lossless-join property.

A test for verifying the lossless-join property of a binary decomposition can be specified as follows. A decomposition D: {R1, R2} of a relation schema, R, is a lossless-join decomposition with respect to a set of FDs, F that holds on R, if and only if F^+ contains:

- either the FD $\qquad$ $(R1 \cap R2) \rightarrow R1$
- or the FD $\qquad$ $(R1 \cap R2) \rightarrow R2$

In other words, the attribute(s) common to R1 and R2 must contain a candidate key of either R1 or R2.[7] In our example, the join attribute in case 1 solution is {**Origin, Destination**} which is the primary key of DISTANCE. Likewise, in the decomposition in case 2, the join attribute, **Flight#**, is the primary key of both FLIGHT_A and DISTANCE_A. Therefore, both these solutions offer lossless-join decompositions, as confirmed by the data in Figure 8.6a. The third solution presented, however, fails the test for lossless-join decomposition prescribed above, because the join attribute in this case, **Mileage**, does not contain a candidate key in either FLIGHT_B or DISTANCE_B. Once again, the spurious tuples in the natural join {FLIGHT_B * DISTANCE_B} (see Figure 8.6a) confirm this.

---

[7] A more general specification of the test is: The FD $(R1 \cap R2) \rightarrow (R1-R2)$ or the FD $(R1 \cap R2) \rightarrow (R2-R1)$.

R: FLIGHT (Flight#, Origin, Destination, Mileage)

FLIGHT

Flight#	Origin	Destination	Mileage
DL507	Seattle	Denver	1537
DL123	Chicago	Dallas	1058
DL723	Boston	St. Louis	1214
DL577	Denver	Los Angeles	1100
DL5219	Minneapolis	St. Louis	580
DL357	Chicago	Dallas	1058
DL555	Denver	Houston	1100
DL5237	Cleveland	St. Louis	580
DL5271	Chicago	Cleveland	300

1 <--- R2 ---> n

D: R1: FLIGHT (Flight#, Origin, Destination); R2: DISTANCE (Origin, Destination, Mileage)

1 R 1

FLIGHT

Flight#	Origin	Destination
DL507	Seattle	Denver
DL123	Chicago	Dallas
DL723	Boston	St. Louis
DL577	Denver	Los Angeles
DL5219	Minneapolis	St. Louis
DL357	Chicago	Dallas
DL555	Denver	Houston
DL5237	Cleveland	St. Louis
DL5271	Chicago	Cleveland

DISTANCE

Origin	Destination	Mileage
Seattle	Denver	1537
Chicago	Dallas	1058
Boston	St. Louis	1214
Denver	Los Angeles	1100
Minneapolis	St. Louis	580
Denver	Houston	1100
Cleveland	St. Louis	580
Chicago	Cleveland	300

{FLIGHT * DISTANCE}

Flight#	Origin	Destination	Mileage
DL507	Seattle	Denver	1537
DL123	Chicago	Dallas	1058
DL723	Boston	St. Louis	1214
DL577	Denver	Los Angeles	1100
DL5219	Minneapolis	St. Louis	580
DL357	Chicago	Dallas	1058
DL555	Denver	Houston	1100
DL5237	Cleveland	St. Louis	580
DL5271	Chicago	Cleveland	300

1 R2 n

D: R1a: FLIGHT_A (Flight#, Origin, Destination); R2a: DISTANCE_A (Flight, Mileage)

1 R 1

FLIGHT_A

Flight#	Origin	Destination
DL507	Seattle	Denver
DL123	Chicago	Dallas
DL723	Boston	St. Louis
DL577	Denver	Los Angeles
DL5219	Minneapolis	St. Louis
DL357	Chicago	Dallas
DL555	Denver	Houston
DL5237	Cleveland	St. Louis
DL5271	Chicago	Cleveland

DISTANCE_A

Flight	Mileage
DL507	1537
DL123	1058
DL723	1214
DL577	1100
DL5219	580
DL357	1058
DL555	1100
DL5237	580
DL5271	300

{FLIGHT_A * DISTANCE_A}

Flight#	Origin	Destination	Mileage
DL507	Seattle	Denver	1537
DL123	Chicago	Dallas	1058
DL723	Boston	St. Louis	1214
DL577	Denver	Los Angeles	1100
DL5219	Minneapolis	St. Louis	580
DL357	Chicago	Dallas	1058
DL555	Denver	Houston	1100
DL5237	Cleveland	St. Louis	580
DL5271	Chicago	Cleveland	300

1 R2 n

D: R1b: FLIGHT_B (Origin, Destination, Mileage); R2b: DISTANCE_A (Flight, Mileage)

1 R 1

FLIGHT_B

Origin	Destination	Mileage
Seattle	Denver	1537
Chicago	Dallas	1058
Boston	St. Louis	1214
Denver	Los Angeles	1100
Minneapolis	St. Louis	580
Denver	Houston	1100
Cleveland	St. Louis	580
Chicago	Cleveland	300

DISTANCE_B

Flight	Mileage
DL507	1537
DL123	1058
DL723	1214
DL577	1100
DL5219	580
DL357	1058
DL555	1100
DL5237	580
DL5271	300

{FLIGHT_B * DISTANCE_B}

Flight#	Origin	Destination	Mileage
DL507	Seattle	Denver	1537
DL123	Chicago	Dallas	1058
DL723	Boston	St. Louis	1214
DL577	Denver	Los Angeles	1100
DL577	Denver	Houston	1100
DL5219	Minneapolis	St. Louis	580
DL5219	Cleveland	St. Louis	580
DL357	Chicago	Dallas	1058
DL555	Denver	Los Angeles	1100
DL555	Denver	Houston	1100
DL5237	Minneapolis	St. Louis	580
DL5237	Cleveland	St. Louis	580
DL5271	Chicago	Cleveland	300

Note: Spurious tuples causing a loss-join decomposition are marked by shading

Figure 8.6a A demonstration of lossless and loss-join decompositions

Normal Forms Based on Functional Dependencies

8.1.5.3 BCNF Design, Dependency-Preservation, and Lossless-Join Decomposition

Decomposition of a relation schema that is in 3NF, but is in violation of BCNF, poses an interesting trade-off. Let us study this using the STU_SUB example from Section 8.1.4. Here, F:

fd1: **{Stu#, Subject}** → **Teacher;**   fd2: **{Stu#, Subject}** → **Ap_score;**   fd3: **Teacher** → **Subject**

is specified over the relation schema:

R:STU_SUB **(Stu#, Subject, Teacher, Ap_score)**

STU_SUB is in 3NF, but violates BCNF since **Teacher**, the determinant in fd3, is not a superkey of STU_SUB.

From an alternative viewpoint, from F, it is possible to infer that:

fd4: **{Stu#, Teacher}** → **Subject;**   fd5: **{Stu#, Teacher}** → **Ap_score;**

exist in F^+. Thus, **{Stu#, Teacher}** is a candidate key of STU_SUB, and if treated as the primary key of STU_SUB renders fd3 a violation of 2NF.

Either way, the decomposition of STU_SUB derived in Section 8.1.4 as the solution is:

D: R1: TEACH_SUB (**Teacher**, Subject);   R2: STU_AP (**Stu#, Teacher**, Ap_score)

First of all, the decomposition is attribute preserving since the union of all attributes in D is exactly the same as the attributes in R. The solution is a lossless-join decomposition because (TEACH_SUB ∩ STU_AP) is the attribute **Teacher** which is the candidate key of TEACH_SUB. This can be verified by doing a natural join of the relations TEACH_SUB and STU_AP that appear in Figure 8.4. This is shown in Figure 8.6b.

Incidentally, it is important to note that while G{fd1, fd2, fd3} is a cover (in fact, a minimal cover) for F^+, G' {fd1, fd3, fd4} is not a cover for F^+.

Is the solution D {R1, R2} dependency-preserving? The union of the FDs that hold on individual relation schemas of D are:

Teacher → **Subject (in R1)** and **{Stu#, Teacher}** → **Ap_score (in R2)**

Clearly, this set of FDs is not a cover for F. In other words, while fd3 (**Teacher** → **Subject**) is preserved, fd1 **{Stu#, Subject}** → **Teacher** and fd2 **{Stu#, Subject}** → **Ap_score** are not preserved in the solution. Thus the solution is not dependency-preserving.

Is there another decomposition that preserves all the FDs specified in F? Clearly, the two-step decomposition procedure that yields a 2NF decomposition and a 3NF decomposition possessing both the dependency-preserving and lossless-join properties, has failed to produce a BCNF solution that preserves both the properties. A second solution is shown here:

D: R1a: TEACH_SUB1 (**Teacher**, Subject);   R2a: STU_AP1 (**Stu#, Subject**, Ap_score)

The decomposition is in BCNF. The union of the FDs that hold on individual relation schemas of D {R1a, R2a} are:

Teacher → **Subject** (in R1a) and **{Stu#, Subject}** → **Ap_score** (in R2a)

R: STU_SUB (Stu#, Subject, Teacher, Ap_score)

STU_SUB

Stu#	Subject	Teacher	Ap_score
IH123	Chemistry	Raturi	4
IH123	English	Stephan	4
IH235	History	Walker	5
IH357	English	Campbell	4
IH571	Chemistry	Raturi	3
IH235	English	Campbell	4

F   fd1: (Stu#, Subject) → Teacher;   fd2: (Stu#, Subject) → Ap_score;
        fd3: Teacher → Subject

Option 1

                                                        1 R1 n
D: R1:   TEACH_SUB (Teacher, Subject);   R2:   STU_AP (Stu#, Teacher, Ap_score)
                                                        1 R 1

STU_AP

Stu#	Teacher	Ap_score
IH123	Raturi	4
IH123	Stephan	4
IH235	Walker	5
IH357	Campbell	4
IH571	Raturi	3
IH235	Campbell	4

TEACH_SUB

Teacher	Subject
Raturi	Chemistry
Stephan	English
Walker	History
Campbell	English

(STU_AP * TEACH_SUB)

Stu#	Subject	Teacher	Ap_score
IH123	Chemistry	Raturi	4
IH123	English	Stephan	4
IH235	History	Walker	5
IH357	English	Campbell	4
IH571	Chemistry	Raturi	3
IH235	English	Campbell	4

Option 2

                                                        1 R1a n
D: R1a: TEACH_SUB1 (Teacher, Subject);   R2a: STU_AP1 (Stu#, Subject, Ap_score)
                                                        1 R 1

STU_AP1

Stu#	Subject	Ap_score
IH123	Chemistry	4
IH123	English	4
IH235	History	5
IH357	English	4
IH571	Chemistry	3
IH235	English	4

TEACH_SUB1

Teacher	Subject
Raturi	Chemistry
Stephan	English
Walker	History
Campbell	English

(STU_AP1 * TEACH_SUB1)

Stu#	Subject	Teacher	Ap_score
IH123	Chemistry	Raturi	4
IH123	English	Stephan	4
IH123	English	Campbell	4
IH235	History	Walker	5
IH357	English	Stephan	4
IH357	English	Campbell	4
IH571	Chemistry	Raturi	3
IH235	English	Stephan	4
IH235	English	Campbell	4

Note: Spurious tuples causing a loss-join decomposition are marked by shading

Figure 8.6b   Lossless-join and dependency preservation in a BCNF resolution

365

Normal Forms Based on Functional Dependencies

Once again, this set of FDs is not a cover for F. In other words, while this solution preserves fd2 and fd3, fd1 is not preserved in this solution. In fact, *there is no BCNF solution that can preserve fd1*. Moreover, this second solution also fails to produce a lossless-join decomposition, as can be seen by applying the prescribed test for lossless-join decomposition. In this case, (TEACH_SUB1 ∩ STU_AP1) is the attribute **Subject** which is neither the candidate key of TEACH_SUB1 nor the candidate key of STU_AP1. Accordingly, the decomposition yields a *loss-join*. This can be verified by constructing the natural join of the relations TEACH_SUB1 and STU_AP1, as shown in Figure 8.6b. The spurious tuples that result from the natural join (TEACH_SUB1 * STU_AP1) are a clear proof for the absence of lossless-join property in the solution.

In sum, since both solutions do not provide full dependency preservation, the designer's choice ought to be the solution that generates at least lossless-join projections, if BCNF design is desired. If one has to aim for one of these two properties, a lossless-join condition is an absolute must (Elmasri and Navathe, 2004, p. 357).

Assuming that we always seek a lossless-join condition, if we are forced to choose between BCNF without preserving dependencies and 3NF with preserved dependencies, it is generally preferable to opt for the latter. After all, if one can't test for dependency preservation efficiently, one either pays a high penalty in system performance or risks the integrity of the data in the database. Neither is an attractive alternative. Thus, the limited amount of redundancy allowed under 3NF is regarded as the lesser of the two evils.

Therefore, the design goals can be expressed in two basic options.

Option 1

BCNF

Lossless-join

Dependency preservation

This is the ideal option and is achieved only when a relational schema is in 3NF and there are no BCNF violations in the relational schema, because then the relational schema is also in BCNF. If the above design cannot be achieved, we may have to settle for:

Option 2

3NF

Lossless-join

Dependency preservation

That said, with the advent of materialized views,[8] it is possible to always achieve option 1 rather cost-effectively. In the absence of BCNF (option 2), the application developer assumes the responsibility to keep redundant data consistent programmatically when modifications to the database occur. If we opt for a BCNF design (option 1), this cost of application programming incurred in option 2 is eliminated. However, we need to supplement the BCNF design with a materialized view for each unpreserved FD in the minimal cover of F. The advantage of this approach is that the DBMS takes care of the maintenance of the materialized views as and when modifications occur in the source relations, thus assuring

---

[8] A *view* defines a 'virtual' relation schema constructed from one or more base relation schemas; unlike base relations, a view does not store data; the value of a view at any given time is a 'derived' relation and results from the evaluation of a specified relational expression at that time. A View is just a logical window to view selected data (attributes and tuples) from one or a set of relation schemas. *Materialized views* (also known as *snapshots*) are also derived like views except that they are stored in the database and refreshed on every modification (i.e., maintained current by the DBMS as and when modifications occur) in the source relations from where the materialized views are generated.

preservation of the associated dependencies. While the DBMS overhead for the maintenance of materialized views requires consideration, often in practice, BCNF violations are few and far between. Costs and inefficiencies associated with application programming in this situation are often far more burdensome than the DBMS overhead.

In conclusion, where a dependency-preserving BCNF design is not possible, it is generally preferable to opt for BCNF (option 1) and supplement it with materialized views to preserve dependencies (Silberschatz, Korth, and Sudarshan 2002).

8.1.6 Summary Notes on Normal Forms

There are several points about normal forms and the normalization process that are noteworthy. Normalization systematically eliminates data redundancies that cause modification anomalies by decomposing the target relation schema that contains undesirable FDs. The decomposition is done in such a way that the FDs that were "undesirable" in the target relation schema are not discarded, but are rendered "desirable" in the resulting relational schema (set of relation schemas). Figure 8.7 summarizes normal form violations due to functional dependencies in a nutshell.

Observe that in the normalization process, there is no specific merit in resolving partial dependencies (2NF violations) before transitive dependencies (3NF violations) are resolved. It simply happens that historically 3NF has been defined with an assumption that a relation schema is tested for 2NF first and then for 3NF. Next, the definition of BCNF is the simplest and yet the strongest and most general among the definitions of normal forms associated with functional dependencies. Interestingly, it can be seen that a 2NF violation is also a BCNF violation. Similarly, a 3NF violation is also a BCNF violation. That is, BCNF, by definition, subsumes 2NF and 3NF. Nonetheless, in practice, it is useful to view the normal form violations in terms of partial dependency (2NF violation) and transitive dependency (3NF violation) and BCNF violation so that the database designer may systematically approach the problem and its solution in incremental steps. Attempting to solve all normal form violations due to functional dependencies as BCNF violations can be overwhelming because the approach to the solution can get tentative. In this context, hereafter we will use the term 'immediate' violation to refer to violations, of 2NF, 3NF, and BCNF so that we do not arbitrarily refer to any undesirable FD as a BCNF violation and be right.

For a database design to be robust, it is not enough if the designer focuses exclusively on the elimination of data redundancies that lead to modification anomalies. The designer must also pay attention to the dependency preservation property and lossless-join property of the design. *A 3NF decomposition D of a relation schema, R, with respect to a set of FDs, F, that possesses the lossless-join property as well as the dependency-preserving property always exists.* (Elmasri and Navathe 2003)

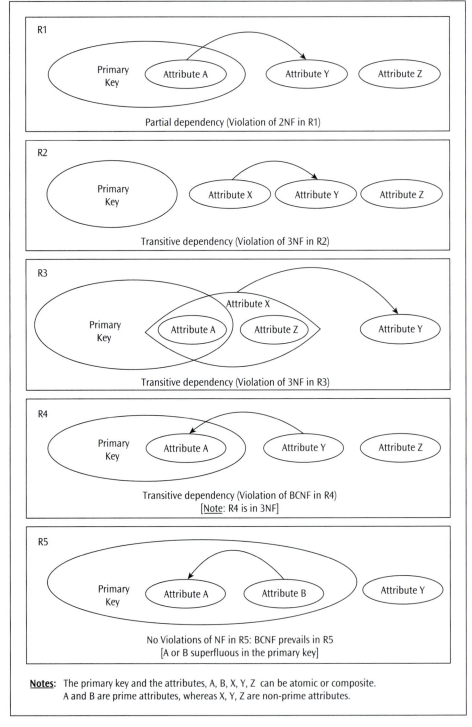

Figure 8.7   Normal forms based on functional dependencies in a nutshell

8.2 The Motivating Exemplar Revisited

At this point, normalization concepts have been presented by analyzing 1NF, 2NF, 3NF, and BCNF in isolation. In practice, however, normal form violations rarely occur in isolation. Therefore, we now explore a comprehensive approach to normalization by studying how, given a set of FDs derived from user-specified business rules, one develops a fully normalized relational schema from the *universal relation schema* that depicts the stated functional dependencies.

To begin with, let us revisit the motivating exemplar in Chapter 7 (see Section 7.1) and understand the process that transpired resulting in the decomposition of STOCK (Figure 7.1c) into STORE, PRODUCT, INVENTORY, and DISC_STRUCTURE (Figure 7.3).

Based on the data in the relation instance of STOCK in Figure 7.1c, we construct the following minimal cover of FDs, F for the relation schema:

R: STOCK (Store, Location, Sq_ft, Manager, Product, Price, Quantity, Discount)

fd1: **Store → Location;**	fd2: **Store → Sq_ft;**	fd3: **Store → Manager;**
fd4: **Product → Price;**	fd5: **{Store, Product} → Quantity**	fd6: **Quantity → Discount**
fd7: **{Manager, Location} → Store**		

The dependency diagram in Figure 8.8a provides a pictorial version of F.

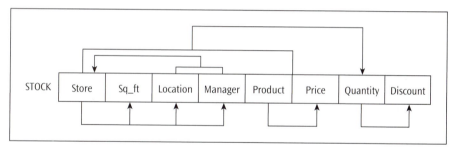

Figure 8.8a Dependency diagram for the relation schema STOCK

STOCK is in 1NF because there are no composite or multi-valued attributes in it.

Using Armstrong's axioms, we can derive **{Store, Product}** as one of the candidate keys and choose it as the primary key of STOCK. The other candidate key is **{Manager, Location, Product}**.

On the basis of **{Store, Product}** as the primary key of STOCK, fd1, fd2, fd3, and fd4 violate 2NF in STOCK; fd6 violates 3NF in STOCK; and fd7 violates BCNF in STOCK.

Neither 2NF nor 3NF nor BCNF is violated in STOCK by fd5.

Applying the two-step normalization process to the partial dependencies fd1, fd2, fd3, and fd4 causing the 2NF violation, we have the decomposition D,

D: {R1, R2, R3, R4, R5}
where:

R1: STORE_LOC (**Store**, **Location**);	R2: STORE_SIZE (**Store**, **Sq_ft**);
R3: STORE_MGR (**Store**, **Manager**);	R4:PRODUCT (**Product**, **Price**);
R5: INVENTORY (**Store, Product**, **Quantity, Discount**)	

and the inclusion dependencies:

INVENTORY.**{Store}** ⊆ STORE_LOC.**{Store}**
INVENTORY.**{Product}** ⊆ PRODUCT.**{Product}**

The decomposed relational schema, D, consisting of the five (arbitrarily named) relation schemas, STORE_LOC, STORE_SIZE, STORE_MGR, PRODUCT, and INVENTORY, does not have any 2NF violations anymore—i.e., D is in 2NF. In addition, the decomposition is lossless and preserves all the FDs in F.

In fact, STORE_LOC, STORE_SIZE, STORE_MGR, and PRODUCT are also in 3NF and BCNF.

However, INVENTORY is in violation of 3NF because fd6: **Quantity → Discount** causes a transitive dependency in INVENTORY. Application of the two step normalization process, once again, leads to the following decomposition of INVENTORY:

R5a: DISC_STRUCTURE (**Quantity, Discount**); R5b: INVENTORY (**Store, Product, Quantity**)

and the inclusion dependency:

INVENTORY.**{Quantity}** ⊆ DISC_STRUCTURE.**{Quantity}**

Both the decomposed relation schema, DISC_STRUCTURE, and the leftover relation schema, INVENTORY (R5b) are in 3NF as well as in BCNF.

Thus, a BCNF solution yields the following result. The relation schema STOCK where 2NF, 3NF, and BCNF violations were present is replaced by the relational schema that contains the set of relation schemas STORE_LOC, STORE_SIZE, STORE_MGR, PRODUCT, INVENTORY, and DISC_STRUCTURE as shown below:

D: {R1, R2, R3, R4, R5a, R5b}
where:

R1: STORE_LOC (**Store**, **Location**);	R2: STORE_SIZE (**Store**, **Sq_ft**);
R3: STORE_MGR (**Store**, **Manager**);	R4: PRODUCT (**Product**, **Price**);
R5a: DISC_STRUCTURE (**Quantity**, **Discount**);	R5b: INVENTORY (**Store, Product**, **Quantity**)

and the inclusion dependencies:

INVENTORY.**{Store}** ⊆ STORE_LOC.**{Store}**
INVENTORY.**{Product}** ⊆ PRODUCT.**{Product}**
INVENTORY.**{Quantity}** ⊆ DISC_STRUCTURE.**{Quantity}**

In addition, the decomposition is lossless and preserves all the FDs in F except fd7.

Since the primary keys of R1, R2, and R3 are the same, the three relations can be consolidated into one without compromising the normal form attained—i.e., BCNF. Observe that, incidentally, this consolidation also preserves fd7, and **{Manager, Location}** is a candidate key of R123. The resulting solution is of the form:

D: {R123, R4, R5a, R5b}

where:

R123: STORE (**Store**, Location, Sq_ft, Manager);	R4: PRODUCT (**Product**, Price);
R5a: DISC_STRUCTURE (**Quantity**, Discount);	R5b: INVENTORY (**Store, Product**, Quantity)

and the inclusion dependencies:

INVENTORY.**{Store}** ⊆ STORE.**{Store}**
INVENTORY.**{Product}** ⊆ PRODUCT.**{Product}**
INVENTORY.**{Quantity}** ⊆ DISC_STRUCTURE.**{Quantity}**

The solution yields a BCNF design, continues to retain the lossless-join property, and preserves all the FDs in F. In addition, the design is parsimonious (tighter) because the number of relation schemas in the design is only four, as opposed to the six relation schemas in the previous solution. Clearly, this is a superior design. This is how the set of tables in Figure 7.3, portraying an instance of this relational schema decomposed from the relation schema, shown in Figure 7.1, is produced.

In Section 7.1, while previewing the transformation of the STOCK relation in Figure 7.1 to the set of relations in Figure 7.3, we asked the following three questions. Now, we can answer these questions with clarity.

- *How do we systematically identify data redundancies?*
  Since we now know that the root cause of data redundancies and the resulting modification anomalies in a relation schema R is undesirable FDs, we first identify the FDs that prevail over R. The only desirable FDs in R are the ones where the determinant is a candidate key of R. The rest, the undesirable FDs, manifest in R as partial dependencies (violation of 2NF) and transitive dependencies (violation of 3NF or BCNF).

- *How do we know how to decompose the base relation schema under investigation?*
  The objective is to render all undesirable FDs in R desirable. This is done by decomposing R into a set of relation schemas D: {R1, R2, R3, ... Rn} such that the FDs captured in each relation schema of D become desirable FDs. The decomposition is done by systematically resolving the 2NF, 3NF, and BCNF violations starting from R and then in the successive versions of D.

- *How do we know that the decomposition is correct and complete (without looking at sample data)?*
  A decomposition is "correct" when:
  i.   the decomposition is attribute-preserving. That is, all attributes of R collectively appear in D (no attribute is lost in the decomposition);
  ii. the join of D: {R1, R2, ... Rn} is strictly equal to R; and

iii. the decomposition does not have any modification anomalies due to FDs. The relational schema D: {R1, R2, ... Rn} when in BCNF assures absence of modification anomalies due to FDs and thus minimal data redundancies.

A decomposition is "complete" when it is a dependency-preserving lossless-join decomposition. Preservation of FDs is a verification process and is accomplished by inspecting the decomposition to see if the union of the FDs that hold on individual relation schemas of D is a cover for F (i.e., can generate F^+). This is demonstrated in Section 8.1. 5.1. One can also test for the lossless-join property. The method of testing is presented in Section 8.1.5.2.

Based on our example above, it appears, at first glance, as if a simple mapping of every FD in F to a relation schema, and consolidation of relation schemas with the same primary key to a single relation schema, produces the solution sought without any concern about even pursuing the normalization process. This apparent simplicity is because the specified F happens to be a minimal cover for the FDs that prevail on STOCK. It is likely that sometimes the set of FDs, F specified on a relation schema R deduced from the business rules contain extraneous attributes and/or extraneous FDs (i.e., F is not a minimal cover of the set of FDs in F). In that case, a simple idea of mapping every FD into a relation schema can also generate extraneous relation schemas in the solution leading to data redundancies and loss-join decompositions. In addition, BCNF violations often present trade-off conditions in the solution—i.e., dependency preservation may have to be sacrificed to achieve a lossless-join design; or, a certain level of data redundancy may have to be tolerated in order to achieve a lossless-join design that is also dependency-preserving. Either way, a systematic method to approach the problem when a set of FDs over a relation schema violates various normal forms is very much needed.

8.3 A Comprehensive Approach to Normalization

In practice, the database designer encounters the normalization task when assessing the quality of a relation schema in the context of the set of FDs that prevails over it. Each relation schema in a relational schema (the set of relation schemas constituting the logical data model) is evaluated individually. Often most of the relation schemas are already in BCNF by virtue of the accurate allocation of attributes to the entity types in the conceptual modeling stage. However, a few relation schemas do suffer from erroneous attribute allocation, often due to inadvertent representation of a set of related entity types as a single entity type in the ER diagram (see Section 7.1 in Chapter 7 for an illustration). In this book we assume that a conceptual design (ER modeling) precedes the logical design, but the issue is even more relevant when a set of FDs is directly derived from the business rules embedded in the user requirements specification. In our case, we start with a single universal relation schema (URS) constructed as a collection of attributes present in the set of FDs, and uses the process of normalization to develop the complete logical schema—i.e., in our case the relational schema.

Given a set of FDs prevailing over a URS either constructed from the set of FDs or mapped from an entity type of an ER diagram, data redundancies and modification anomalies due to undesirable FDs in the URS usually occur in the form of immediate violations of 2NF, 3NF, and/or BCNF. Although a single relation schema may be, in general, the best design from a data retrieval perspective, the normal form violations indicate that a single relation schema is seldom an effective design since modification anomalies (as indicated by the normal form violations) often impose restrictions on effective use and maintenance of the resulting database. Therefore, the ideal goal of database design is to develop a fully normalized relational schema, preferably in BCNF, that is parsimonious—i.e., a tight design containing the least number of relation schemas.

Normalization algorithms do not always yield a parsimonious design, especially if the set of FDs, F provided is not a minimal cover of F. Where multiple minimal covers are possible, these algorithms usually yield a solution based on some minimal cover which need not be the most desirable solution; for instance, not necessarily a parsimonious solution. When the design goal requires evaluation of trade-offs among a BCNF solution, lossless-join decomposition, and dependency preservation, human intervention is almost always necessary. Furthermore, it is rather critical that the database designer has a thorough understanding of the design.

For these reasons, we resort to a heuristic approach based on a few design guidelines for developing a normalized relational schema given a set of FDs. We have already been initiated into this process somewhat informally when we revisited the motivating exemplar from Section 7.1 of Chapter 7 in Section 8.2 above. In this section, we examine a comprehensive approach where trade-off among a BCNF solution, a lossless-join decomposition, and dependency preservation in the normalization process are dealt with. In the interest of continuity and as a means to better learning, we use the same three examples that are used for the derivation of candidate keys in Chapter 7.

8.3.1 Case 1

Let us investigate the example presented in Section 7.3.1 of Chapter 7. The URS and the set of FDs that hold in it are reproduced below:

URS1 (Store, Branch, Location, Sq_ft, Manager, Product, Price, Customer, Address, Vendor, Type)

fd1: {Store, Branch} → Location;	fd5: Product → Price;
fd2: Customer → Address;	fd6: {Store, Branch} → Manager;
fd3: Vendor → Product;	fd7: Manager → {Store, Branch};
fd4: {Store, Branch} → Sq_ft;	fd8: Store → Type;

The set of FDs in F may also be expressed via a dependency diagram as shown in Figure 8.8b.

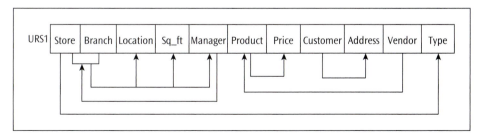

Figure 8.8b   Dependency diagram for the relation schema URS1

Is URS1 normalized?

8.3.1.1 A Simplistic Approach Ignoring Normalization

Before answering this question, let us first ignore normalization and write a relational schema by simply mapping every FD to a relation schema and consolidating the relation schemas with the same primary key into single relation schemas. The result is shown below:

D: {R1, R2, R3, R4, R5, R6}

where:

R1: STORE (**Store**, Type);	R2: BRANCH (**Store, Branch**, Location, Sq_ft, Manager);
R3: CUSTOMER (**Customer**, Address);	R4: VENDOR (**Vendor**, Product);
R5: PRODUCT (**Product**, Price);	R6: MANAGER (**Manager**, Store, Branch)

Is this design free of data redundancies/modification anomalies? The answer is "Yes," because every relation schema in D is in BCNF. Is this design dependency-preserving? Again, the answer is "Yes." Does the design exhibit lossless-join property? Once again, the answer is a resounding "Yes." Is the design parsimonious? A careful inspection of the relational schema reveals that there is a tighter design that does not compromise on the other three properties. R2 and R6 can be consolidated without any adverse effect on the established condition of a dependency-preserving BCNF solution with an enduring lossless-join property. This is possible because R2 and R6 have at least one common candidate key. Consequently, a superior design is of the form:

D: {R1, R26, R3, R4, R5}

where:

R1: STORE (**Store**, Type);	R3: CUSTOMER (**Customer**, Address);
R4: VENDOR (**Vendor**, Product);	R5: PRODUCT (**Product**, Price);
R26: BRANCH (**Store, Branch**, Location, Sq_ft, Manager);	

If the above design is correct and complete, there is no utility in normalization. So, an important question is: *Is the above design correct and complete?* In Section 8.2, three conditions were specified to determine if a design is correct. Accordingly, the above design is attribute-preserving and each relation schema in the relational schema D {R1, R26, R3, R4, R5} is in BCNF. Does the join of D {R1, R26, R3, R4, R5} strictly yield R? The answer is 'No' because while R1 and R26 as well as R4 and R5 can be joined, the results of these joins and R3 are not joinable. So, the design is incorrect. One way to ratify or refute this is to see if we can arrive at the same solution through the normalization process. In order to do that, we return to the original question: *Is URS1 normalized?*

8.3.1.2 The Normalization Approach

The fact that URS1 is a relation schema clarifies that it is in 1NF. Are there any data redundancies in URS1? We know that any violation of 2NF, 3NF, or BCNF implies that data redundancies/modification anomalies attributable to undesirable FDs are present in URS1.

Violations of 2NF, 3NF, and/or BCNF are always with reference to a candidate key—invariably, the primary key. Therefore:

Step 1: Identify the candidate keys of URS1 given the set of FDs, F. In Chapter 7, we derived the candidate keys of URS1 (see Section 7.3.1.) as:

    {Store, Branch, Customer, Vendor} and **{Manager, Customer, Vendor}**

Step 2: Choose a primary key for URS1. Since the choice of primary key from among the candidate keys is essentially arbitrary (see Section 6.3.1 in Chapter 6), using the rules of thumb prescribed in Chapter 7 (see Section 7.3.4), let us choose **{Manager, Customer, Vendor}** as the primary key of URS1. Then, the other candidate key, **{Store, Branch, Customer, Vendor}**, becomes the alternate key of URS1.

Step 3: Record the immediate normal form violated in URS1 with respect to the primary key by each of the FDs in F. The normal form violations are shown in Table 8.1.

Table 8.1 Normal form violations in URS1

URS1 (Store, Branch, Location, Sq_ft, Manager, Product, Price, Customer, Address, Vendor, Type)						
					\multicolumn{2}{}{NF violated in URS1}	
FD	Cover	Role	Determinant	Dependent(s)	PK = CK2	AK = CK1
fd1	F		(Store, Branch)	Location	3NF	2NF
fd2	F		Customer	Address	2NF	2NF
fd3	F		Vendor	Product	2NF	2NF
fd4	F		(Store, Branch)	Sq_ft	3NF	2NF
fd5	F		Product	Price	3NF	3NF
fd6	F		(Store, Branch)	Manager	BCNF	2NF
fd7	F		Manager	(Store, Branch)	2NF	BCNF
fd8	F		Store	Type	3NF	2NF
fdx	F+		(Store, Branch)	Location, Sq_ft, Manager	3NF	2NF
fdy	F+	CK1	(Store, Branch, Customer, Vendor)	Location, Sq_ft, Manager, Address, Product, Price, Type	None	None
fdz	F+	CK2	(Manager, Customer, Vendor)	Location, Sq_ft, Store, Branch, Address, Product, Price, Type	None	None

Note: While we are only interested in the normal form violations predicated upon the primary key, the table above also shows the normal form violations with respect to the alternate key just to demonstrate that the same FD can violate different normal forms depending on the candidate key on which the evaluation is based.

Step 4: Resolve 2NF and 3NF violations in URS1—the sequence of resolution is immaterial.
We use the two-step process prescribed in Sections 8.1.2 and 8.1.3 to resolve the normal form violations. It is important to note that as Step 4 is executed recursively, URS1 ceases to remain a single relation schema. Instead, in each successive execution of this step, the URS1 used is a revised set of decomposed relation schemas as shown below.

Execution 1:

Input: URS1	FD: fd2: **Customer → Address**	Violation: 2NF in URS1

Resolution: Decomposition URS1 {R1, R0} where:

R1:	CUSTOMER (**Customer**, **Address**);	in BCNF
R0:	LORS1 (**Manager, Customer, Vendor**, Store, Branch, Location, Sq_ft, Product, Price, Type)	
#	LORS1.**{Customer}** ⊆ CUSTOMER.**{Customer}**	lossless-join

Note: LORS1 is an acronym for *Leftover Relation Schema 1* The input to the next execution of this step is URS1 {R1, R0}.

Execution 2:

Input: URS1 {R1, R0}	FD: fd3: **Vendor → Product;**	Violation: 2NF in LORS1

Resolution: Decomposition URS1 {R1, R2, R0} where:

R1:	CUSTOMER (**Customer**, **Address**);	in BCNF
R2:	VENDOR (**Vendor**, **Product**)	in BCNF
R0:	LORS2 (**Manager, Customer, Vendor**, Store, Branch, Location, Sq_ft, Price, Type)	
#	LORS2.**{Customer}** ⊆ CUSTOMER.**{Customer}**	lossless-join
#	LORS2.**{Vendor}** ⊆ VENDOR.**{Vendor}**	lossless-join

The input to the next execution of this step is URS1 {R1, R2, R0}.

Execution 3:

Input: URS1 {R1, R2, R0}	FD: fd5: **Product → Price;**	Violation: 3NF in ?

Observe that the attributes in fd5 have been fragmented in previous decompositions. In order to evaluate the effect of fd5 properly, fd5 will have to be restored. This is accomplished by moving the dependent in fd5 (**Price**) to the relation schema R2 where the determinant of fd5 (**Product**) now resides. The revised URS1 {R1, R2, R0} appears below:

R1:	CUSTOMER (<u>Customer</u>, **Address**);	in BCNF
R2:	VENDOR (<u>Vendor</u>, **Product, Price**)	Violates 3NF in R2
R0:	LORS2 (<u>Manager, Customer, Vendor</u>, **Store, Branch, Location, Sq_ft, Type**)	
#	LORS2.**{Customer}** ⊆ CUSTOMER.**{Customer}**	Lossless-join
#	LORS2.**{Vendor}** ⊆ VENDOR.**{Vendor}**	Lossless-join

Accordingly the violation of 3NF by fd5 in URS1 has moved from LORS2 to R2 and is now resolved as follows:

Resolution: Decomposition URS1 {R1, R2, R3, R0} where:

R1:	CUSTOMER (<u>Customer</u>, **Address**);	in BCNF
R2:	VENDOR (<u>Vendor</u>, **Product**);	in BCNF
R3:	PRODUCT (<u>Product</u>, **Price**);	in BCNF
R0:	LORS3 (<u>Manager, Customer, Vendor</u>, **Store, Branch, Location, Sq_ft, Type**)	
#	LORS3.**{Customer}** ⊆ CUSTOMER.**{Customer}**	Lossless-join
#	LORS3.**{Vendor}** ⊆ VENDOR.**{Vendor}**	Lossless-join
#	VENDOR.**{Product}** ⊆ PRODUCT.**{Product}**	Lossless-join

The input to the next execution of this step is URS1 {R1, R2, R3, R0}.

Execution 4:

Input: URS1 {R1, R2, R3, R0}	FD: fd7: **Manager → {Store, Branch}**;	Violation: 2NF in LORS3

Resolution: Decomposition URS1 {R1, R2, R3, R4, R0} where:

R1:	CUSTOMER (<u>Customer</u>, **Address**);	in BCNF
R2:	VENDOR (<u>Vendor</u>, **Product**);	in BCNF
R3:	PRODUCT (<u>Product</u>, **Price**);	in BCNF
R4:	MANAGER (<u>Manager</u>, **Store, Branch**)	in BCNF
R0:	LORS4 (<u>Manager, Customer, Vendor</u>, **Location, Sq_ft, Type**)	
#	LORS4.**{Customer}** ⊆ CUSTOMER.**{Customer}**	Lossless-join

#	LORS4.{Vendor} $\subseteq$ VENDOR.{Vendor}	Lossless-join
#	VENDOR.{Product} $\subseteq$ PRODUCT.{Product}	Lossless-join
#	LORS4.{Manager} $\subseteq$ MANAGER.{Manager}	

Note that the resolution of normal form violation due to fd7 incidentally resolves the BCNF violation in LORS3 due to fd6: **{Store, Branch}** → **Manager** also.

The input to the next execution of this step is URS1 {R1, R2, R3, R4, R0}.

Execution 5:

Input:URS1 {R1, R2, R3, R4, R0}	FD:fd8: **Store** → **Type;**	Violation:3NF in ?

Once again, the attributes in fd8 have been fragmented in previous decompositions. In order to evaluate the effect of fd8 properly, fd8 will have to be restored. This is accomplished by moving the dependent in fd8 (**Type**) to the relation schema R4 where the determinant of fd8 (**Store**) now resides. The affected relation schemas in URS1 {R1, R2, R3, R4, R0) are R4 and R0, as shown below:

R4:	MANAGER (**Manager**, Store, Branch, Type)	Violates 3NF in R4
R0:	LORS4 (**Manager, Customer, Vendor**, Location, Sq_ft)	

{Store, Branch} is a candidate key of R4 based on fd6. Thus, it is possible to designate **{Store, Branch}** as the primary key of R4 in which case **Manager** becomes an alternate key of R4. Then, the same fd6 is seen to violate 2NF in R4. Either way, the resolution of the normal form violation due to fd6 yields the same decomposition given below.

Resolution: Decomposition URS1 {R1, R2, R3, R4, R5, R0} where:

R1:	CUSTOMER (**Customer**, Address);	in BCNF
R2:	VENDOR (**Vendor**, Product);	in BCNF
R3:	PRODUCT (**Product**, Price);	in BCNF
R4:	MANAGER (**Manager**, Store, Branch)	in BCNF
R5:	STORE (**Store**, Type)	in BCNF
R0:	LORS5 (**Manager, Customer, Vendor**, Location, Sq_ft)	
#	LORS5.{Customer} $\subseteq$ CUSTOMER.{Customer}	Lossless-join
#	LORS5.{Vendor} $\subseteq$ VENDOR.{Vendor}	Lossless-join
#	VENDOR.{Product} $\subseteq$ PRODUCT.{Product}	Lossless-join
#	LORS5.{Manager} $\subseteq$ MANAGER.{Manager}	Lossless-join
#	MANAGER.{Store} $\subseteq$ STORE.{Store}	Lossless-join

Since URS1 {R1, R2, R3, R4, R5, R0} is still not fully normalized—for instance, LORS5 is not in BCNF—we continue with the normalization process. Accordingly, the input to the next execution of this step is URS1 {R1, R2, R3, R4, R5, R0}.

Execution 6:

Input: URS1 {R1, R2, R3, R4, R5, R0}	FD: fdx = {fd1 U fd4 U fd6} fdx: {Store, Branch} → {Location, Sq_ft, Manager};	Violation: 3NF in ?

The attributes in fdx have been fragmented in previous decompositions. In order to evaluate the effect of fdx properly, fdx will have to be restored. This is accomplished by moving the dependent in fdx **{Location, Sq_ft, Manager}** to the relation schema R4 where the determinant of fdx **(Store, Branch}** now resides. R4 and R0 are the affected relation schemas in URS1 {R1, R2, R3, R4, R5, R0}:

R4:	MANAGER (**Manager**, Store, Branch, Location, Sq_ft)
R0:	LORS5 (**Manager, Customer, Vendor**)

Since **{Store, Branch}** is a candidate key of R4, fdx does not violate any normal form in R4. Hence R4 is in BCNF. As can be seen, LORS5 is also in BCNF.

Step 5: Resolve all BCNF violations in URS1—the sequence of resolution is immaterial. There are no other BCNF violations in URS1 {R1, R2, R3, R4, R5, R0}.

Thus, the final design URS1 {R1, R2, R3, R4, R5, R0} that is free from modification anomalies is:

R1:	CUSTOMER (**Customer**, Address);	In BCNF
R2:	VENDOR (**Vendor**, Product);	In BCNF
R3:	PRODUCT (**Product**, Price);	In BCNF
R4:	MANAGER (**Manager**, Store, Branch, Location, Sq_ft)	In BCNF
R5:	STORE (**Store**, Type)	In BCNF
R0:	LORS5 (**Manager, Customer, Vendor**)	
#	LORS5.**{Customer}** ⊆ CUSTOMER.**{Customer}**	Lossless-join
#	LORS5.**{Vendor}** ⊆ VENDOR.**{Vendor}**	Lossless-join
#	VENDOR.**{Product}** ⊆ PRODUCT.**{Product}**	Lossless-join
#	LORS5.**{Manager}** ⊆ MANAGER.**{Manager}**	Lossless-join
#	MANAGER.**{Store}** ⊆ STORE.**{Store}**	Lossless-join

Is the above design correct and complete? The above design is attribute-preserving and each relation schema in the relational schema URS1 {R1, R26, R3, R4, R5,R0} is in BCNF.

Normal Forms Based on Functional Dependencies

Does the join of D {R1, R26, R3, R4, R5} strictly yield R? Yes, it does. Therefore, the solution is correct. Since the design also yields a lossless-join decomposition that is also dependency-preserving, the solution is also complete.

It is crucial to observe that LORS5 is part of the final relational schema. Without LORS5 as a part of the final decomposed URS1, the solution is incomplete and therefore incorrect. The initial approach of constructing the relational schema by simply mapping the FDs in F to relation schemas explored at the start of Section 8.3.1 fails to generate LORS5 as a relation schema in the design and therefore is flawed.[9] The solution generated above via normalization demonstrates that the normalization process is indispensable for analyzing a universal relation schema and the set of FDs holding over it, thereby generating a correct and complete relational schema.

8.3.2 Case 2

The next example, first presented in Section 7.3.2.1 of Chapter 7, offers a different variation for a normalization exercise. The URS and the set of FDs that is held in it are reproduced below:

URS2 (Company, Location, Size, President, Product, Price, Sales, Production, Supplier)

fd1: **Company → Location;**	fd2: **Company → Size;**	fd3: **Company → President;**
fd4: **{Product, Company} → Price;**	fd5: **Sales → Production;**	fd6: **{Company, Product} → Sales;**
fd7: **{Product, Company} → Supplier;**	fd8: **Supplier → Product;**	fd9: **President → Company**

The set of FDs in F may also be expressed via a dependency diagram, as shown in Figure 8.8c.

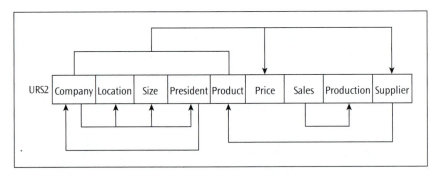

Figure 8.8c   Dependency diagram for the relation schema URS2

8.3.2.1 A Simplistic Approach Ignoring Normalization

Just to convince ourselves that arbitrarily writing relation schemas from FDs yields incomplete and/or incorrect results, let us once again ignore normalization and write a relational

---

[9] This is further clarified when the relational schema is reverse-engineered to an ER diagram (see Section 8.4.1). A simpler decomposition algorithm is available if a 3NF solution is acceptable—see Elmasri and Navathe (2003, p. 342) for details.

schema by simply mapping every FD to a relation schema and consolidating the relation schemas with the same primary key into single relation schemas. Here is the result:

D: {R1, R2, R3, R4, R5}

where:

R1: COMPANY (**Company**, Location, Size, President);   R2: SUPPLIER (**Supplier**, Product);

R3: PRODUCT (**Company, Product**, Price, Sales, Supplier);   R4:SALES (**Sales**, Production);

R5: PRESIDENT (**President**, Company)

Is this design free of data redundancies/modification anomalies? No, because fd8: **Supplier → Product** violates BCNF in R3.

Is this design dependency-preserving? Yes; in fact, it overdoes it. For instance, fd8: **Supplier → Product** is preserved in R2 as well as in R3. Likewise, the FDs fd3: **Company → President** and fd9: **President → Company** are also unnecessarily preserved in two places—R1 and R5.

Does the design possess a lossless-join property? No. While there is a lossless-join navigation path through the relation schemas, a loss-join relationship also exists (i.e., join of R2 and R3 on **Product**). Sometimes, it is impossible to fully eradicate loss-join opportunities and the designer may have to settle for the availability of a lossless-join navigation path through the relation schemas. However, this is not the case here. A close scrutiny of the relational schema above reveals that the relation schema R2 is redundant in the design and elimination of this relation schema from the relational schema, eliminates the loss-join relationship, and renders the design lossless. In addition, elimination of R2 results in a tighter design with one less relation schema.

The design, at this point, is still not parsimonious since R5 is also a redundant relation schema in the design and can be removed without any adverse effects. However, constructing the relation schemas from the FDs, one may not notice such intricacies. At the least, *there is no formal mechanism to ascertain the efficacy of the solution when this approach is used.*

The final design using this ad hoc approach is:

D: {R1, R3, R4}

where:

R1: COMPANY (**Company**, Location, Size, President);

R3: PRODUCT (**Company, Product**, Price, Sales, Supplier);   R4:SALES (**Sales**, Production);

Is this design correct? If it is, then there is no need to bother with normalization. The design is attribute-preserving and the join of D {R1, R3, R4} does strictly yield R. However, fd8: **Supplier → Product** violates BCNF in R3. Therefore, the solution is incorrect; it will have modification anomalies.

8.3.2.2 The Normalization Approach

Let us now employ the normalization procedure to evaluate URS2. Is URS2 normalized? The fact that URS2 is a relation schema clarifies that it is in 1NF. Are there any data redundancies in URS2? In order to know this, we test to see if any violation of 2NF, 3NF, or BCNF is present in URS2 because such violations imply that data redundancies/modification anomalies attributable to undesirable FDs are present in URS2. Violations of 2NF, 3NF, and/or BCNF are always with reference to a candidate key—invariably, the primary key. Therefore:

Step 1: Identify the candidate keys of URS2 given the set of FDs, F. In Chapter 7, we derived the candidate keys of URS2 (see Section 7.3.2) as:

{Company, Product}; {Company, Supplier}; {President, Product}; and {President, Supplier}.

Step 2: Choose a primary key for URS2. Since the choice of a primary key from among the candidate keys is essentially arbitrary, using the rules of thumb prescribed in Chapter 7, let us choose **{Company, Product}** as the primary key of URS2. Then, the other candidate keys, **{Company, Supplier}, {President, Product}, {President, Supplier}** become alternate keys of URS2.

Step 3: Record the immediate normal form violated in URS2 with respect to the primary key by each of the FDs in F. The normal form violations are listed in Table 8.2.

Table 8.2 Normal form violations in URS2

URS2 (Company, Location, Size, President, Product, Price, Sales, Production, Supplier)								
					NF violated in URS1			
FD	Cover Role		Determinant	Dependent(s)	**PK = CK1**	AK1 = CK2	AK2 = CK3	AK3 = CK4
fd1	F		Company	Location	2NF	2NF	3NF	3NF
fd2	F		Company	Size	2NF	2NF	3NF	3NF
fd3	F		Company	President	2NF	2NF	BCNF	BCNF
fd4	F		(Product, Company)	Price	None	None	None	None
fd5	F		Sales	Production	3NF	3NF	3NF	3NF
fd6	F		(Company, Product)	Sales	None	None	None	None
fd7	F		(Product, Company)	Supplier	None	None	None	None
fd8	F		Supplier	Product	BCNF	2NF	BCNF	2NF
fd9	F		President	Company	BCNF	BCNF	2NF	2NF
fdx	F+		Company	Location, Size, President	2NF	2NF	3NF	3NF

Table 8.2 Normal form violations in URS2 (continued)

FD	Cover	Role	Determinant	Dependent(s)	PK = CK1	AK1 = CK2	AK2 = CK3	AK3 = CK4
fdy	F⁺	CK1	(Company, Product)	Location, Size, President, Price, Sales, Production, Supplier	None	None	None	None
fdz	F⁺	CK2	(Company, Supplier)	Location, Size, President, Product, Price, Sales, Production	None	None	None	None
fdu	F⁺	CK3	(President, Product)	Company, Location, Size, Price, Sales, Production, Supplier	None	None	None	None
fdv	F⁺	CK4	(President, Supplier)	Company, Location, Size, Product, Price, Sales, Production	None	None	None	None

Note: While we are only interested in the normal form violations predicated upon the primary key, the table above also shows the normal form violations with respect to the alternate keys to demonstrate that the same FD can violate different normal forms depending on the candidate key on which the evaluation is based.

Step 4. Resolve 2NF and 3NF violations in URS2—the sequence of resolution is immaterial. We use the two-step process prescribed in Sections 8.1.2 and 8.1.3 to resolve the normal form violations. We now know that this step is executed recursively as necessary and in successive execution of this step, the URS2 used is progressively revised to a set of decomposed relation schemas. Since we have already observed the successive, iterative execution of this step for each of the 2NF and 3NF violations in the previous example, we will handle them collectively in one execution of this step in this example. Predicated on **{Company, Product}** as the primary key, fd1, fd2, and fd3 individually (or fdx as a collection of fd1, fd2, and fd3) violate 2NF and fd5 violates 3NF. Solving for these FDs, we have:

Execution 1:

Input: URS2	FDs	Violation: 2NF in URS2;
	fdx: **Company → Location, Size, President;**	
	fd5: **Sales → Production;**	Violation: 3NF in URS2

Resolution: Decomposition URS2 {R1, R2, R0} where:

R1:	COMPANY (**Company**, Location, Size, President);	in BCNF
R2:	SALES (**Sales**, Production)	in BCNF
R0:	LORS1 (**Company, Product**, Sales, Price, Supplier)	
#	LORS1.{Company} ⊆ COMPANY.{Company}	Lossless-join

The relational schema URS2 {R1, R2, R0} is free of any 2NF and 3NF violations.

Step 5. Resolve all BCNF violations in URS2—the sequence of resolution is immaterial. From Table 8.2, we see that there are two BCNF violations (fd8 and fd9) in URS2 when **{Company, Product}** is the primary key. Earlier decompositions have resulted in fd9 displaced from URS2 to R1. In R1, however, fd9: **President → Company** does not violate any normal form since **President** the determinant of fd9 is a candidate key of R1. Thus, at this point, we need to resolve only the BCNF violation that still persists in LORS1 due to fd8: **Supplier → Product.**

Once again, we use the same two-step process of decomposition that is used in the resolution of 2NF and 3NF violations (see Section 8.1.4):

1. Pull out the undesirable FD(s) from the target relation schema as a separate relation schema.
2. Retain the determinant of the pulled-out relation schema as an attribute(s) (foreign key) of the leftover target relation schema to facilitate reconstruction of the original target relation schema.

The input to the next execution of this step is URS2 {R1, R2, R0}.

Execution 2:

Input: URS2 {R1, R2, R0}	FD: fd8: **Supplier → Product;**	Violation: BCNF in LORS1

Resolution: Decomposition URS2 {R1, R2, R3, R0} where:

R1:	COMPANY (**Company**, Location, Size, President);	in BCNF
R2:	SALES (**Sales**, Production)	in BCNF
R3:	SUPPLY (**Supplier**, Product)	in BCNF
R0:	LORS2 (**Company**, Sales, Price, **Supplier**)	in BCNF
#	LORS2.{Company} ⊆ COMPANY.{Company}	Lossless-join
#	LORS2.{Supplier} ⊆ SUPPLY.{Supplier}	Lossless-join

The above design is attribute-preserving and the join of URS2 {R1, R2, R3, R0} does strictly yield R. All the relation schemas in URS2 {R1, R2, R3, R0} are in BCNF. Therefore, the solution is correct. Is the solution complete? The specified conditions for completeness in Section 8.2 are: Lossless-join and dependency-preserving design. Note that in this solution the decomposition possesses the lossless-join property. To verify if the relational schema is dependency-preserving the union of the FDs preserved in individual relation schemas of URS2 {R1, R2, R3, R0} are listed below:

Company → Location;	Company → Size;	Company → President;
{Supplier, Company} → Price;	Sales → Production;	{Company, Supplier} → Sales;
Supplier → Product;	President → Company	

The above set of FDs preserved in this solution is not a cover for F specified over URS2. For instance, it is impossible to deduce fd4: {Product, Company} → Price, fd6: {Company, Product} → Sales and fd7: {Product, Company} → Supplier from the union of the FDs that hold in the *individual* relation schemas of URS2 shown above. In other words, from the union of the *individual* relation schemas of URS2 in the solution above, F^+ cannot be generated. Removal of **Product** from LORS2 appears to have precipitated this situation. An alternative decomposition for the resolution of BCNF violation caused by fd8 in LORS1 is to retain **Product** in LORS2 and pull out **Supplier**:

R3:	SUPPLY (Supplier, Product)	in BCNF
R0:	LORS2 (Company, Product, Sales, Price)	in BCNF

While preserving fd4 and fd6, this solution still does not preserve fd7. Thus, this solution does not provide full dependency preservation either. More importantly, this solution does not yield a lossless-join decomposition. In Section 8.1.5.3, it is stipulated that a lossless-join decomposition is an absolute must. Therefore, this alternative design is rejected. Incidentally, it must be noted that this design yields a loss-join decomposition, because it fails to follow the prescribed two-step algorithm for decomposition. Another alternative discussed in Section 8.1.5.3 is to settle for a relational schema in 3NF and tolerate the BCNF violation by shifting the responsibility of handling the data redundancy to the application programs. In this case, it is possible to achieve lossless-join decomposition as well as dependency preservation in the following solution:
URS2 {R1, R2, R0} where:

R1:	COMPANY (Company, Location, Size, President);	in BCNF
R2:	SALES (Sales, Production)	in BCNF
R0:	LORS1 (Company, Product, Sales, Price, Supplier)	In 3NF
#	LORS1.{Company} ⊆ COMPANY.{Company}	Lossless-join

Observe that the above design is achieved at the conclusion of Step 4 of the normalization process.

With the advent of materialized views, a solution superior to all of the above has become possible as discussed in Section 8.1.5.3. Accordingly, the following solution is presented:

URS2 {R1, R2, R3, R0} where:

R1:	COMPANY (**Company**, Location, Size, President);	in BCNF
R2:	SALES (**Sales**, Production)	in BCNF
R3:	SUPPLY (**Supplier**, Product)	in BCNF
R0:	LORS2 (**Company**, Sales, Price, **Supplier**)	in BCNF
#	LORS2.{Company} ⊆ COMPANY.{Company}	Lossless-join
#	LORS2.{Supplier} ⊆ SUPPLY.{Supplier}	Lossless-join
MV1:	PRODUCT (**Company, Product**, Sales, Price, Supplier)	

Note that this is the solution arrived at the end of Step 5 supplemented by the materialized views for the dependencies not preserved. Since in this case all the three unpreserved dependencies (fd4, fd6, and fd7) have the same determinant, they can be captured with a single materialized view - viz., MV1.

Since **Company → President** (fd3) and **President → Company** (fd9), either **Company** or **President** can be the primary key of R1 and the other becomes the alternate key of R1. Also, using the rule of pseudotransitivity, since **President → Company**, **President** can replace **Company** in LORS2 to generate an equivalent relation schema:

R0:	LORS2 (**President**, Sales, Price, **Supplier**)	in BCNF
#	LORS2.{President} ⊆ COMPANY. {President}	Lossless-join

Finally, while the choice of primary key at the beginning of the normalization process (Step 2) may have some impact on the execution of the normalization steps prescribed, the final solution will be the same or a close equivalent no matter which candidate key is chosen as the primary key. For instance, observe in Table 8.2 that different sets of normal form violations are present in URS2 for the different alternate keys. The solution, however, will be the same as the one that emerges above when CK1 is used as the primary key. Verification of this is left as an exercise for the reader.

8.3.3 Case 3

Let us next review the example presented in Section 7.3.3 of Chapter 7 as a third exercise in normalization. The URS and the set of FDs that prevail over it are reproduced below:

URS3 (Proj_nm, Emp#, Proj#, Job_type, Chg_rate, Emp_nm, Budget, Fund#, Hours, Division)

fd1: **Proj# → Proj_nm;**	fd2: **Job_type → Chg_rate;**	fd3: **Emp# → Emp_nm;**
fd4: **Proj_nm → Budget;**	fd5: **Fund# → Proj_nm;**	fd6: **{Proj_nm, Emp#} → Hours;**
fd7: **Proj_nm → Proj#;**	fd8: **Emp# → Job_type;**	fd9: **{Proj#, Emp#} → Fund#**

The set of FDs in F may also be expressed via a dependency diagram as shown in Figure 8.8d.

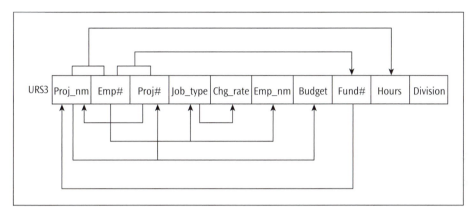

Figure 8.8d   Dependency diagram for the relation schema URS3

Convinced from the previous two examples that arbitrary development of relation schemas from a set of FDs yields an incomplete and incorrect solution, let us reject that approach and proceed with the evaluation of URS3 from the normalization perspective.[10] URS3 is in 1NF by virtue of its definition as a relation schema—i.e., there are no multi-valued attributes or composite attributes in URS3 and all FDs in F are preserved in URS3. Lossless-join property is not an issue since URS3 is a single relation schema. If URS3 is in BCNF, it is guaranteed that there will be no data redundancies/modification anomalies in URS3 due to functional dependencies. In order to check this, we need to know at least one candidate key of URS3 since normal form violations can be checked only with respect to candidate keys (or a primary key, it being one of the candidate keys of the relation schema). So, we start with Step 1 of the normalization heuristic.

Step 1: Identify the candidate keys of URS3 given the set of FDs, F. In Chapter 7, we derived the candidate keys of URS3 (see Section 7.3.3.) as:
  {Proj#, Emp#, Division}, {Proj_nm, Emp#, Division}, and {Fund#, Emp#, Division}

Step 2: Choose a primary key for URS3. Since the choice of primary key from among the candidate keys is essentially arbitrary, using the rules of thumb prescribed in Chapter 7, let us choose **{Proj#, Emp#, Division}** as the primary key of URS3. Then the other two candidate

---

[10] The unconvinced reader may, as an exercise, construct the relational schema by mapping each FD to a relation schema, do some consolidation based on Armstrong's axioms, and compare the results with the normalized solution.

keys, **{Proj_nm, Emp#, Division}** and **{Fund#, Emp#, Division}**, become the alternate keys of URS3.

Step 3: Record the immediate normal form violated in URS3 with respect to the primary key by each of the FDs in F. Based on the *primary key* chosen, viz., **{Proj#, Emp#, Division}**, normal forms are violated in URS3 by:

fd1	fd2	fd3	fd4	fd5	fd6	fd7	fd8	Fd9
2NF	3NF	2NF	3NF	3NF	3NF	BCNF	2NF	2NF

Step 4: Resolve 2NF and 3NF violations in URS2—the sequence of resolution is immaterial. We now know that this step is executed recursively as necessary and in successive execution of this step, the URS3 is progressively revised to a set of decomposed relation schemas. Since we have already observed the successive, iterative execution of this step for each of the 2NF and 3NF violations in the first example, we handle them collectively in one execution of this step in this example. Solving for the 2NF and 3NF violations in URS3, we have:

Decomposition: URS3 {R1, R2, R3, R4, R5, R0} where:

R1:	PROJECT (**Proj#**, Proj_nm, Budget);	in BCNF
R2:	FUND (**Fund#**, Proj_nm)	in BCNF
R3:	EMPLOYEE (**Emp#**, Emp_nm, Job_type)	in BCNF
R4:	JOB (**Job_type**, Chg_rate)	in BCNF
R5:	ASSIGNMENT (**Proj#, Emp#**, Fund#, Hours)	in 3NF (Violates BCNF)*
R0:	LORS1 (**Proj#, Emp#, Division**)	in BCNF
#	LORS1.{Proj#, Emp#} ⊆ ASSIGNMENT.{Proj#, Emp#}	Lossless-join
#	ASSIGNMENT.{Emp#} ⊆ EMPLOYEE.{Emp#}	Lossless-join
#	ASSIGNMENT.{Proj#} ⊆ PROJECT.{Proj#}	Lossless-join
#	ASSIGNMENT.{Fund#} ⊆ FUND.{Fund#}	Lossless-join
#	EMPLOYEE.{Job_type} ⊆ JOB.{Job_type}	Lossless-join
#	FUND.{Proj_nm} ⊆ PROJECT.{Proj_nm}	Lossless-join**

* Violation of BCNF due to the FD in F$^+$ Fund# → Proj#
*** Note that **Proj_nm** is an alternate key of R1.*

While at this stage of the solution, the revised relational schema URS3 is not in BCNF, it possesses the lossless-join property and all FDs in F are preserved. Observe that incidentally the decomposition of URS3 for the resolution of 2NF and 3NF violations eliminated the BCNF violation due to fd7 by the migration of **Proj_nm** to R1. However, the decomposition has seeded a new BCNF violation in R5.

Step 5: Resolve all BCNF violations in URS3—the sequence of resolution is immaterial.
The only BCNF violation in the URS3 {R1, R2, R3, R4, R5, R0} is the one in:

R5:	ASSIGNMENT (**Proj#, Emp#**, Fund#, Hours)	

The resolution of the BCNF violation using the prescribed two step process results in

R5a:	FUND_A (**Fund#**, Proj#)	in BCNF
R5b:	ASSIGNMENT (**Fund#, Emp#**, Hours)	in BCNF

R5a can be consolidated with R2 since the determinant of the FDs preserved in both is the same—viz., **Fund#**. Thus, we have:

R2a:	FUND (**Fund#**, Proj#, Proj_nm)	violates 3NF
R5b:	ASSIGNMENT (**Fund#, Emp#**, Hours)	in BCNF

Since **Proj#** and **Proj_nm** functionally determine each other as per fd1 and fd7, one of the two attributes moves to R1 in the process of decomposing R2a to eliminate the 3NF violation. Thus, we have:

R2:	FUND (**Fund#**, Proj_nm)	in BCNF
R5b:	ASSIGNMENT (**Fund#, Emp#**, Hours)	in BCNF

So, a solution in BCNF can be written as:
Decomposition: URS3 {R1, R2, R3, R4, R5b, R0} where:

R1:	PROJECT (**Proj#**, Proj_nm, Budget);	in BCNF
R2:	FUND (**Fund#**, Proj_nm)	in BCNF
R3:	EMPLOYEE (**Emp#**, Emp_nm, Job_type)	in BCNF
R4:	JOB (**Job_type**, Chg_rate)	in BCNF
R5b:	ASSIGNMENT (**Fund#, Emp#**, Hours)	in BCNF
R0:	LORS1 (**Proj#, Emp#, Division**)	in BCNF
#	LORS1.{Proj#} $\subseteq$ PROJECT.{Proj#}	Lossless-join*
#	LORS1.{Emp#} $\subseteq$ EMPLOYEE.{Emp#}	Lossless-join*
#	ASSIGNMENT.{Emp#} $\subseteq$ EMPLOYEE.{Emp#}	Lossless-join
#	ASSIGNMENT.{Fund#} $\subseteq$ FUND.{Fund#}	Lossless-join

Normal Forms Based on Functional Dependencies

#	EMPLOYEE.{**Job_type**} ⊆ JOB.{**Job_type**}	Lossless-join
#	FUND.{**Proj_nm**} ⊆ PROJECT.{**Proj_nm**}	Lossless-join

* Indicates revision to relationships via changes in inclusion dependencies in order to indicate a lossless-join navigation path.

The relational schema, while eradicating modification anomalies due to functional dependencies, is not dependency-preserving because it is impossible to deduce fd6: {**Proj_nm, Emp#**} → **Hours** and fd9: {**Proj#, Emp#**} → **Fund#** from the union of the FDs that hold in the *individual* relation schemas of URS3. In other words, from the union of the *individual* relation schemas of URS3 in the solution above, F^+ cannot be generated. The earlier lossless-join relationship between R5 and R0 is no longer present; instead, the decomposition has created a loss-join relationship between R5b and R0. A more robust design is to propagate the revision of R5b to R0 and thus retain the originally established lossless-join relationship between R0 and R5.

Step 6: Propagate the revisions to the primary keys in the BCNF resolutions to the primary keys of related relation schemas. Accordingly:

R5b:	ASSIGNMENT (**Fund#, Emp#**, Hours)	in BCNF
R0:	LORS2 (**Fund#, Emp#, Division**)	in BCNF

This design is superior because the lossless-join property is fully restored. Thus, the final form of the BCNF design that retains the lossless-join property is:
Decomposition: URS3 {R1, R2, R3, R4, R5b, R0} where:

R1:	PROJECT (**Proj#**, Proj_nm, Budget);	in BCNF
R2:	FUND (**Fund#**, Proj_nm)	in BCNF
R3:	EMPLOYEE (**Emp#**, Emp_nm, Job_type)	in BCNF
R4:	JOB (**Job_type**, Chg_rate)	in BCNF
R5b:	ASSIGNMENT (**Fund#, Emp#**, Hours)	in BCNF
R0:	LORS2 (**Fund#, Emp#, Division**)	in BCNF
#	LORS2.{**Fund#, Emp#**} ⊆ ASSIGNMENT.{**Fund#, Emp#**}	Lossless-join
#	ASSIGNMENT.{**Emp#**} ⊆ EMPLOYEE.{**Emp#**}	Lossless-join
#	ASSIGNMENT.{**Fund#**} ⊆ FUND.{**Fund#**}	Lossless-join
#	EMPLOYEE.{**Job_type**} ⊆ JOB.{**Job_type**}	Lossless-join
#	FUND.{**Proj_nm**} ⊆ PROJECT.{**Proj_nm**}	Lossless-join

The final solution is attribute-preserving and the join of URS3 {R1, R2, R3, R4, R5b, R0} does strictly yield R. All the relation schemas in URS3 {R1, R2, R3, R4, R5b, R0} are in BCNF. Therefore, the solution is correct. As for completeness, the solution is a lossless-join design, but is not dependency-preserving. The only way to compensate for the dependencies that are not preserved in this solution is to supplement the relational schema with the necessary materialized views for covering the lost FDs. In this case, the two FDs lost happen to have the same determinant and so can be captured in a single materialized view of the form:

MV1: SCHEDULE (Proj#, Emp#, Fund#, Hours)

Once again, irrespective of the primary key of choice from among the candidate keys of URS3, the same final solution will result. The reader is encouraged to try this as an exercise.

8.4 Denormalization

The jovial remark, "Normalize until it hurts, and denormalize until it works," while certainly funny, also indicates that the concept of denormalization is ill-understood. The general case against normalization is that the process results in lots of logically separate relations (tables) leading to lots of physically separate stored files and the consequent data retrieval inefficiencies.

Denormalization entails combining relations so that they are easier to query. The combined relations may lose their normalized status in that they may reintroduce data redundancies eliminated by the normalization process. The general misunderstanding, however, is that denormalization always improves data retrieval performance.

Formally, denormalization may be defined as replacing a set of (often normalized) relation schemas D {R1, R2, ... Rn} by their join R, such that projecting R over the set of attributes of R1, R2, ... Rn respectively is guaranteed to yield the original set D. The objective is to reduce the number of joins that may be required during the run time of queries (data retrieval) by including some of these joins structurally as a part of the database design.

As an example, consider a relation schema:

R: CUSTOMER (**Id, Name, Street, City, State, Zip_code**).

From the semantics of this general scenario, it is obvious that an FD of the form **Zip_code** → **{City, State}** holds on CUSTOMER resulting in a violation of 3NF. In a strict normalization paradigm, one would decompose R to eliminate the 3NF violation leading to the solution D {R1, R2} where:

R1: ZIP (**Zip_code, City, State**); R2: CUSTOMER (**Id, Name, Street, Zip_code**).

While the normalization process eliminates data redundancies and the associated modification anomalies in the design, most queries on CUSTOMER may require joining the two relations R1 and R2, while a denormalized R eliminates the repeated join operation and the attendant retrieval inefficiencies. The semantics of the scenario indicates that a **Zip_code** is rarely deleted, added or updated. Thus the expected modification anomalies are not a serious practical problem. In this case, one may opt for the denormalized design instead of the normalized solution. However, execution of any query that exclusively seeks zip_code data needs to unnecessarily access a relatively larger relation (R) and be accordingly less efficient.

As a second example, consider a scenario where employees of a large corporation have access to a handful of mutual fund companies for their retirement investment. Suppose the relation schema:

R: EMPLOYEE **(Emp#, Name, Age, Salary, M_fund#, Fund_nm, Fund_mgr)**
represents this scenario. The semantics of the scenario would suggest the presence of an FD **M_fund# → {Fund_nm, Fund_mgr}** causing a violation of 3NF in R. In a strict normalization paradigm, one would decompose R to eliminate the 3NF violation leading to the solution D {R1, R2} where:

R1: FUND **(M_fund#, Fund_nm, Fund_mgr)**;

R2: EMPLOYEE **(Emp#, Name, Age, Salary, M_fund#)**

Queries on EMPLOYEE requiring retirement financial data leads to joining R1 and R2, while the unnormalized (or denormalized) R is a better option in such situations. However, suppose there are thousands of employees and just a handful of these retirement mutual funds. Any query requiring only mutual fund data will execute rather inefficiently in the denormalized design. In this case, it may be worthwhile evaluating a normalized design supplemented by a materialized view for the joined relation.

In short, it must be noted that denormalization, contrary to conventional beliefs, need not improve data retrieval performance in all circumstances. Further, denormalization is considered, in general, only during physical database design where sometimes there are other options available to improve data retrieval efficiency.

Denormalization can be a meaningful strategy in the logical data modeling tier also. This application of denormalization is usually not recognized by academics or practitioners. When a relation schema is in immediate violation of BCNF (i.e., the relation schema is in 3NF, but violates BCNF), solving the design for BCNF always fails to preserve some of the functional dependencies in F. Under these circumstances, one is apt to denormalize the design to 3NF in order to preserve all functional dependencies in F. Solution Option 2 discussed in Section 8.1.5.3 is an ideal example for this.

8.5 Role of Reverse Engineering in Data Modeling

The term **reverse engineering** originates in hardware development and refers to the idea of working backwards *in a systematic manner* with a view to discover how a product works. The focus of software engineering over the decades has been in the area of "forward engineering"—i.e., developing *clearly understood* new systems, while one of the major concerns in the practitioners' world has been upgrading *poorly understood* old systems. The critical role of reverse engineering in the software development arena has been acknowledged since the early 1990s (Communications of the ACM, May 1994). The goal of reverse engineering in a software development environment is to understand how existing software systems work. Reverse engineering encompasses a wide array of tasks. Central to these tasks is identifying data and process components of existing software systems and the relationships within and across these components.

Motives suggested for database reverse engineering traditionally include migration from past database paradigms of hierarchical and network data architectures to the contemporary relational or object-oriented paradigms. But the more mundane task of migrating between different implementations within the relational paradigm is equally relevant from a reverse engineering perspective. In addition, reverse engineering sheds light on poorly documented software systems. The starting point of reverse engineering in these cases is an existing operational software system, and the task at hand is to analyze data patterns to distill emerging data structure and behaviors.

The focus in this section of the book is reverse engineering data models in order to generate a high-level conceptual description of the data architecture. Reverse engineering studies have revealed that often the original schemas are fundamentally flawed or violate "good" design practices (Premerlani and Blaha, 1994). This section presents a unique use of reverse engineering in the data modeling task in order to preempt occurrence of such flaws and failures through a better understanding of the design. To that end, the scope of this section is restricted to reverse engineering normalized relational schemas to their conceptual counterparts, namely, ER diagrams.

When a relation schema directly mapped from an entity type of an ERD requires normalization, it simply means that a set of related independent entity types have been erroneously represented as a single entity type in the ERD. Thus, reverse engineering a normalized relational schema to an ERD reveals how the ERD should have been to begin with. Such "discovery" enriches the designers' understanding of the application domain being modeled. Section 7.1 in Chapter 7 provides a case in point. If the source relation schema is not an entity type from an ERD, but, instead a URS constructed from a set of FDs directly arising from the business rules embedded in the user requirements specification, it is all the more valuable to depict the normalized solution as an ERD so that it can serve as a presentation/communication device among the designers and the users.

Unfortunately, reverse engineering is far from being a fully automated technique. Some automation has been tried (e.g., Chiang, et al, 1994) using heuristic approaches. In the following subsections, we use a heuristic approach based on a few guidelines for reverse engineering a relational schema to an ERD. The reader is encouraged to review the example in Section 7.1 of Chapter 7 where a reverse-engineered ERD is shown in Figure 7.4 before continuing with the rest of this section. The reverse engineering heuristic is stated in Figure 8.9.

» **Step 1** Translate the normalized relational schema to an information-preserving logical schema based on available information.

- Denote alternate keys as unique [Q]; Enclose composite attributes in braces [ ]
- Indicate the parent label on top of the foreign key along with the parent side structural constraints of the relationship type—default value for min and max when information not available are 0 and n respectively (see Section 6.7.2 of Chapter 6 for grammar)
- Indicate child side structural constraints of the relationship type right below the foreign key along with the deletion rule—default value for min when information not available is 0 and default deletion rule is R except when the cardinality ratio is 1:1; then, leave unfilled if rule not specified. (see Section 6.7.2 of Chapter 6 for grammar)

» **Step 2** Transform the information-preserving logical schema to a design-specific ER diagram— either coarse or fine granularity or a hybrid thereof depending on the available information.

- Map each logical scheme to an entity type:

  ♦ If the primary key of a logical scheme Lx is a proper subset of the primary key of another logical scheme Ly, map Ly as a weak entity type in the ERD with Lx as its identifying parent
  ♦ If the primary key of a logical scheme Lx is a concatenation of the primary keys of multiple logical schemes Ly, Lz, etc., map Lx as a gerund entity type in the ERD with Ly, Lz, etc. as its identifying parents
  ♦ All other logical schemes are mapped as base entity types

- Represent the foreign key attribute(s) in a logical scheme by a relationship type in the ERD*:

  ♦ Establish the relationship type
  ♦ Connect the relationship type to the parent (referenced) and child (referencing) entity types – if the child is a weak entity type then the relationship type is mapped as an identifying relationship type
  ♦ Map the (min, max) to the appropriate edge of the relationship type

- Map attributes of individual logical schemes to corresponding entity types in the ERD:

  ♦ Map atomic and composite attributes
  ♦ Underline unique identifiers (The primary key and alternate keys of a logical scheme are unique identifiers of the corresponding entity type)
  ♦ Partial key of a weak entity is denoted by a dotted underline

» **Step 3** Abstract the design-specific ER diagram up to a presentation layer ERD.

- Transform gerund entity types to n-way relationship types. Attributes of the gerund entity type remain as attributes of the relationship type
- Any relationship with a gerund entity type is transformed to a relationship type with the cluster entity type that the gerund represents
- A weak entity type not participating in any relationship other than the identifying relationship is transformed to a multi-valued (atomic/composite) attribute of the parent entity type

* The foreign key attribute is removed from the referencing (child) entity type unless the attribute plays some other role in the entity type in which case the attribute is retained in the entity type.

Figure 8.9 Heuristic to reverse-engineer a relational schema to an ER diagram

8.5.1 Reverse Engineering the Normalized Solution of Case 1

To begin with, URS1 in 1NF is of the form:

URS1 (Store, Branch, Location, Sq_ft, Manager, Product, Price, Customer, Address, Vendor, Type)

When fully normalized to BCNF, the resulting relational schema looks like:
URS1 {R1, R2, R3, R4, R5, R0} where:

R1:	CUSTOMER (**Customer**, Address);	in BCNF
R2:	VENDOR (**Vendor**, Product);	in BCNF

R3:	PRODUCT (**Product**, Price);	in BCNF
R4:	MANAGER (**Manager**, Store, Branch, Location, Sq_ft)	in BCNF
R5:	STORE (**Store**, Type)	in BCNF
R0:	LORS5 (**Manager, Customer, Vendor**)	
#	LORS5.{**Customer**} ⊆ CUSTOMER.{**Customer**}	Lossless-join
#	LORS5.{**Vendor**} ⊆ VENDOR.{**Vendor**}	Lossless-join
#	VENDOR.{**Product**} ⊆ PRODUCT.{**Product**}	Lossless-join
#	LORS5.{**Manager**} ⊆ MANAGER.{**Manager**}	Lossless-join
#	MANAGER.{**Store**} ⊆ STORE.{**Store**}	Lossless-join

During the discussion of normalization we suspended the information preservation issue to the background. As we get ready to reverse engineer the normalized URS1 {R1, R2, R3, R4, R5, R0}, the first step prescribed in the reverse engineering heuristic is to translate the normalized relational schema to an information-preserving logical schema *based on available information and assumed default properties where information is not immediately available*. The resulting information preserving logical schema is shown in Figure 8.10a. Next, following Step 2, we construct one entity type for each scheme in the logical schema using base and weak entity types. Since every foreign key in the logical schema represents a relationship, we then systematically replace every foreign key by a relationship type between the referencing entity type (the logical schema carrying the foreign key) and the referenced entity type (the parent entity type in this relationship). At this point, we have a skeletal ER diagram with no attributes allocated to the entity types. The attributes left over in the logical schema after the replacement of foreign key attributes are mapped to the corresponding entity types appropriately indicating non-key attributes with optional property, underlining the unique identifier (primary key and unique attributes in the logical schema) and marking the partial keys by a dotted underline. Figure 8.10b portrays this design-specific reverse engineering.

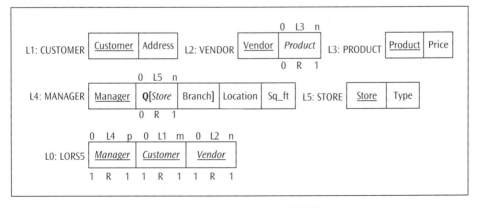

Figure 8.10a   Information-preserving logical schema for URS1

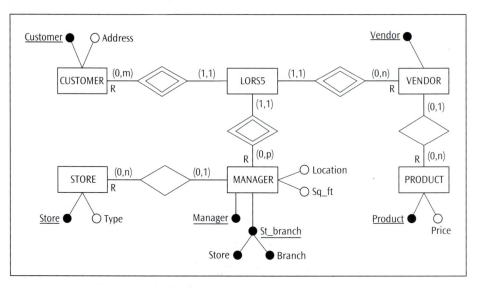

Figure 8.10b   Design-specific ER diagram reverse-engineered from the logical schema in Figure 8.10a

Finally, the design-specific ER diagram is abstracted up one more notch. The gerund entity types are transformed into m:n relationships between the participating entity types, weak entity types with no relationship other than the identifying relationship, and no attributes other than the partial keys are transformed into multi-valued attributes of the identifying parent and the (min, max) grammar, for the structural constraints of relationship types is replaced by cardinality ratio and participation constraint expressed as independent constructs. The presentation layer ERD for this case is shown in Figure 8.10c.

Reverse engineering a relational schema can generate more than one solution—more than one ERD that can produce the same relational schema. From a database implementation perspective the effect is insignificant since the relational schema generated is the same. However, at the conceptual level, the designer/user may relate better to one ERD than the other. While presenting an equivalent ERD for the readers' review, we do not make any comparative assessment of the efficacy of two different reverse-engineered ERDs in this book. Figure 8.11a is an alternative design equivalent to the reverse-engineered design-specific ER diagram shown in Figure 8.10b. By selecting **{Store, Branch}** as the primary key instead of **Manager** in R4, R4 can be depicted as a weak entity child of R5 in the ERD with no impact on the relational schema. The corresponding presentation layer ERD appears in Figure 8.11b. The fact that the attribute **Manager** is a unique identifier of BRANCH does not appear in this ERD. This is because a weak entity type, by definition, does not have an independent unique identifier. Therefore, this information should be carried in the list of semantic integrity constraints that accompanies the ERD.

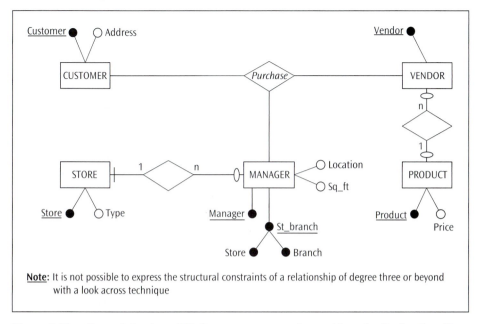

Note: It is not possible to express the structural constraints of a relationship of degree three or beyond with a look across technique

Figure 8.10c Presentation layer ER diagram reverse-engineered from the Design-Specific ER diagram in Figure 8.10b

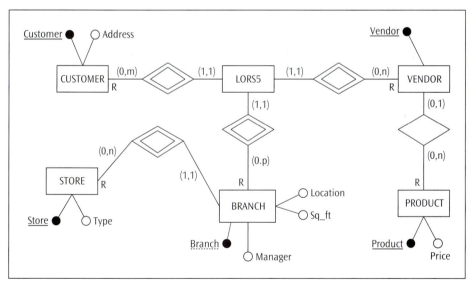

Figure 8.11a An alternative design equivalent to the Design-Specific ER diagram in Figure 8.10b

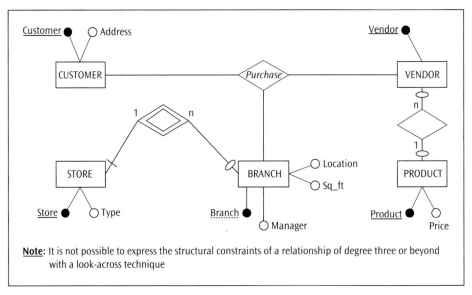

Note: It is not possible to express the structural constraints of a relationship of degree three or beyond with a look-across technique

Figure 8.11b Presentation layer ER diagram reverse-engineered from the Design-Specific ER diagram in Figure 8.11a

The logical schema in Figure 8.12a is the translation of the relational schema constructed by directly mapping the FDs in F to relation schemas instead of following the normalization process. At the end of Section 8.3.1, it is pointed out that the solution from this arbitrary approach fails to capture a relation schema, viz., LORS5 **(Manager, Customer, Vendor)** that the normalized solution yields. The consequence of this error becomes obvious in the ERD reverse-engineered from the logical schema depicted in Figure 8.12a, in that we see three islands of ERDs that are unconnected (see Figure 8.12b). Since the original source of the reengineering task is a single relation schema, viz., URS1, the ERD in Figure 8.12b cannot be correct.

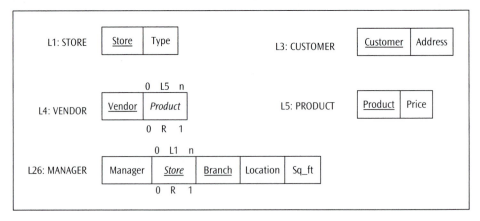

Figure 8.12a Logical schema for URS1 {L1, L3, L4, L5, L26) constructed by directly mapping FDs to relation schemas

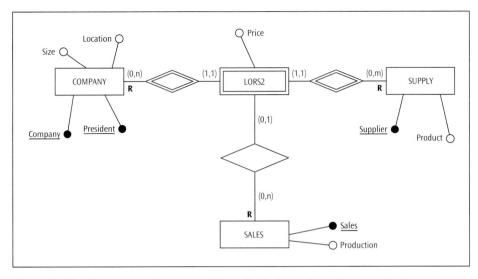

Figure 8.12b   The Presentation layer ERD for the logical schema in Figure 8.12a

8.5.2 Reverse Engineering the Normalized Solution of Case 2

URS2 in 1NF is of the form:

URS2 (Company, Location, Size, President, Product, Price, Sales, Production, Supplier)

When fully normalized to BCNF, the resulting relational schema looks like:
URS2 {R1, R2, R3, R0} where:

R1:	COMPANY (**Company**, Location, Size, President);	in BCNF
R2:	SALES (**Sales**, Production)	in BCNF
R3:	SUPPLY (**Supplier**, Product)	in BCNF
R0:	LORS2 (**Company**, Sales, Price, **Supplier**)	in BCNF
#	LORS2.{Company} ⊆ COMPANY.{Company}	Lossless-join
#	LORS2.{Supplier} ⊆ SUPPLY.{Supplier}	Lossless-join

The information-preserving logical schema based on available information and assumed default properties is shown in Figure 8.13a. Next, following Step 2 in Figure 8.9, the design-specific ER diagram is created. The last step is to abstract up the design-specific ERD to the presentation layer ERD. This is done by following the procedure suggested in Step 3 of the reverse engineering heuristic. The two layers of ERD reverse-engineered from the logical schema shown in Figure 8.13a appear in Figures 8.13b and 8.13c respectively.

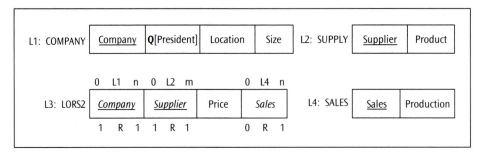

Figure 8.13a Logical schema for URS2

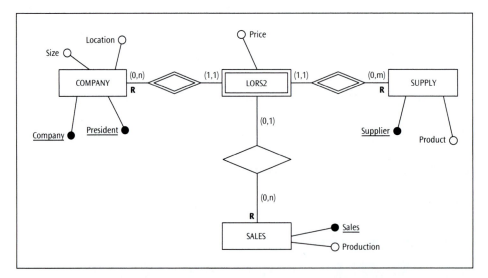

Figure 8.13b Design-Specific ER diagram reverse-engineered from the logical schema in Figure 8.13a

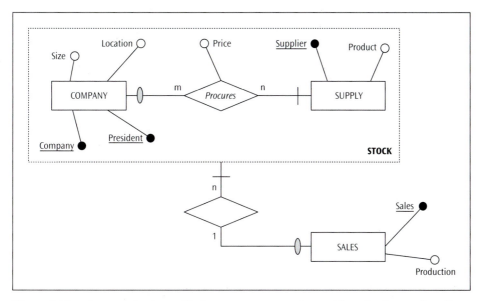

Figure 8.13c Presentation layer ER diagram reverse-engineered from the Design-Specific ER diagram in Figure 8.13b

8.5.3 Reverse Engineering the Normalized Solution of Case 3

URS3 in 1NF is of the form:

URS3 (Proj_nm, Emp#, Proj#, Job_type, Chg_rate, Emp_nm, Budget, Fund#, Hours, Division)

When fully normalized to BCNF, the resulting relational schema looks like:

URS3 {R1, R2, R3, R4, R5b, R0} where:

R1:	PROJECT (**Proj#**, Proj_nm, Budget);	in BCNF
R2:	FUND (**Fund#**, Proj_nm)	in BCNF
R3:	EMPLOYEE (**Emp#**, Emp_nm, Job_type)	in BCNF
R4:	JOB (**Job_type**, Chg_rate)	in BCNF
R5b:	ASSIGNMENT (**Fund#, Emp#**, Hours)	in BCNF
R0:	LORS2 (**Fund#, Emp#, Division**)	in BCNF
#	LORS2.{**Fund#, Emp#**} ⊆ ASSIGNMENT.{**Fund#, Emp#**}	Lossless-join
#	ASSIGNMENT.{**Emp#**} ⊆ EMPLOYEE.{**Emp#**}	Lossless-join
#	ASSIGNMENT.{**Fund#**} ⊆ FUND.{**Fund#**}	Lossless-join
#	EMPLOYEE.{**Job_type**} ⊆ JOB.{**Job_type**}	Lossless-join
#	FUND.{**Proj_nm**} ⊆ PROJECT.{**Proj_nm**}	Lossless-join

The information-preserving logical schema based on available information and assumed default properties is shown in Figure 8.14a. Next, following Step 2, the design-specific ER diagram is created (see Figure 8.14b).

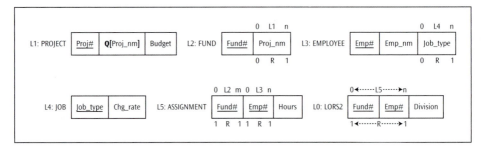

Figure 8.14a   Logical schema for URS3

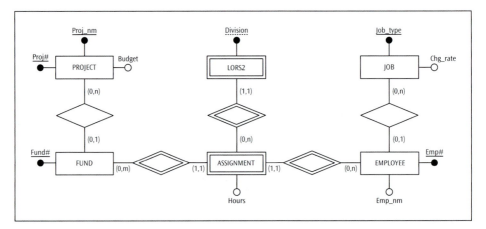

Figure 8.14b   Fine-granular, Design-Specific ER diagram reverse-engineered from the logical schema in Figure 8.14a

The last step is to reverse engineer the design-specific ERD to the presentation layer ERD. This is done by following the procedure suggested in Step 3 of the reverse engineering heuristic in Figure 8.9. The reverse-engineered presentation layer ERD is shown in Figure 8.14c. An additional level of abstraction of Figure 8.14c appears in Figure 8.14d. In fact, only Figure 8.14d represents the presentation layer ERD for the given relational schema.

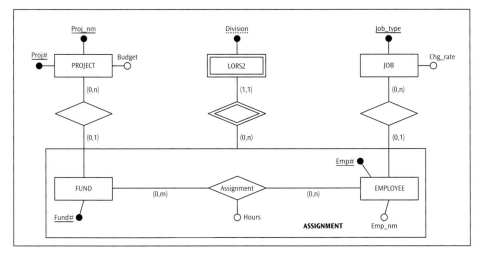

Figure 8.14c Coarse-granular, Design-Specific ER diagram reverse-engineered from the Fine-granular, Design-Specific ER diagram in Figure 8.14b

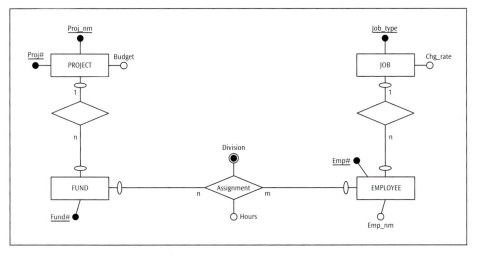

Figure 8.14d Presentation layer ER diagram reverse-engineered from the Design-Specific ER diagram in Figure 8.14c

Once again, a different ERD equivalent to the one in Figures 8.14b–d—equivalent in the sense it generates the same relational schema—is shown in Figure 8.15a–c. The variation in the design arises from the fact that based on the final relational schema in BCNF, R0 can be modeled in the ERD as either the weak entity child of R5b or a weak entity type with two identifying parents, viz., R2 and R3.

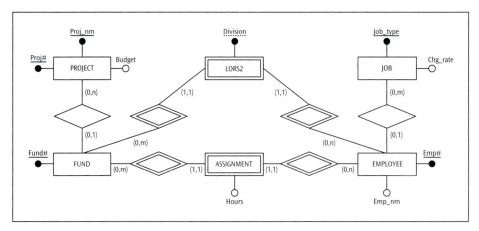

Figure 8.15a An alternative design equivalent to the Fine-granular, Design-Specific ER diagram in Figure 8.15b

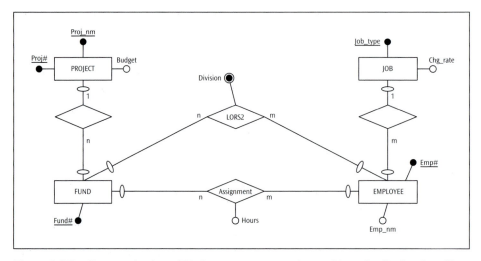

Figure 8.15b Presentation layer ER diagram reverse-engineered from the Design-Specific ER diagram in Figure 8.15a

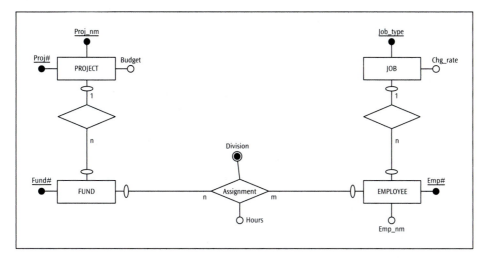

Figure 8.15c Presentation layer ER diagram reverse-engineered from the Design-Specific
ER diagram in Figure 8.15a: An alternative design

Chapter Summary

The root cause of data redundancy and the consequent modification anomalies in a database is the presence of undesirable functional dependencies in a relation schema. Any FD whose determinant is not a candidate key of the relation schema in which the FD holds is an undesirable FD. FDs emerge from business rules of the application domain and so cannot be ignored or discarded when they are undesirable. The set of FDs that prevail over a relation schema is referred to as F. This chapter explores the solution for this problem.

Normalization is presented as the mechanism capable of systematically weeding out undesirable FDs from a relation schema. The undesirable FDs manifest themselves either as partial dependencies or transitive dependencies in a relation schema. Normalization prescribes a method to render the undesirable FDs desirable. This is done by decomposing a relation schema with undesirable FDs to a set of relation schemas such that in each relation schema of the decomposed set, the determinant of the preserved FD is the candidate key of that relation schema.

Normal forms (NFs) provide a stepwise progression towards attaining the goal of a fully normalized relational schema that is guaranteed to be free of data redundancies that cause modification anomalies from a functional dependency perspective. First normal form (1NF) defines a relation schema—i.e., a schema that is not in 1NF is not a relation schema. Elimination of partial dependencies establishes 2NF. Two variations of transitive dependencies are resolved by 3NF and Boyce-Codd normal form (BCNF). A relation schema in BCNF is guaranteed to be free of modification anomalies due to functional dependencies.

A relational schema (i.e., a set of relation schemas) in BCNF does not necessarily result in a good database design. Relational schemas created through the decomposition process should also exhibit the lossless-join and dependency preservation properties. The lossless-join property is critical to ensure that spurious tuples are not generated by a join operation between two relations in the relational schema. The dependency preservation property, which ensures that each functional dependency is represented in some individual relation schema resulting after the decomposition, is sometimes sacrificed.

After introducing first, second, third, and Boyce-Codd normal forms plus the lossless-join and dependency preservation properties in the context of a series of individual examples, the motivating exemplar introduced originally in Chapter 7 is reexamined to illustrate how a set of functional dependencies derived from user-specified business rules can be used to develop a fully normalized relational schema. This discussion answers the three questions presented in Section 7.1: (a) How are data redundancies identified? (b) How is the base relation schema decomposed? and (c) How is the decomposition be evaluated for correctness and completeness?

The chapter includes a presentation of a comprehensive approach to the normalization process that incorporates trade-offs among a Boyce-Codd normal form design, a lossless-join decomposition, and dependency preservation. The approach involves the (a) derivation of the candidate keys along with the primary key of the initial universal relational schema (URS), (b) the identification and resolution of all second and third normal form violations in the URS, (c) the resolution of all Boyce-Codd normal form violations in the URS, and (d) an evaluation of whether the relational schema is dependency-preserving and yields a lossless-join decomposition.

When a relational schema is in BCNF, sometimes it may not be possible to achieve both lossless-join and dependency preservation properties in the design. One alternative in this situation is to accept a relational schema in third normal form and handle the potential data redundancy problems via application programs. Often a superior alternative is to establish a BCNF design that possesses the lossless-join property and use materialized views to capture the few functional dependencies that are not preserved.

Since the early 1990s, reverse engineering has drawn the attention of database researchers. Traditionally, reverse engineering seeks to examine operational software systems with a view to analyze data patterns to extract data structures and behaviors. This chapter presents a unique use of reverse engineering in the data modeling task in order get a better grip on design errors. To that end, the scope of this section is restricted to reverse engineering normalized relational schemas to their conceptual counterparts, viz., ER diagrams. Reverse engineering a normalized relational schema to an ERD reveals how the ERD should have been to begin with. Such discovery enriches the designers' understanding of the application domain being modeled. The chapter concludes with a description and a few demonstrations of a heuristic to reverse engineer a relational schema to a Presentation Layer ER diagram.

Exercises

1. What is the source of functional dependencies?

2. What is the difference between a desirable and an undesirable functional dependency? Describe the nature of the problems caused by undesirable functional dependencies. What prevents us from simply ignoring undesirable functional dependencies?

3. What is the role of normalization in the database design process?

4. Figure 8.1 illustrates how a first normal form violation of ALBUM can be resolved. Suppose that a single album could never have more than four artists. Describe another approach for defining ALBUM that does not violate first normal form.

5. Suppose {A, B} → C in a relation schema R. Under what condition would this not reflect a full functional dependency?

6. Consider the relation instance of the STU-CLASS relation schema.

    STU-CLASS (<u>Snum</u>, Sname, Major, <u>Cname</u>, Time Room)

Snum	Sname	Major	Cname	Time	Room
0110	KHUMAWALA	ACCOUNTING	BA482	MW3	C-150
0110	KHUMAWALA	ACCOUNTING	BD445	TR2	C-213
0110	KHUMAWALA	ACCOUNTING	BA491	TR3	C-141
1000	STEDRY	ANTHROPOLOGY	AP150	MWF9	D-412
1000	STEDRY	ANTHROPOLOGY	BD445	TR2	C-213
2000	KHUMAWALA	STATISTICS	BA491	TR3	C-141
2000	KHUMAWALA	STATISTICS	BD445	TR2	C-213

Normal Forms Based on Functional Dependencies

Snum	Sname	Major	Cname	Time	Room
3000	GAMBLE	ACCOUNTING	BA482	MW3	C-150
3000	GAMBLE	ACCOUNTING	BP490	MW4	C-150

Identify at least one update anomaly, one insertion anomaly, and one deletion anomaly.

7. What are the undesirable functional dependencies in the relation instance of the STU-CLASS relation schema shown in Exercise 6?

8. Consider the following relation instance of the CAR relation schema:

CAR

Model	#cylinders	Origin	Tax	Fee
Camry	4	Japan	15	30
Mustang	6	USA	0	45
Fiat	4	Italy	18	30
Accord	4	Japan	15	30
Century	8	USA	0	45
Mustang	4	Canada	0	30
Monte Carlo	6	Canada	0	45
Civic	4	Japan	15	30
Mustang	4	Mexico	15	30
Mustang	6	Mexico	15	45
Civic	4	Korea	15	30

   a. What is the minimal cover of functional dependencies that exists in CAR?

   b. What is (are) the candidate key(s) of CAR?

   c. If there is more than one candidate key, select a primary key from among the candidate keys.

   d. Based on the primary key, what is the immediate normal form violated in CAR?

   e. Develop a dependency-preserving solution that eliminates this normal form violation. Is your solution a lossless-join decomposition?

9. Consider the following relation instance of the SPORT relation schema.

SPORT

Stu#	Sport	Coach
125	Football	Register
140	Basketball	Lambert

Stu#	Sport	Coach
220	Baseball	Register
246	Basketball	Lambert

a. What is the minimal cover of functional dependencies that exists in SPORT?

b. What is (are) the candidate key(s) of SPORT?

c. If there is more than one candidate key, select a primary key from among the candidate keys.

d. Based on the primary key, what is the normal form violated in SPORT?

e. Develop a dependency-preserving solution that eliminates these normal form violations. Is your solution a lossless-join decomposition?

10. Given a relation schema R (A, B, C) that is in 3NF:

a. State the conditions under which modification anomalies can exist.

b. Express the conditions stated above in terms of functional dependencies.

c. For the conditions stated above in Step a, modification anomalies need not exist. State the functional dependencies for this case.

11. What is the difference between 3NF and Boyce-Codd Normal Form (BCNF)?

12. Consider the relation instance EXAM (Student, Subject, Rank). The meaning of an EXAM tuple is that a specified student examined in a specified subject achieves a specified rank in the class. Further, no two students obtain the same rank in the same subject.

EXAM

Student	Subject	Rank
McGrady	Math	3
Howard	Math	4
McGrady	English	2
Jackson	English	1
Yao	Math	1
Yao	Chemistry	1
Sura	Math	2
Ward	English	3
Taylor	Chemistry	2
Taylor	Math	5
Ewing	Chemistry	3

a. What is the minimal cover of functional dependencies that exists in EXAM?

b. What is (are) the candidate key(s) of EXAM?

c. If there is more than one candidate key, select a primary key from among the candidate keys.

d. Based on your primary key, what normal form violations exist in EXAM?

13. Consider the relation schema PATIENT_VISIT (Patient, Hospital, Doctor) and the relation instance given below:

Patient	Hospital	Doctor
Smith	Methodist	D. Cooley
Lee	St. Luke's	Z. Zhang
Marks	Methodist	D. Cooley
Marks	St. Luke's	W. Lowe
Lou	Hermann	R. Duke

In addition, suppose the following semantic rules exist:

- Each patient may be a patient in several hospitals.
- For each hospital, a patient may have only one doctor.
- Each hospital has several doctors.
- Each doctor uses only one hospital.
- Each doctor treats several patients in one hospital.

a. What is the minimal cover of functional dependencies that exists in PATIENT_VISIT?

b. What is (are) the candidate key(s) of PATIENT_VISIT?

c. If there is more than one candidate key, select a primary key from among the candidate keys.

d. Based on the primary key chosen, what normal form violations exist in PATIENT_VISIT?

e. Develop a solution that eliminates the normal form violations.

f. Is your solution a lossless-join decomposition?

g. Is your solution dependency-preserving? If not, how can dependency preservation be achieved? Is this revised solution in BCNF?

h. Provide a solution that meets all three of the following conditions: (1) is in BCNF, (2) is dependency-preserving, and (3) is a lossless-join decomposition.

14. Why are attribute preservation, dependency preservation, and lossless-join decomposition requirements of a fully normalized database design?

15. Consider the relation schema ACTIVITY (Stu#, Sport, Cost) and associated relation instance shown below:

F: fd1: Stu# $\rightarrow$ Sport; fd2: Stu# $\rightarrow$ Cost; fd3: Sport $\rightarrow$ Cost

Stu#	Sport	Cost
100	SKIING	200
150	TENNIS	50
175	KARATE	50
200	TENNIS	50

    a.   What is the minimal cover of ACTIVITY?

    b.   What, if any, immediate normal form violation exists in ACTIVITY?

    c.   Discuss each of the following three decompositions in the context of attribute preservation, dependency preservation, and lossless-join decomposition.

        D: R1a: STU-SPORT (<u>Stu#</u>, Sport) R2a: SPORT-COST (<u>Sport</u>, Cost)

        D: R1b: STU-SPORT (<u>Stu#</u>, Sport) R2b: STU-COST (<u>Stu#</u>, Cost)

        D: R1c: STU-COST (<u>Stu#</u>, Cost) R2c: SPORT-COST (<u>Sport</u>, Cost)

16.   What is the difference between a loss-join decomposition and a lossless-join decomposition?

17.   Consider relation SUPPLY (S#, Sname, P#, Qty) with supplier names unique such that

    F: fd1: S# → Sname; fd2: Sname → S#; {S#, P#} → Qty

    S# represents a supplier number, Sname represents a supplier name, P# represents a part number, and Qty represents the quantity of a specific part supplied by a specific supplier.

SUPPLY

S#	Sname	P#	Qty
S1	SMITH	P1	300
S1	SMITH	P2	300
S1	SMITH	P3	400
S1	SMITH	P4	200
S1	SMITH	P5	100
S1	SMITH	P6	100
S2	CLARK	P1	300
S2	CLARK	P2	400
S3	MORRIS	P2	200
S4	MCNARY	P2	200
S4	MCNARY	P4	300
S4	MCNARY	P5	400

Normal Forms Based on Functional Dependencies

a. Is there a 3NF violation in SUPPLY? If yes, explain. If no, is there a BCNF violation in SUPPLY? Explain.

b. Decompose SUPPLY if necessary so that the resulting relational schema is in BCNF. Is your design attribute-preserving, dependency-preserving, and a lossless-join decomposition? Explain.

18. Consider the relation schema CLASS with attributes Student, Subject, and Teacher. The meaning of this relation is that the specified student is taught the specified subject by the specified teacher. Assume that semantic rules, depicted by the following functional dependencies, exist:

{Student, Subject} → Teacher

Teacher → Subject

Subject ↛ Teacher

{Student, Teacher} → Subject

a. What do these rules mean in words? Are all these rules necessary? If not, explain which are not needed.

b. Is the following sample data consistent with these rules? Why or why not?

Student	Subject	Teacher
Smith	Math	White
Smith	Physics	Green
Jones	Math	White
Jones	Physics	Brown

c. What causes CLASS to contain a BCNF violation? What anomalies does it exhibit?

d. Decompose CLASS if necessary so that the resulting relational schema is in BCNF. Is your design attribute-preserving, dependency-preserving, and a lossless-join decomposition? Explain.

19. This exercise is a variation of Exercise 18. Consider the following functional dependencies prevailing over the relation schema CLASS (Student, Subject, Teacher):

{Student, Subject} → Teacher

Teacher ↛ Subject

Subject ↛ Teacher

{Student, Teacher} ↛ Subject

a. What do these rules mean in words? Are all these rules necessary? If not, explain which are not needed.

b. Is the following sample data consistent with these rules? Why or why not?

Student	Subject	Teacher
Smith	Math	White
Smith	Physics	White
Jones	Math	Green
Jones	Physics	White

    c.   Does this version of CLASS satisfy the requirements of BCNF? Is it free of modification anomalies triggered by undesirable functional dependencies? Is it free of modification anomalies altogether?

20.   Given the relation schema FLIGHT (Gate#, Flight#, Date, Airport, Aircraft, Pilot) and the constraint set F {fd1, fd2, fd3} where:

    fd1: {Airport, Flight#, Date} → Gate; fd2: {Flight#, Date} → Aircraft

    fd3: {Flight, Date} → Pilot

    a.   List the candidate key(s) of FLIGHT.

    b.   For each candidate key, indicate the immediate normal form violated in FLIGHT by each of the functional dependencies given above.

    c.   If FLIGHT is not in BCNF, design a relational schema that

-     is in BCNF, and

-     yields a lossless-join decomposition

    d.   Are all functional dependencies in F preserved? If not, which are not preserved?

21.   Consider the universal relation schema INVENTORY (Store#, Item, Vendor, Date, Cost, Units, Manager, Price, Sale, Size, Color, Location) and the constraint set F {fd1, fd2, fd3, fd4, fd5, fd6, fd7} introduced originally in Chapter 7, Exercise 13 where:

fd1: {Item, Vendor} → Cost	fd2: {Store#, Date} → {Manager, Sale}
fd3: {Store#, Item, Date} → Units	fd4: Manager → Store#
fd5: Cost → Price	fd6: Item → {Size, Color}
fd7: Vendor → Location	

    a.   Confirm that F is a minimal cover for the set of functional dependencies given above.

    b.   List the candidate key(s) of INVENTORY.

    c.   For each candidate key, indicate the immediate normal form violated in INVENTORY by each of the functional dependencies given above.

    d.   If INVENTORY is not in BCNF, design a relational schema that

-     is in BCNF so that all modification anomalies due to functional dependencies are eradicated, and

-     yields *all* lossless-join decompositions

e. List the functional dependencies in **F** that are not preserved in this design.

f. Show the final design. The design should be parsimonious (i.e., minimal set in BCNF). Also, *clearly indicate entity integrity and referential integrity constraints.*

g. Revise the above design so that <u>all</u> dependencies are preserved in a lossless-join decomposition with the least sacrifice in the achieved level of normal form.

22. Given the set of functional dependencies F {fd1, fd2, fd3, fd4, fd5, fd6, fd7, fd8, fd9, fd10} introduced originally in Chapter 7, Exercise 14:

fd1: **Tenant# ➞ {Name, Job, Phone#, Address}**	fd2: **Job ➞ Salary**
fd3: **Name ➞ Gender**	fd4: **Phone# ➞ Address**
fd5: **{Name, Phone#} ➞ {Tenant, Deposit}**	fd6: **County ➞ Tax_rate**
fd7: **Area ➞ {Rent, County}**	fd8: **Survey# ➞ Lot**
fd9: **{Lot, County} ➞ {Survey#, Area}**	fd10: **{Survey#, Area} ➞ County**

a. Using the universal relation schema (URS) and its associated primary key developed in Chapter 7, Exercise 14, indicate the **immediate** normal form violated in each of the functional dependencies given above and explain how the particular normal form is violated in each case.

b. If the URS is not in BCNF, design a relational schema that
   - is in BCNF so that all modification anomalies due to functional dependencies are eradicated, and
   - yields **all** lossless-join decompositions

c. List the functional dependencies in **F** that are not preserved in this design.

d. Show the final design. The design should be parsimonious (i.e., minimal set in BCNF). Also, *clearly indicate entity integrity and referential integrity constraints.*

e. Revise the above design so that *all* dependencies are preserved in a lossless-join decomposition with the least sacrifice in the achieved level of normal form.

f. Reverse engineer the design to the conceptual level and show it as a Presentation Layer ER diagram.

23. Given the set of functional dependencies F {fd1, fd2, fd3, fd4, fd5, fd6, fd7, fd8, fd9, f10, f11} introduced originally in Chapter 7, Exercise 15:

fd1: **Client ➞ Office**	fd2: **Stock ➞ {Exchange, Dividend}**
fd3: **Broker ➞ Profile**	fd4: **Company ➞ Stock**
fd5: **Client ➞ {Risk_profile, Analyst}**	fd6: **Analyst ➞ Broker**
fd7: **{Stock, Broker} ➞ {Investment, Volume}**	fd8: **Stock ➞ Company**
fd9: **Investment ➞ {Commission, Return}**	fd10: **{Stock, Broker} ➞ Client**
fd11: **Account ➞ Assets**	

a. Using the universal relation schema (URS) and its associated primary key developed in Chapter 7, Exercise 15, indicate the *immediate* normal form violated in each of the functional dependencies given above and explain how the particular normal form is violated in each case.

b. Decompose the URS to arrive at a design/schema that is in BCNF. In each decomposition, identify the primary key and the functional dependencies accounted for. Demonstrate that each decomposition is a lossless-join decomposition.

c. Show the final design. The design should be parsimonious (minimal set in BCNF). *Clearly indicate entity integrity and referential integrity constraints*.

d. Indicate the functional dependencies in F that are not preserved in the BCNF design and specify materialized view(s) for the same.

e. Reverse engineer the design to the conceptual level and show it is a Presentation Layer ER diagram.

f. Revise the design above so that *all* dependencies are preserved in a lossless-join decomposition with the least sacrifice in the achieved level of normal form.

Selected Bibliography

Chiang, R. H. L., Barron, T. M. and Storey, V. C. (1994) "Reverse Engineering of Relational Databases: Extraction of an ER Model from a Relational Database," *Data & Knowledge Engineering*, 12, 107–142.

Chiang, R. H. L., Barron, T. M. and Storey, V. C. (1997) "A Framework for the Design and Evaluation of Reverse Engineering Methods for Relational Databases," *Data & Knowledge Engineering*, 21, 55–77.

Codd, E. F. (1970) "A Relational Model for Large Shared Data Banks," *Communications of the ACM*, 13, 6 (June).

Date, C. J. (2004) *An Introduction to Database Systems*, Eighth Edition, Addison-Wesley.

Elmasri, R., and Navathe, S. B. (2003) *Fundamentals of Database Systems*, Fourth Edition, Addison-Wesley.

Kent, W. (1983) "A Simple Guide to the Five Normal Forms in Relational Database Theory," *Communications of the ACM*, 26, 2 (February), 120–125.

Kifer, M., Bernstein, A., and Lewis, P. M. (2005) *Database Systems: An Application-Oriented Approach*, Second Edition, Addison-Wesley.

Premerlani, W. J. and Blaha, M. R. (1994) "An Approach to Reverse Engineering of Relational Databases," *Communications of the ACM*, 37, 5 (May), 42–49.

Silberschatz, A., Korth, H. F. and Sudarshan, S. (2002) *Database System Concepts*, Fourth Edition, McGraw-Hill.

Teorey, T. J., Yang, D., and Fry, J. P. (1986) "A Logical Design Methodology for Relational Databases Using the Extended Entity-Relationship Model," *Computing Surveys*, 18, 2 (June) 197–222.

Waters, R. C. and Chikofsky, E. (1994) "Reverse Engineering: A Brief Summary of the Papers Presented at the May 1993 ACM/IEEE Computer Society's Working Conference on Reverse Engineering," *Communications of the ACM*, 37, 5 (May), 22–25.

CHAPTER **9**

Higher Normal Forms

The focus of Chapter 8 was data redundancies resulting from undesirable functional dependencies and the aspects of normalization that address this specific problem. This chapter completes the discussion of normalization by addressing normal forms that are beyond the purview of functional dependencies.

This chapter flows as follows. Section 9.1 introduces the concept of multi-valued dependency (MVD) in a relation schema. Beginning with a motivation exemplar in subsection 9.1.1 to help the reader appreciate the import of MVD intuitively, subsections 9.1.2 and 9.1.3 present the formal definition of MVD and the inference rules associated with MVD, respectively. Fourth normal form (4NF) is introduced as the solution to eliminate data redundancies caused by MVDs in Section 9.2. This is followed by a comprehensive example in Section 9.3 describing the occurrence and resolution of 4NF violation in a relation schema. The generality of 4NF is discussed in Section 9.4 by showing how 4NF subsumes all the previously discussed normal forms. The topic of Section 9.5 is the concept of join dependency and the associated normal form called Project/Join normal form (PJNF) or fifth normal form (5NF). A brief note on Domain/Key normal form (DKNF) in Section 9.6 concludes this chapter.

9.1 Multi-valued Dependency

So far, we have examined modification anomalies (insertion, deletion, and update anomalies) due to data redundancies triggered by undesirable functional dependencies in a relation schema and developed a solution to the problem via normalization. We concluded that when a relation schema is normalized to BCNF, the resulting BCNF design (relational schema) is guaranteed to be free of modification anomalies due to functional dependencies. In other words, the basis for 2NF, 3NF, and BCNF has been functional dependencies. However, there are modification anomalies that are not based on functional dependencies, as seen in the following sections.

9.1.1 A Motivating Exemplar for Multi-valued Dependency

Consider an entity type MUSICIAN with attributes **Name, Age, Address, Phone#, and Band.** *Also, assume that a musician can be identified by his/her name. In addition, suppose that a musician may be associated with several types of music (e.g., Jazz, Rock, Classical) and may own several vehicles (e.g., car, Jeep, Truck, Van). In other words, while* **Name, Age, Address, Phone#, and Band** *are single-valued attributes of MUSICIAN,* **Music** *and* **Vehicle** *are multi-valued attributes of MUSICIAN. A schema representing MUSICIAN can be shown as:*

MUSICIAN (**Name, Age, Address, Phone#, {Music}, {Vehicle}, Band**)
where fd1: **Name → Age, Address, Phone#, Band,** and **Music** and **Vehicle** (enclosed in {})are multi-valued attributes.

MUSICIAN, by definition, is not a relation schema because it is not in 1NF due to the presence of multi-valued attributes, **Music** and **Vehicle**.

MUSICIAN can be set in 1NF by reducing **Music** and **Vehicle** to single-valued attributes. This is accomplished by introducing the functional dependency (FD) fd2 in R0 where:

fd2: **(Name, Music, Vehicle)** → **Age, Address, Phone#, Band**

As a result, we have a 1NF schema:

R0: MUSICIAN **(Name, Music, Vehicle, Age, Address, Phone#, Band)**

Since MUSICIAN (R0) is in 1NF, it is a relation schema with **(Name, Music, Vehicle)** as its primary key.

In this relation schema, fd1 violates 2NF. The decomposition D: {R1, R} where:

R1: MUSICIAN **(Name, Age, Address, Phone#, Band)**

R: MUSIC_VEHICLE **(Name, Music, Vehicle)**

is not only in 2NF, but also in 3NF and BCNF.

At this point, let us explore MUSIC_VEHICLE further using an instance of the relation schema shown in Table 9.1. A tuple in this relation indicates that a musician (who has a name) is, say, learning different types of music and owns different kinds of vehicles. Notably, the kinds of vehicles a musician owns are independent of the types of music he or she learns—that is, the attributes **Music** and **Vehicle** are independent of each other in MUSIC_VEHICLE.

Table 9.1 MUSIC_VEHICLE

Name	Music	Vehicle
Kamath	Jazz	Jeep
Kamath	Jazz	Truck
Kamath	Jazz	Van
Kamath	Rock	Truck
Kamath	Rock	Jeep
Kamath	Rock	Van
McKinney	Classical	Van
McKinney	Rock	Van
McKinney	Classical	Car
McKinney	Rock	Car
Barron	Jazz	Jeep
Barron	Jazz	Car

Although MUSIC_VEHICLE is in BCNF, data redundancies exist in the relation. For instance, if Kamath starts learning country music, to keep the relation instance consistent, we need to add the following three tuples to the relation:

Kamath	Country	Jeep
Kamath	Country	Truck
Kamath	Country	Van

If we add only one or two of the above tuples, the semantics of the relation will be altered. For example, if we add only the first tuple from the set above to MUSIC_VEHICLE, it will mean that Kamath learns Country music only when owning a Jeep. In other words, the fact that **Music** and **Vehicle** are independent attributes will be compromised in MUSIC_VEHICLE. Therefore, an insertion anomaly is present. Likewise, if Kamath does not own a Jeep anymore, the following two tuples will have to be deleted in order to keep the relation instance consistent with the implied semantics:

Kamath	Jazz	Jeep
Kamath	Rock	Jeep

This amounts to a deletion anomaly.

In short, MUSIC_VEHICLE is in BCNF and yet modification anomalies persist in the relation schema. It is important to note that these modification anomalies are not due to undesirable FDs. There are no undesirable FDs in MUSIC_VEHICLE. In fact, all FDs in MUSIC_VEHICLE are trivial dependencies, such as (**{Name, Music, Vehicle} → Name; {Name, Music, Vehicle} → {Music, Vehicle}**). A close examination of the relation MUSIC_VEHICLE reveals that for a given value of Name (e.g., Kamath), the same set of **Music** that Kamath learns (Jazz, Rock) occurs for each value of **Vehicle** that Kamath owns (Jeep, Truck, Van). Similarly, for the **Name** McKinney, the set of **Music** values, Classical and Rock, that McKinney learns occurs for each value of **Vehicle**, Van and Car, that McKinney owns. This pattern is due to what is called a multi-valued dependency (MVD) and is a direct consequence of 1NF because 1NF does not permit an attribute to have more than one value in a tuple.

9.1.2 Multi-valued Dependency Defined

Given **R (X, Y, Z)** where X, Y, and Z are atomic or composite attributes and $Z = R - (X \cup Y)$ $X \Rightarrow Y$ means that for a given value of X (say, x_1) in any relation state r of R, the same set of Y values ($y_1, y_2, \ldots y_n$) occurs for each value of Z ($z_1, z_2, z_3, \ldots z_m$). In other words, a multi-valued dependency (MVD) $X \Rightarrow Y$ holds for R (X, Y, Z) **iff** (if and only if) whenever (x, y, z) and (x, y', z') are tuples of any relation state r of R, then so are (x, y', z) and (x, y, z'). Observe that the implication here is that Y and Z are multi-valued attributes "independent" of each other in R.

How can we detect an MVD in a relation schema? In a relation schema **R (X, Y, Z)** that is in BCNF, MVD $X \Rightarrow Y$ exists iff the natural join of r(R1) and r(R2), where **R1 (X, Y)** and

R2 (X, Z) are projections of R, strictly yields r(R) for every relation state r of R. Let us apply this test on the relation state of MUSIC_VEHICLE that appears in Section 9.1.1. Suppose we want to test if the MVD **Name** $\Rightarrow$ **Music** exists in MUSIC_VEHICLE. Then, the projections whose natural join we want to compute and compare with MUSIC_VEHICLE are:

N_MUSIC **(Name, Music)** and N_VEHICLE **(Name, Vehicle)**

For the convenience of the reader the relation state of MUSIC_VEHICLE from Section 9.1.1 along with the two projections, N_MUSIC and N_VEHICLE, are shown in Figure 9.1. Observe that the natural join (N_MUSIC * N_VEHICLE) strictly yields MUSIC_VEHICLE. Therefore, we conclude that the MVD **Name** $\Rightarrow$ **Music** exists in MUSIC_VEHICLE. According to the inference rule of complementation discussed in Section 9.1.3 below, **Name** $\Rightarrow$ **Vehicle** also exists in MUSIC_VEHICLE. In other words, MVDs occur in pairs and are often shown that way, as in **Name** $\Rightarrow$ **Music | Vehicle**.

In essence, if a binary decomposition that has the lossless-join property exists for a relation state of a relation schema R, then:

- *There are non-trivial MVDs in the target relation schema R*, and
- *The MVDs can be inferred from the lossless-join decompositions as the intersection of the two projections multi-determines the differences between the two projections*.

To reinforce understanding, let us test if MVDs **Music** $\Rightarrow$ **Vehicle | Name** exist in MUSIC_VEHICLE. In this case, the projections of MUSIC_VEHICLE whose natural join we want to compute and compare with MUSIC_VEHICLE are:

M_NAME **(Name, Music)** and M_VEHICLE **(Music, Vehicle)**

These two projections of MUSIC_VEHICLE and the natural join of the projections are shown at the bottom of Figure 9.1. Observe that the natural join (M_NAME * M_VEHICLE) does not strictly yield MUSIC_VEHICLE. The shaded tuples in the natural join at the bottom of Figure 9.1 are not present in MUSIC_VEHICLE. Therefore, we conclude that the MVDs **Music** $\Rightarrow$ **Vehicle | Name** do not exist in MUSIC_VEHICLE.

9.1.3 Inference Rules for Multi-valued Dependencies

Similar to the three root inference rules for FDs (the reflexivity rule, the augmentation rule, and the transitivity rule, see Section 7.2.2), there are four inference rules that characterize MVDs.

Given R (A, B, C, D) where A, B, C, and D are arbitrary subsets, atomic or composite, of the set of attributes of a relation schema, R, the following inference rules apply to MVDs:

- **Reflexivity rule**: If B $\subset$ A, then A $\Rightarrow$ B
- **Complementation rule**: If A $\Rightarrow$ B, then A $\Rightarrow$ [R - (A $\cup$ B)]
- **Augmentation rule**: If A $\Rightarrow$ B, and C $\subset$ D, then (A, D) $\Rightarrow$ (B, C)
- **Transitivity rule**: If A $\Rightarrow$ B, and B $\Rightarrow$ C, then A $\Rightarrow$ (C – B)

A few other inference rules can be derived from these four (e.g., union rule, decomposition rule, pseudotransitivity rule). In addition, there is an inference rule that essentially bridges an FD and an MVD[1]:

- **Replication rule**: If A $\rightarrow$ B, then A $\Rightarrow$ B

---

[1] There is a second inference rule called the **coalescence rule** linking FD to MVD which is a little less intuitive: if A $\Rightarrow$ B, and (i) B and D are disjoint, (ii) D $\rightarrow$ C, and (iii) C $\subset$ B, then A $\rightarrow$ C.

MUSIC_VEHICLE

Name	Music	Vehicle
Kamath	Jazz	Jeep
Kamath	Jazz	Truck
Kamath	Jazz	Van
Kamath	Rock	Truck
Kamath	Rock	Jeep
Kamath	Rock	Van
McKinney	Classical	Van
McKinney	Rock	Van
McKinney	Classical	Car
McKinney	Rock	Car
Barron	Jazz	Jeep
Barron	Jazz	Car

Test: Name ↠ Music | Vehicle

N_MUSIC

Name	Music
Kamath	Jazz
Kamath	Rock
McKinney	Classical
McKinney	Rock
Barron	Jazz

N_VEHICLE

Name	Vehicle
Kamath	Jeep
Kamath	Truck
Kamath	Van
McKinney	Van
McKinney	Car
Barron	Jeep
Barron	Car

Test ratified:
Name ↠ Music | Vehicle
holds in MUSIC_VEHICLE

N_MUSIC * N_VEHICLE

Name	Music	Vehicle
Kamath	Jazz	Jeep
Kamath	Jazz	Truck
Kamath	Jazz	Van
Kamath	Rock	Truck
Kamath	Rock	Jeep
Kamath	Rock	Van
McKinney	Classical	Van
McKinney	Rock	Van
McKinney	Classical	Car
McKinney	Rock	Car
Barron	Jazz	Jeep
Barron	Jazz	Car

Test: Music ↠ Vehicle | Name

M_NAME

Name	Music
Kamath	Jazz
Kamath	Rock
McKinney	Classical
McKinney	Rock
Barron	Jazz

M_VEHICLE

Music	Vehicle
Jazz	Jeep
Jazz	Truck
Jazz	Van
Jazz	Car
Rock	Jeep
Rock	Truck
Rock	Van
Rock	Car
Classical	Van
Classical	Car

Test failed:
Music ↠ Vehicle | Name
Does not hold in
MUSIC_VEHICLE

M_NAME * M_VEHICLE

Name	Music	Vehicle
Kamath	Jazz	Jeep
Kamath	Jazz	Truck
Kamath	Jazz	Van
Kamath	Rock	Truck
Kamath	Rock	Jeep
Kamath	Rock	Van
McKinney	Classical	Van
McKinney	Rock	Van
McKinney	Classical	Car
McKinney	Rock	Car
Barron	Jazz	Jeep
Barron	Jazz	Car
Kamath	Jazz	Car
Kamath	Rock	Car
McKinney	Classical	Van
McKinney	Rock	Jeep
McKinney	Rock	Truck
McKinney	Rock	Van
Barron	Jazz	Truck
Barron	Jazz	Van

Figure 9.1   Test to check for the presence of multi-valued dependencies

The replication rule indicates that an MVD is a generalization of an FD in the sense that every FD is an MVD. Note that the converse is not true. More precisely, an FD is an MVD in which the set of dependent values matching a specific determinant value is always a *singleton set*.

From the above set of rules for FDs and MVDs, it is possible to infer the complete set of FDs in F^+ and the MVDs that hold in any relation state r of R that satisfies V (the set of specified MVDs). The closure of V is referred to as V^+.

Another useful rule derived by Catriel, Fagin, and Howard (1977) is:

If $A \Rightarrow B$, and $(A, B) \rightarrow C$, then $A \rightarrow (C - B)$[2]

The property of symmetry in an MVD emerges directly from the complementation rule. Accordingly, if an MVD $X \Rightarrow Y$ holds in the relation schema $R (X, Y, Z)$, then so does the MVD $X \Rightarrow Z$. This is often represented as $X \Rightarrow Y \mid Z$. The MVD $X \Rightarrow Y$ in the relation schema $R (X, Y, Z)$ is a *trivial* MVD if either (i) $Y \subset X$ or (ii) $(X \cup Y) = R$ because in either case the MVD does not convey any additional constraint (meaning). Trivial MVDs are usually removed from V^+ without any consequence. An MVD that satisfies neither (i) nor (ii) is a *non-trivial* MVD and requires attention.

9.2 Fourth Normal Form (4NF)

In Section 9.1.1, we saw that a relation schema in BCNF can still contain data redundancies and associated modification anomalies. However, the data redundancies in this case are not due to functional dependencies. The source of the problem here is what is known as multi-valued dependencies. The only desirable MVD in a relation schema, R, is an MVD whose determinant is a superkey of R. In other words, non-trivial MVDs that are not also FDs are always undesirable because the very presence of these MVDs induces data redundancies in a relation schema, resulting in modification anomalies. This problem is dealt with by **fourth normal form (4NF)**.

DEFINITION

4NF defined: A relation schema R is in 4NF if there are no non-trivial multi-valued dependencies in R, or the determinant of any non-trivial multi-valued dependency in R is a superkey of R.[3]

For instance, let us evaluate the relation schema from Section 9.1.1 for 4NF:
R: MUSIC_VEHICLE **(Name, Music, Vehicle)**

As seen in Section 9.1.2, V^+ contains the MVDs **Name $\Rightarrow$ Music | Vehicle**. Therefore, MUSIC_VEHICLE violates 4NF. The resolution of 4NF violation is accomplished by the decomposition strategy:

- Replace the target relation schema (R) by the projections (R1 and R2) that contain the determinant and dependent present in each of the two MVDs.

---

[2] This rule helps us understand 2NF violation as a special case of 4NF violation.

[3] This is an informal definition of 4NF. More formally, a relation schema R is in 4NF if for every non-trivial MVD $X \Rightarrow Z$ in V^+ in R, X is a superkey of R. Equivalently, it can be stated that R is in 4NF, if it is in BCNF and the dependents in all non-trivial MVDs in R are singleton sets–i.e., the non-trivial MVDs are also FDs whose determinants are candidate keys of R.

Accordingly, we have the decomposition:

Solution 1

D {R1, R2} where:

R1: N_MUSIC **(Name, Music)**; R2: N_VEHICLE **(Name, Vehicle)**

D {R1, R2} is in 4NF because both R1 and R2 are free of any non-trivial MVD. Other decompositions are also possible:

Solution 2

D {R1a, R2a} where:

R1a: N1_MUSIC **(Name, Music)**; R2a: N2_VEHICLE **(Music, Vehicle)**

Does this decomposition resolve the 4NF violation in MUSIC_VEHICLE? Yes, because both R1a and R2a are free of any non-trivial MVD. The difference between Solutions 1 and 2 is that Solution 1 is a lossless-join decomposition while Solution 2 is a loss-join decomposition, as evidenced in Figure 9.1.

How do we detect loss/lossless-join decomposition in a 4NF resolution? With reference to MVDs that hold on a relation schema R, a decomposition D: {R1, R2} is a lossless-join decomposition, if and only if V^+ contains:

- either the MVD (R1 ∩ R2) ⇛ (R1 − R2)
- or the MVD (R1 ∩ R2) ⇛ (R2 − R1)

Accordingly, an examination of the two solutions will reveal that Solution 1 is a lossless-join decomposition, while Solution 2 is not. Note that Solution 2 does not follow the decomposition method prescribed above. The decomposition method prescribed above always yields a lossless-join decomposition.

It seems reasonable to make a semantic conclusion that **Music** and **Vehicle** are *independent* multi-valued attributes of MUSIC_VEHICLE. Let us review another example where the relationship between multi-valued attributes is not semantically obvious.

Suppose, based on user-specified business rules, MVDs **Name** ⇛ **Skill | Music** are specified over the relation schema R: MUSIC_SKILL **(Name, Skill, Music)**. Then it is clear that MUSIC_SKILL is in violation of 4NF. MUSIC_SKILL can be decomposed following the decomposition strategy stated above and results in a 4NF relational schema of the form:

D {R1, R2} where:

R1: N_MUSIC **(Name, Music)**; R2: N_SKILL **(Name, Skill)**

First, observe that the solution (i.e., the projections R1 and R2) yields a lossless-join binary decomposition. Also, the solution suggests that **Music** and **Skill** are *independent* multi-valued attributes of MUSIC_SKILL and that this "independence" is the cause of the MVDs.

What if **Music** and **Skill** are multi-valued attributes that are *not* independent of each other? Will the MVDs **Name** ⇛ **Skill | Music** persist in this case in MUSIC_SKILL? Once again, to get a better understanding, let us review a relation instance of MUSIC_SKILL that appears at the top of Figure 9.2.

The semantics of this relation can be interpreted as: Pixoto, Hathi, and LaMott each have certain skills and each know certain types of music. It also appears that Ms. Pixoto is a composer of jazz, classical, and rock, but works as a critic of *only* classical. Likewise, even though Mr. Hathi is a composer as well as a critic, and is also an exponent of classical and rock, he composes only classical and critiques only rock. Finally, Mr. LaMott is a composer—that is all he does, and since his interest is only jazz, he composes only jazz.

MUSIC_SKILL

Name	Skill	Music
Pixoto	Composer	Jazz
Pixoto	Composer	Classical
Pixoto	Composer	Rock
Pixoto	Critic	Classical
Hathi	Composer	Classical
Hathi	Critic	Rock
LaMott	Composer	Jazz

Three possible binary projections

N_MUSIC

Name	Music
Pixoto	Jazz
Pixoto	Rock
Pixoto	Classical
Hathi	Classical
Hathi	Rock
LaMott	Jazz

N_SKILL

Name	Skill
Pixoto	Composer
Pixoto	Critic
Hathi	Composer
Hathi	Critic
LaMott	Composer

S_MUSIC

Skill	Music
Composer	Jazz
Composer	Classical
Composer	Rock
Critic	Classical
Critic	Rock

Test: Name ⇉ Skill | Music

Test: Music ⇉ Skill | Name

Test: Skill ⇉ Name | Music

N_MUSIC * N_SKILL

Name	Skill	Music
Pixoto	Composer	Jazz
Pixoto	Composer	Rock
Pixoto	Composer	Classical
Pixoto	Critic	Jazz
Pixoto	Critic	Rock
Pixoto	Critic	Classical
Hathi	Composer	Rock
Hathi	Composer	Classical
Hathi	Critic	Rock
Hathi	Critic	Classical
LaMott	Composer	Jazz

N_MUSIC * S_MUSIC

Name	Skill	Music
Pixoto	Composer	Jazz
Pixoto	Composer	Classical
Pixoto	Composer	Rock
Pixoto	Critic	Classical
Pixoto	Critic	Rock
Hathi	Composer	Classical
Hathi	Composer	Rock
Hathi	Critic	Classical
Hathi	Critic	Rock
LaMott	Composer	Jazz

N_SKILL * S_MUSIC

Name	Skill	Music
Pixoto	Composer	Jazz
Pixoto	Composer	Classical
Pixoto	Composer	Rock
Pixoto	Critic	Classical
Pixoto	Critic	Rock
Hathi	Composer	Jazz
Hathi	Composer	Classical
Hathi	Composer	Rock
Hathi	Critic	Classical
Hathi	Critic	Rock
LaMott	Composer	Jazz
LaMott	Composer	Classical
LaMott	Composer	Rock

Test failed:

Name ⇉ Skill | Music

Does not hold in MUSIC_SKILL

Test failed:

Music ⇉ Skill | Name

does not hold in MUSIC_SKILL

Test failed:

Skill ⇉ Name | Music

does not hold in MUSIC_SKILL

Inference: MUSIC_SKILL is in 4NF

Figure 9.2 Testing MUSIC_SKILL for violation of 4NF

Clearly, **Skill** and **Music** appear to be multi-valued attributes with respect to **Name** in a non-1NF schema. However, with **(Name, Skill, Music)** as the primary key, this schema is a relation schema in BCNF (i.e., no immediate violations of 1NF, 2NF, 3NF, or BCNF are present in MUSIC_SKILL). Also, semantically it appears that **Skill** and **Music** are not independent (i.e., **Name's** relationship with **Skill** is not independent of the **Name's** relationship with **Music**). How do we confirm or refute this? Do the MVDs **Name** ⇒ **Skill | Music** exist in MUSIC_SKILL? Applying the method prescribed in Section 9.1.2, we produce the projections **(Name, Skill)** and **(Name, Music)**, join the two back and verify if the natural join of the two strictly produces the target relation instance from which the projections emerged.

A comparison of MUSIC_SKILL with (N_MUSIC * N_SKILL) reveals that this is not the case (see Figure 9.2). So, we reject the premise that MVDs **Name** ⇒ **Skill | Music** exist in MUSIC_SKILL. The next question is: Are there any other MVDs in MUSIC_SKILL? The only other possible MVD pairs in MUSIC_SKILL are **Music** ⇒ **Skill | Name** and **Skill** ⇒ **Name | Music**.

In order to verify these, we have to test if (N_MUSIC * S_MUSIC) or (N_SKILL * S_MUSIC) strictly yield MUSIC_SKILL. An inspection of these two natural joins in Figure 9.2 reveals that neither is true. Thus, we conclude that there are no MVDs in MUSIC_SKILL even when multi-valued attributes are present. Consequently, MUSIC_SKILL is in 4NF and does not require any decomposition—in fact, decomposition of MUSIC_SKILL is incorrect. Note that there is no need to make an "intuitive" judgment about the independence among the multi-valued attributes in a relation schema for inferring the presence or absence of MVDs—it can be scientifically tested and formally inferred based on the test prescribed earlier in this section.

9.3 Resolution of a 4NF Violation—A Comprehensive Example

Consider the schema:
    MUSICIAN (**Name, Age, Ph#, Band, Rate, {Music}, {Skill}, {Dependent}**)
Music, Skill, and **Dependent** are multi-valued attributes.
    The stated set of FDs and MVD that prevail over MUSICIAN are:
    F {fd1, fd2, fd3, fd4} and V{mvd1} where:

fd1: **Name → Age**;	fd2: **Name → Ph#**;
fd3: **Name → Band**;	fd4: **Band → Rate**;
mvd1: **Name** ⇒ **Dependent**	

By the rule of complementation, we can infer that mvd2: **Name** ⇒ **{Music, Skill}**.[4]

MUSICIAN is not in 1NF because of the presence of multi-valued attributes; for this reason, MUSICIAN is not even a relation schema. In order to transform MUSICIAN to a 1NF relation schema, we specify **(Name, Music, Skill, Dependent)** as the primary key of MUSICIAN. Thus we have a 1NF relation schema:
    MUSICIAN (<u>**Name**</u>, **Age, Ph#, Band, Rate, <u>Music, Skill, Dependent</u>**)

---

[4] It is incorrect to conclude that Name ⇒ Music; and Name ⇒ Skill

Eliminating normal form violations due to FDs yields the decomposition:
R1: PERSON (**Name**, **Age, Ph#, Band**)
R2: BAND (**Band**, **Rate**)
R3: NMSD (**Name, Music, Skill, Dependent**)
R1 and R2 above are in BCNF (in fact, in 4NF) and R3 violates 4NF because mvd1l mvd2 persists in R3. Solving R3 for 4NF violation yields:
R3a: FAMILY (**Name, Dependent**)
R3b: NMS (**Name, Music, Skill**)
The Presentation-Layer ER diagram reverse-engineered from this solution enabling a deeper insight into the semantics of the scenario is shown in Figure 9.3.

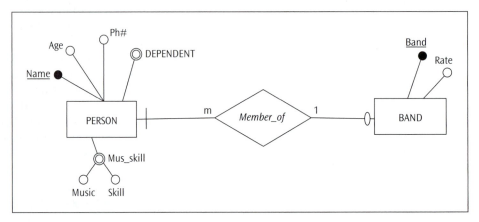

Figure 9.3   Reverse-engineered 4NF design scenario 1
Note: Participation constraints arbitrarily assumed

Suppose we add a few more attributes to the 1NF relation schema MUSICIAN where the following FDs hold:

fd5: **Music → School;**	fd6: **Music → Year;**	fd7: **Skill → Department;**
fd8: **Dependent → Age;**	fd9: **Dependent → Gender;**	fd10: **Dependent → Relationship**

The revised 1NF relation schema will be of the form:
MUSICIAN (**Name**, **Age, Ph#, Band, Rate, Music, Skill, Dependent**, **School, Year, Department, Age, Gender, Relationship**)
Eliminating normal form violations due to the additional set of FDs, {fd5, fd6, fd7, fd8, fd9, fd10} in F, we have a 4NF design:

R1: PERSON (**Name**, **Age, Ph#, Band**);	R2: BAND (**Band**, **Rate**);
R3a: FAMILY (**Name, Dependent**);	R3b: NMS (**Name, Music, Skill**)

The Presentation-Layer ER diagram reverse-engineered from this revised solution appears in Figure 9.4.

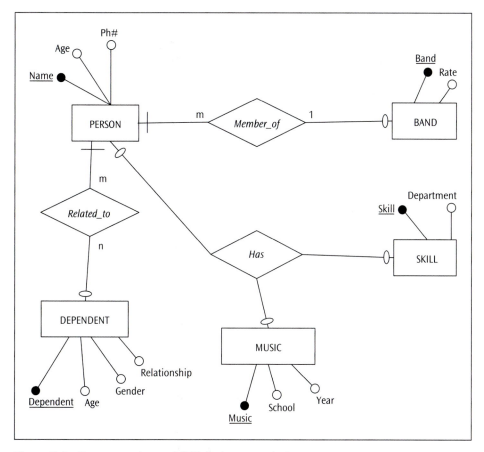

Figure 9.4   Reverse-engineered 4NF design scenario 2
Note: Participation constraints arbitrarily assumed

Comparing the two solutions along with the ERDs in Figures 9.3 and 9.4 provides interesting insights as to how identical relation schemas in the two solutions (R3a and R3b) depict somewhat different scenarios.

9.4 Generality of Multi-valued Dependencies and 4NF

Let us explore the following two assertions from Section 9.1.2 to uncover a broader theme underlying normalization that links the normal forms based on FDs (2NF, 3NF, and BCNF) with the normal form based on MVD (4NF):

- <u>Assertion 1</u>: *If A → B, then A ⇒ B (replication rule)*
- <u>Assertion 2</u>: *If a binary decomposition with the lossless-join property exists for a relation schema R, then non-trivial MVDs are present in R.*

The replication rule simply states that an FD is a special case of an MVD since the FD satisfies the definition of an MVD. However, this occurs only under the special condition that the dependent value is restricted to a *singleton set*—i.e., for a given value of A, B has only one value.

To get a better grip on the concept, let us review the 2NF violation from this perspective using the example studied in Chapter 8 (see Sections 8.1.1 and 8.1.2). The non-1NF schema ALBUM **(Album_no, {Artist_nm}, Price, Stock)** with a multi-valued attribute, **Artist_nm**, and an FD, **Album_no ((Price, Stock)**, is first normalized to 1NF as:

NEW_ALBUM **(Album no, Artist_nm, Price, Stock)**

Since NEW_ALBUM is in 1NF, it is a relation schema and the tenets of the relational theory can be applied to it. Does **Album_no ⇒ Artist_nm** in NEW_ALBUM? This can be determined by comparing the natural join (ALBUM_ARTIST * ALBUM_INFO) with NEW_ALBUM where:

D: ALBUM_INFO **(Album_no, Price, Stock)**; and ALBUM_ARTIST **(Album_no, Artist_nm)** represents a binary decomposition NEW_ALBUM. Figure 8.2 contains the relation instances pertaining to this example. By computing the natural join, (ALBUM_ARTIST * ALBUM_INFO), it can be seen that the natural join indeed strictly yields the relation NEW_ALBUM. Therefore, it is concluded that **Album_no ⇒ Artist_nm**. Then, by the rule of complementation, we have **Album_no ⇒ (Price, Stock)**. In other words, for each value of **Album_no** in NEW_ALBUM, the same set of **(Price, Stock)** occurs for each value of **Artist_nm**. However, **(Price, Stock)** is a singleton set. Because **(Price, Stock)** is a singleton set, the MVD, **Album_no ⇒ (Price, Stock)**, is equivalent to the FD, **Album_no → (Price, Stock)** and that is why the primary key for ALBUM_INFO is **Album_no** instead of **(Album_no, Price, Stock)**. Thus, it is seen that a 2NF violation is a special case of 4NF violation—i.e., 4NF subsumes 2NF. The fact that the binary decomposition of NEW_ALBUM yields a lossless-join decomposition signals the presence of MVD in NEW_ALBUM, thus ratifying Assertion 2 above. The inference rule derived by Catriel, Fagin, and Howard (see Section 9.1.3) essentially conveys the same idea.

In a similar fashion, it can be shown that a 3NF violation and BCNF violation are also special cases of 4NF violation. The mere existence of a lossless-join decomposition of the target relation schema that violates 3NF or BCNF in itself signals the presence of an MVD in the relation schema and that is how a 3NF and a BCNF violation can be seen as a 4NF violation. The reader is encouraged to explore this further using the examples from Chapter 8 (see Sections 8.1.3 and 8.1.4) to gain deeper insight. Identifying the MVDs that prevail in these cases is an interesting exercise to pursue. In short, a relation schema in 4NF also is in 3NF and BCNF.

9.5 Join Dependencies and Fifth Normal Form (5NF)

So far, we were able to resolve normal form violations due to undesirable FDs and MVDs through binary decompositions of the affected relation schemas. Join dependencies (JDs) pertain to conditions where *binary projections* are not sufficient to achieve a lossless-join decomposition and the associated issues.

DEFINITION

Join dependency (JD) defined: JD (R1, R2, R3, . . . Rn) specified over a relation schema, R, states that every legal state r of R has a lossless-join decomposition into R1, R2, R3, . . . Rn. In other words, given (R1, R2, R3, . . . Rn) as projections of R, if the natural join of (R1, R2, R3, . . . Rn) produces every legal state r of R, then JD (R1, R2, R3, . . . Rn) exists in R. A JD (R1, R2, R3, . . . Rn) is trivial if one of the projections, Ri is equal to R because this condition guarantees lossless join property for any state r of R and essentially does not specify any constraint at all on R. Observe that given (R1, R2) as projections of R, the MVDs (R1 ∩ R2) ⇒⇒ (R1 − R2) | (R2 − R1) is equivalent to JD (R1, R2). Thus, MVDs are essentially binary join dependencies possessing a number of algebraic properties similar to those of FDs. Since an FD can be seen as a special case of MVD (i.e., an MVD subsumes an FD) an FD is also subsumed by a JD. Therefore, it can be said that JD is the most general form of dependency constraint that deals with decomposition via projections and re-compositions via natural joins.

Fifth normal form (5NF) is about JDs. In simple terms, presence of JDs in a relation schema R violates 5NF—i.e., if a relation state r of R can be strictly reconstructed from the natural join of all of its projections, (R1, R2, R3, . . . , Rn), then a join dependency is present in R (i.e., a constraint specifying JD has been imposed on R) and R is not in 5NF. In order to establish 5NF, R should be replaced by its decomposition (R1, R2, R3, . . . Rn). The set of relation schemas (R_1, R_2, R_3, . . . R_n) in this case is in 5NF.

DEFINITION

5NF defined: A relation schema R is in 5NF if there are no non-trivial join dependencies in R.[5] A relation schema that cannot be reconstructed by a natural join of all of its projections does not have a JD imposed on it and so is already in 5NF and should not be decomposed to achieve 5NF.

Let us now review an illustration that clarifies the concept. SCHEDULE_X at the top of Figure 9.5 is the relation instance that remains after all the normal form violations due to functional dependencies have been resolved and decomposed from the original relation schema and is represented as:

R: SCHEDULE_X **(Prof_name, Course#, Quarter)**

---

[5] A technically more precise defination is: R is in 5NF if for every non-trival join dependency, JD (R1, R2,...Rn) every Ri is a super key of R (Elmasri and Navathe, 2003). A comprehensive (highly technical) discussion of 4NF and 5NF can be found in Johnson (1997).

Business rule: If an instructor teaches a course, and that course is offered in certain quarters, then the instructor must teach that course in those quarters if s/he is teaching during those quarters.

R: SCHEDULE_X (Prof_name, Course#, Quarter);   R1x: TAUGHT_X (Prof_name, Course#);

R2x: TAUGHT_DURING_X (Prof_name, Quarter);   R3x: COURSE_OFFERING_X (Course#, Quarter)

SCHEDULE_X

Prof_name	Course#	Quarter
Verstrate	IS812	Fall
Verstrate	IS812	Spring
Verstrate	IS832	Winter
Verstrate	IS330	Fall
Verstrate	IS330	Spring
Surendra	IS812	Fall
Surendra	IS812	Spring
Surendra	IS430	Fall
Surendra	IS430	Spring
Kim	IS821	Winter
Kim	IS430	Spring
Kim	IS430	Summer

1. Does the natural join of any two of the three tables below strictly yield the table above? The answer is "No." Therefore, SCHEDULE_X is in 4NF.

TAUGHT_X

Prof_name	Course#
Verstrate	IS812
Verstrate	IS832
Verstrate	IS330
Surendra	IS812
Surendra	IS430
Kim	IS821
Kim	IS430

TAUGHT_DURING_X

Prof_name	Quarter
Verstrate	Fall
Verstrate	Winter
Verstrate	Spring
Surendra	Fall
Surendra	Spring
Kim	Winter
Kim	Spring
Kim	Summer

COURSE_OFFERING_X

Course#	Quarter
IS330	Fall
IS330	Spring
IS430	Fall
IS430	Spring
IS430	Summer
IS812	Fall
IS812	Spring
IS821	Winter
IS832	Winter

2. Does the natural join of all the three tables above strictly yield SCHEDULE_X above? The answer is "Yes," indicating presence of join dependencies. Therefore, SCHEDULE_X violates 5NF. In order to achieve 5NF in the design, SCHEDULE_X must be replaced by the set of relation schemas {TAUGHT_X, TAUGHT_DURING_X, COURSE_OFFERING_X}.

Figure 9.5 An illustration of 5NF violation and resolution

There are no MVDs specified on SCHEDULE_X. In fact, we are able to verify this from the relation instance in Figure 9.5. The three possible decompositions that can suggest presence of non-trivial MVDs in SCHEDULE_X are:

R1x:TAUGHT_X **(Prof_name, Course#)**;

R2x:TAUGHT_DURING_X **(Prof_name, Quarter)**

R3x:COURSE_OFFERING_X **(Course#, Quarter)**

If a natural join of any binary decomposition of R, namely, (R1x * R2x) or (R1x * R3x) or (R2x * R3x), is sufficient to strictly reconstruct R (non-loss composition), then the presence of MVD is evidenced. When verified with the relation instances in Figure 9.5, the reader will find that none of these natural joins yield strictly R. Then, there are no MVDs in R implying that R (i.e., SCHEDULE_X) is in 4NF. Is SCHEDULE_X in 5NF? In order to check this, we need to know if *JD (R1x, R2x, R3x)* exists. If the natural join of {R1x, R2x, R3x} strictly results in R, then we can conclude that *JD (R1x, R2x, R3x)* persists. Then, R, i.e., SCHEDULE_X, violates 5NF. When verified using the relation instances in Figure 9.5, the reader will find that a natural join of {R1x, R2x, R3x} indeed yields strictly R meaning that *JD (R1x, R2x, R3x)* is present. Therefore, SCHEDULE_X violates 5NF. In order to restore the design to 5NF, SCHEDULE_X should be replaced by the set of relation schemas (TAUGHT_X, TAUGHT_DURING_X, COURSE_OFFERING_X).

Let us now review the relation instance SCHEDULE_Y in Figure 9.6. The relation schema that represents this relation instance is:

R: SCHEDULE_Y **(Prof_name, Course#, Quarter)**

SCHEDULE_Y is in 4NF. The verification of this claim is left as an exercise to the reader. Is SCHEDULE_Y in 5NF? In order to check this, we need to know if *JD (R1y, R2y, R3y)* exists where:

R1y: TAUGHT_Y **(Prof_name, Course#)**;

R2y: TAUGHT_DURING_Y **(Prof_name, Quarter)**

R3y: COURSE_OFFERING_Y **(Course#, Quarter)**

Once again, if the natural join of {R1Y, R2Y, R3Y} strictly results in R, then we can conclude that *JD (R1Y, R2Y, R3Y)* persists in R. Then, R, i.e., SCHEDULE_Y, violates 5NF. When verified using the relation instances in Figure 9.6, we find that a natural join of {R1Y, R2Y, R3Y} does not yield strictly R, meaning that *JD (R1Y, R2Y, R3Y)* is not present in R. Therefore, SCHEDULE_Y does not violate 5NF—i.e., SCHEDULE_Y is in 5NF. Replacing SCHEDULE_Y by the set of relation schemas (TAUGHT_Y, TAUGHT_DURING_Y, COURSE_OFFERING_Y) will change the intended semantics of the design and amounts to an erroneous decomposition. Table 9.2 is provided as an aid to get a better understanding of 5NF through a comparative review of SCHEDULE_X (Figure 9.5) and SCHEDULE_Y (Figure 9.6).

Business rule: If an instructor teaches a course, and that course is offered in certain quarters, the instructor need not teach that course in those quarters just because s/he is teaching during those quarters.

R: SCHEDULE_Y (Prof_name, Course#, Quarter);   R1y: TAUGHT_X (Prof_name, Course#);

R2y: TAUGHT_DURING_X (Prof_name, Quarter);   R3y: COURSE_OFFERING_X (Course#, Quarter)

SCHEDULE_Y

Prof_name	Course#	Quarter
~~Verstrate~~	~~IS812~~	~~Fall~~
Verstrate	IS812	Spring
Verstrate	IS832	Winter
Verstrate	IS330	Fall
Verstrate	IS330	Spring
Surendra	IS812	Fall
~~Surendra~~	~~IS812~~	~~Spring~~
Surendra	IS430	Fall
Surendra	IS430	Spring
Kim	IS821	Winter
Kim	IS430	Spring
Kim	IS430	Summer

3. Does the natural join of any two of the three tables below strictly yield the table above? The answer is "No." Therefore, SCHEDULE_Y is in 4NF

TAUGHT_Y

Prof_name	Course#
Verstrate	IS812
Verstrate	IS832
Verstrate	IS330
Surendra	IS812
Surendra	IS430
Kim	IS821
Kim	IS430

TAUGHT_DURING_Y

Prof_name	Quarter
Verstrate	Fall
Verstrate	Winter
Verstrate	Spring
Surendra	Fall
Surendra	Spring
Kim	Winter
Kim	Spring
Kim	Summer

COURSE_OFFERING_Y

Course#	Quarter
IS330	Fall
IS330	Spring
IS430	Fall
IS430	Spring
IS430	Summer
IS812	Fall
IS812	Spring
IS821	Winter
IS832	Winter

4. Does the natural join of all the three tables above strictly yield SCHEDULE_Y above? The answer is "No," indicating absence of join dependencies. Therefore, SCHEDULE_Y is in 5NF. Replacing SCHEDULE_Y by the set of relation schemas {TAUGHT_Y, TAUGHT_DURING_Y, COURSE_OFFERING_Y} changes the intended semantics of the business rule stated above.

Figure 9.6   An illustration of a relation schema in 5NF

Table 9.2 A comparative analysis of the presence and absence of 5NF violation

Question	SCHEDULE_X	SCHEDULE_Y
Who taught IS812 in Winter	Nobody	Nobody
Who taught IS812 in Fall	Verstrate, Surendra	Surendra
Who taught IS812 in Spring	Verstrate, Surendra	Verstrate
Who taught IS430 in Fall	Surendra	Surendra
Who taught IS430 in Spring	Surendra, Kim	Surendra, Kim
Who taught IS430 in Summer	Kim	Kim
Notes	Since Verstrate taught IS812 and he taught during Fall, Winter and Spring, he taught IS812 if it was offered in any of these three quarters—i.e., Verstrate taught IS812 in Fall and Spring. *He didn't teach IS812 in Winter because IS812 was not offered in Winter, even though he taught in Winter.*	Verstrate taught IS812 and Verstrate taught during Fall, Winter and Spring. But, he taught IS812 *only in Spring*—i.e. even though IS812 was offered in Fall and Verstrate taught in Fall, he didn't teach IS812 in Fall. In short, Verstrate didn't teach IS812 whenever it was offered while he was teaching.
	Since Surendra taught IS812 and he taught during Fall and Spring, he taught IS812 if it was offered in any of these two quarters—i.e., Surendra taught IS812 in both Fall and Spring.	Surendra taught IS812 and Surendra taught during Fall and Spring. But, he taught IS812 *only in Fall*—i.e. even though IS812 was offered in Spring and Surendra taught in Spring, he didn't teach IS812 in Spring. In short, Surendra didn't teach IS812 whenever it was offered while he was teaching.
	No MVDs—in 4NF	No MVDs
	But *only* in 4NF; not in 5NF	In 4NF and in 5NF

Table 9.2 A comparative analysis of the presence and absence of 5NF violation (continued)

Question	SCHEDULE_X	SCHEDULE_Y
	SCHEDULE_X can be reconstructed by joining the three of its projections TAUGHT_X, TAUGHT_DURING_X and COURSE_OFFERING_X Therefore, join dependency is **present** in SCHEDULE_X.	SCHEDULE_Y <u>cannot</u> be reconstructed by joining the three of its projections TAUGHT_Y, TAUGHT_DURING_Y and COURSE_OFFERING_Y Therefore, join dependency is **absent** in SCHEDULE_Y.
	SCHEDULE_X cannot be reconstructed by joining any two of its three projections. That is why there are no MVDs in SCHEDULE_X and therefore it is in 4NF.	SCHEDULE_Y cannot be reconstructed by joining any two of its three projections. That is why there are no MVDs in SCHEDULE_Y and therefore it is in 4NF.

The relation instance MUSIC_SKILL in Figure 9.2 is in 4NF. Is it also in 5NF? If not, what can be done to this relation instance to establish 5NF in MUSIC_SKILL. On the other hand, if MUSIC_SKILL is in 5NF, how can the relation instance be altered to depict a 4NF relation that violates 5NF. The reader should find this an interesting exercise.

9.6 A Note on Domain-Key Normal Form (DK/NF)

Fagin (1981) proposed a normal form based exclusively on *domain constraints* and *key constraints*. It is named **domain key normal form (DK/NF)**. DKNF states that if every constraint on a relation schema, R, is a logical consequence of the domain constraints and key constraints that apply to R, then, the relation schema, R, is in DK/NF. In other words, in order for a relation schema to be in DK/NF, all constraints and dependencies that hold on valid relation states (e.g., FD, MVD, JD) are enforced through a set of domain constraints and key constraints. Fagin (1981) showed that a DK/NF relation schema is necessarily in 5NF.

While the rules of DK/NF are rather simple, DK/NF is not always achievable. More importantly, verification of compliance (i.e., when DK/NF is achieved) is difficult. There are no formal methods to systematically verify if a relation schema is in DK/NF. Thus, while theoretically rigorous, DK/NF has limited practical utility.

Chapter Summary

While relation schemas in BCNF are free from modification anomalies due to undesirable functional dependencies, BCNF relation schemas that fail to achieve fourth normal form and fifth normal form are vulnerable to modification anomalies as a result of multi-valued dependencies (in the case of fourth normal form) and join dependencies (in the case of fifth normal form).

A multi-valued dependency is defined as follows: In a relation schema R (X, Y, Z) where X, Y, and Z are atomic or composite attributes, a multi-valued dependency $X \twoheadrightarrow Y$ exists if each value of X in any relation state r of R is associated with a set of Y values independent of the Z values with which X is associated. In a relation schema R (X, Y, Z) in BCNF, a multi-valued dependency $X \twoheadrightarrow Y$ exists if and only if the natural join r (R1) and r (R2), where R1 (X, Y) and R2 (X, Z) are projections of R, strictly yields r (R) for every relation state r of R. A relation schema R is in fourth normal form if it is in BCNF and has no non-trivial multi-valued dependencies.

A join dependency in a relation schema R pertains to conditions where the natural join of all of its projections results in the reconstruction of R. A relation schema that cannot be reconstructed by a natural join of all of its projections does not have a join dependency and is said to be in fifth normal form.

Every normal form up through fifth normal form has considered constraints imposed by functional dependencies, multi-valued dependencies, or join dependencies. Fagin (1981) suggests a generalized constraint that both infers the three kinds of dependencies and allows more generalized constraints. Constraints are explained here in terms of domain constraints of the relation schema's attributes and relation keys. Relation schemas that satisfy these constraints are said to be in domain key normal form (DK/NF). Unfortunately, other than checking to see if each constraint on a relation schema is a logical consequence of the definition of keys of a relation schema or domains of attributes, there are no formal methods to systematically verify if a relation schema is in DK/NF.

Exercises

1. What is a multi-valued dependency?
2. Consider the instance of the relation SHIRT (Shirt#, Color, Size) where Shirt# is equivalent to a style number (e.g., style number 341 might be a shirt with a button-down collar while style number 342 might be a shirt with an open collar, etc.). Observe that each Shirt# comes in a variety of colors and sizes.

Shirt#	Color	Size
341	White	Small
341	White	Medium
341	White	Large
341	Blue	Small
341	Blue	Medium
341	Blue	Large

Shirt#	Color	Size
341	Yellow	Medium
341	Yellow	Large
342	White	Medium
342	White	Large
342	Blue	Large

Does SHIRT possess the multi-valued dependency Shirt# $\twoheadrightarrow$ Color|Size? Why or why not?

3. Which of the inference rules for multi-valued dependencies supports the statement that: _multi-valued_ dependencies (MVDs) are a generalization of functional dependencies (i.e., a functional dependency is a special case of a multi-valued dependency).

4. Consider a revised instance of the SHIRT relation in Exercise 2.

Shirt#	Color	Size
341	White	Large
342	White	Medium
343	White	Large
344	Blue	Large
345	Yellow	Large

Does this version of SHIRT possess the multi-valued dependency Shirt# $\twoheadrightarrow$ Color|Size? Why or why not?

5. Consider the relation schema STUDENT (Sid, Shoe_size, Marital_status)

F: fd1: Sid $\rightarrow$ Shoe_size; fd2: SID$\rightarrow$ Marital_status

   a. Does STUDENT possess a multi-valued dependency? Why or why not?

   b. Does STUDENT possess a fourth normal form violation? Explain your answer.

6. How are multi-valued dependencies incorporated into the definition of fourth normal form (4NF)?

7. Consider the relation instance STUDENT (Major, Stu_id, Activity, Name, Phone)

Major	Stu_id	Activity	Name	Phone
Music	100	Swimming	Costello	444-5456
Accounting	100	Swimming	Costello	444-5456
Music	100	Tennis	Costello	444-5456

Major	Stu_id	Activity	Name	Phone
Accounting	100	Tennis	Costello	444-5456
Mathematics	150	Jogging	Brooks	444-5456
Mathematics	250	Sleeping	Abbott	665-5456
Music	200	Swimming	Costello	665-5456
Mathematics	250	Eating	Abbott	665-5456

    a.   Identify all multi-valued dependencies in STUDENT.

    b.   Describe insertion, deletion, and update anomalies that accompany each of the multi-valued dependencies identified above. Are any of these multi-valued dependencies also functional dependencies?

    c.   Decompose STUDENT into a relational schema (i.e., a set of relation schemas) that exhibit attribute preservation, dependency preservation, and the presence of a lossless join decomposition.

8.   Consider the relation instance COURSE_OFFERED (Course, Teacher, Text)

Course	Teacher	Text
Math	White	Calculus I
Physics	Brown	B. Mechanics
Physics	Black	B. Mechanics
Physics	Black	Prin of Optics
Math	Black	B. Mechanics
Math	White	B. Mechanics
Eng Mech	White	Calculus I

    a.   Does COURSE_OFFERED possess a multi-valued dependency? If the answer is yes, show how this multi-valued dependency occurs. How do you resolve this condition?

    b.   Does COURSE_OFFERED possess a join dependency? If the answer is yes, show how this join dependency occurs. How do you resolve this condition?

9.   What is a join dependency and in what way is it related to a multi-valued dependency?

10. What is the definition of fifth normal form?

11. Is the relation instance SHIRT in Exercise 2 in 5NF? If the answer is yes, explain. If the answer is no, provide a solution.

12. Consider the universal relation schema STOCK (Symbol, Company, Exchange, Investor, Date, Price, Broker, Dividend) and the dependencies set

fd: Symbol → {Company, Exchange, Dividend}

mvd: Symbol ⇒ {Investor, Broker}

   a. Identify the primary key such that STOCK is in first normal form.

   b. Explain the immediate normal forms violated by the functional dependencies and multi-valued dependencies.

   c. Step through the normalization process to obtain a final design that is in Fourth Normal Form. At each level of decomposition, indicate the primary key and dependencies resolved and explain the normal form achieved and violated in each one of the decomposed relation schemas. Explain how each of the decompositions is lossless. Finally, show all integrity constraints in the final relational schema.

   d. Reverse engineer your final design to a Presentation Layer ER diagram.

13. Given the schema SCHEDULE (Prof, Office, Major, {Book}, Course, Quarter) along with

F: fd1: Prof → {Office, Major} and V: mvd: Prof ⇒ Book

   a. Identify the primary key of SCHEDULE such that SCHEDULE is a first normal form relation schema.

   b. Normalize SCHEDULE to fourth normal form.

   c. Indicate the entity integrity and referential integrity constraints.

   d. Reverse engineer the design to a conceptual schema using the ER modeling grammar.

14. This exercise is based on four different inventory relation schemas, P, Q, R, and S. An instance of each relation schema is given below, each of which reflects a different design objective.

P

Part#	Color	Store#
17	Red	A
17	Red	B
17	Red	C
18	White	B
18	White	C
19	Red	B

Q

Part#	Color	Store#
17	Blue	A
17	Green	A
17	Black	A
17	Green	C
18	Green	A
18	Black	C

R

Part#	Color	Store#
17	Blue	A
17	Green	A
17	Blue	B
17	Green	B
17	Blue	C
17	Green	C
18	Green	B
18	Green	C
18	Black	B
18	Black	C

S

Part#	Color	Store#
17	Blue	A
17	Green	A
17	Blue	B
17	Green	B
17	Green	C
18	Green	B
18	Green	C
18	Black	C

Answer each of the following questions separately.

   a. Identify the primary key of each relation schema.

   b. Indicate the functional dependencies and/or multi-valued dependencies present in each design.

   c. Evaluate each design for a fourth or fifth normal form violation. Explain your conclusion in each case.

   d. Where there is a fourth normal form violation, provide a revised lossless join design that is in fourth normal form.

Selected Bibliography

Catriel, B., Fagin, R. and Howard, J.H. (1977) "A Complete Axiomatization for Functional and Multi-valued Dependencies," *Proceedings of the ACM/SIGMOD International Conference on Management of Data*, Toronto, Canada (August) 1977.

Date, C. J. (2004) *An Introduction to Database Systems*, Eighth Edition, Addison-Wesley.

Elmasri, R., and Navathe, S. B. (2003) *Fundamentals of Database Systems*, Fourth Edition, Addison-Wesley.

Fagin, R. (1979) "Normal Forms and Relational Database Operators," *Proceedings of the ACM/SIGMOD International Conference on Management of Data*, (May/June) 1979.

Fagin, R. (1981) "A Normal Form for Relational Databases that is Based on Domains and Keys," *ACM Transactions on Database Systems*, 6, 3 (September).

Hawryszkiewycz, I. T. (1991) *Database Analysis and Design*, Second Edition, Macmillan Publishing Company.

Johnson, J. L. (1997) *Database: Models, Languages, Design*, Oxford university Press.

Kent, W. (1983) "A Simple Guide to the Five Normal Forms in Relational Database Theory," *Communications of the ACM*, 26, 2 (February), 120–125.

Kifer, M., Bernstein, A., and Lewis, P. M. (2005) *Database Systems: An Application-Oriented Approach*, Second Edition, Addison-Wesley.

PART  IV

DATABASE IMPLEMENTATION USING THE RELATIONAL DATA MODEL

INTRODUCTION

The final phase of the data modeling life cycle is physical data modeling. At this point, we have an information-preserving logical schema normalized to the extent we want and ready for implementation. Because contemporary database systems are dominated by relational data models, as we transition from a technology-independent to a technology-dependent paradigm, our mapping of the logical schema to a physical data model will employ the technology-dependent relational data modeling grammar. Accordingly, a relation becomes a *table*, the tuples are the *rows* of the table, and the attributes are the *columns* of the table. Figure IV.1 points out our location in the data modeling journey.

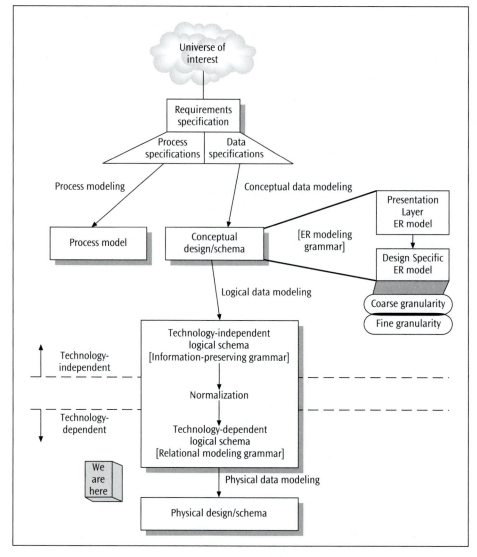

Figure IV.1 Road map for data modeling and database design

Structured Query Language (SQL) is the standard universally accepted by the relational database (RDB) community—the users and the RDBMS vendors—for the implementation of a relational database. The name SQL comes from its predecessor SEQUEL (Structured English Query Language) developed by IBM Research as an experimental product. SQL[1] is a comprehensive relational database language and comprises three sublanguages: (1) a data definition language (SQL/DDL) intended for the creation and alteration of tables

---

[1] Officially pronounced *ess-que-ell* (Date, 2004, p. 4)—not *sequel*.

and other associated structures such as views, domains, and schemas; (2) a data manipulation language (SQL/DML) aimed at data manipulation tasks; and (3) a data control language (SQL/DCL) geared to controlling database access. There are numerous idiosyncratic differences among commercial RDBMS products. Nonetheless, the RDBMS vendors endeavor to meet a certain level of the SQL standards developed jointly by ANSI and ISO. Consequentially, the users of ANSI/ISO standard SQL find migration and interoperability across RDBMS products relatively painless. At this writing, most commercial RDBMS products provide reasonable support to SQL-92 (also known as SQL2). Therefore, most of the discussions in this part are based on SQL-92.[2] Also, we only provide an overview of the salient features of SQL. The reader is directed to the vendor-specific reference manuals for the complete syntax of the language.

Chapter 10 covers portions of SQL/DDL that pertain to creating a database and SQL/DCL. In Chapter 11, relational algebra is introduced as a means to retrieve data from a relational database. A query expressed in relational algebra involves a sequence of operations which when executed in the order specified produces the desired results. Even though SQL uses some of the relational algebra operators explicitly, it is a high-level *declarative* language based on tuple relational calculus.[3] The rest of Chapter 11 is dedicated to an extensive coverage of DML pertaining to data retrieval (querying). Sometimes, this portion of the DML is referred to as the data query language (DQL). In addition, aspects of SQL/DDL that require knowledge of DQL (e.g., assertions, triggers, and views) are covered in Chapter 12 along with some additional features of the SQL language.

---

[2] SQL-99 has already been published [see Gulutzan and Pelzer (1999) for a detailed coverage] and at the time of writing of this book, SQL-03 is available through ISO (*www.iso.org*) and other standards organizations. Significant portions of these standards pertain to object-oriented extensions to SQL. SQL-99-03 is backward-compatible with the (non-object) relational-based SQL-92. While not all commercial RDBMS products fully comply with SQL-92, several of them offer extensions not prescribed in SQL-92. A major SQL construct implemented by many commercial DBMS products called "database trigger" is now fully specified in SQL-99.

[3] For more details on tuple relational calculus, see Elmasri and Navathe (2003).

CHAPTER **10**

Database Creation

At the implementation stage of a database, the principal tasks include creating and modifying the database tables and other related structures, enforcing integrity constraints, populating the database tables, and specifying security and authorizations for access control and transaction processing control.

In this chapter we use the SQL-92 language standard in the code for the SQL statements. It is not necessary that a commercial DBMS implementation include all the standard SQL constructs or follow the standard language syntax verbatim for all SQL constructs. It is highly likely that commercial DBMS products differ in the implementation of at least some of the SQL syntax. Also, some vendors offer additional non-standard SQL constructs. Therefore, the reader using the SQL scripts[1] presented in this chapter may occasionally need to refer to the SQL reference material of the DBMS platform being used. To the extent that a DBMS product does not conform to a common syntactical standard, portability and migration across product platforms might be difficult.

10.1 Data Definition Using SQL

As noted in the introduction of Part IV, a relation is called a *table* in SQL, and attributes and tuples are termed *columns* and *rows*, respectively. In reality, however, the formal object called a "relation" and the informal object called a "table" have several differences. For instance, a table can contain duplicate rows, while a relation is prohibited from having duplicate tuples. SQL tables are allowed to have null values while relations are not. The columns of a table have a left-to-right ordering, but the attributes of a relation in a relational data model are unordered. Columns of a table may end up with duplicate column names (e.g., the result of a join of two tables with the same column name) and sometimes, no column name (a derived column); this, however, is not permissible in a relation. The rows of a table have a top-to-bottom ordering, but the tuples of a relation do not. A table is usually expected to have at least one column, while technically a relation with no attributes is permissible. Notwithstanding many such differences, in the context of a relational data model, a table can represent a concrete picture of an abstract object called a relation—i.e., a table can "represent" a relation, but is not precisely equivalent to a relation. In addition to tables, SQL also supports other structures like *views*, *materialized views*, and *schemas*.

The data definition sublanguage of SQL, known as SQL/DDL, has three major constructs for creating and modifying databases: CREATE, ALTER, and DROP.

---

[1] A script is a command or series of commands usually stored in a file.

10.1.1 Base Table Specification in SQL/DDL

A relation schema (or alternatively a logical scheme) is specified in SQL/DDL by the **CREATE TABLE** statement. The result of a CREATE TABLE statement is referred to as a *base table* to indicate that the table is actually created, populated with rows of data, and stored as a physical file by the DBMS. This facilitates distinguishing base tables from other derived (or virtual) tables such as views or joined tables. The CREATE TABLE construct has a rich syntax that can specify domain constraints on columns, an entity integrity constraint (specification of the primary key), uniqueness constraints (specification of alternate keys), foreign key constraints, deletion and update rules, and more, in addition to the basic table name and column names. We present this SQL construct in gradual increments using appropriate examples. Let us begin with a simple scenario of patients in a clinic placing orders for medications. The ER model (the ER diagram plus a list of semantic integrity constraints that cannot be incorporated in the ER diagram), the relational schema, and the corresponding information-preserving logical schema appear in Figures 10.1a through 10.1d.

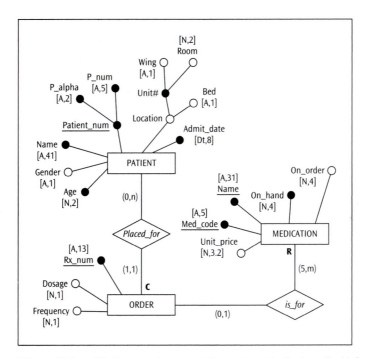

Figure 10.1a ER diagram: An excerpt from a hypothetical medical information system

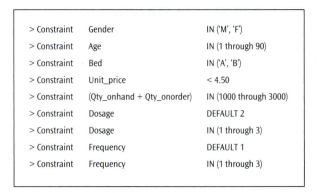

> Constraint	Gender	IN ('M', 'F')
> Constraint	Age	IN (1 through 90)
> Constraint	Bed	IN ('A', 'B')
> Constraint	Unit_price	< 4.50
> Constraint	(Qty_onhand + Qty_onorder)	IN (1000 through 3000)
> Constraint	Dosage	DEFAULT 2
> Constraint	Dosage	IN (1 through 3)
> Constraint	Frequency	DEFAULT 1
> Constraint	Frequency	IN (1 through 3)

Figure 10.1b   Semantic integrity constraints for the Fine-granular Design-Specific ER diagram

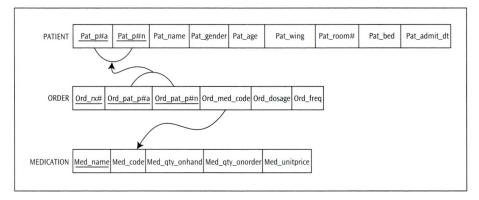

Figure 10.1c   Relational schema for the ERD in Figure 10.1a

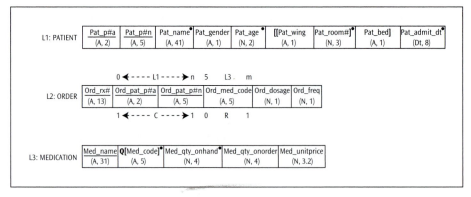

Figure 10.1d   An information-preserving logical schema for the ERD in Figure 10.1a

10.1.1.1 Syntax of the CREATE TABLE Statement

The implementation of the design shown in Figures 10.1a–d in a relational database entails creating the three base table structures first. At the minimum, the table specification includes the table name, the names of the columns in the table, and the data type for each column. The data type of a column is a part of the domain specification for the attribute and is sometimes referred to as the *type constraint* on the attribute. In other words, specifying the data type for a column is equivalent to imposing the type constraint on the attribute. The minimal form of a CREATE TABLE statement for the three base tables is shown in Box 1. SQL statements are *case-insensitive* while string values referenced in the SQL statements are *case-sensitive*.

```
CREATE TABLE patient
(Pat_p#a          char (2),
Pat_p#n           char (5),
Pat_name          varchar (41),
Pat_gender        char (1),
Pat_age           smallint,
Pat_admit_dt      date,
Pat_wing          char (1),
Pat_room#         integer,
Pat_bed           char (1)
);
```

```
CREATE TABLE medication
(Med_code         char (5),
Med_name          varchar (31),
Med_qty_onhand    integer,
Med_qty_onorder   integer,
Med_unitprice     decimal(3,2)
);
```

```
CREATE TABLE order
(Ord_rx#          char (13),
Ord_pat_p#a       char (2),
Ord_pat_p#n       char (5),
Ord_med_code      char (5),
Ord_dosage        smallint,
Ord_freq          smallint
);
```

Box 1

Observe that while it is legal to use the same attribute name in different entity types in an ER diagram, relational database theory stipulates that attribute names must be unique over the entire relational schema. Accordingly, the attribute names in the relational/logical schema in our example follow the naming convention proposed in Chapter 6. The column names of a base table in SQL/DDL are considered to be ordered in the sequence in which they are specified in the CREATE TABLE syntax. However, when populated with data, the rows are not considered to be ordered within a base table. The data types supported by SQL-92 are grouped under *Number*, *String*, *Date/time*, and *Interval*, and are listed in Table 10.1. DBMS vendors often build their own data types based on these four categories.

Table 10.1   Data types supported by SQL-92

Number Data Types	
numeric (p, s) where **p** indicates precision and **s** indicates scale	Exact numeric type—literal representation of number values—*decimal portion exactly the size dictated by the scale*—storage length = (precision + 1) when scale is > 0—most DBMSs have an upper limit on **p** (e.g., 28)
decimal (p, s) where **p** indicates precision and **s** indicates scale	Exact numeric type—literal representation of number values—*decimal portion at least the size of the scale; but expandable to limit set by the specific DBMS*—storage length = (precision + 1) when scale is > 0 —most DBMSs have an upper limit on **p** (e.g., 28)
integer or **integer (p)** where **p** indicates precision	Exact numeric type—binary representation of large whole number values—often precision set by the DBMS vendor (e.g., 2 bytes)
smallint or **smallint (p)** where **p** indicates precision	Exact numeric type—binary representation of small whole number values—often precision set by the DBMS vendor (e.g., 1 byte)
float (p) where **p** indicates precision	Approximate numeric type—represents a given value in an exponential format—precision value represents the minimum size used, up to the maximum set by the DBMS
real	Approximate numeric type—represents a given value in an exponential format—has a default precision value set below that set for double precision data type by the DBMS
double precision	Approximate numeric type—represents a given value in an exponential format—has a default precision value set above that set for real data type by the DBMS
String Data Types	
character (ℓ) or char (ℓ) where ℓ indicates length	Fixed length character strings including blanks from the defined language set SQL_TEXT within a database—can be compared to other columns of the same type with different lengths or varchar type with different maximum lengths—most DBMS have an upper limit on ℓ (e.g., 255)
character varying (ℓ) or **char (ℓ) varying** or **varchar (ℓ)** where ℓ indicates the maximum length	Variable length character strings except trailing blanks from the defined language set SQL_TEXT within a database—DBMS records actual length of column values—can be compared to other columns of the same type with different maximum lengths or char type with different lengths—most DBMS have an upper limit on ℓ (e.g., 2000)

Table 10.1 Data types supported by SQL-92 (continued)

String Data Types (continued)	
bit (ℓ) where ℓ indicates length	Fixed length binary digits (0,1)—can be compared to other columns of the same type with different lengths or bit varying type with different maximum lengths
bit varying (ℓ) where ℓ indicates maximum length	Variable length binary digits (0,1)—can be compared to other columns of the same type with different maximum lengths or bit type with different lengths
Date/Time & Interval Data Types	
date	10 characters long—format: yyyy-mm-dd—can be compared to only other date type columns—allowable dates conform to the Gregorian calendar
time (p)	Format: hh:mi:ss—sometimes precision (p) specified to indicate fractions of a second—the length of a TIME value is 8 characters, if there are no fractions of a second. Otherwise, the length is 8, plus the precision, plus one for the delimiter: hh:mi:ss.p—if no precision is specified, it is 0 by default—TIME can only be compared to other TIME data type columns
timestamp (p)	Format: yyyy:mm:dd hh:mi:ss.p—a timestamp length is nineteen characters, plus the precision, plus one for the precision delimiter—timestamp can only be compared to other timestamp data type columns
interval (q)	Represents measure of time—there are two types of intervals: year-month (yyyy:mm) which stores the year and month; and day-time (dd hh:mi:ss) which stores the days, hours, minutes, and seconds—the qualifier (**q**) known in some databases as the interval lead precision, dictates whether the interval is year-month or day-time—implementation of the qualifier value varies

Note that the "CREATE TABLE order" statement in Box 1 will generate an error; the word "order" (or "ORDER," or any case of the word) cannot be used as a user-defined value for a table, column, or any construct in SQL because ORDER itself is an SQL construct and thus a reserved word. A list of SQL reserved words appears in Appendix C.

The CREATE TABLE statement is a single statement starting at "CREATE" and ending with a semicolon (;). The entire statement could be written on a single line, but spans multiple lines to enhance clarity and readability. The general form of the syntax for the CREATE TABLE statement is:

```
CREATE TABLE table_name (comma-delimited list of table-elements);
```

where:

- *table_name* is a user-supplied name for the base table.
- Each *table-element* in the list is either a column definition or a constraint definition.

There must be at least one column definition in order for a base table to exist. The basic syntax for a column definition is of the form:

```
column_name    representation    [default-definition]    [column-
constraint list]
```

where:

- *column_name* is a user-supplied name for a column.
- *representation* specifies the relevant **data type**, or alternatively, the pre-defined *domain-name*.

The optional *default-definition* (optional elements are indicated by square brackets [ ]) specifies a default value for the column, which overrides any default value specified in the domain definition, if applicable. In the absence of an explicit default definition (directly or via the domain definition), the implicit assumption of a NULL value for the default prevails. The *default-definition* is of the form:

```
DEFAULT ( literal | niladic-function² | NULL )
```

where:

- the | implies "or"
- The optional column-constraint list specifies constraint-definition for the column.

The basic syntax for a *constraint-definition* follows the form:

```
[ CONSTRAINT constraint_name ] constraint-definition³
```

Constraints in a base table are declarative in nature and are imposed either on a single column in the table or on a set of columns in the table. The former is referred to as an *attribute-level* or *column-level constraint* while the latter goes by the name *tuple-level* or *row-level constraint*. A *constraint definition* can be an independent table element (i.e., row-level constraint) or, if applicable to a specific column only, can be part of the column definition (i.e., column-level constraint). Constraint definitions include the following:

- The primary key definition—i.e., specification of an entity integrity constraint

  ```
  PRIMARY KEY (comma-delimited column list)
  ```

  Example:

  ```
  CONSTRAINT pk_pat PRIMARY KEY (Pat_p#a, Pat_p#n)
  ```

- An alternate key definition—i.e., specification of a uniqueness constraint

  ```
  UNIQUE (comma-delimited column list)
  ```

  Example:

  ```
  CONSTRAINT unq_med UNIQUE (Med_code)
  ```

---

² A niladic-function is a built-in function that takes no arguments (Date and Darwen, 1997, p. 55). The niladic-functions that are allowed here are: USER, CURRENT-USER, SESSION-USER, SYSTEM-USER, CURRENT-DATE, CURRENT-TIME, CURRENT-TIMESTAMP.

³ The CONSTRAINT constraint_name phrase enclosed by [ ] is optional. However, in practice, it is *extremely* useful to use this phrase especially for ease of later references to the constraint by a name known to the creator of the table or individual responsible for altering the table. We strongly suggest mandatory use of this phrase in constraint specifications.

- A foreign key constraint—i.e., specification of a referential integrity constraint

```
FOREIGN KEY (comma-delimited column list of referencing table)
REFERENCES table_name
 (comma delimited column list of referenced table)⁴
```

```
[ referential triggered action clause ]
```

The *referential triggered action clause* facilitates specification of the action to be taken when the foreign key constraint is violated upon deletion of a referenced tuple or upon modification of the value of the referenced column list (primary key or alternate key as the case may be). The action options are: CASCADE, SET NULL, SET DEFAULT, and RESTRICT. These actions are referred to as **reactive constraints**, and must be qualified by the referential trigger ON DELETE or ON UPDATE. These action options are described and illustrated with examples in the context of a deletion referential trigger in Section 3.2.3 of Chapter 3 and are equally applicable to an update referential trigger.

Example:

```
CONSTRAINT fk_med FOREIGN KEY (Ord_med_code)
      REFERENCES medication (med_code)
      ON DELETE RESTRICT
      ON UPDATE CASCADE
```

- A check constraint definition in order to restrict column or domain values⁵

```
CHECK (conditional expression)
```

The *conditional expression* here can be of arbitrary complexity; it can even refer to other base tables in the database. Violation of a check constraint occurs when an attempt to create or modify a row in a base table causes the conditional expression stated in the constraint to evaluate as 'false'.

Example:

```
CONSTRAINT chk_gender CHECK Gender IN ('M', 'F')
```

To enhance clarity, it is customary to include the column-level constraint definitions as clauses of the respective column definitions to which they belong, and to list only the row-level constraint definitions at the end of all the column definitions. This style of coding is demonstrated in the DDL code for the MEDICATION and ORDERS tables in Box 2. Notice that the name of the base table ORDER (SQL reserved word) has been changed to ORDERS (not a reserved word). An alternative practice is to code all constraint definitions at the end of all the column definitions, as shown in the DDL code for the PATIENT table in Box 2. Haphazard mixing of column definitions and constraint definitions in the DDL code is strictly discouraged in the interest of best programming practice.

451

---

⁴ If the referenced column list is the primary key of the referenced table, the specification of this column list is optional.

⁵ The CHECK clause is also used to specify conditional expressions in a *general constraint* called ASSERTION which is addressed in Chapter 12.

At this point, we are ready to add other constraints in the CREATE TABLE statement. Let us incorporate the integrity constraints specified in the relational schema (Figure 10.1c) along with the semantic integrity constraints listed in Figure 10.1b in the SQL/DDL code we already have. The resulting script is shown in Box 2.[6]

```
CREATE TABLE patient
(Pat_p#a        char (2),
Pat_p#n         char (5),
Pat_name        varchar (41),
Pat_gender      char (1),
Pat_age         smallint,
Pat_admit_dt    date,
Pat_wing        char (1),
Pat_room#       integer,
Pat_bed         char (1),
CONSTRAINT pk_pat PRIMARY KEY (Pat_p#a, Pat_p#n),
CONSTRAINT chk_gender CHECK (Pat_gender IN ('M', 'F')),
CONSTRAINT chk_age CHECK (Pat_age IN (1 through 90)),
CONSTRAINT chk_bed CHECK (Pat_bed IN ('A', 'B'))
);


CREATE TABLE medication
(Med_code       char (5),
Med_name        varchar (31) CONSTRAINT pk_med  PRIMARY KEY,
Med_unitprice   decimal (3,2) CONSTRAINT chk_unitprice CHECK (Med_unitprice < 4.50),
Med_qty_onhand  integer,
Med_qty_onorder integer,
CONSTRAINT chk_qty CHECK ((Med_qty_onhand + Med_qty_onorder) BETWEEN 1000 AND 3000)
);


CREATE TABLE orders
(Ord_rx#        char (13) CONSTRAINT pk_ord  PRIMARY KEY,
Ord_pat_p#a     char (2),
Ord_pat_p#n     char (5),
Ord_med_code    char (5) CONSTRAINT fk_med FOREIGN KEY REFERENCES medication (med_code),
Ord_dosage      smallint DEFAULT 2 CONSTRAINT chk_dosage CHECK (Ord_dosage BETWEEN 1 AND 3),
Ord_freq        smallint DEFAULT 1 CONSTRAINT chk_freq CHECK (Ord_freq IN (1,2,3)),
CONSTRAINT fk_pat FOREIGN KEY (Ord_pat_p#a, Ord_pat_p#n)
REFERENCES patient (Pat_p#a, Pat_p#n)
);
```

Box 2

Observe that the DDL script produced on the basis of the relational schema (Figure 10.1c) does not fully capture all the information conveyed in the ERD. For instance, the optional property of some attributes, candidate keys of relation schemas, deletion rules, and participation constraints of relationships have not been mapped from the ERD to the relational schema and hence are not reflected in the DDL. An inspection of the information-preserving logical schema (Figure 10.1d) reveals that:

- **Pat_name**, **Pat_age**, **Pat_admit_dt**, **Med_code,** and **Med_qty_onhand** are mandatory attributes, i.e., cannot have null values in any tuple.
- **Med_code** is the alternate key since **Med_name** has been chosen as the primary key of the MEDICATION table.

[6] *Note*: In other chapters in this book, table names are shown in all capital letters; that convention is preserved in the running text of this chapter and other chapters of Part IV. However, in this chapter for clarity in distinguishing SQL keywords from table names, table names are sometimes shown in lower case in SQL code.

- Participation of ORDER in the *Placed_for* relationship is total.
- Participation of PATIENT in the *Placed_for* relationship is partial.
- Participation of ORDER in the *Is_for* relationship is partial.
- Participation of MEDICATION in the *Is_for* relationship is total.
- The deletion rule for the *Is_for* relationship is restrict.
- The deletion rule for the *Placed_for* relationship is cascade.
- **[Pat_wing, Pat_room]** is a composite attribute.
- **[Pat_wing, Pat_room, Pat_bed]** is a composite attribute.

Note: The cardinality ratio of the form (1, **n**) in a relationship type is implicitly captured in the DDL specification via the foreign key constraint. Any (1, **1**) cardinality ratio can be implemented using the UNIQUE constraint definition.

At this point, let us rewrite the SQL/DDL script a third time so as to capture all these constraint definitions. The revised DDL script appears in Box 3, with the added constraint definitions highlighted.

```
CREATE TABLE patient                      Some sort of names
(Pat_p#a        char (2),                 for the constraint
Pat_p#n         char (5),
Pat_name        varchar (41) constraint nn_Patnm not null,
Pat_gender      char (1),
Pat_age         smallint constraint nn_Patage not null,
Pat_admit_dt    date constraint nn_Patadmdt not null,
Pat_wing        char (1),
Pat_room#       integer,
Pat_bed         char (1),
CONSTRAINT pk_pat PRIMARY KEY (Pat_p#a, Pat_p#n),
CONSTRAINT chk_gender CHECK (Pat_gender IN ('M', 'F')),
CONSTRAINT chk_age CHECK (Pat_age IN (1 through 90)),
CONSTRAINT chk_bed CHECK (Pat_bed IN ('A', 'B'))
);
                                                  alternate key
CREATE TABLE medication
(Med_code       char (5) CONSTRAINT nn_medcd not null CONSTRAINT unq_med UNIQUE,
Med_name        varchar (31) CONSTRAINT pk_med  PRIMARY KEY,
Med_unitprice   decimal (3,2) CONSTRAINT chk_unitprice CHECK (Med_unitprice < 4.50),
Med_qty_onhand  integer CONSTRAINT nn_medqty not null,
Med_qty_onorder integer,
CONSTRAINT chk_qty CHECK ((Med_qty_onhand + Med_qty_onorder) BETWEEN 1000 AND 3000)
);


CREATE TABLE orders
(Ord_rx#        char (13) CONSTRAINT pk_ord  PRIMARY KEY,
Ord_pat_p#a     char (2) CONSTRAINT nn_ord_pat_p#a not null,
Ord_pat_p#n     char (5) CONSTRAINT nn_ord_pat_p#n not null,
Ord_med_code    char (5) CONSTRAINT fk_med REFERENCES medication (Med_code)
ON DELETE RESTRICT ON UPDATE RESTRICT,
Ord_dosage      smallint DEFAULT 2 CONSTRAINT chk_dosage CHECK (Ord_dosage BETWEEN 1 AND 3),
Ord_freq        smallint DEFAULT 1 CONSTRAINT chk_freq CHECK (Ord_freq IN (1, 2, 3)),
CONSTRAINT fk_pat FOREIGN KEY (Ord_pat_p#a, Ord_pat_p#n)
REFERENCES patient (Pat_p#a, Pat_p#n) ON DELETE CASCADE ON UPDATE CASCADE
);
```

ON UPDATE → oracle doesn't support this

Box 3

453

Database Creation

By default, a column is allowed to have null values in the rows. The "not null" constraint definitions take care of the mandatory attribute value specification in the corresponding columns in the associated base tables. Likewise, unless explicitly prohibited, a column can contain the same value in multiple rows. An alternate key (i.e., a candidate key not chosen as the primary key of a base table) is required to enforce the uniqueness constraint. This is accomplished by the UNIQUE constraint definition for the **Med_code** column in the MEDICATION table.

Partial participation of a parent as well as a child in a relationship (i.e., min = 0) exists by default. Total participation of a parent in a relationship (i.e., min > 0) cannot be enforced using any of the constraint definitions discussed so far. SQL-92 offers another mechanism called *declarative assertion* to specify broader constraints at the database schema level. This is presented in Chapter 12. Total participation of a child in a relationship (i.e., min = 1) is enforced by specifying a "not null" constraint on the foreign key attribute(s) (e.g., not null constraint definition for (**Ord_pat_p#a, Ord_pat_p#n**) in the ORDERS table). The deletion rules (referential triggered action clause) are incorporated using the ON DELETE clause of the foreign key constraint definition.[7] Since, by definition, a relation schema has only atomic attributes, SQL/DDL does not provide for the specification of composite columns—all columns in a table are atomic.

10.1.1.2 Syntax of the ALTER TABLE Statement

At this point, we have successfully created the table structures for the logical schema constituting the base tables PATIENT, MEDICATION, and ORDERS. Suppose we want to make some corrections or changes to one or more of these base table structures. The **ALTER TABLE** statement in SQL/DDL is used to accomplish this. The general form of the syntax for the ALTER TABLE statement is:

```
ALTER TABLE table_name action;
```

where *table_name* is the name of the base table being altered and *action* is one of the following:

- Adding a column or altering a column's *default-definition* or removing the existing *default-definition* via the syntax:

  ```
  ADD [ COLUMN ] column_definition

  ALTER [ COLUMN ] column_name
  { SET default-definition | DROP DEFAULT }[8]
  ```
- Removing an existing column via the syntax:

  ```
  DROP [ COLUMN ] column_name { RESTRICT | CASCADE }
  ```

---

[7] Similar to the ON DELETE clause, SQL/DDL offers an ON UPDATE clause for referential triggered action which is intended for specifying action to be taken when the referenced attribute(s) value (primary key or alternate key value) in a foreign key constraint is changed. Since in our example this is not specified, we defaulted to an assumption of same as ON DELETE clause specifications.

[8] Braces { } are used to specify that one of the items from the list of items separated by the vertical bar must be chosen.

- Adding to an existing set of constraints via the syntax:

```
ADD table_constraint_definition
```
- Removing a named constraint via the syntax:

```
DROP CONSTRAINT constraint_name { RESTRICT | CASCADE }
```

Suppose we want to add a column to the base table PATIENT to store the phone number of every patient. The SQL/DDL code to do this is:

```
ALTER TABLE patient  ADD Pat_phone#     char (10);
```

Now the rows of the base table PATIENT are capable of receiving values for the **Pat_phone#** column. Since no default has been specified for the new column added to the table, the rows of PATIENT will, by default, have "null" value for the column, **Pat_phone#**. Clearly, it is not possible to specify a "not null" constraint on the column until either a non-null default value is specified for the column or the column is populated with non-null values in all rows of the base table.

The column can be removed from the base table by either of these two statements:

```
ALTER TABLE patient  DROP Pat_phone# CASCADE;
```

or:

```
ALTER TABLE patient  DROP Pat_phone# RESTRICT;
```

Observe the definition of DROP behavior (i.e., CASCADE or RESTRICT) in the SQL/DDL statement. The SQL-92 standard requires the DROP behavior definition. The CASCADE option implies that all constraints and derived tables that reference the column also be dropped from the database schema. Likewise, the RESTRICT option prevents the deletion of the column should any schema element that references the column exist. Also, SQL-92 provides for the deletion of only one column per ALTER statement.

Suppose we want to specify a default value of $3.00 for the unit price of all medications. This can be done as follows:

```
ALTER TABLE medication  ALTER Med_unitprice SET DEFAULT 3.00;
```

The default clause can be removed by:

```
ALTER TABLE medication  ALTER Med_unitprice DROP DEFAULT;
```

10.1.1.3 A Best Practice Hint

Let us take a look at two methods of specifying the domain constraints on the column **Pat_age** in the base table PATIENT. Here is the first method:

```
Pat_age       smallint CONSTRAINT nn_Patage not null,
CONSTRAINT chk_age CHECK (Pat_age IN (1 through 90))
```

Since the naming of the constraint definition is optional, it is possible to write the above DDL using a second method:

```
Pat_age       smallint not null CHECK (Pat_age IN (1 through 90))
```

Clearly, the second method code appears simpler and more concise. However, if we decide to permit null values for **Pat_age**, in method 2, the entire column definition has to be re-specified. On the other hand, using method 1, we simply drop the "not null" constraint as shown below:

```
ALTER TABLE patient  DROP CONSTRAINT nn_patage CASCADE;
```

or:

```
ALTER TABLE patient  DROP CONSTRAINT nn_patage RESTRICT;
```

This is possible only because we named the constraint. While the DBMS names every constraint when we don't, finding out the constraint name given by the DBMS is an inefficient task; and the constraint name given by the DBMS is not generally a user-friendly name. Thus, the coding technique of method 1 offers greater flexibility and is strongly recommended.

10.1.1.4 Syntax of the DROP TABLE Statement

Just as we can create a base table and make certain changes to it, it is also possible to remove a base table (structure and content) from the database using the **DROP TABLE** SQL/DDL statement:

```
DROP TABLE table_name drop behavior;
```

where:

- *table_name* is the name for the base table being deleted.
- The *drop behaviors* possible are CASCADE and RESTRICT.

For example,

```
DROP TABLE medication CASCADE;
```

not only deletes the MEDICATION table (all rows and the table definition) from the database, but also removes all schema elements (constraints and derived tables) that reference any column of the MEDICATION table. For instance, the constraint definition, **fk_med** in the base table ORDERS, defines the foreign key constraint that references a candidate key of the MEDICATION table. The CASCADE option in the DROP TABLE statement automatically drops the constraint **fk_med** in ORDERS when the MEDICATION table is dropped. The RESTRICT option, on the other hand, disallows the deletion of the MEDICATION table, because the constraint definition, **fk_med**, exists in the schema.

10.1.1.5 A Comprehensive Example

Figure 10.2 contains a Fine-granular Design-Specific ER diagram for Madeira College which emerged as a result of use of the conceptual modeling process described in Part I of this book. This database is used to keep track of the courses offered by departments, the enrollment of students in courses, and the teaching-related activities of professors.

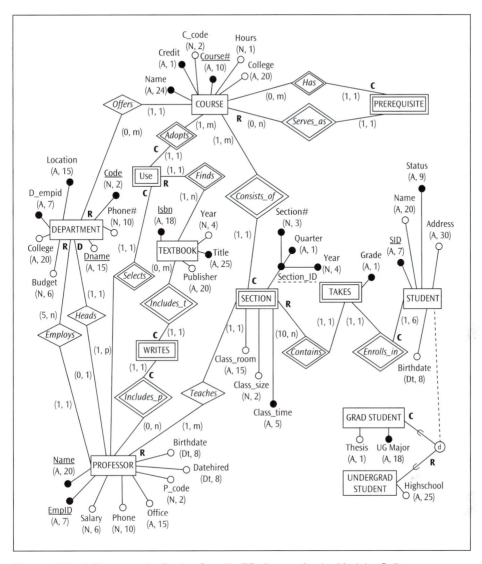

Figure 10.2 A Fine-granular Design-Specific ER diagram for the Madeira College registration system

A brief description of the various relationships follows:

- Departments can offer a variety of courses with each offered by a single department.
- Each department must have at least five professors, one of whom serves as the department head.
- Each course consists of at least one section and may serve as a prerequisite of other courses as well as have other courses as its prerequisite.
- Madeira College has both graduate and undergraduate students. Each student must enroll in (i.e., take) at least one but no more than six courses.

- In order for a section of a course to be offered (i.e., exist), it must have a minimum of ten students.
- Each professor selects at least one textbook to be used in each course he or she teaches, at least one textbook is adopted for each course offered, and each textbook is used in at least one course.
- Some professors write textbooks used in courses and some of the textbooks used in courses are written by professors.

The relational schema (information-reducing logical schema) shown in Figure 10.3 and the information-preserving logical schema that appears in Figure 10.4 result from the logical data modeling process discussed in Chapter 6. The implementation of this logical schema takes the form of the SQL/DDL script for the Madeira College registration system in Box 4. Observe how the physical schema in Box 4 captures the constraints defined in the information-preserving logical schema.

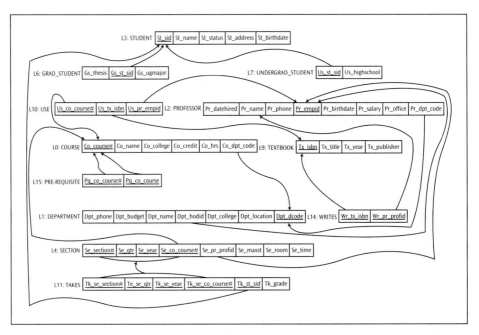

Figure 10.3 Information-reducing logical schema for the Madeira College registration system

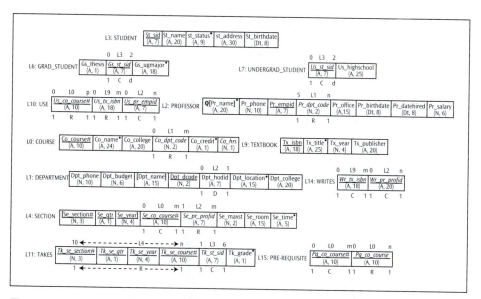

Figure 10.4 Information-preserving logical schema for the Madeira College registration system

```
CREATE TABLE PROFESSOR
(PR_NAME  CHAR(20)  CONSTRAINT NN_3NAME UNIQUE,
PR_EMPID VARCHAR(7)   CONSTRAINT PK_3COU PRIMARY KEY,
PR_PHONE INTEGER(10),
PR_OFFICE CHAR(15),
PR_BIRTHDATE DATE,
PR_DATEHIRED DATE,
CONSTRAINT CK_DATES CHECK (PR_DATEHIRED - PR_BIRTHDATE >= 0),
PR_DPT_DCODE  NUMBER(2) CONSTRAINT NN_3DCOD NOT NULL,
PR_SALARY NUMBER(6));


CREATE TABLE DEPARTMENT
(DPT_NAME   CHAR(15),
DPT_DCODE     INTEGER(2) CONSTRAINT PK_2DEPT PRIMARY KEY,
               CONSTRAINT CK_2COD CHECK(DPT_DCODE BETWEEN 1 AND 10),
DPT_COLLEGE  CHAR(20),
DPT_PHONE    INTEGER(10) CONSTRAINT UNQ_PHONE UNIQUE,
DPT_BUDGET   DECIMAL(6,0),
               CONSTRAINT CK_2BUD CHECK(DPT_BUDGET BETWEEN 100000 AND 500000),
DPT_LOCATION CHAR(15)  CONSTRAINT NN_2LOC NOT NULL,
DPT_HODID  VARCHAR(7)  CONSTRAINT NN_2HOD NOT NULL,
               CONSTRAINT UNQ_2HOD UNIQUE(DPT_HODID),
               CONSTRAINT FK_2DEPT FOREIGN KEY(DPT_HODID) REFERENCES PROFESSOR(PR_EMPID),
               CONSTRAINT UNQ_2DNMCOL UNIQUE(DPT_NAME,DPT_COLLEGE));


ALTER TABLE PROFESSOR ADD
CONSTRAINT FK_3DEPT FOREIGN KEY(PR_DPT_DCODE) REFERENCES DEPARTMENT(DPT_DCODE);


CREATE TABLE COURSE
(CO_NAME   CHAR(24)  CONSTRAINT NN_1NAME NOT NULL,
CO_COURSE# CHAR(10)  CONSTRAINT PK_1COU PRIMARY KEY,
CO_CREDIT  CHAR(1)   CONSTRAINT NN_1CRED NOT NULL,
               CONSTRAINT CK_1CRED CHECK(UPPER(CO_CREDIT) IN ('U','G')),
CO_COLLEGE CHAR(20)  CONSTRAINT CK_1COL CHECK(CO_COLLEGE IN ('ARTS AND SCIENCES','EDUCATION',
'ENGINEERING','BUSINESS')),
CO_HRS   SMALLINT,
CO_DPT_DCODE   INTEGER(2) CONSTRAINT NN_1DCOD NOT NULL,
CONSTRAINT UNQ_1COLNAM UNIQUE(CO_NAME,CO_COLLEGE),
CONSTRAINT FK_1DEP FOREIGN KEY(CO_DPT_DCODE) REFERENCES DEPARTMENT(DPT_DCODE),
CONSTRAINT CK_1HRS CHECK(((UPPER(CO_CREDIT) IN ('U')) AND (CO_HRS<=4)) OR
                  ((UPPER(CO_CREDIT) IN ('G')) AND (CO_HRS=3))));


CREATE TABLE STUDENT
(ST_SID        VARCHAR(7)   CONSTRAINT PK_4STD PRIMARY KEY,
ST_NAME        CHAR(20),
ST_ADDRESS     CHAR(30),
ST_STATUS      CHAR(9),
ST_BIRTHDATE DATE,
               CONSTRAINT CK_4STAT CHECK(ST_STATUS IN ('FULL TIME','PART TIME')));


CREATE TABLE SECTION
(SE_SECTION#     INTEGER(3),
SE_QTR        CHAR(1)   CONSTRAINT CK_5QTR CHECK(SE_QTR IN ('A','W','S','U')),
SE_YEAR         INTEGER(4),
SE_TIME      CHAR(5)   CONSTRAINT NN_5TIM NOT NULL,
SE_MAXST     INTEGER(2) CONSTRAINT CK_5SIZ CHECK(SE_MAXST <= 35),
SE_ROOM      VARCHAR(15),
SE_CO_COURSE#       CHAR(10),
         CONSTRAINT FK_5COU FOREIGN KEY(SE_CO_COURSE#) REFERENCES COURSE(CO_COURSE#) ON DELETE
CASCADE,
SE_PR_PROFID        VARCHAR(7) CONSTRAINT NN_5PROF NOT NULL,
               CONSTRAINT PK_5SEC PRIMARY KEY(SE_SECTION#,SE_QTR,SE_YEAR,SE_CO_COURSE#),
               CONSTRAINT FK_5DEPT FOREIGN KEY(SE_PR_PROFID) REFERENCES PROFESSOR(PR_EMPID),
               CONSTRAINT UNQ_5SECT UNIQUE(SE_QTR,SE_YEAR,SE_TIME,SE_ROOM));
```

Box 4

```
CREATE TABLE GRAD_STUDENT
(GS_ST_SID    VARCHAR(7)  CONSTRAINT PK_7SID PRIMARY KEY,
GS_THESIS CHAR(1) CONSTRAINT CK_7THES CHECK(GS_THESIS IN ('Y','N')),
GS_UGMAJOR CHAR(18) CONSTRAINT NN_7MAJ NOT NULL,
        CONSTRAINT FK_7STD FOREIGN KEY(GS_ST_SID) REFERENCES STUDENT(ST_SID) ON DELETE CASCADE);


CREATE TABLE UNDERGRAD_STUDENT
(US_ST_SID           VARCHAR(7) CONSTRAINT PK_8SID PRIMARY KEY,
US_HIGHSCHOOL      CHAR(25),
        CONSTRAINT FK_8STD FOREIGN KEY(US_ST_SID) REFERENCES STUDENT(ST_SID) ON DELETE CASCADE);


CREATE TABLE TEXTBOOK
(TX_ISBN     VARCHAR(18) CONSTRAINT PK_10TB PRIMARY KEY,
TX_TITLE     VARCHAR(25)   CONSTRAINT NN_10TIT NOT NULL,
TX_YEAR      INTEGER(4),
TX_PUBLISHER CHAR(20)   CONSTRAINT CK_10PUB CHECK(TX_PUBLISHER IN ('Thomson','Springer',
'Prentice-Hall')));


CREATE TABLE PREREQUISITE
(PQ_CO_COURSE#  CHAR(10),
        CONSTRAINT FK_11APRE FOREIGN KEY(PQ_CO_COURSE#) REFERENCES COURSE(CO_COURSE#),
PQ_CO_COURSE    CHAR(10),
        CONSTRAINT FK_11BPRE FOREIGN KEY(PQ_CO_COURSE) REFERENCES COURSE(CO_COURSE#),
        CONSTRAINT PK_11PRE PRIMARY KEY(PQ_CO_COURSE#,PQ_CO_COURSE),
        CONSTRAINT CK_11PRE CHECK(PQ_CO_COURSE# <> PQ_CO_COURSE));


CREATE TABLE USE
(US_CO_COURSE#       CHAR(10),
US_TX_ISBN          VARCHAR(18),
US_PR_EMPID         VARCHAR(7),
CONSTRAINT PK_12USE PRIMARY KEY(US_CO_COURSE#,US_TX_ISBN,US_PR_EMPID),
CONSTRAINT FK_12AUSE FOREIGN KEY(US_CO_COURSE#) REFERENCES COURSE(CO_COURSE#),
CONSTRAINT FK_12BUSE FOREIGN KEY(US_TX_ISBN) REFERENCES TEXTBOOK(TX_ISBN),
CONSTRAINT FK_12CUSE FOREIGN KEY(US_PR_EMPID) REFERENCES PROFESSOR(PR_EMPID));


CREATE TABLE WRITES
(WR_TX_ISBN          VARCHAR(18),
WR_PR_PROFID         VARCHAR(7),
CONSTRAINT PK_13WRT PRIMARY KEY(WR_TX_ISBN,WR_PR_PROFID),
CONSTRAINT FK_13AWRT FOREIGN KEY(WR_TX_ISBN) REFERENCES TEXTBOOK(TX_ISBN) ON DELETE CASCADE,
CONSTRAINT FK_13BWRT FOREIGN KEY(WR_PR_NAME) REFERENCES PROFESSOR(PR_NAME));


CREATE TABLE TAKES
(TK_SE_SECTION# INTEGER(3),
TK_SE_QTR    CHAR(1),
TK_SE_YEAR      INTEGER(4),
TK_SE_CO_COURSE#   CHAR(10),
TK_GRADE       CHAR(1)    CONSTRAINT NN_14GRADE NOT NULL,
                CONSTRAINT CK_14GRADE CHECK(UPPER(TK_GRADE) IN ('A','B','C')),
TK_ST_SID       VARCHAR(7),
                CONSTRAINT FK_14SID FOREIGN KEY(TK_ST_SID) REFERENCES STUDENT(ST_SID) ON
DELETE        CASCADE,
                CONSTRAINT FK_14TAK FOREIGN
KEY(TK_SE_SECTION#,TK_SE_QTR,TK_SE_YEAR,TK_SE_CO_COURSE#)  REFERENCES
SECTION(SE_SECTION#,SE_QTR,SE_YEAR,SE_CO_COURSE#),
                CONSTRAINT PK_14TAK PRIMARY
KEY(TK_SE_SECTION#,TK_SE_QTR,TK_SE_YEAR,TK_SE_CO_COURSE#,TK_ST_SID));
```

Box 4 (continued)

10.1.2 Specification of User-Defined Domains

The SQL-92 standard provides for the formal specification of a *domain*. A domain specification can be used to define a constraint over one or more columns of a table with a formal name so that the domain name can be used wherever that constraint is applicable. In other words, a domain may be regarded as a named collection of data values that can be treated as a user-defined data type in a column definition. This approach lends the ability for the design to be modular. The **CREATE DOMAIN** statement is of the form:

```
CREATE DOMAIN domain_name [ AS ] data type [ default-definition ]
    [ domain-constraint-definition list ];
```

where:

- *domain_name* is a user-supplied name for the domain.
- The optional *default-definition* specifies a default value for the domain. In the absence of an explicit default definition, the domain has *no* default value.
- The optional *domain-constraint-definition* is of the form:

```
[ CONSTRAINT constraint_name ] CHECK (VALUE (conditional-expression))
```

Note: The keyword VALUE is required here; it is not used in any other SQL/DDL statement.

Several examples of creating a domain follow.

Example 1. Specify a domain to capture integer values 1, 2, and 3, with a default value of 2.

```
CREATE DOMAIN measure smallint DEFAULT 2 CONSTRAINT chk_measure
CHECK (VALUE IN (1,2,3));
```

Since [CONSTRAINT *constraint_name*] is optional, the above statement can also be stated as:

```
CREATE DOMAIN measure smallint DEFAULT 2 CHECK (VALUE IN (1,2,3));
```

Although the latter definition appears short and crisp, the former expression is better practice since it provides a direct referencing mechanism to the constraint definition within the domain definition; for instance, it gives the ability to delete the CHECK constraint while retaining the rest of the domain definition. Now, any column definition that is a number in the range of 1—3 and can accept a default value of 2 can use this domain name as the *representation* for the column instead of the *data type* and a *constraint on the data type* for the range of values. For instance, the column definition in the ORDERS table:

```
Ord_dosage smallint DEFAULT 2 CONSTRAINT chk_dosage
CHECK (Ord_dosage BETWEEN 1 AND 3)
```

can be coded as:

```
Ord_dosage  measure
```

Likewise,

```
Ord_freq    smallint DEFAULT 1 CONSTRAINT chk_freq
CHECK (Ord_freq IN (1, 2, 3))
```

can also be coded as:

```
Ord_freq    measure DEFAULT 1
```

Accordingly, **Ord_dosage** has a default value of 2 propagated from the domain **measure** and honors the acceptable range of values, 1 through 3. **Ord_freq**, on the other hand, honors the acceptable range of values, 1 through 3 based on the domain **measure** it is referencing; but the default value of 2 propagated from **measure** is overridden by the local default value of 1 specified as part of its column definition. The revised SQL/DDL is highlighted in Box 5.

```
CREATE TABLE orders
 (Ord_rx#           char (13) CONSTRAINT pk_ord  PRIMARY KEY,
 Ord_pat_p#a        char (2), CONSTRAINT nn_ord_pat_p#a not null,
 Ord_pat_p#n        char (5), CONSTRAINT nn_ord_pat_p#n not null,
 Ord_med_code       char (5) CONSTRAINT fk_med REFERENCES medication (Med_code)
 ON DELETE RESTRICT ON UPDATE RESTRICT,
 Ord_dosage         measure,
 Ord_freq           measure DEFAULT 1,
 CONSTRAINT fk_pat FOREIGN KEY (Ord_pat_p#a, Ord_pat_p#n)
 REFERENCES patient (Pat_p#a, Pat_p#n) ON DELETE CASCADE ON UPDATE CASCADE
 );
```

Box 5

Example 2. Specify a domain with mandatory values for the U.S. Postal Service abbreviation for the list of states OH, PA, IL, IN, KY, WV, and MI. Also, designate OH as the default state.

```
CREATE DOMAIN valid_states CHAR (2) DEFAULT 'OH'
   CONSTRAINT nn_states CHECK (VALUE IS NOT NULL)
   CONSTRAINT chk_states CHECK (VALUE IN ('OH', 'PA', 'IL', 'IN', 'KY',
   'WV', 'MI'));
```

This domain name can now be used as the *representation* for any column name in the database that represents a 2-character attribute—hopefully the **State** attributes in different base tables. Obviously, the default designation cannot contradict the constraint definition on the DOMAIN—i.e., the CREATE DOMAIN operation will fail if the specified DEFAULT value is TX since TX is not present in the list of valid states shown in the constraint definition. Note that:

```
CREATE DOMAIN valid_states CHAR (2) NOT NULL;
```

is incorrect syntax, while:

```
CREATE DOMAIN valid_states CHAR (2) CHECK (VALUE IS NOT NULL);
```

is correct syntax.

A predefined *domain name* comes in handy when it can be used in several tables without having to repeat the complete data type and column constraint specification. A domain is a schema element (object) of a database schema like a base table. Thus, changes to the domain definition automatically propagate to all tables in which the domain name is referenced. To demonstrate this, let us follow up on Example 2. While most columns in a database that represent the attribute **State** cannot have a missing value for **State**, suppose a few are allowed to have null values. One way to handle this situation is to let the DOMAIN **valid_states** include a null value and let the individual column definitions specify "not null" as needed. This way, any constraint specification in a column definition overrides the corresponding constraint in the domain definition referenced by the column for that particular column; the rest of the domain definition continues to apply for the referencing column. Therefore, now we need to remove the "not null" constraint from the DOMAIN **valid_states**.

Example 3. Remove the "not null" constraint from the domain specification for **valid_states**. The SQL/DDL statement that accomplishes this task is **ALTER DOMAIN**:

```
ALTER DOMAIN valid_states DROP CONSTRAINT nn_states;
```

Note: Had we not named the constraint, the only recourse we would have is to drop the complete domain definition and create it over again.

The effect of the above ALTER DOMAIN statement is that all columns in the database schema which reference the domain **valid_states** as their column representation will now accept a missing value except, of course, the ones that enforce a local column constraint explicitly prohibiting a missing value.

The general form of the syntax for the ALTER DOMAIN statement is:

```
ALTER DOMAIN domain_name action;
```

where *domain_name* is the name of the DOMAIN being altered and the *actions* possible are:

- Adding *default-definition* or replacing an existing *default-definition* via the syntax:

  ```
  SET default-definition
  ```

- Copying *default-definition* to the columns defined on the domain which do not have explicitly specified default values of their own, then removing *default-definition* from the domain definition via the syntax:

  ```
  DROP DEFAULT
  ```

- Adding to the existing set of constraints, if any, via the syntax:

  ```
  ADD domain_constraint_definition
  ```

- Removing the named constraint via the syntax:

  ```
  DROP CONSTRAINT constraint_name
  ```

Note: Only one action per ALTER DOMAIN statement is allowed.

Example 4. Add Maryland (MD) and Virginia (VA) to the domain **valid_states**. This is done by adding a new constraint to the domain **valid_states** as shown here:

```
ALTER DOMAIN valid_states ADD CONSTRAINT chk_2more
CHECK (VALUE IN ('MD', 'VA'));
```

Note that this constraint labeled **chk_2more** does not replace the existing **chk_states** constraint in the domain **valid_states**. Instead, it is included as an additional constraint of the domain to allow 'MD' and 'VA' as valid states of the domain. Having altered the domain **valid_states** thus, what if we decide that 'VA' is not a valid state, but 'MD' is? In this case, the only recourse is to remove the constraint **chk_2more** and add a new constraint checking only for 'MD' in the **valid_states** domain.

Example 5. Assuming that the scripts in Example 1 and Box 5 have been executed, remove the default value for the DOMAIN **measure**. As of now, the default for **measure** is 2, the default for **Ord_dosage** is 2, and the default for **Ord_freq** is 1. The SQL/DDL syntax follows:

```
ALTER DOMAIN measure DROP DEFAULT;
```

The execution of the above code yields the following result. The DOMAIN **measure** no longer has a default value—*not even null*; **Ord_dosage** continues to have a default value of 2; and **Ord_freq** continues to have a default value of 1. The verification of this result is left as an exercise for the reader.

Example 6. Change the default value of **State** in the **valid_states** domain to PA.

```
ALTER DOMAIN valid_states SET DEFAULT 'PA';
```

A domain definition can be eliminated using the **DROP DOMAIN** SQL/DDL syntax:

```
DROP DOMAIN domain_name CASCADE | RESTRICT
```

The rules for the DROP behavior in this case are somewhat different and should be noted carefully. With the RESTRICT option any attempt to drop a domain definition fails if any column in the database tables, views, and/or integrity constraints references the domain name. With the CASCADE option, however, dropping a domain entails dropping of any referencing views and integrity constraints only. The columns referencing the domain are *not* dropped. Instead the domain constraints are effectively converted into base table constraints and are attached to every base table that has a column(s) defined on the domain.

Example 7. Suppose the domain named **measure** is no longer needed and the following DROP DOMAIN statement is used:

```
DROP DOMAIN measure RESTRICT;
```

In this case, since the columns **Ord_dosage** and **Ord_freq** are defined on the DOMAIN **measure**, the DROP DOMAIN operation will fail with the DROP behavior specification of RESTRICT.

On the other hand, suppose the following DROP DOMAIN statement is used:

```
DROP DOMAIN measure CASCADE;
```

In this case, since the columns **Ord_dosage** and **Ord_freq** are defined on the DOMAIN **measure**, the DROP DOMAIN operation first creates row-level constraints in the base table ORDERS equivalent to the domain constraint **chk_measure**. The default definition of the DOMAIN **measure** is mapped to the column **Ord_dosage** since it does not have a column-level default definition. The column-level default definition of **Ord_freq** stays intact.

10.1.3 Schema and Catalog Concepts in SQL/DDL

An **SQL-schema** represents a named collection of schema elements (objects) such as base table definitions, domain definitions, view definitions, constraint definitions, and so on under the control of a single user (or application) who alone is authorized to create and access objects within it. This concept is introduced in the SQL-92 standard. Before SQL-92, the database objects of all users belonged to the same schema/database. A named collection of SQL-schema in an SQL environment is referred to as a **catalog** in SQL-92—i.e., the DBMS partitions the catalog into SQL-schemas. The purpose of a catalog is to better organize database elements into groups by application/user. In general, only a few users, such as a database administrator (DBA), are authorized to create an SQL-schema. The privilege to create an SQL-schema and other schema elements must be explicitly granted to other users by the DBA.

10.1.3.1 CREATE SCHEMA

SQL-schemas are created by the **CREATE SCHEMA** statement using the following syntax:

```
CREATE SCHEMA { [ schema_name ] [AUTHORIZATION user_name ] }
        [ schema element list ];
```

where:

- *schema_name* is a user supplied name for the SQL-schema.
- *user_name* is intended to identify the owner of the SQL-schema.
- *schema element list* includes objects such as base table definitions, domain definitions, view definitions, constraint definitions, etc.

Note: At least one of *schema_name* and *user_name* must be present in the definition; however, it is a good practice to include both.

Example:

```
CREATE SCHEMA clinic AUTHORIZATION Debakey;
```

Observe that the inclusion of *schema element list* in the CREATE SCHEMA script is optional. In other words, an SQL-schema can be created without any schema elements in which case the schema elements will be added as and when required because the creation of schema elements as independent operations is valid in SQL. A sample script for the creation of an SQL-schema containing an assortment of schema elements appears in Box 6.

```
CREATE SCHEMA clinic AUTHORIZATION Debakey
CREATE TABLE patient
(Pat_p#a        char (2),
Pat_p#n         char (5),
Pat_name        varchar (41) constraint nn_Patnm not null,
Pat_gender      char (1),
Pat_age         smallint constraint nn_Patage not null,
Pat_admit_dt    date constraint nn_Patadmdt not null,
Pat_wing        char (1),
Pat_room#       integer,
Pat_bed         char (1),
CONSTRAINT pk_pat PRIMARY KEY (Pat_p#a, Pat_p#n),
CONSTRAINT chk_gender CHECK (Pat_gender IN ('M', 'F')),
CONSTRAINT chk_age CHECK (Pat_age IN (1 through 90)),
CONSTRAINT chk_bed CHECK (Pat_bed IN ('A', 'B'))
)

CREATE VIEW senior_citizen AS
  SELECT patient.Pat_name, patient.Pat_age, patient.Pat_gender
    FROM patient
  WHERE patient.Pat_age > 64

CREATE VIEW senior_stat (V_gender, V_#ofpats) AS
  SELECT patient.Pat_gender, count (*)
    FROM patient
  WHERE patient.Pat_age > 64
  GROUP BY patient.Pat_gender

CREATE TABLE medication
(Med_code        char (5) CONSTRAINT nn_medcd not null CONSTRAINT unq_med UNIQUE,
Med_name         varchar (31) CONSTRAINT pk_med  PRIMARY KEY,
Med_unitprice    decimal (3,2) CONSTRAINT chk_unitprice CHECK (Med_unitprice < 4.50),
Med_qty_onhand   integer CONSTRAINT nn_medqty not null,
Med_qty_onorder  integer,
CONSTRAINT chk_qty CHECK ((Med_qty_onhand + Med_qty_onorder) BETWEEN 1000 AND 3000)
)

CREATE VIEW unused_med AS
  SELECT medication.Med_name, medication.Med_code, medication.Med_qty_onhand
    FROM medication
  WHERE medication.Med_code NOT IN
    (SELECT orders.Ord_med_code FROM orders)
WITH CHECK OPTION

CREATE TABLE orders
 (Ord_rx#       char (13) CONSTRAINT pk_ord  PRIMARY KEY,
Ord_pat_p#a    char (2),
Ord_pat_p#n    char (5),
Ord_med_code   char (5) CONSTRAINT fk_med REFERENCES medication (Med_code)
ON DELETE RESTRICT ON UPDATE RESTRICT,
Ord_dosage       measure,
Ord_freq         measure,
CONSTRAINT fk_pat FOREIGN KEY (Ord_pat_p#a, Ord_pat_p#n)
REFERENCES patient (Pat_p#a, Pat_p#n) ON DELETE CASCADE ON UPDATE CASCADE
 )
```

Box 6

```
CREATE DOMAIN measure AS smallint CHECK (measure > 0 and measure < 4)

CREATE VIEW used_med AS
  SELECT orders.Ord_pat_p#a, orders.Ord_pat_p#n, medication.Med_name,
orders.Ord_dosage,
             orders.Ord_frequency
    FROM medication, orders
  WHERE medication.Med_code = orders.Ord_med_code

CREATE ASSERTION Chk_orders
           CHECK (SELECT COUNT (*) FROM orders >= 100)

CREATE ASSERTION Chk_ordr_per_med
CHECK (NOT EXISTS
(SELECT * FROM medication
 WHERE medication.Med_code NOT IN
      (SELECT medx.Med_code   FROM medication medx
          WHERE medx.Med_code IN
          (SELECT Ord_med_code FROM orders
             GROUP BY Ord_med_code
                                      HAVING COUNT (*) >= 5)))
                    )

CREATE ASSERTION Chk_unit#
CHECK (NOT EXISTS
(SELECT * FROM patient
 WHERE Pat_wing IS NULL AND Pat_room# IS NULL
                        )

CREATE ASSERTION Chk_no_ordr_pats
CHECK (NOT EXISTS
(SELECT * FROM patient
 WHERE (Pat_p#a, Pat_p#n) NOT IN
      (SELECT Ord_pat_p#a, Ord_pat_p#n   FROM orders))
                        )

CREATE TRIGGER Pat_discharge_dt
AFTER DELETE ON patient
FOR EACH ROW
INSERT INTO PATIENT_AUDIT VALUES (Pat_p#a, Pat_p#a, SYSDATE);

);
```

Box 6   (continued)

Note: Assertions, triggers, and views are covered in Chapter 12 because they require some knowledge of the SQL querying language.

10.1.3.2 DROP SCHEMA

The SQL-92 **DROP SCHEMA** syntax for deleting an SQL-schema is:

```
DROP SCHEMA schema_name { CASCADE | RESTRICT };
```

Example:

```
DROP SCHEMA clinic RESTRICT;
```

If RESTRICT is specified as the drop behavior, the DROP operation will fail unless the SQL-schema is empty. If CASCADE is specified as the drop behavior, the SQL-schema and all the objects contained in it will be eradicated from the database.

10.1.3.3 The INFORMATION_SCHEMA

The SQL-92 standard does not stipulate the structure of a catalog or the format of the SQL-schemas in a catalog. However, an idealized catalog structure for a DBMS product to emulate has been proposed. Known as the **definition schema** in the SQL-92 standard, this idealized catalog structure defines a set of system-level tables (see Groff and Weinberg, 2002). While not requiring a DBMS product to support the idealized catalog, SQL-92 does require that every system catalog contain one particular schema named INFORMATION_ SCHEMA in order to claim adherence to intermediate or full SQL-92 compliance level. The INFORMATION_SCHEMA essentially defines a series of views on the idealized catalog tables that identify database objects accessible to current users in the catalog. A list of these catalog views in the INFORMATION_SCHEMA is also available in Groff and Weinberg (2002). All object definitions from all the SQL-schemas in a catalog are also captured in the INFORMATION_SCHEMA via the defined catalog views. Only authorized users may access the INFORMATION_SCHEMA of a catalog. Also, SQL-schemas within the same catalog are allowed to share many of the schema elements. Finally, SQL-92 also presents the concept of a **cluster** of catalogs in the SQL-92 environment. The CLUSTER identifies the set of databases a single SQL program can access.

10.2 Data Population Using SQL

Every database is subject to continual change. Three SQL statements (INSERT, DELETE, and UPDATE) are used for data insertion, deletion, and modification. This section discusses these three statements in the context of the PATIENT, MEDICATION, and ORDERS tables introduced Section 10.1.1. For convenience of the reader, the SQL/DDL script from Box 3 in Section 10.1.1.1 is reproduced as Box 7.

```
CREATE TABLE patient
(Pat_p#a       char (2),
Pat_p#n        char (5),
Pat_name       varchar (41) constraint nn_Patnm not null,
Pat_gender     char (1),
Pat_age        smallint constraint nn_Patage not null,
Pat_admit_dt   date constraint nn_Patadmdt not null,
Pat_wing       char (1),
Pat_room#      integer,
Pat_bed        char (1),
CONSTRAINT pk_pat PRIMARY KEY (Pat_p#a, Pat_p#n),
CONSTRAINT chk_gender CHECK (Pat_gender IN ('M', 'F')),
CONSTRAINT chk_age CHECK (Pat_age IN (1 through 90)),
CONSTRAINT chk_bed CHECK (Pat_bed IN ('A', 'B'))
);


CREATE TABLE medication
(Med_code        char (5) CONSTRAINT nn_medcd not null CONSTRAINT unq_med UNIQUE,
Med_name         varchar (31) CONSTRAINT pk_med  PRIMARY KEY,
Med_unitprice    decimal (3,2) CONSTRAINT chk_unitprice CHECK (Med_unitprice < 4.50),
Med_qty_onhand   integer CONSTRAINT nn_medqty not null,
Med_qty_onorder integer,
CONSTRAINT chk_qty CHECK ((Med_qty_onhand + Med_qty_onorder) BETWEEN 1000 AND 3000)
);


CREATE TABLE orders
(Ord_rx#        char (13) CONSTRAINT pk_ord  PRIMARY KEY,
Ord_pat_p#a     char (2) CONSTRAINT nn_ord_pat_p#a not null,
Ord_pat_p#n     char (5) CONSTRAINT nn_ord_pat_p#n not null,
Ord_med_code    char (5) CONSTRAINT fk_med REFERENCES medication (Med_code)
ON DELETE RESTRICT ON UPDATE RESTRICT,
Ord_dosage      smallint DEFAULT 2 CONSTRAINT chk_dosage CHECK (Ord_dosage BETWEEN 1 AND 3),
Ord_freq        smallint DEFAULT 1 CONSTRAINT chk_freq CHECK (Ord_freq IN (1, 2, 3)),
CONSTRAINT fk_pat FOREIGN KEY (Ord_pat_p#a, Ord_pat_p#n)
REFERENCES patient (Pat_p#a, Pat_p#n) ON DELETE CASCADE ON UPDATE CASCADE
);
```

Box 7

10.2.1 The INSERT Statement

The SQL-92 standard provides two types of **INSERT** statements to add new rows of data to a database:

- Single-row INSERT—adds a single row of data to a table.
- Multi-row INSERT—extracts rows of data from another part of the database and adds them to a table.

These statements take the following forms:

```
INSERT INTO <table-name> [(column-name {, column-name})]
VALUES (expression {, expression})
```

```
INSERT INTO <table-name> [(column-name {, column-name})]
<select-statement>⁹
```

---

⁹ The SQL SELECT statement is used to retrieve (i.e., query) data from tables. In its simplest form, SELECT * FROM *table_name*, all columns from the table_name listed are retrieved. The remainder of this chapter contains several examples that make use of this form of the SQL SELECT statement. Chapters 11 and 12 contain an extensive discussion of the SQL SELECT statement.

The values should be listed in the same order in which they are specified in the CREATE TABLE statement. The following three INSERT statements add one row to the PATIENT, MEDICATION, and ORDERS tables. Note that SQL allows us to omit the column names from the INSERT statement when assigning a value to each column in the table.

```
INSERT INTO PATIENT VALUES ('DB','77642','Davis, Bill', 'M', 27, '2007-07-
07', 'B', 108, 'B');10
```

```
1 row created.
```

```
INSERT INTO MEDICATION VALUES ('TAG', 'Tagament', 3.00, 3000, 0);
```

```
1 row created.
```

```
INSERT INTO ORDERS VALUES ('104', 'DB', '77642', 'TAG', 3, 1);
```

```
1 row created.
```

Each of these INSERT statements is successful only because each honors the declarative constraints established in the respective CREATE TABLE statements.

It is also permissible for an INSERT statement to specify explicit column names that correspond to the values provided in the INSERT statement. This is useful if a table has a number of columns but only a few columns are assigned values in a particular new row. Example 1 inserts a row into the PATIENT table that contains only the patient number, patient name, age, and date of admission. In this case, specification of the column names is required.

Example 1.

```
INSERT INTO PATIENT (PATIENT.PAT_P#A, PATIENT.PAT_P#N, PATIENT.PAT_
NAME, PATIENT.PAT_AGE, PATIENT.PAT_ADMIT_DT) VALUES ('GD','72222','Grimes,
David', 44, '2007-07-12');
```

```
1 row created.
```

This INSERT statement was successful because each of the columns not listed in the INSERT statement permits null values.[11] For example, had an attempt been made to insert a patient without specifying a date of admission, the INSERT statement would have failed. Thus every row in the PATIENT table must contain a patient name, age, and date of admission. In addition, since the patient number consisting of the combination of **PATIENT.Pat_p#a** and **PATIENT.Pat_p#n** constitutes the primary key, these two columns must be defined as well.

---

[10] The '2007-07-07' character string represents the date of admission of the patient. For a date data type, SQL-92 uses a default date format where the first four digits represent the year component, the next two digits (1-12) represent the month component, and final two digits (as constrained by the rules of the Gregorian calendar) represent the day of month (see Table 10.1). Other formats for representing dates are covered in Chapter 12.

[11] Although not illustrated here, it is also permissible to omit columns with a DEFAULT value. If a DEFAULT exists for a column not explicitly listed in the INSERT statement, the default value will also be included for this column when the row is inserted.

Each order must involve both an existing patient and existing medication. Observe what happens in Example 2 when an attempt is made to insert an order for an existing patient (David Grimes) but nonexistent medication (KEF).

Example 2.

```
INSERT INTO ORDERS VALUES ('109', 'GD', '72222', 'KEF', 1, 1);
 integrity constraint FK_MED violated—parent key not found
```

Note that FK_MED (see Box 7) is the name of the referential integrity constraint requiring each medication code in the ORDERS table to exist in the MEDICATION table.

The multi-row INSERT statement adds multiple rows of data to a table via the execution of a query. In this form of the INSERT statement, the data values for the new rows appear in a SELECT statement specified as part of the INSERT statement. Suppose, for example, separate patient tables exist for different hospitals within the same hospital system. The INSERT statement in Example 3 inserts all rows in the PATIENT_SUGARLAND table into the PATIENT table. Since the PATIENT_SUGARLAND table has only six columns while the PATIENT table has nine columns, the column names are specified in the INSERT statement.

Example 3.

```
INSERT INTO PATIENT
(PATIENT.PAT_P#A,PATIENT.PAT_P#N,PATIENT.PAT_NAME,PATIENT.PAT_GENDER,PATIENT.PAT_AGE,
PATIENT.PAT_ADMIT_DT)
      SELECT * FROM PATIENT_SUGARLAND;

3 rows created.

SELECT * FROM PATIENT;
Pat_p#a  Pat_p#n  Pat_name      Pat_gender  Pat_age Pat_admit_dt Pat_wing  Pat_room#  Pat_bed
------------------------------------------------ -----------------------------------------------
DB       77642    Davis, Bill   M               27 2007-07-07  B            108  B
GD       72222    Grimes, David                 44 2007-07-12
LH       97384    Lisauckis, Hal M              69 2008-06-06
HJ       99182    Hargrove, Jan  F              21 2008-05-25
RN       31678    Robins, Nancy  F              57 2008-06-01
```

Had there been an interest in inserting only those rows in the PATIENT_SUGARLAND table with a date of admission after June 1, 2008, a WHERE clause referencing the appropriate column name in the PATIENT_SUGARLAND table could have been added to the SELECT statement in the INSERT statement given above.

10.2.2 The DELETE Statement

The **DELETE** statement removes selected rows of data from a single table and takes the following form:

```
DELETE FROM <table-name> [WHERE <search-condition>]
```

Since the WHERE clause in a DELETE statement is optional, a DELETE statement of the form DELETE FROM <table-name> can be used to delete all rows in a table. When used in this manner, while the target table has no rows after execution of the deletion, the table still exists and new rows can still be inserted into the table with the INSERT statement. To erase the table definition from the database, the DROP TABLE statement (described in Section 10.1.1.3) must be used.

Care must be exercised when using the DELETE statement. While rows are deleted from only one table at a time, the deletion may propagate to rows in other tables if actions are specified in the referential integrity constraints. For example, the constraint in the ORDERS table:

```
constraint fk_pat foreign key (Ord_pat_p#a, Ord_pat_p#n)
references patient (Pat_p#a, Pat_p#n) ON DELETE CASCADE ON UPDATE CASCADE
```

results in the deletion of all orders for a particular patient when that patient is deleted from the PATIENT table. The DELETE statement in Example 1 illustrates the propagation of a deletion to another table by deleting the first patient inserted into the PATIENT table, Bill Davis. Observe the content of the PATIENT and ORDERS tables before and after the deletion.

Content of Tables Prior to Deletion

```
SELECT * FROM PATIENT;
Pat_p#a  Pat_p#n  Pat_name            Pat_gender  Pat_age Pat_admit_dt Pat_wing  Pat_room# Pat_bed
-------- -------- ------------------- ---------- ---------- ------------ -------- ---------- ---
DB       77642    Davis, Bill         M              27 2007-07-07   B              108 B
GD       72222    Grimes, David                      44 2007-07-12

SELECT * FROM MEDICATION;
Med_code Med_name                        Med_unitprice Med_qty_onhand Med_qty_onorder
-------- ------------------------------- ------------- -------------- ---------------
TAG      Tagament                                    3           3000               0

SELECT * FROM ORDERS;
Ord_rx       Ord_pat_p#a  Ord_pat_p#n  Ord_med_code Ord_dosage   Ord_freq
------------ ------------ ------------ ------------ ---------- ----------
 104         DB           77642        TAG                   3          1
```

Example 1.

```
DELETE FROM PATIENT WHERE PATIENT.PAT_NAME LIKE '%Davis, Bill%';[12]
```

1 row deleted.

Content of Tables After Deletion

```
SELECT * FROM PATIENT;
Pat_p#a  Pat_p#n  Pat_name          Pat_gender  Pat_age Pat_admit_dt Pat_wing  Pat_room# Pat_bed
-------- -------- ----------------- ---------- ---------- ------------ -------- ---------- ------
GD       72222    Grimes, David                    44 2007-07-12

SELECT * FROM MEDICATION;
Med_code Med_name                        Med_unitprice Med_qty_onhand Med_qty_onorder
-------- ------------------------------- ------------- -------------- ---------------
TAG      Tagament                                    3           3000               0

SELECT * FROM ORDERS;
no rows selected
```

[12] The LIKE operator and the percent character (%) are used for pattern matching. Pattern matching in SQL is discussed in Chapter 11.

Database Creation

On the other hand, observe the effect of the constraint in the ORDERS table:

```
Ord_med_code char(5) constraint fk_med references medication (Med_code)
ON DELETE RESTRICT ON UPDATE RESTRICT
```

when the attempt is made in Example 2 below to delete a medication for which one or more orders exists. Assume that the rows previously deleted from the PATIENT and ORDERS tables have been reinserted prior to the execution of the DELETE statement which attempts to delete the Tagament medication from the MEDICATION table.

Example 2.

```
DELETE FROM MEDICATION WHERE MEDICATION.MED_CODE = 'TAG';
integrity constraint (FK_MED) violated - child record found
```

Note that FK_MED is the name of the referential integrity constraint requiring each medication code in the ORDERS table to exist in the MEDICATION table and restricting the deletion of a medication with one or more orders.

10.2.3 The UPDATE Statement

The **UPDATE** statement modifies the values of one or more columns in selected rows of a single table. The UPDATE statement takes the following form:

```
UPDATE <table-name>
SET column-name = expression
        {, column-name = expression}
[WHERE <search-condition>]
```

The SET clause specifies which columns are to be updated and calculates the new values for the columns.

It is important that an UPDATE statement not violate any existing constraints. See Example 1.

Example 1.

```
UPDATE MEDICATION SET MEDICATION.MED_UNITPRICE = 5.00
WHERE MEDICATION.MED_CODE = 'TAG';
```

The UPDATE statement in this example violates the check constraint CHK_UNITPRICE and thus generates the following message:

```
check constraint (CHK_UNITPRICE) violated
```

Several rows can be modified by a single UPDATE statement. As an example, suppose the MEDICATION table now contains the following six rows:

```
SELECT * FROM MEDICATION;
Med_code Med_name              Med_unitprice    Med_qty_onhand Med_qty_onorder
-------- --------------------- ---------------- -------------- ---------------
TAG      Tagament              3                3000           0
VIB      Vibramycin            1.5              1700           300
KEF      Keflin                2.5              900            410
ASP      Aspirin               .02              3000           0
PCN      Penicillin            .4               2700           0
VAL      Valium                .75              2100           0
```

Observe the effect of the UPDATE statement in Example 2 designed to add 500 to the quantity on hand for each medication with a unit price greater than 0.50.

Example 2.

```
UPDATE MEDICATION
SET MEDICATION.MED_QTY_ONHAND = MEDICATION.MED_QTY_ONHAND + 500
WHERE MEDICATION.MED_UNITPRICE > 0.50;

check constraint (CHK_QTY) violated

SELECT * FROM MEDICATION;
Med_code Med_name                        Med_unitprice Med_qty_onhand Med_qty_onorder
-------- ------------------------------- ------------- -------------- ---------------

TAG      Tagament                                    3           3000               0
VIB      Vibramycin                                1.5           1700             300
KEF      Keflin                                    2.5            900             410
ASP      Aspirin                                   .02           3000               0
PCN      Penicillin                                 .4           2700               0
VAL      Valium                                    .75           2100               0
```

While the CHK_QTY constraint requiring the quantity on hand plus the quantity on order is violated for only one of the four otherwise qualifying rows, none of the four rows is updated. When the WHERE clause excludes Tagament, as shown in Example 3, the UPDATE is successful.

Example 3.

```
UPDATE MEDICATION
SET MEDICATION.MED_QTY_ONHAND = MEDICATION.MED_QTY_ONHAND + 500
WHERE MEDICATION.MED_UNITPRICE > 0.50 AND MEDICATION.MED_CODE <> 'TAG';

3 rows updated.

SELECT * FROM MEDICATION;
Med_code Med_name                        Med_unitprice Med_qty_onhand Med_qty_onorder
-------- --------------------- --------------- ------------- -------------- ---------------

TAG      Tagament                                    3           3000               0
VIB      Vibramycin                                1.5           2200             300
KEF      Keflin                                    2.5           1400             410
ASP      Aspirin                                   .02           3000               0
PCN      Penicillin                                 .4           2700               0
VAL      Valium                                    .75           2600               0
```

10.3 Access Control in the SQL-92 Standard

In a database environment, the DBMS must ensure that only authenticated users[13] are authorized to access the database and that they are allowed to access only the information that has been specifically made available to them. **Privileges** constitute the set of actions a user is permitted to carry out on a table or a view. The privileges defined by the SQL-92 standard are:

- SELECT—Permission to retrieve data from a table or view
- INSERT—Permission to insert rows into a table or view

[13] An authenticated user is one who has demonstrated, by providing a password and perhaps meeting a series of other requirements, the right to access information in a database.

- UPDATE—Permission to modify column values of a row in a table or view
- DELETE—Permission to delete rows of data in a table or view
- REFERENCES—Permission to reference columns of a table named in integrity constraints
- USAGE—Permission to use domains, collation sequences, character sets, and translations[14]

The UPDATE privilege can be restricted to specific columns of a table, making it possible to change these columns but disallowing changes to any other columns. A similar column list associated with the INSERT privilege restricts the grantee to supply values only for the listed columns while inserting a row. Likewise, the REFERENCES privilege can be restricted to specific columns of a table, allowing these columns to be referenced in foreign key constraints and check constraints but disallowing other columns from being referenced.

Many DBMS products offer additional privileges beyond those specified in the SQL-92 standard. For example, Oracle and DB2 support ALTER and INDEX privileges for tables. With the ALTER privilege on a given table, a user can use the ALTER TABLE statement to modify the definition (i.e., structure) of a table. A user with the INDEX privilege on a given table can create an index for that table with the CREATE INDEX statement. DBMS products without the ALTER and INDEX privileges only allow the user who created the table to use the ALTER TABLE and CREATE INDEX statements.

10.3.1 The GRANT and REVOKE Statements

The **GRANT** statement is used to grant privileges on database objects to specific users. The format of the GRANT statement is:

```
GRANT {Privilege-list |ALL PRIVILEGES}
ON Object-name
TO  {User-list | PUBLIC}
[WITH GRANT OPTION]
```

Privilege-list consists of one or more of the following privileges, separated by commas:

```
SELECT
DELETE
INSERT [(Column-name [, ... ])]
UPDATE [(Column-name [, ... ])]
REFERENCES [(Column-name [, ... ])]
USAGE
```

Object-name can be the name of a base table, view, domain, character set, collation sequence, or translation.

The keywords ALL PRIVILEGES and PUBLIC are shortcuts that can be used when granting all privileges to all users. Note that when privileges are given to the PUBLIC they are granted to all present and future authorized users, not just to the users currently known to the DBMS. The WITH GRANT OPTION clause allows for the horizontal propagation of privileges by permitting the user(s) in the designated *User-list* to pass on the privileges they have been given for the named object to other users.

---

[14] Collation sequences, character sets, and translations are not covered in this book and thus the USAGE privilege is not included in the remainder of this discussion. The interested reader may wish to reference Cannan and Otten (1993) for a discussion of these database structures.

The **REVOKE** statement is used to take away all or some of the privileges previously granted to another user or users. The format of the REVOKE statement is:

```
REVOKE [GRANT OPTION FOR] {Privilege-list | ALL PRIVILEGES}
ON Object-name
FROM [{User-list | PUBLIC} RESTRICT | CASCADE]
```

The keyword ALL PRIVILEGES refers to all the privileges granted to a user by the user revoking the privileges. The optional GRANT OPTION FOR clause allows privileges passed on via the WITH GRANT OPTION of the GRANT statement to be revoked separately from the privileges themselves. CASCADE means that if USER_A, whose user name appears in the list *User-list*, has granted those privileges to USER_B, the privileges granted to USER_B are also revoked. If USER_B has granted those privileges to USER_C, those privileges are revoked as well, and so on. The option RESTRICT indicates that if any such dependent privileges exist, the REVOKE statement will fail.

10.3.2 Some Examples of Granting and Revoking Privileges[15]

Suppose that the six users (USER_A, USER_B, USER_C, USER_D, USER_E, and USER_F) exist and that USER_A owns the PATIENT, MEDICATION, and ORDERS tables. Let's experiment by observing what happens as a result of USER_A's granting and subsequently revoking privileges on these three tables.

Example 1. USER_A grants an assortment of privileges on the ORDERS, PATIENT, and MEDICATION tables to USER_B, USER_C, and USER_D. Note the information about these privileges recorded in the System Catalog that contains a record of USER_A's grants.

```
GRANT SELECT, INSERT, DELETE, UPDATE
ON ORDERS
TO USER_B;

Grant succeeded.

GRANT SELECT, INSERT
ON PATIENT
TO USER_C;

Grant succeeded.

GRANT INSERT, DELETE
ON MEDICATION
TO USER_D;

Grant succeeded.
```

USER_A's table privileges granted as recorded in the System Catalog

GRANTEE	OWNER	TABLE_NAME	GRANTOR	PRIVILEGE	GRANTABLE
USER_B	USER_A	ORDERS	USER_A	DELETE	NO
USER_B	USER_A	ORDERS	USER_A	INSERT	NO
USER_B	USER_A	ORDERS	USER_A	SELECT	NO
USER_B	USER_A	ORDERS	USER_A	UPDATE	NO

---

[15] The examples in this section adhere to the syntax in the SQL-92 standard and may require some slight modifications to run on certain DBMS platforms. In addition, some of the error messages are abridged versions of those that might be generated by a particular DBMS product.

Database Creation

USER_C	USER_A	PATIENT	USER_A	INSERT	NO
USER_C	USER_A	PATIENT	USER_A	SELECT	NO
USER_D	USER_A	MEDICATION	USER_A	DELETE	NO
USER_D	USER_A	MEDICATION	USER_A	INSERT	NO

Example 2. USER_A grants all privileges on the MEDICATION table to the PUBLIC. This means that USER_D has received the DELETE and INSERT privileges on the MEDICA-TION table twice (once explicitly and once implicitly as a member of the PUBLIC). Thus should USER_A choose to revoke all privileges on the MEDICATION table from the PUB-LIC, USER_D would still maintain the DELETE and INSERT privileges.

```
GRANT ALL PRIVILEGES
ON MEDICATION
TO PUBLIC;

Grant succeeded.
```

USER_A's table privileges granted as recorded in the System Catalog (PUBLIC's privileges on the MEDICATION table have been added)

GRANTEE	OWNER	TABLE_NAME	GRANTOR	PRIVILEGE	GRANTABLE
PUBLIC	USER_A	MEDICATION	USER_A	DELETE	NO
PUBLIC	USER_A	MEDICATION	USER_A	INSERT	NO
PUBLIC	USER_A	MEDICATION	USER_A	SELECT	NO
PUBLIC	USER_A	MEDICATION	USER_A	UPDATE	NO
PUBLIC	USER_A	MEDICATION	USER_A	REFERENCES	NO
USER_B	USER_A	ORDERS	USER_A	DELETE	NO
USER_B	USER_A	ORDERS	USER_A	INSERT	NO
USER_B	USER_A	ORDERS	USER_A	SELECT	NO
USER_B	USER_A	ORDERS	USER_A	UPDATE	NO
USER_C	USER_A	PATIENT	USER_A	INSERT	NO
USER_C	USER_A	PATIENT	USER_A	SELECT	NO
USER_D	USER_A	MEDICATION	USER_A	DELETE	NO
USER_D	USER_A	MEDICATION	USER_A	INSERT	NO

Example 3. USER_A has not granted USER_B any privileges on the PATIENT table. The following GRANT statement allows USER_B to retrieve (i.e., select) rows from USER_A's PATIENT table and also grant the SELECT privilege to other users.

```
GRANT SELECT
ON PATIENT
TO USER_B
WITH GRANT OPTION;

Grant succeeded.
```

Example 4. At this point assume USER_B is connected to the database and attempts to grant the SELECT privilege received from USER_A to USER_D. Note how the first attempt fails because USER_B failed to qualify the table name (PATIENT) with the name of its owner (USER_A). After the successful grant, the system catalog now records the fact that USER_B has granted the SELECT privilege to USER_D.

```
GRANT SELECT
ON PATIENT
TO USER_D;

ERROR at line 2:  table PATIENT does not exist
```

```
GRANT SELECT
ON USER_A.PATIENT
TO USER_D;

Grant succeeded.
```

USER_B's table privileges granted as recorded in the System Catalog

GRANTEE	OWNER	TABLE_NAME	GRANTOR	PRIVILEGE	GRANTABLE
USER_D	USER_A	PATIENT	USER_B	SELECT	NO

As a result of the GRANT statement in Example 3, the System Catalog now records USER_A's granting of the SELECT privilege to USER_B on the PATIENT table as grantable. In addition, it also records USER_B's granting of the SELECT privilege on the PATIENT table to USER_D.

USER_A's table privileges granted as recorded in the System Catalog

GRANTEE	OWNER	TABLE_NAME	GRANTOR	PRIVILEGE	GRANTABLE
PUBLIC	USER_A	MEDICATION	USER_A	DELETE	NO
PUBLIC	USER_A	MEDICATION	USER_A	INSERT	NO
PUBLIC	USER_A	MEDICATION	USER_A	SELECT	NO
PUBLIC	USER_A	MEDICATION	USER_A	UPDATE	NO
PUBLIC	USER_A	MEDICATION	USER_A	REFERENCES	NO
USER_B	USER_A	PATIENT	USER_A	SELECT	YES
USER_B	USER_A	ORDERS	USER_A	DELETE	NO
USER_B	USER_A	ORDERS	USER_A	INSERT	NO
USER_B	USER_A	ORDERS	USER_A	SELECT	NO
USER_B	USER_A	ORDERS	USER_A	UPDATE	NO
USER_C	USER_A	PATIENT	USER_A	INSERT	NO
USER_C	USER_A	PATIENT	USER_A	SELECT	NO
USER_D	USER_A	PATIENT	USER_B	SELECT	NO
USER_D	USER_A	MEDICATION	USER_A	DELETE	NO
USER_D	USER_A	MEDICATION	USER_A	INSERT	NO

Another table in the System Catalog records the five table privileges received by USER_B from USER_A.

USER_B's table privileges received as recorded in the System Catalog

OWNER	TABLE_NAME	GRANTOR	PRIVILEGE	GRANTABLE
USER_A	PATIENT	USER_A	SELECT	YES
USER_A	ORDERS	USER_A	DELETE	NO
USER_A	ORDERS	USER_A	INSERT	NO
USER_A	ORDERS	USER_A	SELECT	NO
USER_A	ORDERS	USER_A	UPDATE	NO

Example 5. Privileges can be granted on specific columns of a table as well as on all columns. Here USER_A grants USER_E the UPDATE privilege on three columns of the PATIENT table. Notice that, as shown below, a record of column privileges granted by USER_A is recorded in a different table in the System Catalog.

```
GRANT UPDATE (PAT_WING, PAT_ROOM#, PAT_BED)
ON PATIENT
TO USER_E;
```

Grant succeeded.

USER_A's table privileges granted as recorded in the System Catalog

GRANTEE	OWNER	TABLE_NAME	GRANTOR	PRIVILEGE	GRANTABLE
PUBLIC	USER_A	MEDICATION	USER_A	DELETE	NO
PUBLIC	USER_A	MEDICATION	USER_A	INSERT	NO
PUBLIC	USER_A	MEDICATION	USER_A	SELECT	NO
PUBLIC	USER_A	MEDICATION	USER_A	UPDATE	NO
PUBLIC	USER_A	MEDICATION	USER_A	REFERENCES	NO
USER_B	USER_A	PATIENT	USER_A	SELECT	YES
USER_B	USER_A	ORDERS	USER_A	DELETE	NO
USER_B	USER_A	ORDERS	USER_A	INSERT	NO
USER_B	USER_A	ORDERS	USER_A	SELECT	NO
USER_B	USER_A	ORDERS	USER_A	UPDATE	NO
USER_C	USER_A	PATIENT	USER_A	INSERT	NO
USER_C	USER_A	PATIENT	USER_A	SELECT	NO
USER_D	USER_A	PATIENT	USER_B	SELECT	NO
USER_D	USER_A	MEDICATION	USER_A	DELETE	NO
USER_D	USER_A	MEDICATION	USER_A	INSERT	NO

Observe how the record of the column privileges granted by USER_A to USER_E is recorded in a separate part of the System Catalog.

USER_A's column privileges granted as recorded in the System Catalog

GRANTEE	OWNER	TABLE_NAME	COLUMN_NAME	GRANTOR	PRIVILEGE	GRANTABLE
USER_E	USER_A	PATIENT	PAT_WING	USER_A	UPDATE	NO
USER_E	USER_A	PATIENT	PAT_ROOM#	USER_A	UPDATE	NO
USER_E	USER_A	PATIENT	PAT_BED	USER_A	UPDATE	NO

When connected as USER_E, we can observe how the System Catalog also records the column privileges on the PATIENT table received from (i.e., granted by) USER_A. Note that at this point USER_E has not received any explicit table privileges but only the three column privileges on the PATIENT table. Of course, USER_E does have all privileges on the MEDICATION table received by virtue of being a member of the PUBLIC.

USER_E's column privileges received as recorded in the System Catalog

OWNER	TABLE_NAME	COLUMN_NAME	GRANTOR	PRIVILEGE	GRANTABLE
USER_A	PATIENT	PAT_WING	USER_A	UPDATE	NO
USER_A	PATIENT	PAT_ROOM#	USER_A	UPDATE	NO
USER_A	PATIENT	PAT_BED	USER_A	UPDATE	NO

Example 6. Here USER_A is granting the UPDATE privilege on all columns of the PATIENT table to USER_F. Observe how information about this grant is recorded in the System Catalog table that contains a record of table privileges granted (i.e., made). Likewise, information about this grant is recorded when USER_F looks at the System Catalog table to view the table privileges received.

```
GRANT UPDATE
ON PATIENT
TO USER_F;

Grant succeeded.
```

USER_A's table privileges granted as recorded in the System Catalog (note addition of USER_F's UPDATE privilege)

GRANTEE	OWNER	TABLE_NAME	GRANTOR	PRIVILEGE	GRANTABLE
PUBLIC	USER_A	MEDICATION	USER_A	DELETE	NO
PUBLIC	USER_A	MEDICATION	USER_A	INSERT	NO
PUBLIC	USER_A	MEDICATION	USER_A	SELECT	NO
PUBLIC	USER_A	MEDICATION	USER_A	UPDATE	NO
PUBLIC	USER_A	MEDICATION	USER_A	REFERENCES	NO
USER_B	USER_A	PATIENT	USER_A	SELECT	YES
USER_B	USER_A	ORDERS	USER_A	DELETE	NO
USER_B	USER_A	ORDERS	USER_A	INSERT	NO
USER_B	USER_A	ORDERS	USER_A	SELECT	NO
USER_B	USER_A	ORDERS	USER_A	UPDATE	NO
USER_C	USER_A	PATIENT	USER_A	INSERT	NO
USER_C	USER_A	PATIENT	USER_A	SELECT	NO
USER_D	USER_A	PATIENT	USER_B	SELECT	NO
USER_D	USER_A	MEDICATION	USER_A	DELETE	NO
USER_D	USER_A	MEDICATION	USER_A	INSERT	NO
USER_F	**USER_A**	**PATIENT**	**USER_A**	**UPDATE**	**NO**

USER_F's table privileges received as recorded in the System Catalog

OWNER	TABLE_NAME	GRANTOR	PRIVILEGE	GRANTABLE
USER_A	PATIENT	USER_A	UPDATE	NO

Figure 10.5 summarizes all privileges granted at this point by USER_A and USER_B to other users as well as to the PUBLIC.

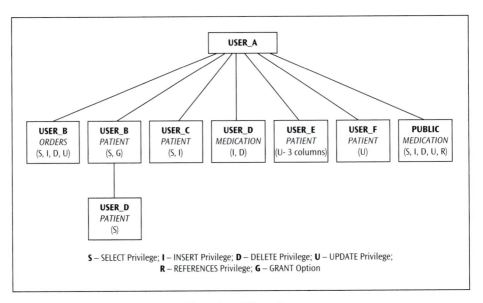

Figure 10.5 Privileges granted by User_A and User_B

Example 7. As shown below, the System Catalog records the fact that USER_B along with the entire public has been granted all privileges by USER_A on the MEDICATION table. Here USER_B is creating the table MY_MEDICATION which references the **MED_CODE** column in USER_A's MEDICATION table in a referential integrity constraint. Note that it is important that the references constraint in the MY_MEDICATION table specify that the MEDICATION table is owned by USER_A.

```
USER_B's table privileges received as a result of being a member of the PUBLIC
as recorded in the System Catalog
GRANTEE     OWNER      TABLE_NAME GRANTOR     PRIVILEGE          GRANTABLE
-------------------------------------------------------------------------
PUBLIC      USER_A     MEDICATION USER_A      DELETE             NO
PUBLIC      USER_A     MEDICATION USER_A      INSERT             NO
PUBLIC      USER_A     MEDICATION USER_A      SELECT             NO
PUBLIC      USER_A     MEDICATION USER_A      UPDATE             NO
PUBLIC      USER_A     MEDICATION USER_A      REFERENCES         NO


CREATE TABLE MY_MEDICATION
(My_name char(20),
My_med_code CHAR(5) REFERENCES USER_A.MEDICATION(MED_CODE));

Table created.
```

Observe what happens when USER_B attempts to insert a row into the MY_MEDICATION table that honors and fails to honor the referential integrity constraint.

```
INSERT INTO MY_MEDICATION VALUES ('Thomas Jones', 'TAG');

1 row created.

INSERT INTO MY_MEDICATION VALUES ('Sally Fields', 'XXX');
integrity constraint (USER_B.SYS_C004527) violated - parent key not found
```

Example 8. Observe that the REVOKE statement issued here by USER_A revoking all privileges on the PATIENT table from USER_C is successful because no dependent privileges have been based on this grant.

```
REVOKE ALL PRIVILEGES
ON PATIENT
FROM USER_C;

Revoke succeeded.
```

```
USER_A's table privileges granted as recorded in the System Catalog
(note the removal of USER_C's two privileges)
GRANTEE      TABLE_NAME  GRANTOR    PRIVILEGE            GRANTABLE
----------   ----------  ---------- -------------------- ----------
PUBLIC       MEDICATION  USER_A     DELETE               NO
PUBLIC       MEDICATION  USER_A     INSERT               NO
PUBLIC       MEDICATION  USER_A     SELECT               NO
PUBLIC       MEDICATION  USER_A     UPDATE               NO
PUBLIC       MEDICATION  USER_A     REFERENCES           NO
USER_D       MEDICATION  USER_A     DELETE               NO
USER_D       MEDICATION  USER_A     INSERT               NO
USER_B       ORDERS      USER_A     DELETE               NO
USER_B       ORDERS      USER_A     INSERT               NO
USER_B       ORDERS      USER_A     SELECT               NO
USER_B       ORDERS      USER_A     UPDATE               NO
USER_B       PATIENT     USER_A     SELECT               YES
USER_F       PATIENT     USER_A     UPDATE               NO
USER_D       PATIENT     USER_B     SELECT               NO
```

Example 9. USER_A now wishes to revoke all privileges on the MEDICATION table previously given to the PUBLIC. Since USER_B has used the REFERENCES privilege received as a member of the PUBLIC in the creation of the MY_MEDICATION table, the default RESTRICT option causes the REVOKE statement issued by USER_A to fail.

```
REVOKE ALL PRIVILEGES
ON MEDICATION
FROM PUBLIC;

ERROR:  CASCADE option must be specified to perform this revoke

REVOKE ALL PRIVILEGES
ON MEDICATION
FROM PUBLIC
CASCADE;

Revoke succeeded.
```

Database Creation

USER_A's table privileges granted as recorded in the System Catalog

GRANTEE	TABLE_NAME	GRANTOR	PRIVILEGE	GRANTABLE
USER_D	MEDICATION	USER_A	DELETE	NO
USER_D	MEDICATION	USER_A	INSERT	NO
USER_B	ORDERS	USER_A	DELETE	NO
USER_B	ORDERS	USER_A	INSERT	NO
USER_B	ORDERS	USER_A	SELECT	NO
USER_B	ORDERS	USER_A	UPDATE	NO
USER_B	PATIENT	USER_A	SELECT	YES
USER_F	PATIENT	USER_A	UPDATE	NO
USER_D	PATIENT	USER_B	SELECT	NO

Observe that none of the PUBLIC's privileges on the MEDICATION table exists, including the REFERENCES constraint. Thus, as shown below, it is possible for USER_B to insert a row into the MY_MEDICATION table that refers to a non-existent **My_med_code** (i.e., XXX).

```
SELECT * FROM MY_MEDICATION;
```

My_name	My_med_code
Thomas Jones	TAG

```
INSERT INTO MY_MEDICATION VALUES ('Chet Gladchuk', 'TAG');

1 row created.

INSERT INTO MY_MEDICATION VALUES ('Sally Fields', 'XXX');

1 row created.
```

Example 10. At this point USER_D is still able to make use of its SELECT privilege to query the PATIENT table owned by USER_A (see the query given immediately below). Observe what happens when USER_A revokes the SELECT privilege on the PATIENT table from USER_B who had subsequently granted this privilege to USER_D. USER_D is now unable to query the PATIENT table owned by USER_A.

```
SELECT * FROM USER_A.PATIENT;
```

Pat_p#a	Pat_p#n	Pat_name	Pat_gender	Pat_age	Pat_admit_dt	Pat_wing	Pat_room#	Pat_bed
DB	77642	Davis, Bill	M	27	2007-07-07	B	108	B

USER_D's table privileges received as recorded in the System Catalog

OWNER	TABLE_NAME	GRANTOR	PRIVILEGE	GRANTABLE
USER_A	PATIENT	USER_B	SELECT	NO
USER_A	MEDICATION	USER_A	DELETE	NO
USER_A	MEDICATION	USER_A	INSERT	NO

USER_A revokes all privileges on the PATIENT table from USER_B.
```
REVOKE ALL PRIVILEGES
ON PATIENT
FROM USER_B;
```

Revoke succeeded.

USER_A's table privileges granted as recorded in the System Catalog (note USER_B's privileges on the PATIENT table no longer exist; although not recorded in this portion of the System Catalog, the SELECT privilege on the PATIENT table granted by USER_B to USER_D no longer exists as well).

GRANTEE	TABLE_NAME	GRANTOR	PRIVILEGE	GRANTABLE
USER_D	MEDICATION	USER_A	DELETE	NO
USER_D	MEDICATION	USER_A	INSERT	NO
USER_B	ORDERS	USER_A	DELETE	NO
USER_B	ORDERS	USER_A	INSERT	NO
USER_B	ORDERS	USER_A	SELECT	NO
USER_B	ORDERS	USER_A	UPDATE	NO
USER_F	PATIENT	USER_A	UPDATE	NO

USER_B's table privileges received as recorded in the System Catalog (compare with comparable System Catalog information in Example 4)

OWNER	TABLE_NAME	GRANTOR	PRIVILEGE	GRANTABLE
USER_A	ORDERS	USER_A	DELETE	NO
USER_A	ORDERS	USER_A	INSERT	NO
USER_A	ORDERS	USER_A	SELECT	NO
USER_A	ORDERS	USER_A	UPDATE	NO

USER_D's table privileges received as recorded in the System Catalog (compare with comparable System Catalog immediately prior to REVOKE statement immediately above)

OWNER	TABLE_NAME	GRANTOR	PRIVILEGE	GRANTABLE
USER_A	MEDICATION	USER_A	DELETE	NO
USER_A	MEDICATION	USER_A	INSERT	NO

USER_D is no longer able to query the PATIENT table owned by USER_A.
```
SELECT * FROM USER_A.PATIENT;
```

ERROR: table or view does not exist

Chapter Summary

The SQL-92 standard contains a data definition language (SQL/DDL) that allows database objects to be created. This chapter focuses primarily on the CREATE TABLE statement as the vehicle for defining required data by specifying the proper data type and using, where appropriate, the NOT NULL clause, domain constraints (defined by either the CHECK clause or by explicitly defining domains using the CREATE DOMAIN statement), entity integrity (via the PRIMARY KEY clause), referential integrity (via the FOREIGN KEY clause along with, as appropriate, update and deletion rules), and row-level constraints (defined by the CHECK and UNIQUE clauses).

The discussion highlights the efficacy of using a logical schema based on the information-preserving grammar in the SQL-92 SQL/DDL to create tables that fully capture all information contained in the Fine-granular Design-Specific ER model.

The SQL-92 DDL includes an ALTER TABLE statement as well as a DROP TABLE statement. The ALTER TABLE statement is used to add or drop a column, add or drop a constraint, or to modify a column definition. The DROP TABLE statement is used to remove a table (structure and content) from the database. Both the ALTER TABLE statement, if it involves some kind of drop action, and the DROP TABLE statement must specify a drop behavior associated with the action (i.e., dropping a column or constraint in the alteration of a table or the dropping of an entire table). The options available in both cases are RESTRICT or CASCADE. RESTRICT implies that the action is rejected if any other object referencing the base table that is the subject of the ALTER TABLE statement or DROP TABLE statement exists. On the other hand, the CASCADE option deletes the object (column, constraint, or base table) along with all references to the object.

The SQL-92 standard also includes a CREATE DOMAIN statement as a means to factor out column data type specifications that can be shared by any number of column definitions in any number of base tables. Use of this statement requires that the domain be given a name, a data type, an optional default definition, and an optional constraint definition. In an existing domain, the default definition as well as constraint definitions can be altered using the ALTER DOMAIN statement. A domain can be removed from the database by using the DROP DOMAIN statement. It is important to note that dropping a domain effectively maps the domain default definition and other domain constraint definitions to base tables which contain columns defined on the domain in question before actually dropping the domain.

An SQL-schema represents a named collection of schema elements (objects) usually under the ownership of an authorized user. An SQL-schema is created by a CREATE SCHEMA statement. The creation of an SQL-schema can also include the creation of base tables, domain definitions, constraint definitions, assertions, views, and triggers.[16] A DROP SCHEMA statement can be used to drop an SQL-schema. Specification of a drop behavior (RESTRICT or CASCADE) is mandatory. With the RESTRICT option, the drop operation fails unless the SQL-schema is empty. If CASCADE is specified, all schema elements of the SQL-schema are purged along with the SQL-schema.

---

[16] Assertions, views and triggers are discussed in Chapter 12.

In SQL, three statements can be used to modify the database: INSERT, DELETE, and UPDATE. Two types of INSERT statements exist. One allows for the addition of a single row to a table while the other allows for multiple rows to be added to a table. Only one type of DELETE and UPDATE statement exists. However, with each statement it is possible to delete (in the case of the DELETE statement) or update (in the case of the UPDATE statement) one or more rows in the table.

In a database environment, it is important that the DBMS ensures that only authorized users are allowed to access the database, and that they are allowed to access only information that has been specifically made available to them. Privileges constitute the various actions that a user is permitted to carry out on a table or a view. The privileges discussed in this chapter that are defined by the SQL-92 standard are: SELECT, INSERT, UPDATE, DELETE, and REFERENCES. The SELECT and DELETE privileges apply only to a table, whereas the INSERT and UPDATE privileges can apply to both a base table or a column or group of columns within a table. Although the REFERENCES privilege can be used in the context of a table, it is applied most often to a column or group of columns in a base table. Privileges are granted to and revoked from other users by the GRANT and REVOKE statements.

Exercises

1. Discuss the differences between a relation and a table.
2. What are the minimum elements that must be included in the CREATE TABLE statement in defining the structure of a table?
3. What is the difference between a column-level constraint and a row-level constraint?
4. Describe the SQL clauses used in the definition of:
   a. a primary key constraint
   b. an alternate key constraint
   c. a foreign key constraint
   d. a check constraint
5. With which of the four types of constraints in Exercise 4 is a requirement that an attribute not contain a null value associated?
6. With which type of constraint is the specification of a deletion rule associated?
7. Describe the types of modifications that can be made to a table using the ALTER TABLE statement.
8. Suppose that a CREATE DOMAIN statement of the following form had been used to create a domain SSN_TYPE:

   ```
   CREATE DOMAIN ssn_type char(9);
   ```
   How would this statement impact the definition of attributes such as **Ssn** and **Essn** for the tables EMPLOYEE and DEPENDENT in the relational schema for Bearcat Incorporated in Chapter 6?
9. Instructors at the University of Houston are allowed to award grades of A, A–, B+, B, B–, C+, C, C–, D+, D, D–, F+, F, W, I, S, and U. Use a CREATE DOMAIN statement to specify these values as constituting the domain grades.

10. Use an ALTER DOMAIN statement to modify the domain grades to include the recently adopted Q grade.

11. Consider a relational schema with three relation schemas: DRIVER (Dr_license_no, Dr_name, Dr_city, Dr_state), TICKET_TYPE (Ttp_offense, Ttp_fine) and TICKET (Tic_ticket_no, Tic_ticket_date, Tic_dr_license_no, Tic_ttp_offense). Assume that Dr_license_no is a character data type of size 7, Dr_name, Dr_city and Ttp_offense are character data types of size 20, Dr_state is a character data type of size 2, Tic_ticket_no is a character data type of size 5, and Ttp_fine is an integer between 15 and 150. Each tuple in the TICKET relation corresponds to a ticket received by a driver for having been issued a specific type of ticket on a given date for committing a given offense. Use SQL/DDL to create a schema of these relations, including the appropriate primary key, integrity (e.g., not null), and foreign key constraints. In addition, include the definition of domains for each of the distinct attribute types.

12. The series of tasks in this exercise is based on the following ER diagram and its associated logical schema.

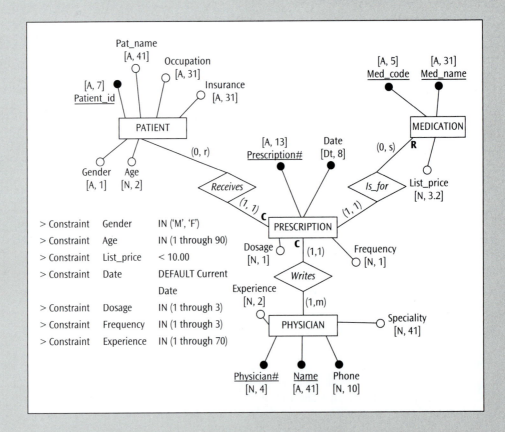

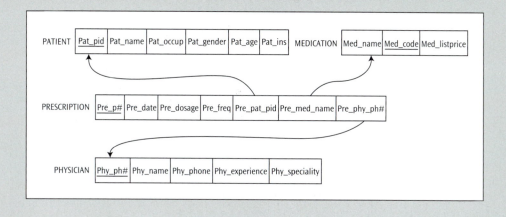

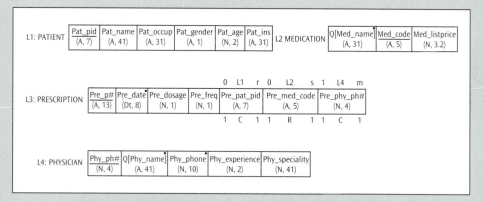

a. Write appropriate CREATE TABLE statements for the logical schema. Be sure to define all appropriate constraints.

b. Write an ALTER TABLE statement to add to the MEDICATION table the attribute unit cost that represents the per unit cost of the medication. The unit cost of a medication can range from $0.50 to $7.50.

c. Write an ALTER TABLE statement that imposes the business rule that the list price of a medication must be at least 20 percent higher than its unit cost.

d. Write an ALTER TABLE statement to drop the occupation attribute from the PATIENT table.

Database Creation

13. Suppose that the two users (USER_1 and USER_2) exist and that USER_1 owns the DRIVER, TICKET_TYPE, and TICKET tables described in Exercise 11. Write the appropriate SQL Access Control statements to implement the following tasks.

   a. As USER_1, grant USER_2 all privileges on the TICKET table. Would USER_2 be allowed to insert a row into the TICKET table that violates either of its referential integrity constraints? Would USER_2 be allowed to insert a row into the TICKET table as long as it does not violate either of its referential integrity constraints?

   b. As USER_1, revoke all privileges on the TICKET table from USER_2.

   c. As USER_1, grant USER_2 all privileges on the TICKET, TICKET_TYPE, and DRIVER tables. Would USER_2 be allowed to insert a row into the TICKET table that violates either of its referential integrity constraints? Would USER_2 be allowed to insert a row into the TICKET table as long as it does not violate either of its referential integrity constraints?

   d. As USER_1, revoke all privileges on the TICKET, TICKET_TYPE, and DRIVER tables from USER_2.

   e. As USER_1, grant USER_2 all privileges on the TICKET table except the references privilege. Would USER_2 be allowed to insert a row into the TICKET table that violates either of its referential integrity constraints? Would USER_2 be allowed to insert a row into the TICKET table as long as it does not violate either of its referential integrity constraints?

   f. As USER_1, revoke all privileges on the TICKET table from USER_2.

   g. As USER_1, grant USER_2 all privileges on the TICKET, TICKET_TYPE, and DRIVER tables except for the references privilege on the TICKET_TYPE and DRIVER tables. Would USER_2 be allowed to insert a row into the TICKET table that violates either of its referential integrity constraints? Would USER_2 be allowed to insert a row into the TICKET table as long as it does not violate either of its referential integrity constraints?

   h. As USER_1, revoke all privileges on the TICKET, TICKET_TYPE, and DRIVER tables from USER_2.

   i. As USER_1, grant USER_2 all privileges on the TICKET table except the references privilege. Would USER_2 be able to create a table with the same name and structure as the TICKET_TYPE table that belongs to USER_1?

   j. As USER_1, revoke all privileges on the TICKET table from USER_2.

   k. As USER_1, grant USER_2 all privileges on the TICKET table. Would USER_2 be able to create a table with the same name and structure as the TICKET_TYPE table that belongs to USER_1?

   l. As USER_1, revoke all privileges on the TICKET table from USER_2.

   m. As USER_1, grant USER_2 all privileges on the TICKET_TYPE and DRIVER tables except the references privileges. Would USER_2 be able to create a table with the same name and structure as the TICKET table that belongs to USER_1?

   n. As USER_1, revoke all privileges on the TICKET_TYPE and DRIVER tables from USER_2.

o.   As USER_1, grant USER_2 all privileges on the TICKET_TYPE and DRIVER tables. Would USER_2 be able to create a table with the same name and structure as the TICKET table that belongs to USER_1?

p.   As USER_1, revoke all privileges on the TICKET_TYPE and DRIVER tables from USER_2.

q.   After analyzing the results of tasks a through p, summarize the effect of the references privilege on the insertion of rows in tables and the creation of tables.

Selected Bibliography

Date, C. J. (2004) *An Introduction to Database Systems*, Eighth Edition, Addison-Wesley.

Date, C. J. and Darwen, H. (1997) *A Guide to the SQL Standard*, Fourth Edition, Addison-Wesley.

Elmasri, R. and Navathe, S. B. (2003) *Fundamentals of Database Systems*, Fourth Edition, Addison-Wesley.

Groff, J. R. and Weinberg, P. N. (2002) *SQL: The Complete Reference*, McGraw-Hill/Osborne, Second Edition.

Gulutzan, P. and Pelzer, T. (1999) *SQL-99 Complete, Really*, R&D Books.

Kifer, M.; Bernstein, A.; and Lewis, P. M. (2005) *Databases and Transactions Processing: An Application-Oriented Approach*, Second Edition, Addison-Wesley.

Sunderraman, R. (2003) *Oracle9i Programming: A Primer*, Addison-Wesley.

11

Data Manipulation: Relational Algebra and SQL

From the SQL/DDL for database creation in the previous chapter, we now move on to learn about data retrieval. Relational algebra, a mathematical expression of data retrieval methods prescribed by E. F. Codd, is introduced first as a means to specify the logic for data retrieval from a relational database. A query expressed in relational algebra involves a sequence of operations which, when executed in the order specified, produces the desired results. SQL is the most common way that relational algebra is implemented for data retrieval operations in a relational database.

Chapter 11 is divided into two major sections. Section 11.1 discusses relational algebra with examples. Section 11.2 presents the syntax for SQL and enumerates various ways in which SQL can be used for data manipulation, with copious examples. Included as part of the discussion are illustrations of how SQL treats missing data (or what are called null values), and a discussion of different types of subqueries.

11.1 Relational Algebra[1]

The relational data model includes a group of basic data manipulation operations. As a result of its theoretical foundation in set theory, the relational data model's operations include Union, Intersection, and Difference. Five other relational operators also exist: Select, Project, Cartesian Product, Join, and Divide. Collectively these eight operators comprise relational algebra. This section discusses each of these eight operators and gives examples of their use in the formulation of queries. The examples are based on the Madeira College registration system introduced in Chapter 10. The Fine-granular Design-Specific ER diagram for Madeira College appears in Figure 10.2, and its information-reducing and information-preserving logical schema are shown in Figures 10.3 and 10.4, respectively. Figure 11.1 contains representative data for the DEPARTMENT, PROFESSOR, COURSE, and SECTION relations used in the relational algebra examples in Section 11.1 along with representative data for other Madeira College relations used in conjunction with the SQL examples that begin in Section 11.2.

The discussion of these relational algebra operators begins with the two that operate on a single relation (**unary operators**) followed by those that operate on two relations (**binary operators**). Figure 11.2 shows the fundamental relational algebra operators, along with their symbolic representation. In Figure 11.2, the fundamental operators are indicated by a double asterisk (**). The rest can be defined from the fundamental operators. Nonetheless, these operators are usually included in relational algebra as a matter of convenience.

---

[1] The discussion of and the notation used for various relational algebra operations in this section is based on R. Elmasri and S.B. Navathe (2003), *Fundamental of Database Systems*, Addison-Wesley. The reader is encouraged to refer to Elmasri and Navathe (2003) and C. J. Date (1995), *An Introduction to Database Systems*, Addison-Wesley for a more in-depth discussion of relational algebra.

DEPARTMENT Relation

Dpt_name	Dpt_dcode	Dpt_college	Dpt_phone	Dpt_budget	Dpt_location	Dpt_hodid
Economics	1	Arts and Sciences	5235567654	433545	123 McMicken	FM49276
QA/QM	3	Business	5235566656	134556	333 Lindner	BA54325
Economics	4	Education	5235569978	400000	336 Dyer	CM65436
Mathematics	6	Engineering	5235564379	433567	728 Old Chem	RR79345
IS	7	Business	5235567489	400000	333 Lindner	CC49234
Philosophy	9	Arts and Sciences	5235565546	333333	272 McMicken	CM87659

COURSE Relation

Co_name	Co_course#	Co_credit	Co_college	Co_hrs	Co_dpt_dcode
Intro to Economics	15ECON112	U	Arts and Sciences	3	1
Operations Research	22QA375	U	Business	2	3
Intro to Economics	18ECON123	U	Education	4	4
Supply Chain Analysis	22QA411	U	Business	3	3
Principles of IS	22IS270	G	Business	3	7
Programming in C++	20ECES212	G	Engineering	3	6
Optimization	22QA888	G	Business	3	3
Financial Accounting	18ACCT801	G	Education	3	4
Database Concepts	22IS330	U	Business	4	7
Database Principles	22IS832	G	Business	3	7
Systems Analysis	22IS430	G	Business	3	7

STUDENT Relation

St_sid	St_name	St_address	St_status	St_birthdate
BE76598	Elijah Baley	2920 Scioto Street	Part time	
OD76578	Daniel Olive	338 Bishop Street	Full time	1982-05-12
SW56547	Wanda Seldon	3138 Probasco	Full time	1970-03-03
BG66765	Gladis Bale	356 Vine Street	Full time	1977-10-23
GS76775	Shweta Gupta	356 Probasco	Full time	1979-05-21
HT67657	Troy Hudson		Part time	
FR45545	Rick Fox	314 Clifton	Full time	1983-10-09
FV67733	Vanessa Fox	314 Clifton	Full time	1983-10-20
HJ45633	Jenna Hopp	2930 Scioto Street	Full time	1970-03-03
SD23556	David Sane	245 University Avenue	Part time	1984-07-14
DT87656	Tim Duncan		Part time	1975-05-21
KJ56656	Joumana Kidd	2920 Scioto Street	Part time	
AJ76998	Jenny Aniston	88 MLK	Full time	
KP78924	Poppy Kramer	437 Love Lane	Full time	1980-11-11
KS39874	Sweety Kramer	748 Hope Avenue	Full time	1980-11-11
JD35477	Diana Jackson	2920 Scioto Street	Part time	1976-02-20

GRAD_STUDENT Relation

Gs_st_sid	Gs_thesis	Gs_ugmajor
BE76598	Y	Marketing
SW56547	Y	Finance
BG66765	N	Archeology
GS76775	N	Archeology
HJ45633	Y	History
DT87656	N	Physics
KJ56656	Y	History
AJ76998	Y	Child Care
JD35477	N	Mathematics

SECTION Relation

Se_section#	Se_qtr	Se_year	Se_time	Se_maxst	Se_room	Se_co_course#	Se_pr_profid
101	A	2007	T1015	25		22QA375	HT54347
901	A	2006	W1800	35	Rhodes 611	22IS270	SK85977
902	A	2006	H1700	25	Lindner 108	22IS270	SK85977
101	S	2006	T1045	29	Lindner 110	22IS330	SK85977
102	S	2006	H1045	29	Lindner 110	22IS330	CC49234
701	W	2007	M1000	33	Braunstien 211	22IS832	CC49234
101	A	2007	W1800		Baldwin 437	20ECES212	RR79345
101	U	2007	T1015	33		22QA375	HT54347
101	A	2007	H1700	29	Lindner 108	22IS330	SK85977
101	S	2007	T1015	30		22QA375	HT54347
101	W	2007	T1015	20		22QA375	HT54347

Figure 11.1 Madeira College relations

TAKES Relation

Tk_se_section#	Tk_se_qtr	Tk_se_year	Tk_se_co_course#	Tk_grade	Tk_st_sid
101	A	2007	22QA375	A	KP78924
101	A	2007	22QA375	A	KS39874
101	A	2007	22QA375	B	BG66765
101	S	2006	22IS330	C	BE76598
101	A	2007	22IS330	B	KJ56656
101	A	2007	22IS330	A	KP78924
101	A	2007	22IS330	A	KS39874
701	W	2007	22IS832	A	KS39874
101	A	2007	22IS330	A	BE76598
701	W	2007	22IS832	B	BG66765
101	A	2007	22IS330	C	GS76775

PROFESSOR Relation

Pr_name	Pr_empid	Pr_phone	Pr_office	Pr_birthdate	Pr_datehired	Pr_dpt_dcode	Pr_salary
John Smith	SJ89324	5235567645	223 McMicken	1966-10-12	2001-06-23	1	45000
Mike Faraday	FM49276	5235568492	249 McMicken	1960-08-26	1996-05-01	1	92000
Kobe Bryant	BK68765	5235568522	322 McMicken	1968-03-02	1998-05-01	1	66000
Ram Raj	RR79345	5235567244	822 Old Chem	1970-02-06	2001-06-23	6	44000
John B Smith	SJ65436	5235567556	838 Old Chem			6	
Prester John	JP77869	5235567244	822 Old Chem	1955-08-25	1995-08-25	6	44000
Chelsea Bush	BC65437	5235567777	227 Lindner	1946-09-03	1993-05-01	3	77000
Tony Hopkins	HT54347	5235569977	324 Lindner	1949-11-24	1997-01-20	3	77000
Alan Brodie	BA54325	5235569876	238 Lindner	1944-01-14	2000-05-16	3	76000
Jessica Simpson	SJ67543	5235565567	324 Lindner	1955-08-25	1995-08-25	3	67000
Laura Jackson	JL65436	5235565436	336 Lindner	1973-10-16	2000-09-23	3	43000
Marie Curie	CM65436	5235569899	331 Dyer	1972-02-29	1999-10-22	4	99000
Jack Nicklaus	NJ33533	5235566767		1976-01-01	1999-12-31	4	67000
John Nicholson	NJ43728	5235569999	324 Dyer	1966-05-01	2003-06-22	4	99000
Sunil Shetty	SS43278	5235566764	526 Lindner		1993-06-28	7	64000
Katie Shef	SK85977	5235568765	572 Lindner	1948-08-08	1997-06-06	7	65000
Cathy Cobal	CC49234	5235565345	544 Lindner	1968-02-28	2001-01-23	7	45000
Jeanine Troy	TJ76546	5235565545	423 McMicken	1968-01-16		9	45000
Tiger Woods	WT65487	5235565563		1975-11-14	2003-11-14	9	
Mike Crick	CM87659	5235565569	444 McMicken	1970-05-31	2002-05-30	9	69000

TEXTBOOK Relation

Tx_isbn	Tx_title	Tx_year	Tx_publisher
000-66574998	Database Management	1999	Thomson
003-6679233	Linear Programming	1997	Prentice-Hall
001-55-435	Simulation Modeling	2001	Springer
118-99898-67	Systems Analysis	2000	Thomson
77898-8769	Principles of IS	2002	Prentice-Hall
0296748-99	Economics For Managers	2001	
0296437-1118	Programming in C++	2002	Thomson
012-54765-32	Fundamentals of SQL	2004	
111-11111111	Data Modeling	2006	

USES Relation

Us_co_course#	Us_tx_isbn	Us_pr_empid
22IS832	000-66574998	SS43278
22IS270	000-66574998	SS43278
22IS270	77898-8769	SS43278
22IS270	77898-8769	SK85977
22IS270	77898-8769	CC49234
20ECES212	0296437-1118	CC49234
22QA375	0296437-1118	SJ65436
22IS330	003-6679233	BC65437
18ECON123	0296748-99	CM65436
22IS330	118-99898-67	SS43278
22IS832	118-99898-67	SK85977
22QA888	001-55-435	HT54347

Figure 11.1 Madeira College relations (continued)

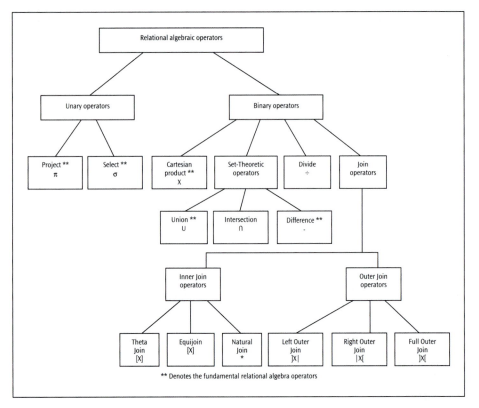

Figure 11.2 Classification of relational algebra operators

11.1.1 Unary Operators

Unary operators "operate" on a single relation. There are two relational algebra unary operators: the Select operator and the Project operator.

11.1.1.1 The Select Operator

The **Select operator** is used to select a horizontal subset of the tuples that satisfy a selection condition from a relation. The general form of the Selection operation is:

$$\sigma_{\text{<selection condition>}}(R)$$

where:

- the symbol σ (sigma) designates the Select operator, and
- the selection condition is a Boolean expression specified on the attributes of relation schema R.

R is generally a *relational algebra expression* whose result is a relation; the simplest expression is the name of a single relation. The relation resulting from the Selection operation has the same attributes as R. The Boolean expression specified as <selection condition> is composed of a number of clauses of the form:

```
<attribute name><comparison operator><constant value>
```

or:

```
<attribute name><comparison operator><attribute name>
```

where:

- <attribute name> is the name of an attribute of R,
- <comparison operator> is normally one of the operators $\{=, \neq, <, \leq, >, \geq\}$, and
- <constant value> is a constant value from the domain of the attribute.

Following are three examples of Selection operations performed on the COURSE relation shown in Figure 11.1.

Selection Example 1. Which courses are three-hour courses?

Relational Algebra Syntax:

$\sigma_{(Co\_hrs\ =\ 3)}(\text{COURSE})$

Observe that the result is an unnamed relation that contains a subset of the tuples in the COURSE relation.

Result:

Co_name	Co_course#	Co_credit	Co_college	Co_hrs	Co_dpt_dcode
Supply Chain Analysis	22QA411	U	Business	3	3
Principles of IS	22IS270	G	Business	3	7
Programming in C++	20ECES212	G	Engineering	3	6
Optimization	22QA888	G	Business	3	3
Financial Accounting	18ACCT801	G	Education	3	4
Database Principles	22IS832	G	Business	3	7
Systems Analysis	22IS430	G	Business	3	7

The Boolean operators AND, OR, and NOT can be used to form a general selection condition.

Selection Example 2. Which courses offered by department 7 are three-hour courses?

Relational Algebra Syntax:

$\sigma_{(Co\_dpt\_dcode\ =\ 7\ and\ Co\_hrs\ =\ 3)}(\text{COURSE})$

Use of the logical operator AND requires that *both* conditions (**Co_dpt_dcode** = 7 and **Co_hrs** = 3) be satisfied.

Result:

Co_name	Co_course#	Co_credit	Co_college	Co_hrs	Co_dpt_dcode
Principles of IS	22IS270	G	Business	3	7
Database Principles	22IS832	G	Business	3	7
Systems Analysis	22IS430	G	Business	3	7

Selection Example 3. Which courses are offered in either the College of Arts and Sciences or the College of Education?

Relational Algebra Syntax:

$\sigma_{\text{(Co\_college = 'Arts and Sciences' or Co\_college = 'Education')}}$ (COURSE)

Use of the logical operator OR allows either condition (**Co_college** = 'Arts and Sciences' or **Co_college** = 'Education') to be satisfied.

Result:

Co_name	Co_course#	Co_credit	Co_college	Co_hrs	Co_dpt_dcode
Intro to Economics	15ECON112	U	Arts and Sciences	3	1
Intro to Economics	18ECON123	U	Education	4	4
Financial Accounting	18ACCT801	G	Education	3	4

11.1.1.2 The Project Operator

While the Select operator selects some of the *tuples* from the relation while eliminating unwanted tuples, the **Project operator** selects certain *attributes* from the relation and eliminates unwanted attributes. In other words, a Selection operation forms a new relation by taking a *horizontal* subset of an existing relation, whereas a Projection operation forms a new relation by taking a *vertical* subset of an existing relation. The difference between Selection and Projection is shown pictorially in Figure 11.3.

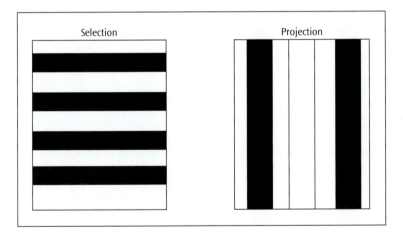

Figure 11.3 Selection compared to projection

The general form of the Projection operation is:

$\pi_{\text{<attribute list>}}$ (R)

where:

- the symbol π (pi) is used to represent the Project operator, and
- <attribute list> is a subset of the attributes of the relation schema R.

As was the case in the Selection operation, R, in general, is a *relational algebra expression* whose result is a relation, which in the simplest case is just the name of a single relation. The result of the Projection operation contains only the attributes specified in the <attribute list> in the same order as they appear in the list.

In cases where the attribute list produced as a result of the Projection operation is not a superkey of R, duplicate tuples are likely to occur in the result. Since relations are sets and do not allow duplicate tuples, only one copy of each group of identical tuples is included in the result of a Projection. Following are a couple of examples of Projection operations based on the COLLEGE and DEPARTMENT relations in Figure 11.1.

Projection Example 1. Which colleges offer courses?

Relational Algebra Syntax:

$\pi_{(Co\_college)}$ (COURSE)

Since **Co_college** is not a superkey of COURSE, the number of tuples in the result is less than the number of tuples in COURSE.

Result:

```
Co_college
-------------------
Arts and Sciences
Business
Education
Engineering
```

Projection Example 2. What is the name and college of each department?

Relational Algebra Syntax:

$\pi_{(Dpt\_name, Dpt\_college)}$ (DEPARTMENT)

Since **[Dpt_name, Dpt_college]** is a superkey of DEPARTMENT, the number of tuples resulting from a Projection operation on DEPARTMENT is equal to the number of tuples in DEPARTMENT.

Result:

```
Dpt_name          Dpt_college
---------------   ------------------
Economics         Arts and Sciences
QA/QM             Business
Economics         Education
Mathematics       Engineering
IS                Business
Philosophy        Arts and Sciences
```

11.1.2 Binary Operators

The relational algebra binary operators "operate" on two relations. There are four binary operators:

- The Cartesian Product operator
- Set theoretic operators
- Join operators
- The Divide operator

11.1.2.1 The Cartesian Product Operator

The **Cartesian Product operator** (often referred to as the Product or Cross-Product operation), denoted by **X**, is used to combine tuples from any two relations in a combinatorial fashion. The Cartesian Product of relations R and S is created by (a) concatenating the attributes of R and S together, and (b) attaching to each tuple in R each of the tuples in S. Thus if R has n_R tuples and S has n_S tuples, then the Cartesian Product Q will have n_R times n_S tuples and take the form shown in Figure 11.4.

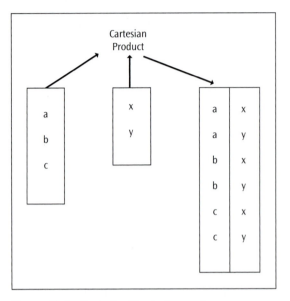

Figure 11.4 Cartesian Product operation

Cartesian Product Example 1. What is the product of the DEPARTMENT and COURSE relations?

Relational Algebra Syntax:

COURSE **X** DEPARTMENT

Since there are six tuples in DEPARTMENT and eleven tuples in COURSE, a total of 66 tuples that contain 13 attributes per tuple is produced when the Cartesian Product of DEPARTMENT and COURSE is formed.

In order to illustrate the result obtained by a Cartesian Product operation, consider two relations—D and C derived from the DEPARTMENT and COURSE relations. Relation D contains six tuples with the attributes **D_name**, **D_dcode**, and **D_college**; relation C contains eleven tuples with the attributes **C_name**, **C_course#**, **C_credit**, and **C_d_dcode**. The content of relations C and D is shown in Figure 11.5.

```
┌─────────────────────────────────────────────────────────────────────┐
│  C Relation                                                           │
│  C_name                    C_course#  C_credit  C_d_dcode             │
│  ───────────────────       ─────────  ────────  ─────────             │
│  Intro to Economics        15ECON112  U              1                │
│  Operations Research       22QA375    U              3                │
│  Intro to Economics        18ECON123  U              4                │
│  Supply Chain Analysis     22QA411    U              3                │
│  Principles of IS          22IS270    G              7                │
│  Programming in C++        20ECES212  G              6                │
│  Optimization              22QA888    G              3                │
│  Financial Accounting      18ACCT801  G              4                │
│  Database Concepts         22IS330    U              7                │
│  Database Principles       22IS832    G              7                │
│  Systems Analysis          22IS430    G              7                │
│                                                                       │
│                                                                       │
│  D Relation                                                           │
│  D_name                 D_dcode  D_college                            │
│  ───────────            ───────  ─────────────────                    │
│  Economics                    1  Arts and Sciences                    │
│  QA/QM                        3  Business                             │
│  Economics                    4  Education                            │
│  Mathematics                  6  Engineering                          │
│  IS                           7  Business                             │
│  Philosophy                   9  Arts and Sciences                    │
│                                                                       │
└─────────────────────────────────────────────────────────────────────┘
```

Figure 11.5   The C and D relations

Relational Algebra Syntax:

C **X** D

The first 18 of the 66 tuples produced by the Cartesian Product of relations C and D follow.
Result (First 18 tuples of the Cartesian Product):

D_name	D_dcode	D_college	C_name	C_course#	C_credit	C_d_dcode
Economics	1	Arts and Sciences	Intro to Economics	15ECON112	U	1
QA/QM	3	Business	Intro to Economics	15ECON112	U	1
Economics	4	Education	Intro to Economics	15ECON112	U	1
Mathematics	6	Engineering	Intro to Economics	15ECON112	U	1
IS	7	Business	Intro to Economics	15ECON112	U	1
Philosophy	9	Arts and Sciences	Intro to Economics	15ECON112	U	1
Economics	1	Arts and Sciences	Operations Research	22QA375	U	3
QA/QM	3	Business	Operations Research	22QA375	U	3
Economics	4	Education	Operations Research	22QA375	U	3
Mathematics	6	Engineering	Operations Research	22QA375	U	3
IS	7	Business	Operations Research	22QA375	U	3
Philosophy	9	Arts and Sciences	Operations Research	22QA375	U	3
Economics	1	Arts and Sciences	Intro to Economics	18ECON123	U	4
QA/QM	3	Business	Intro to Economics	18ECON123	U	4
Economics	4	Education	Intro to Economics	18ECON123	U	4
Mathematics	6	Engineering	Intro to Economics	18ECON123	U	4
IS	7	Business	Intro to Economics	18ECON123	U	4
Philosophy	9	Arts and Sciences	Intro to Economics	18ECON123	U	4

Note that the result shown above is the concatenation of the first three tuples of relation C (see columns 4-7) with all six tuples of relation D (see columns 1-3, rows 1-6, 7-12, and 13-18).

The Cartesian Product operation by itself is generally of little value. It is useful when followed first by a Selection operation that matches values of attributes coming from the component relations (technically, a Cartesian Product operation followed by a Selection operation is equivalent to a Join operation) and sometimes by a Projection operation that selects certain columns from the selected set of tuples.

Cartesian Product Example 2. What are the names of the departments and associated colleges that offer a four-hour course?

Relational Algebra Syntax:

$$\pi_{(Dpt\_name,\ Dpt\_college)}\ (\sigma_{(Co\_hrs\ =\ 4\ and\ Co\_dpt\_dcode\ =\ Dpt\_dcode)}\ (COURSE\ \textbf{X}\ DEPARTMENT))$$

As we will see in Section 11.1.2.3, $\sigma_{(Co\_hrs\ =\ 4\ and\ Co\_dpt\_dcode\ =\ Dpt\_dcode)}$ (COURSE **X** DEPARTMENT) represents a JOIN operation with COURSE **X** DEPARTMENT serving as its fundamental building block.

Result:

```
Dpt_name          Dpt_college
----------------  ------------------
Economics         Education
IS                Business
```

11.1.2.2 Set Theoretic Operators

Three **set theoretic operators**—Union, Intersection, and Difference—are used to combine the tuples from two relations. These are binary operations as each is applied to two sets. When adapted to relational databases, the two relations on which any of the above three operations are applied must be union compatible. Two relations $R(A_1, A_2, ..., A_n)$ and $S(B_1, B_2, ..., B_n)$ are said to be **union compatible** if (a) they have the same degree (i.e., have the same number of attributes), and (b) each pair of corresponding attributes in R and S share the same domain. Venn diagrams illustrating Union, Intersection, and Difference are shown in Figure 11.6.

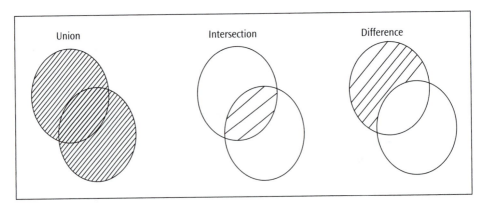

Figure 11.6 The Union, Intersection, and Difference operations

The remainder of this section consists of the definition and examples of the three set theoretic operators.

Union: The result of this operation, denoted by R ∪ S, is a relation that includes all tuples that belong to either R or S or to both R and S. Duplicate tuples are eliminated.

Union Example. Let relations R and S be derived from the SECTION relation. R contains tuples indicating Fall quarter sections (**Se_qtr** = 'A') and S contains tuples listing sections offered in a Lindner classroom (**Se_room** = 'Lindner').

RELATION R

Se_section#	Se_qtr	Se_year	Se_time	Se_maxst	Se_room	Se_co_course#	Se_pr_profid
902	A	2006	H1700	25	Lindner 108	22IS270	SK85977
901	A	2006	W1800	35	Rhodes 611	22IS270	SK85977
101	A	2007	H1700	29	Lindner 108	22IS330	SK85977
101	A	2007	T1015	25		22QA375	HT54347
101	A	2007	W1800		Baldwin 437	20ECES212	RR79345

RELATION S

Se_section#	Se_qtr	Se_year	Se_time	Se_maxst	Se_room	Se_co_course#	Se_pr_profid
902	A	2006	H1700	25	Lindner 108	22IS270	SK85977
101	S	2006	T1045	29	Lindner 110	22IS330	SK85977
102	S	2006	H1045	29	Lindner 110	22IS330	CC49234
101	A	2007	H1700	29	Lindner 108	22IS330	SK85977

Relational Algebra Syntax and Result: R∪S

Se_section#	Se_qtr	Se_year	Se_time	Se_maxst	Se_room	Se_co_course#	Se_pr_profid
101	A	2007	H1700	29	Lindner 108	22IS330	SK85977
101	A	2007	T1015	25		22QA375	HT54347
101	A	2007	W1800		Baldwin 437	20ECES212	RR79345
101	S	2006	T1045	29	Lindner 110	22IS330	SK85977
102	S	2006	H1045	29	Lindner 110	22IS330	CC49234
901	A	2006	W1800	35	Rhodes 611	22IS270	SK85977
902	A	2006	H1700	25	Lindner 108	22IS270	SK85977

The Union R ∪ S contains the sections that are offered either exclusively in a fall quarter or exclusively in a Lindner classroom, or offered in a Lindner classroom during a fall quarter (see the first and seventh tuples).

Intersection: The result of this operation, denoted by R ∩ S, is a relation that includes all tuples that are in both R and S.

Intersection Example. Using the data in the relations R and S given above, form the intersection of R and S.

Relational Algebra Syntax and Result: R ∩ S

Se_section#	Se_qtr	Se_year	Se_time	Se_maxst	Se_room	Se_co_course#	Se_pr_profid
101	A	2007	H1700	29	Lindner 108	22IS330	SK85977
902	A	2006	H1700	25	Lindner 108	22IS270	SK85977

Observe that the sections shown above are the only sections offered in a Lindner classroom during a fall quarter.

Difference: The result of this operation, denoted by R – S, is a relation that includes all tuples that are in R but not in S.

Difference Example 1. Using the data in the relations R and S given above, form the difference R minus S.

Relational Algebra Syntax and Result: R – S

Se_section#	Se_qtr	Se_year	Se_time	Se_maxst	Se_room	Se_co_course#	Se_pr_profid
101	A	2007	T1015	25		22QA375	HT54347
101	A	2007	W1800		Baldwin 437	20ECES212	RR79345
901	A	2006	W1800	35	Rhodes 611	22IS270	SK85977

Note that the result obtained by subtracting S from R is equal to all sections offered during a fall quarter in a classroom other than Lindner. The classroom associated with the section in the first tuple is not available (i.e., is a null value) and satisfies the condition that the section be offered in a classroom other than Lindner.

Difference Example 2. Using the data in the relations R and S given above, form the difference S minus R.

Relational Algebra Syntax and Result: S – R

Se_section#	Se_qtr	Se_year	Se_time	Se_maxst	Se_room	Se_co_course#	Se_pr_profid
101	S	2006	T1045	29	Lindner 110	22IS330	SK85977
102	S	2006	H1045	29	Lindner 110	22IS330	CC49234

Observe that the result obtained by subtracting R from S is equal to all sections offered in a classroom located in Lindner during a quarter other than the fall quarter.

11.1.2.3 Join Operators

Joins come in several varieties. Basically in each, the **Join operation**, denoted by $[X]^2$, is used to combine related tuples from two relations into single tuples. The example given previously in Cartesian Product Example 2 where the names of the departments and their associated colleges that offer a four-hour course is requested can be specified using the Join operation by replacing:

$$\pi_{(Dpt\_name, \, Dpt\_college)} \left(\sigma_{(Co\_hrs \, = \, 4 \, and \, Co\_dpt\_dcode \, = \, Dpt\_dcode)} \, (COURSE \; X \; DEPARTMENT) \right)$$

with a Join operation followed by a Projection operation as follows:

$$\pi_{(Dpt\_name, \, Dpt\_college)} \left(COURSE \; [X]_{\, Co\_hrs \, = \, 4 \, and \, Co\_dpt\_dcode \, = \, Dpt\_dcode} \, DEPARTMENT \right)$$

The general form of a Join operation on two relations $R(A_1, A_2, ..., A_n)$ and $S(B_1, B_2, ..., B_m)$ is:

R [X] <join condition> S

The result of the Join operation is a relation Q with n + m attributes $Q(A_1, A_2, ..., A_n, B_1, B_2, ... B_m)$, in that order. Q has one tuple for each combination of tuples—one from R and one from S—whenever the combination satisfies the join condition. This is the main difference between Cartesian Product and Join; in Join only combinations of tuples satisfying the join

---

[2] In this book, [X] is used to represent the Equijoin and Theta Join operations as well as all Outer Join operations while * is used to represent the Natural Join operation.

condition appear in the result, while in the Cartesian Product all combinations of tuples are included in the result. The join condition is specified on attributes from the two relations R and S and is evaluated for each combination of tuples. Each tuple combination for which the join condition evaluates to true is included in the resulting relation Q as a single combined tuple. In order to join the two relations R and S, they must be **join compatible**, that is, the join condition must involve attributes from R and S which share the same domain.

A general join condition is of the form:

```
<condition> AND <condition> AND ... AND <condition>
```

where each condition is of the form $A_i \Theta B_j$, A_i is an attribute of R, B_j is an attribute of S, A_i and B_j have the same domain, and Θ (theta) is one of the comparison operators $\{=, <, , >, , \neq \}$. A Join operation with such a general join condition is called a Theta join. Tuples whose join attributes are null do not appear in the result.

The Equijoin Operator The most common join involves join conditions where the comparison operator is "=". This type of join is called the **Equijoin**. The result of an Equijoin includes all attributes from both relations participating in the Join operation. This implies duplication of the joining attributes in the result.

Equijoin Example. Join the C and D relations (derived from the COURSE and DEPARTMENT relations) over their common attribute department code (**D_dcode** in relation D and **C_d_dcode** in relation C).

Relational Algebra Syntax:

C [**x**] $_{\text{C\_d\_dcode} = \text{D\_dcode}}$ D

Result:

C_name	C_course#	C_credit	C_d_dcode	D_name	D_dcode	D_college
Intro to Economics	15ECON112	U	1	Economics	1	Arts and Sciences
Operations Research	22QA375	U	3	QA/QM	3	Business
Optimization	22QA888	G	3	QA/QM	3	Business
Supply Chain Analysis	22QA411	U	3	QA/QM	3	Business
Intro to Economics	18ECON123	U	4	Economics	4	Education
Financial Accounting	18ACCT801	G	4	Economics	4	Education
Programming in C++	20ECES212	G	6	Mathematics	6	Engineering
Principles of IS	22IS270	G	7	IS	7	Business
Database Concepts	22IS330	U	7	IS	7	Business
Database Principles	22IS832	G	7	IS	7	Business
Systems Analysis	22IS430	G	7	IS	7	Business

Observe that the name (**D_name**) and college (**D_college**) of the department associated with each course appears in the result shown above.

Had the Equijoin involved the complete COURSE and DEPARTMENT relations joined on the **Co_dpt_dcode** and **Dpt_dcode** attributes instead of just relations C and D, the result would have also included 11 tuples with each tuple containing 13 attributes.

While it is possible to join COURSE and DEPARTMENT over the **Co_college** attribute from COURSE and the **Dpt_college** attribute from DEPARTMENT, the result of such a join would be meaningless—each tuple in COURSE would be concatenated not only with the tuple from DEPARTMENT that offers the course, but also with the tuples from DEPARTMENT with the same **Dpt_college** but associated with a different department code than that associated with the course. Using the data in the COURSE and DEPARTMENT relations in

Figure 11.1, it is left as an exercise for the reader to demonstrate that an Equijoin of COURSE and DEPARTMENT on the attributes **Co_college** and **Dpt_college** yields a result that contains 19 tuples.

The Natural Join Operator   Because the result of an Equijoin results in pairs of attributes with identical values in all the tuples (see **C_d_dcode** and **D_dcode** in the result shown above), a new relational algebra operation called a **Natural Join**, denoted by *, was created to omit the second (and unnecessary) attribute in an Equijoin condition. The Natural Join of two relations with the common attribute b is illustrated in Figure 11.7.

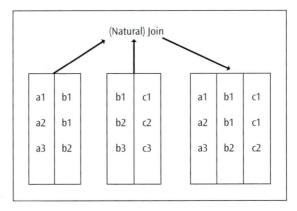

Figure 11.7   The Natural Join operation

Natural Join Example. Join the C and D relations over their common attribute department code (**D_dcode** in relation D and **C_d_dcode** in relation C).
Relational Algebra Syntax:[3]

$$C \; * \; _{C\_d\_dcode \; = \; D\_dcode} \; D$$

Result:

C_name	C_course#	C_credit	D_name	D_dcode	D_college
Intro to Economics	15ECON112	U	Economics	1	Arts and Sciences
Operations Research	22QA375	U	QA/QM	3	Business
Optimization	22QA888	G	QA/QM	3	Business
Supply Chain Analysis	22QA411	U	QA/QM	3	Business
Intro to Economics	18ECON123	U	Economics	4	Education
Financial Accounting	18ACCT801	G	Economics	4	Education
Programming in C++	20ECES212	G	Mathematics	6	Engineering
Principles of IS	22IS270	G	IS	7	Business
Database Concepts	22IS330	U	IS	7	Business
Database Principles	22IS832	G	IS	7	Business
Systems Analysis	22IS430	G	IS	7	Business

[3] Contrary to the requirement that attributes have unique names over the entire relational schema, the standard definition of Natural Join requires that the two join attributes (or each pair of join attributes) have the same name. If this is not the case, a renaming operation must be applied first. See Elmasri and Navathe (2004) for a discussion of the use of the renaming operation in representing a Natural Join.

Observe that only one (**D_dcode**) of the two join attributes common to both relations (**D_dcode** and **C_d_dcode**) appears in the result shown above.

The Join operation is used to combine data from multiple relations so that related information can be presented in a single relation. As illustrated in Cartesian Product Example 2, Join operations are typically followed by a Projection operation.

The Theta Join Operator   While occurring infrequently in practical applications, **Theta Joins** that do not involve equality conditions are possible as long as the join condition involves attributes that share the same domain.

Theta Join Example. Instead of doing an Equijoin of C and D when an equality condition involving their two common attributes (**C_d_dcode** and **D_dcode**) exists, do a join of C and D when an inequality condition exists for these common attributes.

Relational Algebra Syntax:

C [X] C_d_dcode <> D_dcode D

In such a Theta Join, a tuple from D is concatenated with a tuple from C only when an inequality exists for each of the join conditions. Thus the first tuple in D is not concatenated with the first tuple in C because the join condition is not satisfied. Observe, however, that the join condition is satisfied when the first tuple of D is evaluated against all other tuples in C, thus resulting in the first ten tuples of the 55 tuples in the result. Using the data for relations C and D shown in Figure 11.5, the reader is encouraged to verify why this Theta Join produces a total of 55 tuples.

Outer Join Operators   The Join operations discussed to this point are **Inner Join** operations, meaning that for the relations R and S, only tuples from R that have matching tuples in S (and vice versa) appear in the result. In other words, tuples without a matching (or related) tuple are eliminated from the join result. Tuples with null values in the join attributes are also eliminated. A set of operations called **Outer Joins**, can be used when we want to keep all the tuples in R, or those in S, or those in both relations in the result of the join, whether or not they have matching tuples in the other relation.

The **Left Outer Join** operation, denoted by **]X|**, keeps every tuple in the *first or left* relation R in R]X| S. If no matching tuple is found in S, then the attributes of S in the join result are filled or "padded" with null values. Thus, in effect a tuple of null values is added to relation S, and any tuple in relation R without a matching tuple in relation S is concatenated with the tuple of null values in relation S. On the other hand, a **Right Outer Join**, denoted by **|X[**, keeps every tuple in the *second or right* relation S in the result R |X[S. If for a particular tuple in relation S there is no matching tuple in relation R, then for that particular tuple of relation S attributes from relation R are "padded" with null values. A third operation, **Full Outer Join**, denoted by **]X[**, keeps all tuples in both the left and right relations when no matching tuples are found, padding them with null values as needed.

Left Outer Join Example. Do a Left Outer Join of relations D and C over their common attributes (**D_dcode** in relation D and **C_d_dcode in relation C**).

Relational Algebra Syntax:

D]X| D_dcode = C_d_dcode C

Result:

D_name	D_dcode	D_college	C_name	C_course#	C_credit	C_d_dcode
Economics	1	Arts and Sciences	Intro to Economics	15ECON112	U	1
QA/QM	3	Business	Operations Research	22QA375	U	3
Economics	4	Education	Intro to Economics	18ECON123	U	4
QA/QM	3	Business	Supply Chain Analysis	22QA411	U	3
IS	7	Business	Principles of IS	22IS270	G	7
Mathematics	6	Engineering	Programming in C++	20ECES212	G	6
QA/QM	3	Business	Optimization	22QA888	G	3
Economics	4	Education	Financial Accounting	18ACCT801	G	4
IS	7	Business	Database Concepts	22IS330	U	7
IS	7	Business	Database Principles	22IS832	G	7
IS	7	Business	Systems Analysis	22IS430	G	7
Philosophy	9	Arts and Sciences				

Observe how the sixth tuple of D is concatenated with the tuple of null values from the C relation to produce the twelfth tuple in the result, revealing that the Philosophy Department has yet to offer a course.

Right Outer Join Example. In order to conveniently illustrate the result of a Right Outer Join operation, the relation SS has been created from the SECTION relation. Like SECTION, SS contains 11 tuples, but only four of the eight attributes found in SECTION. These four attributes are: **Ss_section#, Ss_qtr, Ss_year**, and **Ss_c_course#**. The content of relation SS is shown in Figure 11.8.

Ss_section#	Ss_qtr	Ss_year	Ss_c_course#
101	A	2007	22QA375
901	A	2006	22IS270
902	A	2006	22IS270
101	S	2006	22IS330
102	S	2006	22IS330
701	W	2007	22IS832
101	A	2007	20ECES212
101	U	2007	22QA375
101	A	2007	22IS330
101	S	2007	22QA375
101	W	2007	22QA375

Figure 11.8   Relation SS

The following shows a Right Outer Join of relations SS and C over their common attributes (**Ss_c_course#** in relation SS and **C_course#** in relation C).

Relational Algebra Syntax:

SS |X[Ss_c_course# = C_course# C

Result:

Ss_section#	Ss_qtr	Ss_year	Ss_c_course#	C_name	C_course#	C_credit	C_d_dcode
101	A	2007	22QA375	Operations Research	22QA375	U	3
901	A	2006	22IS270	Principles of IS	22IS270	G	7
902	A	2006	22IS270	Principles of IS	22IS270	G	7
101	S	2006	22IS330	Database Concepts	22IS330	U	7
102	S	2006	22IS330	Database Concepts	22IS330	U	7
701	W	2007	22IS832	Database Principles	22IS832	G	7
101	A	2007	20ECES212	Programming in C++	20ECES212	G	6
101	U	2007	22QA375	Operations Research	22QA375	U	3
101	A	2007	22IS330	Database Concepts	22IS330	U	7
101	S	2007	22QA375	Operations Research	22QA375	U	3
101	W	2007	22QA375	Operations Research	22QA375	U	3
				Financial Accounting	18ACCT801	G	4
				Careers Colloquium	06US100	U	
				Systems Analysis	22IS430	G	7
				Intro to Economics	18ECON123	U	4
				Supply Chain Analysis	22QA411	U	3
				Intro to Economics	15ECON112	U	1
				Optimization	22QA888	G	3

Observe how each tuple of C (including those for which a section has not been offered) appears in the result shown above.

Full Outer Join Example. Join the D and C relations, making sure that each tuple from each relation appears in the result.

In order to illustrate a Full Outer Join, assume that it is possible for a course to exist without being affiliated with a department or college. The revised C relation shown below reflects the addition of a one-hour Careers Colloquium course. *For purposes of this example, assume that the business rule requiring each course to be affiliated with a department has been temporarily disabled.*

Revised C Relation

C_name	C_course#	C_credit	C_d_dcode
Intro to Economics	15ECON112	U	1
Operations Research	22QA375	U	3
Intro to Economics	18ECON123	U	4
Supply Chain Analysis	22QA411	U	3
Principles of IS	22IS270	G	7
Programming in C++	20ECES212	G	6
Optimization	22QA888	G	3
Financial Accounting	18ACCT801	G	4
Database Concepts	22IS330	U	7
Database Principles	22IS832	G	7
Systems Analysis	22IS430	G	7
Careers Colloquium	06US100	U	

This example uses the D relation and the revised C relation to illustrate the distinction between a Left, Right, and Full Outer Join. The result of a Left Outer Join of D and C, expressed as:

D]X| $_{\text{D\_dcode = C\_d\_dcode}}$ C

adds a blank tuple to the revised C relation to ensure that each tuple in relation D is reflected in the result shown below.

D_name	D_dcode	D_college	C_name	C_course#	C_credit	C_d_dcode
Economics	1	Arts and Sciences	Intro to Economics	15ECON112	U	1
QA/QM	3	Business	Operations Research	22QA375	U	3
Economics	4	Education	Intro to Economics	18ECON123	U	4
QA/QM	3	Business	Supply Chain Analysis	22QA411	U	3
IS	7	Business	Principles of IS	22IS270	G	7
Mathematics	6	Engineering	Programming in C++	20ECES212	G	6
QA/QM	3	Business	Optimization	22QA888	G	3
Economics	4	Education	Financial Accounting	18ACCT801	G	4
IS	7	Business	Database Concepts	22IS330	U	7
IS	7	Business	Database Principles	22IS832	G	7
IS	7	Business	Systems Analysis	22IS430	G	7
Philosophy	9	Arts and Sciences				

A Right Outer Join of the relation D and the revised C relation, expressed as:

D |X[$_{\text{D\_dcode = C\_d\_dcode}}$ C

adds a blank tuple to relation D to ensure that each tuple in the revised C relation is reflected in the result shown below.

D_name	D_dcode	D_college	C_name	C_course#	C_credit	C_d_dcode
Economics	1	Arts and Sciences	Intro to Economics	15ECON112	U	1
QA/QM	3	Business	Optimization	22QA888	G	3
QA/QM	3	Business	Supply Chain Analysis	22QA411	U	3
QA/QM	3	Business	Operations Research	22QA375	U	3
Economics	4	Education	Financial Accounting	18ACCT801	G	4
Economics	4	Education	Intro to Economics	18ECON123	U	4
Mathematics	6	Engineering	Programming in C++	20ECES212	G	6
IS	7	Business	Systems Analysis	22IS430	G	7
IS	7	Business	Database Principles	22IS832	G	7
IS	7	Business	Database Concepts	22IS330	U	7
IS	7	Business	Principles of IS	22IS270	G	7
			Careers Colloquium	06US100	U	

A Full Outer Join of D and the revised C relation, expressed as:

D]X[$_{\text{D\_dcode = C\_d\_dcode}}$ C

adds a blank tuple to both relation D and the revised C relation to insure that each tuple in each relation is reflected in the result. Observe how each tuple in each relation appears in the result shown below.

D_name	D_dcode	D_college	C_name	C_course#	C_credit	C_d_dcode
Economics	1	Arts and Sciences	Intro to Economics	15ECON112	U	1
QA/QM	3	Business	Operations Research	22QA375	U	3
Economics	4	Education	Intro to Economics	18ECON123	U	4

QA/QM	3	Business	Supply Chain Analysis	22QA411	U	3
IS	7	Business	Principles of IS	22IS270	G	7
Mathematics	6	Engineering	Programming in C++	20ECES212	G	6
QA/QM	3	Business	Optimization	22QA888	G	3
Economics	4	Education	Financial Accounting	18ACCT801	G	4
IS	7	Business	Database Concepts	22IS330	U	7
IS	7	Business	Database Principles	22IS832	G	7
IS	7	Business	Systems Analysis	22IS430	G	7
Philosophy	9	Arts and Sciences				
			Careers Colloquium	06US100	U	

It is left as an exercise for the reader to verify that the union of the results of a Left Outer Join and a Right Outer Join is equal to the result of a Full Outer Join.

11.1.2.4 The Divide Operator

The **Divide operator** is useful when there is a need to identify tuples in one relation that match *all* tuples in another relation. For example, let R be a relation schema with attributes $A_1, A_2, ..., A_n, B_1, B_2, ..., B_m$ and S be a relation schema with attributes $B_1, B_2, ..., B_m$. In other words, the set of attributes of S (*the divisor*) is a subset of the attributes of R (*the dividend*). The division of R by S can be expressed as R/S. Figure 11.9 contains a schematic view of the Division operation. Observe how the two tuples in relation T represent a subset of the tuples in relation R that match all three tuples in relation S. In order to divide R by S, relations R and S must be division compatible in that the set of attributes of S must be a subset of the attributes of R. In other words, if relation S in Figure 11.9 were also to contain the attribute p, then division compatibility would not exist.

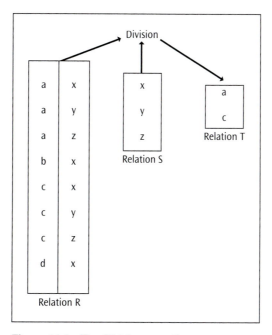

Figure 11.9 The Division operation

For ease of understanding, Kifer, Bernstein, and Lewis (2005) illustrate how the Division operation can be decomposed into a sequence of Projection, Cartesian Product, and Difference operations as shown here:

$T_1 = \pi_A (R) \times S$	All possible associations between the A values in R and B values in S.
$T_2 = \pi_A(T_1 - R)$	All those A values in R that are not associated in R with every B value in S. These are those A values that should not be in the answer.
$T_3 = \pi_A (R) - T_2$	The quotient: all those A values in R that are associated in R with all B values in S.

Divide Example. List the course numbers of courses that are offered in all quarters during which course sections are offered.

Relational Algebra Syntax:

```
R = π (Se_co_course#,Se_qtr) (SECTION)
S = π (Se_qtr) (SECTION)
R ÷ S
```

Result:

```
Se_co_course#
-------------
22QA375
```

Expressing the result as a sequence of projection, Cartesian Product, and Difference operations yields the following:

Relation R

Se_co_course#	Se_qtr
22QA375	A
22IS270	A
22IS330	S
22IS832	W
20ECES212	A
22QA375	U
22IS330	A
22QA375	S
22QA375	W

Relation S

Se_qtr
A
S
W
U

$T_1 = \pi_{(Se\_co\_course\#,Se\_qtr)}$ (R) X S — All possible associations between courses and quarters during which sections of courses are offered:

$\pi_{Se\_co\_course\#}$(R)	S
22QA375	A
22IS270	S
20IS330	W
22IS832	U
20ECES212	

T_1	
22QA375	A
22QA375	S
22QA375	W
22QA375	U
22IS270	A
22IS270	S
22IS270	W
22IS270	U
20IS330	A
20IS330	S
20IS330	W
20IS330	U
22IS832	A
22IS832	S
22IS832	W
22IS832	U
20ECES212	A
20ECES212	S
20ECES212	W
20ECES212	U

$T_2 = \pi_{Se\_co\_course\#}$ $(T_1 - R)$ — All courses that are not offered during each quarter:

$(T_1 - R)$	
22IS270	S
22IS270	W
22IS270	U
22IS832	A
22IS832	S
22IS832	U
20ECES21	S
20ECES212	W
20ECES212	U

T_2
22IS270
20IS330
20IS832
20ECES212

$T_3 = \pi_{Se\_co\_course\#}$ (R) $- T_2$ — All courses offered during each quarter:

$\pi_{Se\_co\_course\#}$ (R)	T_2	T_3
22QA375	20IS270	
22IS270	20IS330	22QA375
20IS330	22IS832	
22IS832	20ECES212	
20ECES212		

Table 11.1 summarizes the basic relational algebra operators and notation.

Table 11.1 Basic relation algebra operators[4]

Relational Algebra Operator	Purpose	Notation
Select	Selects all tuples that satisfy the selection condition from a relation R.	$\sigma_{<\text{selection condition}>} (R)$
Project	Produces a new relation with only some of the attributes of R, and removes duplicate tuples.	$\pi_{<\text{attribute list}>} (R)$
Cartesian Product	Produces a relation that has the attributes of R_1 and R_2 and includes as tuples all possible combinations of tuples from R_1 and R_2.	$R_1 \times R_2$
Union	Produces a relation that includes all the tuples in R_1 or R_2 or both R_1 and R_2; R_1 and R_2 must be union compatible.	$R_1 \cup R_2$
Intersection	Produces a relation that includes all the tuples in both R_1 and R_2; R_1 and R_2 must be union compatible.	$R_1 \cap R_2$
Difference	Produces a relation that includes all the tuples in R_1 that are not in R_2; R_1 and R_2 must be union compatible.	$R_1 - R_2$
Equijoin	Produces all the combinations of tuples from R_1 and R_2 that satisfy a join condition with only equality comparisons. R_1 and R_2 must be join compatible.	$R_1 [X]_{<\text{join condition}>} R_2$
Natural Join	Produces all the combinations of tuples from R_1 and R_2 that satisfy a join condition with only equality comparisons, except that one of the join attributes is not included in the resulting relation. R_1 and R_2 must be join compatible.	$R_1 *_{<\text{join condition}>} R_2$
Theta Join	Produces all the combinations of tuples from R_1 and R_2 that satisfy a join condition which does not have to involve equality comparisons. R_1 and R_2 must be join compatible.	$R_1 [X]_{<\text{join condition}>} R_2$
Divide	Returns every tuple from R_1 that match all tuples in R_2; R_1 and R_2 must be division compatible.	$R_1 \div R_2$

11.1.2.5 Additional Relational Operators

Some common database requests cannot be performed with the basic relational algebra operations described previously. This section defines three additional operations to express these requests.

The Semi-Join Operation The **Semi-Join operation** defines a relation that contains the tuples of R that participate in the join of R with S. In other words, a Semi-Join is defined to

---

[4] Source: R. Elmasri and S.B. Navathe (2003), *Fundamentals of Database Systems*, Addison-Wesley.

be equal to the join of R and S, projected back on the attributes of R. The Semi-Join operation can be expressed using the Projection and Join operations as follows:

$$\Pi_R(R \ |X_{A=B} \ S)$$

where A and B are domain-compatible attributes of R and S, respectively.

Semi-Join Example. List the complete details of all courses for which sections are being offered in the fall (**Se_qtr = 'A'**) 2007 quarter.

Relational Algebra Syntax:

Π COURSE(COURSE |X $_{(Co\_course\#=Se\_co\_course\#\ and\ Se\_qtr\ =\ 'A'\ and\ Se\_year\ =\ 2007)}$ SECTION)

Result:

Co_name	Co_course#	Co_credit	Co_college	Co_hrs	Co_dpt_code
Database Concepts	22IS330	U	Business	4	7
Operations Research	22QA375	U	Business	2	3
Programming in C++	20ECES212	G	Engineering	3	6

The Semi-Minus Operation   The semi-difference between R and S (in this order) is defined to be equivalent to:

$$R - \Pi_R(R \ |X_{A=B} \ S)$$

The result of the **Semi-Minus operation** is thus the tuples of R that have no counterpart in S.

Semi-Minus Example. List complete details of all courses for which sections are not offered in the fall 2007 quarter.

Relational Algebra Syntax:

COURSE - Π_{COURSE}(COURSE |X $_{(Co\_course\#=Se\_co\_course\#\ and\ Se\_qtr\ =\ 'A'\ and\ Se\_year\ =\ 2007)}$ SECTION)

Result:

Co_name	Co_course#	Co_credit	Co_college	Co_hrs	Co_dpt_dcode
Database Principles	22IS832	G	Business	3	7
Financial Accounting	18ACCT801	G	Education	3	4
Intro to Economics	15ECON112	U	Arts and Sciences	3	1
Intro to Economics	18ECON123	U	Education	4	4
Optimization	22QA888	G	Business	3	3
Principles of IS	22IS270	G	Business	3	7
Supply Chain Analysis	22QA411	U	Business	3	3
Systems Analysis	22IS430	G	Business	3	7

Aggregate Functions and Grouping   One type of request that cannot be expressed in the basic relational algebra involves mathematical aggregate functions on collections of values from the database. Examples of such functions include retrieving the sum, average, maximum, and minimum of a series of numeric values. Another type of request involves the grouping of tuples in a relation by the value of some of their attributes and then applying an aggregate function independently to each group. An example would be to group the courses taken by course number and then count the number of students taking each course during the year 2007.

The **Aggregate Function** operation can be defined using the symbol $\mathcal{F}$ to specify these types of requests as follows:

<grouping attributes> $\mathcal{F}$ <function list> (R)

Data Manipulation: Relational Algebra and SQL

where <grouping attributes> is a list of attributes of the relation specified in R, and <function list> is a list of (<function> <attribute>) pairs where:

- <function> is one of the allowed functions, such as SUM, AVERAGE, MAXIMUM, MINIMUM, COUNT, and
- <attribute> is an attribute of the relation specified by R.

The resulting relation has the grouping attributes plus one attribute for each element in the function list.

Aggregate Functions and Grouping Example. Compute the average number of hours for the courses offered by each college.

Relational Algebra Syntax:

$$Co\_college \ \mathcal{F} \ \text{AVERAGE Co\_hrs} \ (COURSE)$$

Result:

```
Co_college              Average
------------------- ---------------
Arts and Sciences          3
Business                   3
Education                3.5
Engineering                3
```

11.2 Structured Query Language (SQL)

SQL is the standard language for manipulating relational databases. The language was created to facilitate implementation of relational algebra in a database. Like relational algebra, SQL uses one or more relations as input and produces a single relation as output.[5] This section and Chapter 12 contain an informal overview of the use of the SQL SELECT statement for information retrieval. Through the use of numerous example queries, Section 11.2 introduces the SELECT statement's features gradually beginning with queries based on a single table followed by queries that retrieve data from several tables. As was the case in Chapter 10, except where indicated, the examples in this chapter as well as in Chapter 12 are based on the syntax associated with the SQL-92 standard.

Before proceeding further, it is important to note that space does not permit a thorough and complete discussion of all features of the SQL SELECT statement as well as other features associated with SQL's data definition language,[6] data manipulation language, and data control language. Hundreds of books and reference manuals are available on SQL, and the interested reader is encouraged to consider such sources for more comprehensive syntax than that shown in this book. In addition, many Web sites have information about standard SQL and its implementation under various database platforms. For example, considerable online documentation is available from the Microsoft developers site (*http://msdn.microsoft.com*) and the Oracle TechNet site (*http://technet.oracle.com*). The O'Reilly site (*http://www.ora.com*) also publishes information online about SQL.

---

[5] SQL uses the terms *table*, *row*, and *column* for relation, tuple, and attribute, respectively. Thus, when the discussion focuses on SQL the terms table, row, and column will be used. When the discussion focuses on relational algebra operations, the terms relation, tuple, and attribute will be used.

[6] Three more advanced features of SQL-92 are assertions, triggers, and views. Assertions and triggers are SQL/DDL facilities for capturing constraints while views are "virtual tables" created using SQL/DDL. These features are discussed in Chapter 12.

As mentioned previously, the SQL SELECT statement is used to retrieve (i.e., query) data from tables (i.e., relations) and is used in conjunction with all relational algebra operations. For example, using a SELECT statement, it is possible to view all the columns and rows within a table (i.e., to execute a relational algebra Selection operation) or specify that only certain columns and rows be viewed (execute a Selection operation followed by a Projection operation).

The syntax for an SQL statement gives the basic structure, or rules, required to execute the statement. The basic form of the SQL SELECT statement is called a **select-from-where block** and contains the three clauses SELECT, FROM, and WHERE in the following form:

```
SELECT <column list>
FROM <table list>
WHERE <condition>
```

where:

- <column list> is a list of column names whose values are to be retrieved by the query,
- <table list> is a list of the table names required to process the query,[7] and
- <condition> is a conditional (Boolean) expression that identifies the rows to be retrieved by the query.

Kifer, Bernstein, and Lewis (2005) provide a useful description of the basic algorithm for evaluating an SQL query.

1. The FROM clause is evaluated. It produces a table that is the Cartesian Product of the tables listed as arguments. If a table occurs more than once in the FROM clause, then this table occurs as many times in the Cartesian Product.
2. The WHERE clause is evaluated by taking the table produced in Step 1 and processing each row individually. Values from the row are substituted for the column names in the condition and the condition is evaluated. The table produced by the WHERE clause contains exactly those rows for which the condition evaluates to true.
3. The SELECT clause is evaluated. It takes the table produced in Step 2 and retains only those columns that are listed as arguments. The resulting table is output by the SELECT statement.

Additional features of the language result in the addition of steps to this algorithm. Also, certain steps need not be present in a particular evaluation. For example, Step 2 above is not evaluated if there is no WHERE clause.[8] Likewise, if the FROM clause refers to only a single table, the Cartesian Product is not formed.

Three other clauses can also be used in an SQL SELECT statement:

```
GROUP BY group_by_expression
HAVING group_condition
ORDER BY column name(s)
```

---

[7] As we will see, the table list may include both database views as well as inline views.

[8] While the WHERE clause is optional, the SELECT and FROM clauses are not.

where:

- *group_by_expression* forms groups of rows with the same value,
- *group_condition* filters the groups subject to some condition, and
- *column name(s)* specifies the order of the output.

In this book, a semicolon (;) follows the last clause in an SQL SELECT statement.[9] In addition, while the keywords and clauses (e.g., SELECT, FROM, WHERE, GROUP BY, etc.) in an SQL SELECT statement appear capitalized here, only conditions in the WHERE clause that involve non-numeric (i.e., character or string) literals are actually case sensitive.

Many of the illustrations in this chapter make use of the DEPARTMENT, COURSE, STUDENT, GRAD_STUDENT, SECTION, TAKES, PROFESSOR, TEXTBOOK, and USES tables from the Madeira College registration system described in Chapter 10. The initial form of these nine tables is shown in Figure 11.1 at the beginning of this chapter. Each illustration has been tested using Oracle. Thus please note that not all syntax shown will work on other platforms such as IBM DB2, Microsoft SQL Server, and MySQL. For more comprehensive syntax than that shown in conjunction with the examples in this book, refer to the SQL reference material of the respective database platform.

11.2.1 SQL Queries Based on a Single Table

This section considers several SQL features in the context of queries that involve a single table. Included in the discussion are illustrations of (a) the Select and Project operators, the two unary relational algebra operators based on a single relation, (b) the hierarchy of operations, (c) the handling of null values, and (d) pattern matching.

The examples in the remainder of this chapter are labeled in accordance with the following scheme.

- The first two elements of the section number are ignored (e.g., in Section 11.2.1.1, the 11.2 portion is ignored)
- The examples within the section are numbered from 1 to n beginning with the third element of the section number.

Thus you will find that the three examples in Section 11.2.1.1 are labeled 1.1.1 through 1.1.3; the ten examples in Section 11.2.1.6 are labeled 1.6.1 through 1.6.10; and the three examples in Section 11.2.3.1 are labeled 3.1.1.1 through 3.1.1.3.

11.2.1.1 Examples of the Selection Operation

Three examples of the use of the relational algebra Select operator in a Selection operation appear in Section 11.1.1.1. SQL SELECT statements that correspond to each of these examples follow.

Example 1.1.1 (Corresponds to Selection Example 1 in Section 11.1.1.1). Which courses are three-hour courses?

SQL SELECT Statements:

```
SELECT COURSE.CO_NAME, COURSE.CO_COURSE#, COURSE.CO_CREDIT, COURSE.CO_
COLLEGE, COURSE.CO_HRS, COURSE.CO_DPT_DCODE
FROM COURSE
WHERE COURSE.CO_HRS = 3;
```

---

[9] The SQL-92 standard actually prescribes the semicolon as a statement terminator only in the case of embedded SQL.

or:

```
SELECT *
FROM COURSE
WHERE COURSE.CO_HRS = 3;
```

Result:

```
Co_name                Co_course# Co_credit Co_college            Co_hrs Co_dpt_dcode
---------------------  ---------- --------- --------------------  ------ ------------
Intro to Economics     15ECON112  U         Arts and Sciences          3            1
Supply Chain Analysis  22QA411    U         Business                   3            3
Principles of IS       22IS270    G         Business                   3            7
Programming in C++     20ECES212  G         Engineering                3            6
Optimization           22QA888    G         Business                   3            3
Financial Accounting   18ACCT801  G         Education                  3            4
Database Principles    22IS832    G         Business                   3            7
Systems Analysis       22IS430    G         Business                   3            7
```

The asterisk (*) means that all columns from the table are to be selected. The WHERE clause tells SQL to search the rows in the COURSE table and to return (i.e., display) only those rows where the value of **COURSE.Co_hrs** is equal to exactly 3. In addition, while not required, in an SQL SELECT statement it is a good idea to prefix each column name with its table name in order to minimize ambiguity and confusion.[10]

Since SQL SELECT statements are not case sensitive, the two SQL statements shown above could also have been written as follows:

```
select co_name, co_course#, co_credit, co_college, co_hrs, co_dpt_dcode
from course
where co_hrs = 3;

select *
from course
where co_hrs = 3;
```

While each of these statements is semantically correct, once again we recommend that each column name referenced be preceded (i.e., prefixed) by the table name COURSE in either lower or upper case.

The ORDER BY clause is used to specify how the results of a query are to be sorted. The default sort is an ascending sort. The keywords, ASCENDING and DESCENDING (abbreviated ASC and DESC), are used to control the sorting of each column. Thus the query:

```
SELECT *
FROM COURSE
WHERE COURSE.CO_HRS = 3
ORDER BY COURSE.CO_DPT_DCODE DESC;
```

---

[10] The convention of prefixing each column name with its table name is used throughout this chapter. SQL however, permits duplicate column names as long as they appear in different tables. When two tables involved in a SELECT statement share a common column name or names, the qualification of the column name(s) with the appropriate table name is required if the column name appears in the <attribute list> or in the WHERE condition. Otherwise ambiguity will exist and an error message of the form "column ambiguously defined" will appear. Since the column names used in the tables for Madeira College adhere to the convention for developing attribute names described in Chapter 6 (see Section 6.2), ambiguous column names will not occur in the examples in this book.

lists the three-hour courses in descending order by the department number in which the course is offered. The query:

```
SELECT *
FROM COURSE
WHERE COURSE.CO_HRS = 3
ORDER BY COURSE.CO_DPT_DCODE DESC,COURSE.CO_NAME;
```

lists the three-hour courses in descending order by the department number in which the course is offered and in ascending order by course name within each department.

Example 1.1.2 (Corresponds to Selection Example 2 in Section 11.1.1.1). Which courses offered by department 7 are three-hour courses?

SQL SELECT Statement:

```
SELECT *
FROM COURSE
WHERE COURSE.CO_DPT_DCODE = 7 AND COURSE.CO_HRS = 3;
```

Result:

Co_name	Co_course#	Co_credit	Co_college	Co_hrs	Co_dpt_dcode
Principles of IS	22IS270	G	Business	3	7
Database Principles	22IS832	G	Business	3	7
Systems Analysis	22IS430	G	Business	3	7

Use of the logical operator AND requires that *both* conditions (i.e., **COURSE.Co_dpt_dcode** = 7 and **COURSE.Co_hrs** = 3) be satisfied.

Example 1.1.3. Which sections have a maximum number of students greater than 30 or are offered in Lindner 110?

SQL SELECT Statement:

```
SELECT *
FROM SECTION
WHERE SECTION.SE_MAXST > 30 OR SECTION.SE_ROOM = 'Lindner 110';
```

Result:

Se_section#	Se_qtr	Se_year	Se_time	Se_maxst	Se_room	Se_co_course#	Se_pr_profid
901	A	2006	W1800	35	Rhodes 611	22IS270	SK85977
101	S	2006	T1045	29	Lindner 110	22IS330	SK85977
102	S	2006	H1045	29	Lindner 110	22IS330	CC49234
701	W	2007	M1000	33	Braunstien 211	22IS832	CC49234
101	U	2007	T1015	33		22QA375	HT54347

Use of the logical operator OR requires that either one or both conditions (i.e., **SECTION.Se_maxst** > 30 OR **SECTION.Se_room** = 'Lindner 110') be satisfied. Observe that in Example 1.1.3 above, none of the rows in the SECTION table satisfy both conditions of the WHERE clause. Also, whenever a character or string literal (e.g., Lindner 110) is used as part of a condition, the value must be enclosed within *single quotation marks*. Character or string literals enclosed in single quotation marks are case sensitive. Thus the WHERE condition will not be true if anything other than Lindner 110 appears inside the single quotation marks.

11.2.1.2 Use of Comparison and Logical Operators

Combining AND or OR in the same logical expression must be done with care. When AND and OR appear in the same WHERE clause, all the ANDs are performed first; then all the ORs are performed. In this way, AND is said to have a higher precedence than OR.

For example, the WHERE clause:

```
WHERE COURSE.CO_CREDIT = 'U' AND COURSE.CO_COLLEGE = 'Business' OR
COURSE.CO_COLLEGE = 'Engineering'
```

is evaluated in the following manner:

1. Each of the component expressions is evaluated yielding a value that is either "true" or "false."
2. The results of **COURSE.Co_credit** = 'U' AND **COURSE.Co_college** = 'Business' are combined, yielding a result that is "true" if both expressions are true, and otherwise is "false."
3. The result of **COURSE.Co_college** = 'Engineering' is evaluated as being either "true" or "false."
4. The results of Steps 2 and 3 are combined, yielding a result that is "true" if either result is "true" and "false" if both results are "false."

Applying this evaluation scheme to the first row of the COURSE table yields the following results:

1. **COURSE.Co_credit** = 'U' — True
    COURSE.Co_college = 'Business' — False
    COURSE.Co_college = 'Engineering' — False
2. **COURSE.Co_credit** = 'U' AND **COURSE.Co_college** = 'Business' — False
3. **COURSE.Co_college** = 'Engineering' — False
4. Since both steps 2 and 3 are false, the overall result is *false* and the first row fails to satisfy the WHERE clause.

In SQL, all operators are arranged in a hierarchy that determines their precedence. In any expression, operations are performed in order of their precedence, from highest to lowest. When operators of equal precedence are used next to each other, they are performed from left to right. The precedence of common logical operators in SQL is:

1. All of the comparison operators have equal precedence
    =, <>, <=, <, >=, >
2. NOT
3. AND
4. OR

When the normal rules of operator precedence do not fit the needs, one can override them by placing part of an expression in parentheses. That part of the expression will be evaluated first; then the rest of the expression will be evaluated.

The following examples illustrate the incorporation of logical operators in a SELECT statement. Note that Example 1.2.1 is based on the WHERE clause discussed above.

Example 1.2.1

```
SELECT *
FROM COURSE
WHERE COURSE.CO_CREDIT = 'U'
```

521

Data Manipulation: Relational Algebra and SQL

```
AND COURSE.CO_COLLEGE = 'Business'
OR COURSE.CO_COLLEGE = 'Engineering';
```

Result:

Co_name	Co_course#	Co_credit	Co_college	Co_hrs	Co_dpt_dcode
Operations Research	22QA375	U	Business	2	3
Supply Chain Analysis	22QA411	U	Business	3	3
Programming in C++	20ECES212	G	Engineering	3	6
Database Concepts	22IS330	U	Business	4	7

In Example 1.2.2, the rows with **COURSE.Co_college** equal to Business or Engineering are selected first as a result of the presence of the parentheses. However, in order to be included in the result, **COURSE.Co_credit** must be equal to U. This eliminates the graduate Programming in C++ course offered in the College of Engineering.

Example 1.2.2

```
SELECT *
FROM COURSE
WHERE COURSE.CO_CREDIT = 'U'
AND (COURSE.CO_COLLEGE = 'Business'
OR COURSE.CO_COLLEGE = 'Engineering');
```

Result:

Co_name	Co_course#	Co_credit	Co_college	Co_hrs	Co_dpt_dcode
Operations Research	22QA375	U	Business	2	3
Supply Chain Analysis	22QA411	U	Business	3	3
Database Concepts	22IS330	U	Business	4	7

In Example 1.2.3, each of the 11 rows satisfies at least one of the conditions in parentheses (**COURSE.Co_college** <> 'Business' OR **COURSE.Co_college** <> 'Engineering'). Five of these rows are also associated with a course that is offered for undergraduate credit.

Example 1.2.3

```
SELECT *
FROM COURSE
WHERE (COURSE.CO_COLLEGE <> 'Business'
OR COURSE.CO_COLLEGE <> 'Engineering')
AND CO_CREDIT = 'U';
```

Result:

Co_name	Co_course#	Co_credit	Co_college	Co_hrs	Co_dpt_dcode
Intro to Economics	15ECON112	U	Arts and Sciences	3	1
Operations Research	22QA375	U	Business	2	3
Intro to Economics	18ECON123	U	Education	4	4
Supply Chain Analysis	22QA411	U	Business	3	3
Database Concepts	22IS330	U	Business	4	7

In Example 1.2.4, the condition **COURSE.Co_college** <> 'Engineering' AND **COURSE.Co_credit** = 'U' is evaluated first and is satisfied by the rows for the five undergraduate courses. The condition **COURSE.Co_college** <> 'Business' is evaluated next and satisfied by the rows for the two Intro to Economics courses, the graduate Programming in C++

course, and the graduate Financial Accounting course. The rows for the two Intro to Economics courses satisfy both conditions, while each of the other five rows satisfies one condition.

Example 1.2.4

```
SELECT *
FROM COURSE
WHERE COURSE.CO_COLLEGE <> 'Business'
OR COURSE.CO_COLLEGE <> 'Engineering'
AND CO_CREDIT = 'U';
```

Result:

Co_name	Co_course#	Co_credit	Co_college	Co_hrs	Co_dpt_dcode
Intro to Economics	15ECON112	U	Arts and Sciences	3	1
Operations Research	22QA375	U	Business	2	3
Intro to Economics	18ECON123	U	Education	4	4
Supply Chain Analysis	22QA411	U	Business	3	3
Programming in C++	20ECES212	G	Engineering	3	6
Financial Accounting	18ACCT801	G	Education	3	4
Database Concepts	22IS330	U	Business	4	7

The keywords IN or NOT IN can also be used as comparison operators. IN is evaluated in the context of being "equal to any member of" a set of values. Thus Example 1.2.5 is equivalent to Example 1.2.2.

Example 1.2.5

```
SELECT *
FROM COURSE
WHERE COURSE.CO_CREDIT = 'U'
AND COURSE.CO_COLLEGE IN ('Business', 'Engineering');
```

Result:

Co_name	Co_course#	Co_credit	Co_college	Co_hrs	Co_dpt_dcode
Operations Research	22QA375	U	Business	2	3
Supply Chain Analysis	22QA411	U	Business	3	3
Database Concepts	22IS330	U	Business	4	7

The logical operator NOT reverses the result of a logical expression. NOT can be used to precede any of the comparison operators =, <>, <=, <, >=, > as well as the word IN. Thus Example 1.2.6 displays all courses offered for undergraduate credit that are not offered at either the College of Business or the College of Engineering.

Example 1.2.6

```
SELECT *
FROM COURSE
WHERE COURSE.CO_CREDIT = 'U'
AND COURSE.CO_COLLEGE NOT IN ('Business', 'Engineering');
```

Result:

Co_name	Co_course#	Co_credit	Co_college	Co_hrs	Co_dpt_dcode
Intro to Economics	15ECON112	U	Arts and Sciences	3	1
Intro to Economics	18ECON123	U	Education	4	4

So far, the conditions in the WHERE clause have involved comparisons where the operands were of the form:

```
<column name><comparison operator><constant value>
```

and the constant value was either a numeric constant or a character constant (i.e., character string). However, as we will see, the operands of a comparison operator can be expressions, not just a simple column name or constant. For example, suppose we are interested in identifying all professors with a monthly salary that exceeds $6,000.

Example 1.2.7

```
SELECT *
FROM PROFESSOR
WHERE PROFESSOR.PR_SALARY/12 > 6000;
```

Result:

Pr_name	Pr_empid	Pr_phone	Pr_office	Pr_birthdate	Pr_datehired	Pr_dpt_dcode	Pr_salary
Mike Faraday	FM49276	5235568492	249 McMicken	1960-08-26	1996-05-01	1	92000
Chelsea Bush	BC65437	5235567777	227 Lindner	1946-09-03	1993-05-01	3	77000
Tony Hopkins	HT54347	5235569977	324 Lindner	1949-11-24	1997-01-20	3	77000
Alan Brodie	BA54325	5235569876	238 Lindner	1944-01-14	2000-05-16	3	76000
Marie Curie	CM65436	5235569899	331 Dyer	1972-02-29	1999-10-22	4	99000
John Nicholson	NJ43728	5235569999	324 Dyer	1966-05-01	2003-06-22	4	99000

If we follow this Selection operation with a Projection operation, the query in Example 1.2.7 could be rewritten to display just the name and monthly salary of each qualifying professor.

```
SELECT PROFESSOR.PR_NAME, PROFESSOR.PR_SALARY/12 AS "Monthly Salary"
FROM PROFESSOR
WHERE PROFESSOR.PR_SALARY/12 > 6000;
```

Result:

Pr_name	Monthly Salary
Mike Faraday	76611.66667
Chelsea Bush	64111.66667
Tony Hopkins	64111.66667
Alan Brodie	6333.33333
Marie Curie	8250
John Nicholson	8250

Rather than displaying the heading of the second column as **PROFESSOR.Pr_salary**/12, SQL allows the column name to be changed with the use of the keyword AS[11] and thus provide a more descriptive column heading as a column alias. It should be noted that this column alias cannot be used in other clauses associated with the SELECT statement. The TRUNC function of the form TRUNC (**PROFESSOR.Pr_salary**/12,2), or the ROUND function of the form ROUND (**PROFESSOR.Pr_salary**/12,2), could be used to either truncate or round each monthly salary to two places to the right of the decimal point.

---

[11] The keyword, AS, is optional and is often used in the column list to distinguish between the column name and column alias.

WHERE clauses can also refer to a range of values through use of the comparison operator BETWEEN, which searches for rows in a specific range of values. For example, suppose we are interested in identifying the name and monthly salary of all professors whose monthly salary is between $6,000 and $7,000.

Example 1.2.8

```
SELECT PROFESSOR.PR_NAME, PROFESSOR.PR_SALARY/12 AS "Monthly Salary"
FROM PROFESSOR
WHERE PROFESSOR.PR_SALARY/12 BETWEEN 6000 AND 7000;
```

Result:

```
Pr_name          Monthly Salary
---------------  --------------
Chelsea Bush        64111.66667
Tony Hopkins        64111.66667
Alan Brodie         6333.33333
```

The BETWEEN operator is inclusive (i.e., professors with a monthly salary of exactly $6,000 or $7,000 would also be included in the result) and can be applied to all data types. Further, NOT can also be used with the BETWEEN operator. For example, the query in Example 1.2.9 identifies all professors whose monthly salary is outside the range of from $6,000 to $7,000. A close inspection of the results indicates that the two professors without a salary (John B. Smith and Tiger Woods) do not appear. *Note*: Professor John Smith has a $45,000 annual salary.

Example 1.2.9

```
SELECT PROFESSOR.PR_NAME, ROUND(PROFESSOR.PR_SALARY/12,0) AS
"Monthly Salary"
FROM PROFESSOR
WHERE PROFESSOR.PR_SALARY/12 NOT BETWEEN 6000 AND 7000;
```

Result:

```
Pr_name          Monthly Salary
---------------  --------------
John Smith               3750
Mike Faraday             7667
Kobe Bryant              5500
Ram Raj                  3667
Prester John             3667
Jessica Simpson          5583
Laura Jackson            3583
Marie Curie              8250
Jack Nicklaus            5583
John Nicholson           8250
Sunil Shetty             5333
Katie Shef               5417
Cathy Cobal              3750
Jeanine Troy             3750
Mike Crick               5750
```

Section 11.2.1.5, "Handling Null Values," explains why John B. Smith and Tiger Woods are not included in the results for either Example 1.2.8 or 1.2.9. Section 11.2.2 illustrates how the WHERE and ON clauses have comparisons where operands of the form:

```
<column name><comparison operator><column name>
```

are used to express join conditions.

11.2.1.3 Examples of the Projection Operation

Examples illustrating use of the relational algebra Project operator in a Projection operation appear in Section 11.1.1.2. SQL SELECT statements that correspond to one of these examples follow along with other examples illustrating how a Projection operation typically follows a Selection operation.

Example 1.3.1 (Corresponds to Projection Example 1 in Section 11.1.1.2). Which colleges offer courses?

SQL SELECT Statement:

```
SELECT COURSE.CO_COLLEGE
FROM COURSE;
```

Execution of the SELECT statement given above, since it refers to only one of the six columns in the COURSE table, actually generates duplicate rows and thus the result shown below does not constitute a relation.

Result:

```
Co_college
--------------------
Arts and Sciences
Business
Education
Business
Business
Engineering
Business
Education
Business
Business
Business
```

However, if the qualifier **DISTINCT** is specified as part of the SELECT statement, duplicate rows are removed. In other words, in SQL, without use of the qualifier DISTINCT, the general form of the SELECT statement is:

```
SELECT ALL <column list>
FROM   <table list>
WHERE <condition>
```

where ALL retains duplicate values in queries (note that ALL is the default as compared to DISTINCT).

Thus the revised SELECT statement with the DISTINCT qualifier produces the following result:

```
SELECT DISTINCT COURSE.CO_COLLEGE
FROM COURSE;
```

Result:

```
Co_college
--------------------
Arts and Sciences
Business
Education
Engineering
```

Example 1.3.2. What is the name and status of each student?
SQL SELECT Statement:

```
SELECT STUDENT.ST_NAME, STUDENT.ST_STATUS
FROM STUDENT;
```

Result:

```
St_name         St_status
--------------  ---------
Elijah Baley    Part time
Daniel Olive    Full time
Wanda Seldon    Full time
Gladis Bale     Full time
Shweta Gupta    Full time
Troy Hudson     Part time
Rick Fox        Full time
Vanessa Fox     Full time
Jenna Hopp      Full time
David Sane      Part time
Tim Duncan      Part time
Joumana Kidd    Part time
Jenny Aniston   Full time
Poppy Kramer    Full time
Sweety Kramer   Full time
Diana Jackson   Part time
```

In most cases, several relational algebra operations are applied one after the other (e.g., a Selection operation is followed by a Projection operation).

Example 1.3.3. What are the names of the courses offered in the College of Business?
SQL SELECT Statement:

```
SELECT COURSE.CO_NAME
FROM COURSE
WHERE COURSE.CO_COLLEGE = 'Business';
```

Result:

```
Co_name
--------------------
Operations Research
Supply Chain Analysis
Principles of IS
Optimization
Database Concepts
Database Principles
Systems Analysis
```

Data Manipulation: Relational Algebra and SQL

Example 1.3.4. What are the names and addresses of all part time students?
SQL SELECT Statement

```
SELECT STUDENT.ST_NAME, STUDENT.ST_ADDRESS
FROM STUDENT
WHERE ST_STATUS = 'Part time';
```

Result:

```
St_name         St_address
-------------   ----------------------
Elijah Baley    2920 Scioto Street
Troy Hudson
David Sane      245 University Avenue
Tim Duncan
Joumana Kidd    2920 Scioto Street
Diana Jackson   2920 Scioto Street
```

Examples 1.2.7 – 1.2.9 in Section 11.2.1.2 illustrate how an expression can be included in column-list of a query. The following query in Example 1.3.5 illustrates that an expression that involves dates can also be part of a column-list of a query.

Example 1.3.5. What was the age (in years) of each professor in department 3 when hired?
SQL SELECT Statement:

```
SELECT PROFESSOR.PR_NAME,
TRUNC((PROFESSOR.PR_DATEHIRED - PROFESSOR.PR_BIRTHDATE)/365.25,0)
"Age When Hired"
FROM PROFESSOR
WHERE PROFESSOR.PR_DPT_DCODE = 3;
```

Result:

```
Pr_name         Age When Hired
-------------   --------------
Chelsea Bush                46
Tony Hopkins                47
Alan Brodie                 56
Jessica Simpson             40
Laura Jackson               26
```

Date arithmetic allowing one date to be subtracted from another date is discussed in Chapter 12.

11.2.1.4 Grouping and Summarizing

SQL allows the grouping of rows into sets; it then can summarize data in such a way that one row is returned for each set. The GROUP BY and HAVING clauses facilitate the grouping, while built-in aggregate functions are used in conjunction with grouping.

Aggregate functions take as input a set of values, one from each row in a group of rows, and return one value as output. Common aggregate functions include:

- COUNT (x) — counts the number of non-null values in a set of values
- SUM (x) — sums all numbers in a set of values
- AVG (x) — computes the average of a set of numbers in a set of values

- MAX (x) — computes the maximum of a set of numbers in a set of values
- MIN (x) — computes the minimum of a set of numbers in a set of values

Note: x represents a numeric column name.

Example 1.4.1. Count the number of salaries in the PROFESSOR table and, at the same time, display the sum of the salaries, the average salary, the maximum salary, and the minimum salary.

SQL SELECT Statement:

```
SELECT COUNT(PR_SALARY), SUM(PR_SALARY),AVG(PR_SALARY),
MAX(PR_SALARY), MIN(PR_SALARY)
FROM PROFESSOR;
```

Result:

COUNT(Pr_salary)	SUM(Pr_salary)	AVG(Pr_salary)	MAX(Pr_salary)	MIN(Pr_salary)
18	1184000	65777.7778	99000	43000

Note that while there are 20 rows in the PROFESSOR table, the salaries of two professors (John B. Smith and Tiger Woods) are unknown and thus a null value is stored in the respective **Pr_salary** column for these professors. Section 11.2.1.5 discusses the impact of null values on aggregate functions.

Aggregate functions are often used in conjunction with groups of rows rather than with all rows in a table. The column or columns on which the grouping takes place are called the grouping column(s). Doing this requires the application of the GROUP BY clause. The GROUP BY clause divides data into sets (i.e., groups) based on the contents of specified columns. The general form of the GROUP BY clause is:

```
GROUP BY column name, [,column name,...]
```

The grouping column(s) must also appear in the SELECT clause so that the value from applying each function to a group of rows appears along with the value of the grouping column(s).

Example 1.4.2. Count the number full-time and part-time students.

SQL SELECT Statement:

```
SELECT STUDENT.ST_STATUS, COUNT(*)
FROM STUDENT
GROUP BY STUDENT.ST_STATUS;
```

Result:

St_status	COUNT(*)
Full time	10
Part time	6

Example 1.4.3. Calculate the average salary of the professors in each department.
SQL SELECT Statement:

```
SELECT PROFESSOR.PR_DPT_DCODE, AVG(PROFESSOR.PR_SALARY)
FROM PROFESSOR
GROUP BY PROFESSOR.PR_DPT_DCODE;
```

Result:

```
Pr_dpt_dcode AVG(PROFESSOR.Pr_salary)
------------ -----------------------
           1              67666.6667
           3                    68000
           4              88333.3333
           6                    44000
           7                    58000
           9                    57000
```

As mentioned above, grouping can be done on a combination of columns. Example 1.4.4 illustrates the use of the GROUP BY clause to count the number of sections offered during each year and quarter combination.

Example 1.4.4. Count the number of sections offered during each quarter of each year.
SQL SELECT Statement:

```
SELECT SECTION.SE_YEAR, SECTION.SE_QTR, COUNT(*)
FROM SECTION
GROUP BY SECTION.SE_YEAR, SECTION.SE_QTR;
```

Result:

```
Se_year Se_qtr   COUNT(*)
------- ------ ----------
   2006 A             2
   2006 S             2
   2007 A             3
   2007 S             1
   2007 U             1
   2007 W             2
```

The HAVING clause is used in conjunction with the GROUP BY clause to place restrictions on the rows returned by the GROUP BY clause in a query. A condition in a HAVING clause must always involve an aggregation. In addition, a HAVING clause cannot be used apart from an associated GROUP BY clause.

Example 1.4.5. Calculate the average salary of the professors in each department for those departments where the minimum salary of a professor is $45,000 or more.
SQL SELECT Statement:

```
SELECT PROFESSOR.PR_DPT_DCODE, AVG(PROFESSOR.PR_SALARY)
FROM PROFESSOR
GROUP BY PROFESSOR.PR_DPT_DCODE
HAVING MIN(PROFESSOR.PR_SALARY) >= 45000;
```

Result:

```
Pr_dpt_dcode AVG(PROFESSOR.Pr_salary)
------------ -----------------------
           1              67666.6667
           4              88333.3333
```

```
          7                    58000
          9                    57000
```

Note that departments 3 and 6 do not appear in the results because there is at least one professor in each department with a salary less than $45,000. In addition, observe that the calculation of the average salary of the professors in department 9 ignores Tiger Woods since his salary is a null value (i.e., is not available). The following section goes into more detail on the handling of null values. In addition, Section 11.2.2.3 describes how to join tables in SQL, which would represent one way to revise the query in Example 1.4.5 in order to display department names instead of department numbers.

11.2.1.5 Handling Null Values

It is important to understand how null values are handled in SQL. A data field without a value in it is said to contain a **null value**. A null value can occur in two situations: (a) where a value is unknown (e.g., the salary of an employee is unknown or unavailable), and (b) where a value is not meaningful (e.g., in a column representing "commission" for an employee who is not a salesperson and thus is not eligible for a commission).

A number field containing a null value is different from one that contains a value of zero. Null values are displayed as blanks while zero values are displayed as numeric zeros. A null value will evaluate to null in any expression. For example, a null value added to 15 results in a null value; a null value minus a null value yields a null value, not zero. The aggregate function COUNT(*) counts all null and not null rows in a table; COUNT (attribute) counts all rows whose attribute value is not null. Other SQL aggregate functions ignore null values in their computation.

This section consists of examples that illustrate the impact of null values in SQL. Each of the examples is based on the textbook table whose structure and initial contents follow.

```
CREATE TABLE TEXTBOOK
(TX_ISBN      VARCHAR(14)    CONSTRAINT PK_10TB PRIMARY KEY,
TX_TITLE      VARCHAR(22)    CONSTRAINT NN_10TIT NOT NULL,
TX_YEAR       INTEGER(4),
TX_PUBLISHER CHAR(13));


SELECT *
FROM TEXTBOOK;

Tx_isbn          Tx_title                   Tx_year Tx_publisher
--------------   ----------------------     ---------- -------------
000-66574998     Database Management          1999 Thomson
003-6679233      Linear Programming           1997 Prentice-Hall
001-55-435       Simulation Modeling          2001 Springer
118-99898-67     Systems Analysis             2000 Thomson
77898-8769       Principles of IS             2002 Prentice-Hall
0296748-99       Economics For Managers       2001
0296437-1118     Programming in C++           2002 Thomson
012-54765-32     Fundamentals of SQL          2004
111-11111111     Data Modeling                2006
```

The behavior of null values in SQL can on occasion be surprising or unintuitive. In order to illustrate differences between how SQL handles a null value (defined as a character string zero characters long) versus a character string that consists of a single blank space, a null value appears in the **TEXTBOOK.Tx_publisher** column for both the titles Economics For Managers and Fundamentals of SQL[12] while the value in the **TEXTBOOK. Tx_publisher** column for the title Data Modeling consists of a single blank space.

As expected, both the previous query and the following query (labeled Example 1.5.1) display the **TEXTBOOK.Tx_publisher** column for Economics For Managers, Fundamentals of SQL, and Data Modeling as blank values in the **TEXTBOOK.Tx_publisher** column.

Example 1.5.1

```
SELECT TEXTBOOK.TX_PUBLISHER
FROM TEXTBOOK;
```

Result:

```
Tx_publisher
-------------
Thomson
Prentice-Hall
Springer
Thomson
Prentice-Hall

Thomson



9 rows selected.
```
[13]

A query (see Example 1.5.2) that displays the textbooks where the **TEXTBOOK. Tx_publisher** column contains a not null value excludes the rows associated with the titles Economics For Managers and Fundamentals of SQL, since the **TEXTBOOK.Tx_publisher** column for each of these textbooks contains a null value. The row associated with the title Data Modeling is not excluded since the value in the **TEXTBOOK.Tx_publisher** column for this title is not null (i.e., consists of a single blank space).

Example 1.5.2

```
SELECT TEXTBOOK.TX_TITLE, TEXTBOOK.TX_PUBLISHER
FROM TEXTBOOK
WHERE TEXTBOOK.TX_PUBLISHER IS NOT NULL;
```

Result:

Tx_title	Tx_publisher
Database Management	Thomson
Linear Programming	Prentice-Hall
Simulation Modeling	Springer

---

[12] In SQL-92, a character value that is zero characters long is treated as a null value.

[13] Most versions of SQL contain a system variable that allows for the number of rows retrieved by the execution of a query to be displayed. The value of this variable will be shown in conjunction with various examples when appropriate. It is useful here as a way of showing that all rows from the TEXTBOOK table were selected.

```
Systems Analysis        Thomson
Principles of IS        Prentice-Hall
Programming in C++      Thomson
Data Modeling
```

```
7 rows selected.
```

The only comparison operators that can be used with null values are IS NULL and IS NOT NULL. If any other operator (e.g., =, >, <>, etc.) is used with a null value, the result is always unknown.[14] In addition, since a NULL represents a lack of data, a null value cannot be equal or unequal to any other value, even another NULL. Examples 1.5.3, 1.5.4, and 1.5.5 illustrate the fact that only IS NULL and IS NOT NULL can be used as comparison operators with null values.

Example 1.5.3

```
SELECT *
FROM TEXTBOOK
WHERE TEXTBOOK.TX_PUBLISHER = NULL;
```

Result:

```
no rows selected
```

Example 1.5.4

```
SELECT *
FROM TEXTBOOK
WHERE TEXTBOOK.TX_PUBLISHER <> NULL;
```

Result:

```
no rows selected
```

While conditional expressions of the form "WHERE X = NULL" and "WHERE X <> NULL" are illegal, SQL does not generate a syntax error. This can create serious problems in cases where "WHERE X IS NULL" would otherwise cause one or more rows to be selected.

Example 1.5.5

```
SELECT *
FROM TEXTBOOK
WHERE TEXTBOOK.TX_PUBLISHER IS NULL;
```

Result:

Tx_isbn	Tx_title	Tx_year	Tx_publisher
0296748-99	Economics For Managers	2001	
012-54765-32	Fundamentals of SQL	2004	

The SELECT statement in Example 1.5.6 demonstrates that there are five distinct publishers in the TEXTBOOK table and reflects the fact that a null value in a column can be distinguished from a column that contains a single blank space. The SELECT statement in Example 1.5.7 (with the is not null condition) indicates that the publisher whose name

---

[14] SQL-92 treats conditions evaluating to unknown as FALSE.

consists of a single blank space can be distinguished from the publisher with a null value for its name.

Example 1.5.6

```
SELECT DISTINCT TEXTBOOK.TX_PUBLISHER
FROM TEXTBOOK;
```

Result:

```
Tx_publisher
-------------
Prentice-Hall
Springer
Thomson


5 rows selected.
```

Example 1.5.7

```
SELECT DISTINCT TEXTBOOK.TX_PUBLISHER
FROM TEXTBOOK
WHERE TEXTBOOK.TX_PUBLISHER IS NOT NULL;
```

Result:

```
Tx_publisher
-------------
Prentice-Hall
Springer
Thomson

4 rows selected.
```

Example 1.5.8 uses the COUNT function to count the number of rows in the TEXTBOOK table and returns the result in a single table with a single column. Recall that when COUNT(*) is used, SQL focuses on the presence of rows rather than values appearing in a column.

Example 1.5.8

```
SELECT COUNT(*)
FROM TEXTBOOK;
```

Result:

```
  COUNT(*)
----------
         9

1 row selected.
```

When the COUNT function refers to a column, it behaves like the other aggregate functions and ignores null values (see Section 11.2.1.4). Thus the query in Example 1.5.9 displays the number of rows in the TEXTBOOK table with something other than a null value in the **TEXTBOOK.Tx_publisher** column.

Example 1.5.9

```
SELECT COUNT(TEXTBOOK.TX_PUBLISHER)
FROM TEXTBOOK;
```

Result:

```
COUNT(TEXTBOOK.Tx_publisher)
---------------------------
                          7
```

```
1 row selected.
```

Numeric functions, such as the COUNT function, are associated with columns of output whose width is equal to the number of characters required to display the name of the function plus its argument(s). Often, a column alias, such as the one used in Example 1.5.10, is used to provide a more descriptive column heading. Observe how the width of the column has been adjusted to display the entire column alias (i.e., heading).

Example 1.5.10

```
SELECT COUNT(TEXTBOOK.TX_PUBLISHER) AS "Number of Publishers"
FROM TEXTBOOK;
```

Result:

```
Number of Publishers
--------------------
                   7
```

In Example 1.5.10, the two rows in the TEXTBOOK table with null publishers are ignored by the COUNT function. This can be verified by the query in Example 1.5.11 which counts the number of distinct not null values in the **TEXTBOOK.Tx_publisher** column.

Example 1.5.11

```
SELECT COUNT(DISTINCT TEXTBOOK.TX_PUBLISHER) AS "Number of Distinct
Publishers"
FROM TEXTBOOK;
```

Result:

```
Number of Distinct Publishers
-----------------------------
                            4
```

```
1 row selected.
```

Aggregate functions are frequently applied to groups of rows in a table rather than to all rows in a table. The SELECT statement in Example 1.5.12 reflects the fact that there are five distinct publishers in the TEXTBOOK table and counts the number of rows (i.e., the COUNT (*) function is focusing on the presence of a row) associated with each distinct publisher.

Example 1.5.12

```
SELECT TEXTBOOK.TX_PUBLISHER, COUNT(*)
FROM TEXTBOOK
GROUP BY TEXTBOOK.TX_PUBLISHER;
```

Result:

```
Tx_publisher    COUNT(*)
------------- ----------
                       1
Prentice-Hall          2
Springer               1
Thomson                3
                       2
```

5 rows selected.

Observe that the last row displayed is associated with the two textbooks with a null value in the **TEXTBOOK.Tx_publisher** column.

The SELECT statement in Example 1.5.13 also recognizes that there are five distinct publishers, but since group functions ignore the presence of null values in the **TEXTBOOK.Tx_publisher** column, the accumulator set up for the publisher whose value is a null value never has its initial value of zero incremented.

Example 1.5.13

```
SQL> SELECT TEXTBOOK.TX_PUBLISHER, COUNT(TEXTBOOK.TX_PUBLISHER)
FROM TEXTBOOK
GROUP BY TEXTBOOK.TX_PUBLISHER;
```

Result:

```
Tx_publisher   COUNT(TEXTBOOK.Tx_publisher)
------------- ----------------------------
                                         1
Prentice-Hall                            2
Springer                                 1
Thomson                                  3
                                         0
```

5 rows selected.

The SELECT statement in Example 1.5.14 works as expected, since the WHERE clause places the condition that the value in the **TEXTBOOK.Tx_publisher** column must be not null before that publisher can be used to form a group. The SELECT statement in Example 1.5.15 also works as expected since the WHERE clause focuses only on the publisher whose name consists of a single blank space. Finally, the SELECT statement in Example 1.5.16 serves as a reminder that the only publishers whose name consists of something other than a single blank space are Thomson, Prentice-Hall, and Springer. Recall, using a <> comparison operator in the evaluation of whether a null value differs from a single blank space produces an unknown result (which is treated as false).

Example 1.5.14

```
SELECT TEXTBOOK.TX_PUBLISHER, COUNT(*)
FROM TEXTBOOK
WHERE TEXTBOOK.TX_PUBLISHER IS NOT NULL
GROUP BY TEXTBOOK.TX_PUBLISHER;
```

Result:

```
Tx_publisher     COUNT(*)
-------------    ----------
                         1
Prentice-Hall            2
Springer                 1
Thomson                  3.
```

4 rows selected.

Example 1.5.15

```
SELECT TEXTBOOK.TX_PUBLISHER, COUNT(*)
FROM TEXTBOOK
WHERE TEXTBOOK.TX_PUBLISHER = ' '
GROUP BY TEXTBOOK.TX_PUBLISHER;
```

Result:

```
Tx_publisher     COUNT(*)
-------------    ----------
                         1
```

1 row selected.

Example 1.5.16

```
SELECT TEXTBOOK.TX_PUBLISHER, COUNT(*)
FROM TEXTBOOK
WHERE TEXTBOOK.TX_PUBLISHER <> ' '
GROUP BY TEXTBOOK.TX_PUBLISHER;
```

Result:

```
Tx_publisher     COUNT(*)
-------------    ----------
Prentice-Hall            2
Springer                 1
Thomson                  3
```

3 rows selected.

The SELECT statements in Examples 1.5.17, 1.5.18, and 1.5.19 illustrate the impact of a null value in queries that involve multiple conditions. In Example 1.5.17, the first condition selects the rows associated with the four not null publishers, while the second condition selects the rows associated with the three publishers Thomson, Prentice-Hall, and Springer. Since OR is used to connect the two conditions, groups are formed and counts accumulated for the four not null publishers.

Example 1.5.17

```
SELECT TEXTBOOK.TX_PUBLISHER, COUNT(*)
FROM TEXTBOOK
WHERE TEXTBOOK.TX_PUBLISHER IS NOT NULL
OR TEXTBOOK.TX_PUBLISHER <> ' '
GROUP BY TEXTBOOK.TX_PUBLISHER;
```

Result:

```
Tx_publisher    COUNT(*)
------------- ----------
                        1
Prentice-Hall           2
Springer                1
Thomson                 3
```

4 rows selected.

In Example 1.5.18, since AND is used to connect the two conditions, groups are formed only for those publishers that satisfy both conditions (i.e., Thomson, Prentice-Hall, and Springer).

Example 1.5.18

```
SELECT TEXTBOOK.TX_PUBLISHER, COUNT(*)
FROM TEXTBOOK
WHERE TEXTBOOK.TX_PUBLISHER IS NOT NULL
AND TEXTBOOK.TX_PUBLISHER <> ' '
GROUP BY TEXTBOOK.TX_PUBLISHER;
```

Result:

```
Tx_publisher    COUNT(*)
------------- ----------
Prentice-Hall           2
Springer                1
Thomson                 3
```

3 rows selected.

The SELECT statement in Example 1.5.19 illustrates a situation where the first condition selects the rows associated with the three publishers other than the single-space publisher and the second condition selects rows associated with the **TEXTBOOK.Tx_publisher** column containing a null value. Since OR is used to connect the two conditions, groups are formed and counts accumulated for Thomson, Prentice-Hall, and Springer, and the rows associated with the **TEXTBOOK.Tx_publisher** column containing a null value.

Example 1.5.19

```
SELECT TEXTBOOK.TX_PUBLISHER, COUNT(*)
FROM TEXTBOOK
WHERE TEXTBOOK.TX_PUBLISHER <> ' '
OR TEXTBOOK.TX_PUBLISHER IS NULL
GROUP BY TEXTBOOK.TX_PUBLISHER;
```

Result:

```
Tx_publisher    COUNT(*)
------------- ----------
Prentice-Hall           2
Springer                1
Thomson                 3
                        2
```

4 rows selected.

The SELECT statement in Example 1.5.20 introduces the use of the operator IN to test whether a value is contained within a set of values.

Example 1.5.20

```
SELECT *
FROM TEXTBOOK
WHERE TEXTBOOK.TX_PUBLISHER IN ('Thomson', 'Springer');
```

Result:

```
Tx_isbn          Tx_title                Tx_year Tx_publisher
-------------    ----------------------  ------- -------------
000-66574998     Database Management        1999 Thomson
001-55-435       Simulation Modeling        2001 Springer
118-99898-67     Systems Analysis           2000 Thomson
0296437-1118     Programming in C++         2002 Thomson
```

4 rows selected.

Prefaced with the word NOT, NOT IN tests for whether a value does not appear within a set of values. Example 1.5.21 illustrates how null values only satisfy a WHERE clause that uses IS NULL or IS NOT NULL since the titles Economics For Managers and Fundamentals of SQL do not appear in the results but Prentice-Hall and the single-space publisher do.

Example 1.5.21

```
SELECT * FROM TEXTBOOK
WHERE TEXTBOOK.TX_PUBLISHER NOT IN ('Thomson', 'Springer');
```

Result:

```
Tx_isbn          Tx_title                Tx_year Tx_publisher
-------------    ----------------------  ------- -------------
003-6679233      Linear Programming         1997 Prentice-Hall
77898-8769       Principles of IS           2002 Prentice-Hall
111-11111111     Data Modeling              2006
```

3 rows selected.

11.2.1.6 Pattern Matching in SQL

In the context of the TEXTBOOK table, pattern matching would be useful if we were trying to find all textbooks used in Madeira College with a specific word or phrase in their titles. Examples include (a) all textbooks with the word "Introduction" in the title, (b) all textbooks whose title includes the words "Information System" or "Accounting System," and (c) all textbooks with titles that include "Programming." SQL-92 supports pattern matching through the use of the LIKE operator in conjunction with the two wildcard characters—the percent character (%) and the underscore character (_).[15] The percent character represents *a series of one or more unspecified characters* while the underscore character represents *exactly one character*. Let us continue to use the data associated with the TEXTBOOK table as a vehicle to explore use of the LIKE operator. Given the number of rows and columns associated with

---

[15] Although the SQL-92 standard specifies use of the underscore character(_) and percent character (%), some implementations of SQL use other characters to represent a single character or a series of characters.

Data Manipulation: Relational Algebra and SQL

the TEXTBOOK table, the examples that follow have limited practical application but instead are intended to simply illustrate how the LIKE operator works.

The SELECT statement in Example 1.6.1 searches for all textbooks such that the first two characters of their ISBN begin with the digit 1 and is followed by any alphanumeric character. Observe that while the textbooks with the titles Systems Analysis and Data Modeling seem to satisfy this request, no rows are selected. Likewise, the SELECT statement in Example 1.6.2 searches for all textbooks whose title contains the letter "i" in the second character position. While two textbooks would appear to satisfy this request, no rows are selected.

Example 1.6.1

```
SELECT TEXTBOOK.TX_TITLE, TEXTBOOK.TX_YEAR
FROM TEXTBOOK
WHERE TEXTBOOK.TX_ISBN LIKE '1_';
```

Result:

```
No rows selected.
```

Example 1.6.2

```
SELECT TEXTBOOK.TX_TITLE
FROM TEXTBOOK
WHERE TEXTBOOK.TX_TITLE LIKE '_i';
```

Result:

```
No rows selected.
```

The SELECT statements in Examples 1.6.1 and 1.6.2 return no rows because no textbook has an ISBN number that is exactly two characters long nor is there any textbook title that has the letter "i" in the second character position and is exactly two characters long. As illustrated in Example 1.6.3, revising Example 1.6.2 by adding the % sign as a wildcard character searches for textbook titles with the letter "i" in the second character position but allows the title to have as many as twenty-two characters (the width of the **TEXTBOOK.Tx_title** column).

Example 1.6.3

```
SELECT TEXTBOOK.TX_TITLE
FROM TEXTBOOK
WHERE TEXTBOOK.TX_TITLE LIKE '_i%';
```

Result:

```
Tx_title
---------------------
Linear Programming
Simulation Modeling

2 rows selected.
```

Examples 1.6.4 and 1.6.5 illustrate that the LIKE Operator is case sensitive.

Example 1.6.4

```
SELECT TEXTBOOK.TX_TITLE
FROM TEXTBOOK
WHERE TEXTBOOK.TX_TITLE LIKE 'P%';
```

Result:

```
Tx_title
---------------------
Principles of IS
Programming in C++

2 rows selected.
```

Example 1.6.5

```
SELECT TEXTBOOK.TX_TITLE
FROM TEXTBOOK
WHERE TEXTBOOK.TX_TITLE LIKE 'p%';
```

Result:

```
No rows selected.
```

The SELECT statements in Examples 1.6.6 and 1.6.7 represent rather unusual uses of the LIKE Operator. Example 1.6.6 searches for the titles of all textbooks that contain the letter "e" while Example 1.6.7 displays the titles of all textbooks.

Example 1.6.6

```
SELECT TEXTBOOK.TX_TITLE
FROM TEXTBOOK
WHERE TEXTBOOK.TX_TITLE LIKE '%e%';
```

Result:

```
Tx_title
---------------------
Database Management
Linear Programming
Simulation Modeling
Systems Analysis
Principles of IS
Economics For Managers
Fundamentals of SQL
Data Modeling

8 rows selected.
```

Example 1.6.7

```
SELECT TEXTBOOK.TX_TITLE
FROM TEXTBOOK
WHERE TEXTBOOK.TX_TITLE LIKE '%';
```

Result:

```
Tx_title
---------------------
Database Management
Linear Programming
```

```
Simulation Modeling
Systems Analysis
Principles of IS
Economics For Managers
Programming in C++
Fundamentals of SQL
Data Modeling

9 rows selected.
```

As shown in the SELECT statement in Example 1.6.8, the string '%' cannot match a null value as the two publishers with a null value in the **TEXTBOOK.Tx_publisher** column do not appear in the results.

Example 1.6.8

```
SELECT *
FROM TEXTBOOK
WHERE TEXTBOOK.TX_PUBLISHER LIKE '%';
```

Result:

```
Tx_isbn          Tx_title               Tx_year Tx_publisher
-------------    ---------------------  --------- -------------
000-66574998     Database Management     1999 Thomson
003-6679233      Linear Programming      1997 Prentice-Hall
001-55-435       Simulation Modeling     2001 Springer
118-99898-67     Systems Analysis        2000 Thomson
77898-8769       Principles of IS        2002 Prentice-Hall
0296437-1118     Programming in C++      2002 Thomson
111-11111111     Data Modeling           2006
```

7 rows selected.

CHAR(ℓ) and VARCHAR(ℓ) are string data types (see Table 10.1 in Chapter 10) commonly used to define fixed-length and variable-length character data. A CHAR(ℓ) data type, where ℓ represents the length of the column, has a maximum size of 255 characters in most DBMSs; blank characters are added to the data should the number of characters be less than ℓ. A VARCHAR(ℓ) data type, where ℓ also represents the length of the column, has a maximum size of 2,000 characters in most DBMSs. A VARCHAR(ℓ) data type does not append blank characters to the data if the number of characters is less than ℓ. CHAR(ℓ) and VARCHAR(ℓ) data types can produce different results in some comparisons that involve the LIKE Operator. For example, the query in Example 1.6.9 searches for and locates all titles that end with the letter "s".

Example 1.6.9

```
SELECT *
FROM TEXTBOOK
WHERE TEXTBOOK.TX_TITLE LIKE '%s';
```

Result:

```
Tx_isbn          Tx_title               Tx_year Tx_publisher
-------------    ---------------------  --------- -------------
118-99898-67     Systems Analysis        2000 Thomson
0296748-99       Economics For Managers  2001
```

2 rows selected.

However, had the **TEXTBOOK.Tx_title** column been defined as a CHAR(22) data type instead of as a VARCHAR(22) data type, only one of the titles, Economics For Managers (a title with an "s" in the 22nd character position) would have been located since the 22nd character position in the title, Systems Analysis, would have contained a single blank space character (i.e., the rightmost "s" in Systems Analysis would have been in the 16th character position with character positions 17-22 containing blank spaces).

SQL allows for the definition of an escape character in cases where either a percent character (%) or an underscore character (_) stand for themselves as part of the search. For example, suppose you need to identify the name of each table in the Madeira College database with an underscore character as part of its table name. This is possible with the following SELECT statement:

Example 1.6.10

```
SELECT TABLE_NAME
FROM USER_TABLES16
WHERE TABLE_NAME LIKE '%/_%'
ESCAPE '/';
```

Result:

```
Table_name
------------------------------
GRAD_STUDENT
```

ESCAPE is used here to declare the slash (/) as an escape character so that it can be prefixed to the underscore character in the character string expression used in the LIKE operator.

11.2.2 SQL Queries Based on Binary Operators

Recall that binary operators "operate" on two relations and are of four types: (a) the Cartesian Product operator, (b) set theoretic operators, (c) join operators, and (d) the divide operator. The examples in this section are based on the Madeira College tables shown in Figure 11.1.

11.2.2.1 The Cartesian Product Operation

An SQL SELECT statement that references two or more tables and does not include a WHERE clause always involves a Cartesian Product operation. The Cartesian Product operation by itself is generally meaningless. However, all joins actually begin as a Cartesian Product followed by a WHERE condition that selects only the rows that make sense in the context of the question. In other words, a Cartesian Product is useful when followed first by a Selection operation that matches values of attributes coming from the component relations (technically, a Cartesian Product operation followed by a Selection operation is a Join operation) and then by a Projection operation that selects certain columns from the combined result.

---

[16] USER_TABLES is the name of a data dictionary table comprised of columns that contain data about the various tables that comprise the Madeira College database. Included in this table is a column that records the name of each table.

Example 2.1.1. What is the product of the T and SS relations? *Note:* In order to illustrate some of the SQL queries based on binary operators, relations SS and T have been created from the larger SECTION and TAKES relations. Relation SS (introduced earlier in conjunction with the Right Outer Join example in Section 11.1.2.3) contains four attributes: **Ss_section#, Ss_qtr, Ss_year**, and **Ss_c_course#** and relation T contains **T_ss_c_course#, T_grade, and T_s_sid**.

SQL SELECT Statement:

```
SELECT * FROM T CROSS JOIN SS;
```

In the SQL-92 standard, the CROSS keyword, combined with the JOIN keyword, is used in the FROM clause to create a Cartesian Product. Sometimes a Cartesian Product is referred to as a Cross Join. The first 22 rows of the result of this Cartesian Product appear below. Since there are 11 rows in SS and 11 rows in T, a total of 121 rows are produced as a result of the concatenation of SS and T. The result shown constitutes the concatenation of the first row of SS (observe how the final four columns remain the same) with the 11 rows in T. Observe that the next 11 rows of the result shows second row of SS with each of the 11 rows of T.

Result:

T_se_c_course#	T_grade	T_s_sid	Ss_section#	Ss_qtr	Ss_year	Ss_c_course#
22QA375	A	KP78924	101	A	2007	22QA375
22QA375	A	KS39874	101	A	2007	22QA375
22QA375	B	BG66765	101	A	2007	22QA375
22IS330	C	BE76598	101	A	2007	22QA375
22IS330	B	KJ56656	101	A	2007	22QA375
22IS330	A	KP78924	101	A	2007	22QA375
22IS330	A	KS39874	101	A	2007	22QA375
22IS832	A	KS39874	101	A	2007	22QA375
22IS330	A	BE76598	101	A	2007	22QA375
22IS832	B	BG66765	101	A	2007	22QA375
22IS330	C	GS76775	101	A	2007	22QA375
22QA375	A	KP78924	901	A	2006	22IS270
22QA375	A	KS39874	901	A	2006	22IS270
22QA375	B	BG66765	901	A	2006	22IS270
22IS330	C	BE76598	901	A	2006	22IS270
22IS330	B	KJ56656	901	A	2006	22IS270
22IS330	A	KP78924	901	A	2006	22IS270
22IS330	A	KS39874	901	A	2006	22IS270
22IS832	A	KS39874	901	A	2006	22IS270
22IS330	A	BE76598	901	A	2006	22IS270
22IS832	B	BG66765	901	A	2006	22IS270
22IS330	C	GS76775	901	A	2006	22IS270

When a Cartesian Product operation is accompanied by a Selection operation, an Inner Join results. As shown below, using the SQL-92 standard, the words INNER JOIN[17] are used in the FROM clause instead of the words CROSS JOIN, and the ON clause is used to specify the join condition(s).

---

[17] Use of the word, INNER, is optional.

Example 2.1.2. What are the student IDs of those students who took a section of a course during the year 2006? *Note*: Since the result here explicitly calls for only one column to be displayed, the SECTION and TAKES tables are used in lieu of the abridged versions SS and T used in Example 2.1.1.

SQL SELECT Statement:

```
SELECT TAKES.TK_ST_SID
FROM TAKES INNER JOIN SECTION
ON SECTION.SE_YEAR = 2006
AND TAKES.TK_SE_YEAR = 2006
AND SECTION.SE_SECTION# = TAKES.TK_SE_SECTION#;
```

Result:

```
Tk_st_sid
---------
BE76598

1 row selected.
```

Observe that the ON clause given above effectively constrains the concatenation of a row from TAKES with a row from SECTION to the condition where both the year in TAKES (**TAKES.Tk_se_year**) and the year in SECTION (**SECTION.Se_year**) are equal to 2006, and the section# in the SECTION row (**SECTION.Se_section#**) matches the section# in the TAKES row (**TAKES.Tk_se_section#**).

Reviewing the content of the SECTION and TAKES tables provides an explanation of this result. Since there are 11 rows in the SECTION table and 11 rows in the TAKES tables, the Cartesian Product operation results in an unnamed table with 121 rows and 14 columns (there are eight columns in SECTION and six columns in TAKES). Since only one row in TAKES contains the year 2006, this Cartesian Product operation generates only four of these 121 rows which have 2006 in both the **TAKES.Tk_se_year** and **SECTION.Se_year** columns. Each of these four rows contains the same value in the **TAKES.Tk_se_section#** column, 101, and the same value in the **TAKES.Tk_st_sid** column, BE76598. On the other hand, only one of these four rows contains the section number 101 in the **SECTION.Se_section#** column. Hence, the portion of the Selection operation which requires that **TAKES.Tk_se_section#** = **SECTION.Se_section#** "selects" only one of these four rows, and the SQL Projection operation displays the **TAKES.Tk_st_sid** value BE76598.[18]

11.2.2.2 SQL Queries Involving Set Theoretic Operations

Union, Intersection, and Difference are three set theoretic operators used to merge elements of two union compatible sets together. The examples shown below illustrate the incorporation of these operators in SQL statements that involve the tables R and S

---

[18] The actual execution of this SELECT statement would differ from this description. For example, a Selection operation on the TAKES table would occur first and result in "selecting" the one row where **TAKES.Tk_se_year** = 2006. A Cartesian Product operation would follow concatenating this one row with the 11 rows in the SECTION table. This would be followed by a second Select operation which would "select" the four rows in the SECTION table where **SECTION.Se_year** = 2006 and where **TAKES.Tk_se_section#** = **SECTION.Se_section#**. The Projection operation on the **TAKES.Tk_st_sid** column and the imposition of the DISTINCT qualifier is the final operation executed.

derived from the SECTION table, where R contains those sections offered during a Fall Quarter and S contains those sections offered in a room located in Lindner Hall.[19]

RELATION R

Se_section#	Se_qtr	Se_year	Se_time	Se_maxst	Se_room	Se_co_course#	Se_pr_profid
902	A	2006	H1700	25	Lindner 108	22IS270	SK85977
901	A	2006	W1800	35	Rhodes 611	22IS270	SK85977
101	A	2007	H1700	29	Lindner 108	22IS330	SK85977
101	A	2007	T1015	25		22QA375	HT54347
101	A	2007	W1800		Baldwin 437	20ECES212	RR79345

RELATION S

Se_section#	Se_qtr	Se_year	Se_time	Se_maxst	Se_room	Se_co_course#	Se_pr_profid
902	A	2006	H1700	25	Lindner 108	22IS270	SK85977
101	S	2006	T1045	29	Lindner 110	22IS330	SK85977
102	S	2006	H1045	29	Lindner 110	22IS330	CC49234
101	A	2007	H1700	29	Lindner 108	22IS330	SK85977

Example 2.2.1 (Corresponds to Union Example in Section 11.1.2.2). Display the union of R and S (i.e., those sections offered in either the fall quarter or in a room located in Lindner Hall, or offered in both the fall quarter and in a room located in Lindner Hall).

SQL SELECT Statement:

```
SELECT *
FROM R
    UNION
SELECT *
FROM S;
```

Note that when duplicate rows exist in tables, only one row per set of duplicates is displayed in the result when using the UNION operator unless UNION ALL has been used.

Result:

Se_section#	Se_qtr	Se_year	Se_time	Se_maxst	Se_room	Se_co_course#	Se_pr_profid
101	A	2007	H1700	29	Lindner 108	22IS330	SK85977
101	A	2007	T1015	25		22QA375	HT54347
101	A	2007	W1800		Baldwin 437	20ECES212	RR79345
101	S	2006	T1045	29	Lindner 110	22IS330	SK85977
102	S	2006	H1045	29	Lindner 110	22IS330	CC49234
901	A	2006	W1800	35	Rhodes 611	22IS270	SK85977
902	A	2006	H1700	25	Lindner 108	22IS270	SK85977

Example 2.2.2 (Corresponds to Intersection Example in Section 11.1.2.2). Display the intersection of R and S (i.e., those sections offered in both the fall quarter and in a room located in Lindner Hall).

SQL SELECT Statement:

```
SELECT *
FROM R
```

[19] The tables R and S were created simply as a means to illustrate use of the UNION, INTERSECT and DIF-FERENCE operators in an SQL query. Each of the examples in this section can be expressed by a query that refers to just the SECTION table.

```
        INTERSECT
SELECT *
FROM S;
```

Result:

```
Se_section# Se_qtr Se_year Se_time Se_maxst Se_room        Se_co_course# Se_pr_profid
----------- ------ ------- ------- -------- -------------- ------------- ------------
        101 A         2007 H1700        29 Lindner 108     22IS330       SK85977
        902 A         2006 H1700        25 Lindner 108     22IS270       SK85977
```

Example 2.2.3 (Corresponds to Difference Example 1 in Section 11.1.2.2). Display the difference R minus S (i.e., those sections offered in the fall quarter but not in a room located in Lindner Hall).

SQL SELECT Statement:

```
SELECT *
FROM R
    MINUS
SELECT *
FROM S;
```

Result:

```
Se_section# Se_qtr Se_year Se_time Se_maxst Se_room        Se_co_course# Se_pr_profid
----------- ------ ------- ------- -------- -------------- ------------- ------------
        101 A         2007 T1015        25                 22QA375       HT54347
        101 A         2007 W1800           Baldwin 437     20ECES212     RR79345
        901 A         2006 W1800        35 Rhodes 611      22IS270       SK85977
```

The word MINUS is used in Oracle's SQL. The word EXCEPT is part of the SQL-92 standard.

Example 2.2.4 (Corresponds to Difference Example 2 in Section 11.1.2.2). Using the data in the relations R and S given above, form the difference S minus R (i.e., those sections offered in a room located in Lindner Hall but not in the fall quarter).

SQL SELECT Statement:

```
SELECT *
FROM S
    MINUS
SELECT *
FROM R;
```

Result:

```
Se_section# Se_qtr Se_year Se_time Se_maxst Se_room        Se_co_course# Se_pr_profid
----------- ------ ------- ------- -------- -------------- ------------- ------------
        101 S         2006 T1045        29 Lindner 110     22IS330       SK85977
        102 S         2006 H1045        29 Lindner 110     22IS330       CC49234
```

11.2.2.3 Join Operations

Four types of relational algebra joins were discussed earlier in Section 11.1.2.3. The Equijoin involves join conditions with equality comparisons only and produces a new relation which contains all columns associated with the join condition. On the other hand, a Natural Join omits one of each pair of columns associated with the join condition, thus eliminating the possibility of obtaining a result that contains columns with the same set of values. A Theta Join, the third type of join, is based on a join condition that does not involve an equality comparison.

Data Manipulation: Relational Algebra and SQL

A fourth type of join is the Outer Join. An Outer Join is used when it is desired to include each row in a relation in the result even if the row does not contain an attribute value satisfying the join condition. Examples of these four types of joins using SQL are illustrated here in the context of the SECTION and TAKES tables.

In the SECTION table, a Section# is assigned to each course offered during a particular quarter and year. This allows Section# 101 to be assigned to the course 22QA375 offered during the Fall quarter in 2007. It is also possible for Section# 101 to be assigned to other courses offered during the Fall quarter in 2007 (e.g., 20ECES212 and 22IS330). Observe that **(Se_section#, Se_qtr, Se_year,** and **Se_co_course#)** constitute the concatenated primary key of SECTION.

The TAKES table records the sections of the courses taken by various students. Observe that the concatenated primary key of TAKES is **(Tk_se_section#, Tk_se_qtr, Tk_se_year, Tk_se_co_course#,** and **Tk_st_sid)**, which makes it possible for a student to take two different courses with the same Section# during a particular quarter and year (e.g., **Tk_st_sid** KS39874 takes both 22QA375 and 22IS330 during the fall quarter in 2007).

While it is possible to join TAKES and SECTION over the two common attributes **SECTION.Se_section#** and **TAKES.Tk_se_section#** sharing the same domain, the result of such a join would be meaningless. This is because each row in TAKES would be concatenated not only with the row from SECTION that describes the section of the course taken by the student, but also with the rows from SECTION with the same **SECTION.Se_section#** but associated with a different course than that taken by the student. Using the data in the SECTION and TAKES tables in Figure 11.1, it is left as an exercise for the reader to demonstrate that an Equijoin of TAKES and SECTION on just the common attributes **SECTION.Se_section#** and **TAKES.Tk_se_section#** yields a result that contains 65 rows.

An Example of an Equijoin Operation The most meaningful Equijoin of TAKES and SECTION would involve all of the attributes that share the same domains in TAKES and SECTION. This allows the result of the join to include only rows that match information on a section taken by a student (recorded in TAKES) with the information about that section (recorded in SECTION).

Example 2.3.1.1. Join the SECTION and TAKES tables over their common attributes.
SQL SELECT Statement:

```
SELECT *
FROM SECTION JOIN TAKES
ON SECTION.SE_SECTION# = TAKES.TK_SE_SECTION#
AND SECTION.SE_QTR = TAKES.TK_SE_QTR
AND SECTION.SE_YEAR = TAKES.TK_SE_YEAR
AND SECTION.SE_CO_COURSE# = TAKES.TK_SE_CO_COURSE#;
```

Had the result of this Equijoin been displayed for each section taken by a student, a complete description of the section would have appeared as well. In addition, the section number taken, the quarter during which the section taken was offered, the year during which the section taken was offered, and the time the section taken was offered would have appeared twice in each row as a result of the Equijoin operation. The following example illustrates how the combination of a Natural Join operation and a Projection operation allow a less cluttered result to be displayed.

Examples of Natural Join Operations Since the result of an Equijoin results in pairs of attributes with identical values, a Natural Join operation was created to omit the second (and superfluous) attribute(s) in an Equijoin condition. In direct violation of the requirement of the relational data model that attributes have unique names over the entire relational schema, the standard definition of NATURAL JOIN requires that the two join attributes (or each pair of join attributes) have the same name in both relations.

Example 2.3.2.1. Join the SECTION and TAKES tables over their common attributes (in our case attributes with different names but sharing the same domain).

SQL-92 supports three approaches for representing a Natural Join. One uses the NATURAL JOIN keyword, another uses JOIN ... USING, and the third uses JOIN ... ON. The first two approaches can only be used if the requirement that attributes have unique names over the entire relational schema is not enforced and tables that contain columns with the same name are allowed to exist. Thus in the context of this example, use of the first two approaches would require that in addition to the section#, quarter, year, and course# sharing the same domain, they must also have exactly the same column names in the SECTION and TAKES tables. Since this book adheres strictly to the requirement that attribute (and hence column) names be unique over the entire relational schema, the first two approaches can only be illustrated in general. In addition, in order to display just the distinct column names, the third approach based on the SQL-92 syntax requires that each column name appear in the <column list>.

Approach 1

```
SELECT * FROM tablename1 NATURAL JOIN tablename2
```

Approach 2

```
SELECT * FROM tablename1
JOIN tablename2 USING (columnname_a, columnname_b, ..., columnname_n)
```

Approach 3

```
SELECT SECTION.*, TAKES.TK_GRADE, TAKES.TK_ST_SID
FROM SECTION JOIN TAKES ON
SECTION.SE_SECTION# = TAKES.TK_SE_SECTION#
AND SECTION.SE_QTR = TAKES.TK_SE_QTR
AND SECTION.SE_YEAR = TAKES.TK_SE_YEAR
AND SECTION.SE_CO_COURSE# = TAKES.TK_SE_CO_COURSE#;
```

Result:

Se_section#	Se_qtr	Se_year	Se_time	Se_maxst	Se_room	Se_co_course#	Se_pr_profid	Tk_grade	Tk_st_sid
101	A	2007	T1015	25		22QA375	HT54347	A	KP78924
101	A	2007	T1015	25		22QA375	HT54347	A	KS39874
101	A	2007	T1015	25		22QA375	HT54347	B	BG66765
101	S	2006	T1045	29	Lindner 110	22IS330	SK85977	C	BE76598
101	A	2007	H1700	29	Lindner 108	22IS330	SK85977	B	KJ56656
101	A	2007	H1700	29	Lindner 108	22IS330	SK85977	A	KP78924
101	A	2007	H1700	29	Lindner 108	22IS330	SK85977	A	KS39874
701	W	2007	M1000	33	Braunstien 211	22IS832	CC49234	A	KS39874
101	A	2007	H1700	29	Lindner 108	22IS330	SK85977	A	BE76598
701	W	2007	M1000	33	Braunstien 211	22IS832	CC49234	B	BG66765
101	A	2007	H1700	29	Lindner 108	22IS330	SK85977	C	GS76775

Data Manipulation: Relational Algebra and SQL

Since most queries that involve one or more Join operations also include some sort of Projection operation, in effect virtually all joins take the form of this "variation" of a Natural Join.

Sometimes a join may be specified between a relation and itself. This type of join is often referred to as a Self Join.

Example 2.3.2.2. List the student IDs of those students recorded as having taken more than one course.

```
SELECT X.TK_ST_SID
FROM TAKES X JOIN TAKES Y
ON X.TK_ST_SID = Y.TK_ST_SID
AND X.TK_SE_CO_COURSE# <> Y.TK_SE_CO_COURSE#;
```

In this SELECT statement the TAKES table is referenced twice. To prevent ambiguity, each use of TAKES in the FROM clause has been assigned a temporary name (called a table alias). X and Y serve as the table aliases in this SELECT statement. Whenever a table alias is associated with a table name in the FROM clause, it must also be used any time the table is referenced in the SELECT statement (i.e., in this example when referencing the table in the WHERE condition and also in the column list that follows use of the word SELECT). Whenever a table alias has been introduced in the FROM clause, we cannot use the full table name anywhere else in the SELECT statement.

Since, as we have seen, duplicate rows are not automatically removed in SQL, the execution of the SELECT statement given above generates ten rows (six with an **X.Tk_st_sid** value of KS39874, two with an **X.Tk_st_sid** value of KP78924, and two with an **X.Tk_st_sid** value of BG66765):

```
Tk_st_sid
---------
BG66765
BG66765
KP78924
KP78924
KS39874
KS39874
KS39874
KS39874
KS39874
KS39874
```

This is due to the fact that the join condition is satisfied two times for each of the three rows in TAKES that involve **X.Tk_st_sid** KS39874, one time for each of the two rows in TAKES that involve **X.Tk_st_sid** KP78924, and one time for each of the two rows in TAKES that involve **X.Tk_st_sid** BG66765. Observe that while **X.Tk_st_sid** BE76598 appears in two rows in TAKES, both rows involve the same course (22IS330) taken once in the summer quarter of the year 2006 and again in the fall quarter of the year 2007. The fact that the join condition in the relational algebra expression and in the SQL SELECT statement is not satisfied is the cause of **X.Tk_st_sid** BE76598 not being displayed.

Use of the qualifier DISTINCT in the SELECT statement (see below) eliminates the duplicate rows and produces a result that corresponds to the relational algebra result.

SQL SELECT Statement Revised:

```
SELECT DISTINCT X.TK_ST_SID
FROM TAKES X JOIN TAKES Y
ON X.TK_ST_SID = Y.TK_ST_SID
AND X.TK_SE_CO_COURSE# <> Y.TK_SE_CO_COURSE#;
```

Result:

```
Tk_st_sid
----------
BG66765
KP78924
KS39874
```

The Natural Join or equijoin operation can also be specified among multiple tables, leading to what is sometimes referred to as an *n-way join*.

Example 2.3.2.3. Instead of listing the student IDs of those students having taken more than one course, list the names of those students having taken more than one course.

SQL SELECT Statement:

```
SELECT DISTINCT STUDENT.ST_NAME
FROM (TAKES X JOIN TAKES Y
ON X.TK_ST_SID = Y.TK_ST_SID
AND X.TK_SE_CO_COURSE# <> Y.TK_SE_CO_COURSE#)
JOIN STUDENT ON X.TK_ST_SID = STUDENT.ST_SID;
```

This SELECT statement uses the result of the Self Join from Example 2.3.2.2 and joins it with the STUDENT table. Without the use of the qualifier DISTINCT to eliminate duplicate rows, the three names shown below would be displayed a total of ten times (i.e., Gladis Bale twice, Poppy Kramer twice, and Sweety Kramer six times).

Result:

```
St_name
--------------
Gladis Bale
Poppy Kramer
Sweety Kramer
```

Example 2.3.2.4. For each student taking a course in the fall quarter of 2007, list the student's name, classroom where the course is offered, and course number.

SQL SELECT Statement:

```
SELECT STUDENT.ST_NAME, SECTION.SE_ROOM, TAKES.TK_SE_CO_COURSE#
FROM (STUDENT JOIN TAKES
ON STUDENT.ST_SID = TAKES.TK_ST_SID)
JOIN SECTION ON TAKES.TK_SE_CO_COURSE# = SECTION.SE_CO_COURSE#
AND SECTION.SE_QTR = TAKES.TK_SE_QTR
AND SECTION.SE_YEAR = TAKES.TK_SE_YEAR
AND TAKES.TK_SE_QTR = 'A' AND TAKES.TK_SE_YEAR = 2007;
```

Result:

```
St_name         Se_room         Tk_se_co_course#
-------------   --------------  ----------------
Poppy Kramer                    22QA375
Sweety Kramer                   22QA375
Gladis Bale                     22QA375
```

```
Joumana Kidd    Lindner 108    22IS330
Poppy Kramer    Lindner 108    22IS330
Sweety Kramer   Lindner 108    22IS330
Elijah Baley    Lindner 108    22IS330
Shweta Gupta    Lindner 108    22IS330
```

In effect, prior to the execution of the Projection operation, the SQL SELECT statement first links (i.e., concatenates) each row of TAKES to the corresponding row in SECTION and then links each row in the combined result to the corresponding row in STUDENT. Should both the course number and course name need to be displayed, the COURSE table must be included in the join. As shown below, the SELECT statement required to generate this result takes the result of joining the STUDENT, TAKES, and SECTION tables and joins it with the COURSE table.

SQL SELECT Statement:

```
SELECT STUDENT.ST_NAME, SECTION.SE_ROOM, TAKES.TK_SE_CO_COURSE#, COURSE.CO_
NAME
FROM ((STUDENT JOIN TAKES
ON STUDENT.ST_SID = TAKES.TK_ST_SID)
JOIN SECTION ON TAKES.TK_SE_CO_COURSE# = SECTION.SE_CO_COURSE#
AND SECTION.SE_QTR = TAKES.TK_SE_QTR
AND SECTION.SE_YEAR = TAKES.TK_SE_YEAR
AND TAKES.TK_SE_QTR = 'A' AND TAKES.TK_SE_YEAR = 2007)
JOIN COURSE ON SECTION.SE_CO_COURSE# = COURSE.CO_COURSE#;
```

Result:

St_name	Se_room	Tk_se_co_course#	Co_name
Poppy Kramer		22QA375	Operations Research
Sweety Kramer		22QA375	Operations Research
Gladis Bale		22QA375	Operations Research
Joumana Kidd	Lindner 108	22IS330	Database Concepts
Poppy Kramer	Lindner 108	22IS330	Database Concepts
Sweety Kramer	Lindner 108	22IS330	Database Concepts
Elijah Baley	Lindner 108	22IS330	Database Concepts
Shweta Gupta	Lindner 108	22IS330	Database Concepts

The Theta Join Operation While occurring infrequently in practical applications, Theta Joins (or Non-Equijoins) that do not involve equality conditions are possible as long as the join condition involves attributes that share the same domain.

Example 2.3.3.1. Instead of doing an Equijoin of SECTION and TAKES when an equality condition involving each of their common attributes exists, do a Join of SECTION and TAKES when an inequality condition exists for each of their common attributes.

The following SQL SELECT statement displays the columns in SECTION and TAKES involved in the join condition. On each row, observe how adjacent pairs of columns reflect inequalities for the four columns involved in the join condition.

SQL SELECT Statement:

```
SELECT SECTION.SE_SECTION#, TAKES.TK_SE_SECTION#,
       SECTION.SE_YEAR, TAKES.TK_SE_YEAR,
       SECTION.SE_QTR, TAKES.TK_SE_QTR,
       SECTION.SE_CO_COURSE#, TAKES.TK_SE_CO_COURSE#
FROM SECTION JOIN TAKES
ON SECTION.SE_SECTION# <> TAKES.TK_SE_SECTION#
```

```
AND SECTION.SE_YEAR <> TAKES.TK_SE_YEAR
AND SECTION.SE_QTR <> TAKES.TK_SE_QTR
AND SECTION.SE_CO_COURSE# <> TAKES.TK_SE_CO_COURSE#
ORDER BY SECTION.SE_SECTION#;
```

Result:

Se_section#	Tk_se_section#	Se_year	Tk_se_year	Se_qtr	Tk_se_qtr	Se_co_course#	Tk_se_co_course#
101	701	2006	2007 S	W		22IS330	22IS832
101	701	2006	2007 S	W		22IS330	22IS832
102	101	2006	2007 S	A		22IS330	22QA375
102	101	2006	2007 S	A		22IS330	22QA375
102	101	2006	2007 S	A		22IS330	22QA375
102	701	2006	2007 S	W		22IS330	22IS832
102	701	2006	2007 S	W		22IS330	22IS832
701	101	2007	2006 W	S		22IS832	22IS330
901	701	2006	2007 A	W		22IS270	22IS832
901	701	2006	2007 A	W		22IS270	22IS832
902	701	2006	2007 A	W		22IS270	22IS832
902	701	2006	2007 A	W		22IS270	22IS832

11.2.2.4 Outer Join Operations

In relational algebra, Outer Joins can be used when there is a need to keep all the tuples in R, or those in S, or those in both relations in the result of the Join, whether or not they have matching tuples in the other relation.

Left Outer Join The Left Outer Join operation is used when we want to retain all rows in the leftmost table (i.e., the first table to be listed) regardless of whether corresponding rows exist in the other table. SQL uses the LEFT OUTER JOIN[20] keywords to create a Left Outer Join.

Example 2.4.1.1. For each section taken by a student, display the section number, quarter, year, course number, grade, student ID, and student name. Include the names of all students (i.e., even those who have never taken a course).

SQL SELECT Statement:

```
SELECT TAKES.*, ST_NAME
FROM STUDENT LEFT OUTER JOIN TAKES
ON STUDENT.ST_SID = TAKES.TK_ST_SID;
```

Result:

Tk_se_section#	Tk_se_qtr	Tk_se_year	Tk_se_co_course#	Tk_grade	Tk_st_sid	St_name
101	A	2007	22QA375	A	KP78924	Poppy Kramer
101	A	2007	22QA375	A	KS39874	Sweety Kramer
101	A	2007	22QA375	B	BG66765	Gladis Bale
101	S	2006	22IS330	C	BE76598	Elijah Baley
101	A	2007	22IS330	B	KJ56656	Joumana Kidd
101	A	2007	22IS330	A	KP78924	Poppy Kramer

---

[20] Use of the word OUTER is not required to create either a LEFT, RIGHT, or FULL outer join.

101	A	2007	22IS330	A	KS39874	Sweety Kramer
701	W	2007	22IS832	A	KS39874	Sweety Kramer
101	A	2007	22IS330	A	BE76598	Elijah Baley
701	W	2007	22IS832	B	BG66765	Gladis Bale
101	A	2007	22IS330	C	GS76775	Shweta Gupta
						Tim Duncan
						Troy Hudson
						Rick Fox
						Jenny Aniston
						Daniel Olive
						Diana Jackson
						Jenna Hopp
						Wanda Seldon
						David Sane
						Vanessa Fox

In SQL-92, the words LEFT OUTER JOIN are used to designate a Left Outer Join operation. The use of the LEFT JOIN keywords means that if the table listed on the left side of the join condition given in the ON clause has an unmatched row, it should be matched with a null row and displayed in the results.

Observe how the first three rows in TAKES are concatenated with those rows in STU-DENT that correspond to the students taking the course 22QA375 to produce the first three rows in the result shown above. On the other hand, the second row in STUDENT (the one for Daniel Olive) has been concatenated with the row of null values appended to the end of the TAKES table to produce the 16th row in the result. The result reveals that ten of the students have never taken a course.

One way to verify that the ten rows with null values for the attributes in TAKES represent the ten students who have not taken a course is to take the difference between the STUDENT and the TAKES relations over those attributes that share the same domain.

```
SELECT STUDENT.ST_SID
FROM STUDENT
    MINUS
SELECT TAKES.TK_ST_SID
FROM TAKES;
```

Result:

```
St_sid
-------
AJ76998
DT87656
FR45545
FV67733
HJ45633
HT67657
JD35477
OD76578
SD23556
SW56547
```

The student IDs shown above can be replaced by a list of student names through the use of the following nested subquery. Subqueries are discussed in Section 11.2.3.

```
SELECT STUDENT.ST_NAME
FROM STUDENT
WHERE STUDENT.ST_SID IN
    (SELECT STUDENT.ST_SID
    FROM STUDENT
        MINUS
    SELECT TAKES.TK_ST_SID
    FROM TAKES);
```

Result:

```
St_name
--------------
Jenny Aniston
Tim Duncan
Rick Fox
Vanessa Fox
Jenna Hopp
Troy Hudson
Diana Jackson
Daniel Olive
David Sane
Wanda Seldon
```

Right Outer Join Semantically, a Right Outer Join is the same as a Left Outer Join. The difference is that the required table is the rightmost table listed.

Example 2.4.2.1. For each textbook, display all available information about the book and its usage in courses. Include textbooks that have never been used.

SQL SELECT Statement:

```
SELECT *
FROM USES RIGHT OUTER JOIN TEXTBOOK
ON USES.US_TX_ISBN = TEXTBOOK.TX_ISBN;
```

Result:

Us_co_course#	Us_tx_isbn	Us_pr_empid	Tx_isbn	Tx_title	Tx_year	Tx_publisher
22IS832	000-66574998	S43278	000-66574998	Database Management	1999	Thomson
22IS270	000-66574998	SS43278	000-66574998	Database Management	1999	Thomson
22IS270	77898-8769	SS43278	77898-8769	Principles of IS	2002	Prentice-Hall
22IS270	77898-8769	SK85977	77898-8769	Principles of IS	2002	Prentice-Hall
22IS270	77898-8769	CC49234	77898-8769	Principles of IS	2002	Prentice-Hall
20ECES212	0296437-1118	CC49234	0296437-1118	Programming in C++	2002	Thomson
22QA375	0296437-1118	SJ65436	0296437-1118	Programming in C++	2002	Thomson
22IS330	003-6679233	BC65437	003-6679233	Linear Programming	1997	Prentice-Hall
18ECON123	0296748-99	CM65436	0296748-99	Economics For Managers	2001	
22IS330	118-99898-67	SS43278	118-99898-67	Systems Analysis	2000	Thomson
22IS832	118-99898-67	SK85977	118-99898-67	Systems Analysis	2000	Thomson
22QA888	001-55-435	HT54347	001-55-435	Simulation Modeling	2001	Springer
			111-11111111	Data Modeling	2006	
			012-54765-32	Fundamentals of SQL	2004	

In SQL-92, the words RIGHT OUTER JOIN are used to designate a Right Outer Join operation. The use of the RIGHT JOIN keywords means that if the table listed on the right side of the join condition given in the ON clause has an unmatched row, it should be matched with a null row and displayed in the results.

Full Outer Join A Full Outer Join operation is used when we want all rows in both the left and right tables included even though no corresponding rows exist in the other table.

Example 2.4.3.1. Join the GRAD_STUDENT and TAKES tables making sure that each row from each table appears in the result.

Only syntaxes beginning with SQL-92 support the full outer join operation directly. This is accomplished through use of the FULL OUTER JOIN keywords.

SQL SELECT Statement:

```
SELECT *
FROM GRAD_STUDENT FULL OUTER JOIN TAKES
ON GRAD_STUDENT.GS_ST_SID = TAKES.TK_ST_SID;
```

Result:

Gs_st_sid	Gs_thesis	Gs_ugmajor	Tk_se_section#	Tk_se_qtr	Tk_se_year	Tk_se_co_course#	Tk_grade	Tk_st_sid
BG66765	N	Archeology	101	A	2007	22QA375	B	BG66765
BE76598	Y	Marketing	101	S	2006	22IS330	C	BE76598
KJ56656	Y	History	101	A	2007	22IS330	B	KJ56656
BE76598	Y	Marketing	101	A	2007	22IS330	A	BE76598
BG66765	N	Archeology	701	W	2007	22IS832	B	BG66765
GS76775	N	Archeology	101	A	2007	22IS330	C	GS76775
DT87656	N	Physics						
SW56547	Y	Finance						
AJ76998	Y	Child Care						
JD35477	N	Mathematics						
HJ45633	Y	History						
			101	A	2007	22QA375	A	KP78924
			101	A	2007	22QA375	A	KS39874
			101	A	2007	22IS330	A	KP78924
			101	A	2007	22IS330	A	KS39874
			701	W	2007	22IS832	A	KS39874

Note that the union of a Left Outer Join and a Right Outer Join is equal to the results of a Full Outer Join.

11.2.2.5 SQL and the Semi-Join and Semi-Minus Operations

Recall that the Semi-Join operation defines a relation that contains the tuples of R that participate in the Join of R with S. In other words, a Semi-Join is defined to be equal to the Join of R and S, projected back on the attributes of R. As such, a Semi-Join can be handled in SQL by using the Projection and Join operations.

Example 2.5.1 (Corresponds to Semi-Join Example 1 in Section 11.1.2.5). List the complete details of all courses for which sections are being offered in the fall 2007 quarter.

SQL SELECT Statement:

```
SELECT COURSE.*
FROM COURSE JOIN SECTION
ON COURSE.CO_COURSE# = SECTION.SE_CO_COURSE#
AND SE_YEAR = 2007 AND SE_QTR = 'A';
```

Result:

Co_name	Co_course#	Co_credit	Co_college	Co_hrs	Co_dpt_dcode
Database Concepts	22IS330	U	Business	4	7
Operations Research	22QA375	U	Business	2	3
Programming in C++	20ECES212	G	Engineering	3	6

A Semi–Minus operation occurs in cases where there are tuples in relation R that have no counterpart in relation S. A Semi-Minus operation can be handled in SQL through use of a combination of Join, Projection, and Minus operations.

Example 2.5.2 (Corresponds to Semi-Minus Example 1 in Section 11.1.2.5). List complete details of all courses for which sections are not offered in the fall 2007 quarter.

SQL SELECT Statement:

```
SELECT * FROM COURSE
MINUS
SELECT COURSE.* FROM COURSE JOIN SECTION
ON COURSE.CO_COURSE# = SECTION.SE_CO_COURSE#
AND SECTION.SE_YEAR = 2007 AND SECTION.SE_QTR = 'A';
```

Result:

Co_name	Co_course#	Co_credit	Co_college	Co_hrs	Co_dpt_dcode
Database Principles	22IS832	G	Business	3	7
Financial Accounting	18ACCT801	G	Education	3	4
Intro to Economics	15ECON112	U	Arts and Sciences	3	1
Intro to Economics	18ECON123	U	Education	4	4
Optimization	22QA888	G	Business	3	3
Principles of IS	22IS270	G	Business	3	7
Supply Chain Analysis	22QA411	U	Business	3	3
Systems Analysis	22IS430	G	Business	3	7

11.2.3 Subqueries

A complete SELECT statement embedded within another SELECT statement is called a subquery. In data retrieval, subqueries may be used (a) in the SELECT list of a SELECT statement, (b) in the FROM clause of a SELECT statement, (c) in the WHERE clause of a SELECT statement, and (d) in the ORDER BY clause of a SELECT statement. As illustrated in Section 10.2.1, subqueries can also be used in an INSERT...SELECT...FROM statement as well as the SET clause of an UPDATE statement. The output of a subquery can consist of a single value (a single-row subquery) or several rows of values (a multiple-row subquery). There are two types of subqueries: (a) uncorrelated subqueries where the subquery is executed first and passes one or more values to the outer query, and (b) correlated subqueries where the subquery is executed once for every row retrieved by the outer query.

11.2.3.1 Multiple-Row Uncorrelated Subqueries

Multiple-row subqueries are nested queries that can return more than one row of results to the parent query. Three multiple-row operators are used with multiple-row queries (IN, ALL, and ANY).

The IN and NOT IN Operators When used in conjunction with a subquery, the IN operator evaluates if rows processed by the outer query are equal to any of the values returned by

Data Manipulation: Relational Algebra and SQL

the subquery (i.e., it creates an OR condition). Example 3.1.1.1 illustrates how the IN operator can be used to display the course number, course name, and college of those courses for which sections have been offered. In this query, the subquery is executed first and returns the set of values (22QA375, 22IS270, 22IS330, 22IS832, and 20ECES212). The main query then displays the course number, name, and college for these courses. Note that a subquery of the form SELECT DISTINCT SECTION.CO_COURSE# FROM SECTION would have produced exactly the same result.

Example 3.1.1.1

```
SELECT COURSE.CO_COURSE#, COURSE.CO_NAME, COURSE.CO_COLLEGE
FROM COURSE
WHERE COURSE.CO_COURSE# IN
      (SELECT SECTION.SE_CO_COURSE#
       FROM SECTION);
```

Result:

```
Co_course# Co_name                Co_college
---------- ---------------------- --------------------
20ECES212  Programming in C++     Engineering
22IS270    Principles of IS       Business
22IS330    Database Concepts      Business
22IS832    Database Principles    Business
22QA375    Operations Research    Business
```

The NOT IN operator is the opposite of the IN operator and indicates that the rows processed by the outer query are not equal to any of the values returned by the subquery. Example 3.1.1.2 displays the course number, course name, and college of those courses for which sections have not been offered.

Example 3.1.1.2

```
SELECT COURSE.CO_COURSE#, COURSE.CO_NAME, COURSE.CO_COLLEGE
FROM COURSE
WHERE COURSE.CO_COURSE# NOT IN
      (SELECT SECTION.SE_CO_COURSE#
       FROM SECTION);
```

Result:

```
Co_course# Co_name                Co_college
---------- ---------------------- --------------------
15ECON112  Intro to Economics     Arts and Sciences
18ECON123  Intro to Economics     Education
22QA411    Supply Chain Analysis  Business
22QA888    Optimization           Business
18ACCT801  Financial Accounting   Education
22IS430    Systems Analysis       Business
```

The comparison operators =, <>, >, >=, <, and <= are *single-row operators*. Observe the error message generated when the multiple-row operator IN is replaced by the single-row operator = (equals sign) in Example 3.1.1.3. This error is caused by the fact that there are several course number-section number combinations in the TAKES table associated with a grade of 'A'. Observe what happens, however, when the single-row operator = is replaced by IN.

The purpose of Example 3.1.1.3 is to display the section number and course number for which at least one grade of "A" has been assigned.

Example 3.1.1.3

```
SELECT DISTINCT SECTION.SE_SECTION#, SECTION.SE_CO_COURSE#
FROM SECTION
WHERE (SECTION.SE_SECTION#, SECTION.SE_CO_COURSE#) =
      (SELECT TAKES.TK_SE_SECTION#,TAKES.TK_SE_CO_COURSE#
      FROM TAKES
      WHERE TAKES.TK_GRADE = 'A');

(SELECT TAKES.TK_SE_SECTION#,TAKES.TK_SE_CO_COURSE#
 *
ERROR at line 4:  single-row subquery returns more than one row
```

Example 3.1.1.3 (Corrected)

```
SELECT DISTINCT SECTION.SE_SECTION#, SECTION.SE_CO_COURSE#
FROM SECTION
WHERE (SECTION.SE_SECTION#, SECTION.SE_CO_COURSE#) IN
      (SELECT TAKES.TK_SE_SECTION#, TAKES.TK_SE_CO_COURSE#
      FROM TAKES
      WHERE TAKES.TK_GRADE = 'A');
```

Result:

```
Se_section# Se_co_course#
----------- -------------
        101 22IS330
        101 22QA375
        701 22IS832
```

The remaining examples in this section illustrate the use of the ANY and ALL operators in the context of the PROFESSOR table.

The ALL and ANY Operators The ALL and ANY operators can be combined with the comparison operators =, <>, >, >=, <, and <= to treat the results of a subquery as a set of values, rather than as individual values. ANY specifies that the condition be true for *at least one value* from the set of values. ALL, on the other hand, specifies that the condition be true for *all values* in the set of values. Table 11.2 summarizes the use of the ALL and ANY operators in conjunction with other comparison operators.

Table 11.2 Use of ANY and ALL operators in subqueries

Operator	Description
> ALL	Greater than the highest value returned by the subquery
>= ALL	Greater than or equal to the highest value returned by the subquery
< ALL	Less than the lowest value returned by the subquery
<= ALL	Less than or equal to the lowest value returned by the subquery
> ANY	Greater than the lowest value returned by the subquery

Table 11.2 Use of ANY and ALL operators in subqueries (continued)

Operator	Description
>= ANY	Greater than or equal to the lowest value returned by the subquery
< ANY	Less than the highest value returned by the subquery
<= ANY	Less than or equal to the highest value returned by the subquery
= ANY	Equal to any value returned by the subquery (same as the IN operator)

Example 3.1.2.1. Display the names and salaries of those professors who earn more than all professors in department number 3.

SQL SELECT Statement:

```
SELECT PROFESSOR.PR_NAME, PROFESSOR.PR_SALARY
FROM PROFESSOR
WHERE PROFESSOR.PR_SALARY > ALL
      (SELECT PROFESSOR.PR_SALARY
       FROM PROFESSOR
       WHERE PROFESSOR.PR_DPT_DCODE = 3);
```

Result:

```
Pr_name            Pr_salary
---------------    ----------
Mike Faraday          92000
Marie Curie           99000
John Nicholson        99000
```

The following query includes Chelsea Bush and Tony Hopkins in the result since their salary is equal to the highest value returned by the subquery.

```
SELECT PROFESSOR.PR_NAME, PROFESSOR.PR_SALARY
FROM PROFESSOR
WHERE PROFESSOR.PR_SALARY >= ALL
      (SELECT PROFESSOR.PR_SALARY
       FROM PROFESSOR
       WHERE PROFESSOR.PR_DPT_DCODE = 3);
```

Result:

```
Pr_name            Pr_salary
---------------    ----------
Mike Faraday          92000
Chelsea Bush          77000
Tony Hopkins          77000
Marie Curie           99000
John Nicholson        99000
```

Example 3.1.2.2. Display the names and salaries of those professors who earn less than all professors in department number 7.

```
SELECT PROFESSOR.PR_NAME, PROFESSOR.PR_SALARY
FROM PROFESSOR
WHERE PROFESSOR.PR_SALARY < ALL
      (SELECT PROFESSOR.PR_SALARY
       FROM PROFESSOR
       WHERE PROFESSOR.PR_DPT_DCODE = 7);
```

Result:

```
Pr_name          Pr_salary
--------------   ----------
Ram Raj               44000
Prester John          44000
Laura Jackson         43000
```

Example 3.1.2.3. Revise the query in Example 3.1.2.2 and display the names and salaries of those professors with a salary that is less than or equal to that of the lowest paid professor in department number 7.

```
SELECT PROFESSOR.PR_NAME, PROFESSOR.PR_SALARY
FROM PROFESSOR
WHERE PROFESSOR.PR_SALARY <= ALL
      (SELECT PROFESSOR.PR_SALARY
       FROM PROFESSOR
       WHERE PROFESSOR.PR_DPT_DCODE = 7);
```

Result:

```
Pr_name          Pr_salary
--------------   ----------
John Smith            45000
Ram Raj               44000
Prester John          44000
Laura Jackson         43000
Cathy Cobal           45000
Jeanine Troy          45000
```

Example 3.1.2.4. Revise the query in Example 3.1.2.3 to exclude display of any employees in department number 7.

```
SELECT PROFESSOR.PR_NAME, PROFESSOR.PR_SALARY
FROM PROFESSOR
WHERE  PROFESSOR.PR_DPT_DCODE <> 7
AND PROFESSOR.PR_SALARY <= ALL
      (SELECT PROFESSOR.PR_SALARY
       FROM PROFESSOR
       WHERE PROFESSOR.PR_DPT_DCODE = 7);
```

Result:

```
Pr_name          Pr_salary
--------------   ----------
John Smith            45000
Ram Raj               44000
Prester John          44000
Laura Jackson         43000
Jeanine Troy          45000
```

Since < ANY returns all rows with a salary less than highest salary associated with department 3, the query in Example 3.1.2.5 displays the rows for all professors with a salary less than $77,000.

Example 3.1.2.5

```
SELECT PR_NAME, PR_SALARY
FROM PROFESSOR
WHERE PROFESSOR.PR_SALARY < ANY
```

Data Manipulation: Relational Algebra and SQL

```
(SELECT PROFESSOR.PR_SALARY
FROM PROFESSOR
WHERE PROFESSOR.PR_DPT_DCODE = 3);
```

Result:

```
Pr_name           Pr_salary
--------------- ----------
John Smith           45000
Kobe Bryant          66000
Ram Raj              44000
Prester John         44000
Alan Brodie          76000
Jessica Simpson      67000
Laura Jackson        43000
Jack Nicklaus        67000
Sunil Shetty         64000
Katie Shef           65000
Cathy Cobal          45000
Jeanine Troy         45000
Mike Crick           69000
```

On the other hand, since <= ANY returns all rows with a salary less than or equal to the highest salary associated with department 3, the query in Example 3.1.2.6 also displays the rows for both Chelsea Bush and Tony Hopkins.

Example 3.1.2.6

```
SELECT PR_NAME, PR_SALARY
FROM PROFESSOR
WHERE PROFESSOR.PR_SALARY <= ANY
      (SELECT PROFESSOR.PR_SALARY
       FROM PROFESSOR
       WHERE PROFESSOR.PR_DPT_DCODE = 3);
```

Result:

```
Pr_name           Pr_salary
--------------- ----------
John Smith           45000
Kobe Bryant          66000
Ram Raj              44000
Prester John         44000
Chelsea Bush         77000
Tony Hopkins         77000
Alan Brodie          76000
Jessica Simpson      67000
Laura Jackson        43000
Jack Nicklaus        67000
Sunil Shetty         64000
Katie Shef           65000
Cathy Cobal          45000
Jeanine Troy         45000
Mike Crick           69000
```

Since > ANY returns all rows with a salary greater than the lowest salary associated with professors who work in department 3, the query in Example 3.1.2.7 displays all rows except for the professor with the lowest salary (i.e., Laura Jackson) and the two professors who have a null salary.

Example 3.1.2.7

```
SELECT * PR_NAME, PR_SALARY
FROM PROFESSOR
WHERE PROFESSOR.PR_SALARY > ANY
      (SELECT PROFESSOR.PR_SALARY
       FROM PROFESSOR
       WHERE PROFESSOR.PR_DPT_DCODE = 3);
```

Result:

Pr_name	Pr_salary
John Smith	45000
Mike Faraday	92000
Kobe Bryant	66000
Ram Raj	44000
Prester John	44000
Chelsea Bush	77000
Tony Hopkins	77000
Alan Brodie	76000
Jessica Simpson	67000
Marie Curie	99000
Jack Nicklaus	67000
John Nicholson	99000
Sunil Shetty	64000
Katie Shef	65000
Cathy Cobal	45000
Jeanine Troy	45000
Mike Crick	69000

As expected, since >= ANY returns all rows with a salary greater than or equal to the lowest salary associated with department 3, all rows are returned in Example 3.1.2.8 except for those associated with the professors who have null salaries.

Example 3.1.2.8

```
SELECT PR_NAME, PR_SALARY
FROM PROFESSOR
WHERE PROFESSOR.PR_SALARY >= ANY
      (SELECT PROFESSOR.PR_SALARY
FROM PROFESSOR
WHERE PROFESSOR.PR_DPT_DCODE = 3);
```

Result:

Pr_name	Pr_salary
John Smith	45000
Mike Faraday	92000
Kobe Bryant	66000
Ram Raj	44000
Prester John	44000
Chelsea Bush	77000
Tony Hopkins	77000
Alan Brodie	76000
Jessica Simpson	67000
Laura Jackson	43000
Marie Curie	99000
Jack Nicklaus	67000

```
John Nicholson         99000
Sunil Shetty           64000
Katie Shef             65000
Cathy Cobal            45000
Jeanine Troy           45000
Mike Crick             69000
```

As illustrated in Example 3.1.2.9, = ANY produces the same result as the IN operator. Note how the query in Example 3.1.2.9a restricts the rows displayed to those not associated with department 3.

Example 3.1.2.9

```
SELECT PR_NAME, PR_SALARY
FROM PROFESSOR
WHERE PROFESSOR.PR_SALARY = ANY
      (SELECT PROFESSOR.PR_SALARY
      FROM PROFESSOR
      WHERE PROFESSOR.PR_DPT_DCODE = 3);
```

Result:

```
Pr_name            Pr_salary
--------------- ----------
Laura Jackson         43000
Jessica Simpson       67000
Jack Nicklaus         67000
Alan Brodie           76000
Chelsea Bush          77000
Tony Hopkins          77000
```

Example 3.1.2.9a

```
SELECT PR_NAME, PR_SALARY
FROM PROFESSOR
WHERE PROFESSOR.PR_SALARY = ANY
      (SELECT PROFESSOR.PR_SALARY
      FROM PROFESSOR
      WHERE PROFESSOR.PR_DPT_DCODE = 3)
      AND PROFESSOR.PR_DPT_DCODE <> 3;
```

Result:

```
Pr_name            Pr_salary
--------------- ----------
Jack Nicklaus         67000
```

Although ANY and ALL are most commonly used with subqueries that return a set of numeric values, it is also possible to use them in conjunction with subqueries that return a set of character values.

The MAX and MIN Functions The MAX and MIN functions can be used in place of > ANY, < ANY, > ALL, and < ALL when the WHERE clause of the outer query involves a numeric value. Examples 3.1.3.1 through 3.1.3.3 are equivalent to Examples 3.1.2.1 through 3.1.2.3 but use the MAX and MIN function instead of ANY or ALL.

Example 3.1.3.1 (Compare with Example 3.1.2.1)

```
SELECT PR_NAME, PR_SALARY
FROM PROFESSOR
```

```
WHERE PROFESSOR.PR_SALARY >
        (SELECT MAX(PROFESSOR.PR_SALARY)
        FROM PROFESSOR
        WHERE PROFESSOR.PR_DPT_DCODE = 3);
```

Result:

```
Pr_name          Pr_salary
---------------  ----------
Mike Faraday         92000
Marie Curie          99000
John Nicholson       99000
```

Example 3.1.3.2 (Compare with Example 3.1.2.2)

```
SELECT PR_NAME, PR_SALARY
FROM PROFESSOR WHERE
PROFESSOR.PR_SALARY <
        (SELECT MIN(PROFESSOR.PR_SALARY)
        FROM PROFESSOR
        WHERE PROFESSOR.PR_DPT_DCODE = 7);
```

Result:

```
Pr_name          Pr_salary
---------------  ----------
Ram Raj              44000
Prester John         44000
Laura Jackson        43000
```

Example 3.1.3.3 (Compare with Example 3.1.2.3)

```
SELECT PR_NAME, PR_SALARY
FROM PROFESSOR
WHERE PROFESSOR.PR_SALARY <=
        (SELECT MIN(PROFESSOR.PR_SALARY)
        FROM PROFESSOR
        WHERE PROFESSOR.PR_DPT_DCODE = 7);
```

Result:

```
Pr_name          Pr_salary
---------------  ----------
John Smith           45000
Ram Raj              44000
Prester John         44000
Laura Jackson        43000
Cathy Cobal          45000
Jeanine Troy         45000
```

Subqueries in the FROM Clause It is also possible to have nested subqueries in the FROM clause and in effect treat the subquery itself as if it were the name of a table.[21] This approach is often used when the subquery is a multiple-column subquery. As an example, suppose we are interested in listing all professors with a salary that is equal to or exceeds the average salary of all professors in their department.

---

[21] Such a "temporary table" is more formally called an inline view.

Example 3.1.4.1. Display all professors with a salary that is equal to or exceeds the average salary of all professors in their department.

SQL SELECT Statement:

```
SELECT A.PR_NAME, A.PR_DPT_DCODE, A.PR_SALARY, B."Department Average"
FROM PROFESSOR A
    JOIN(SELECT PROFESSOR.PR_DPT_DCODE, AVG(PROFESSOR.PR_SALARY) AS "Department Average"
    FROM PROFESSOR
    GROUP BY PROFESSOR.PR_DPT_DCODE) B
    ON A.PR_DPT_DCODE = B.PR_DPT_DCODE
    AND A.PR_SALARY >= B."Department Average";
```

Observe how the shaded subquery in essence creates a temporary table that records the average salary of the professors in each department. The syntax calls for the table alias B to be located outside the parenthetical expression of the subquery since the execution of the subquery yields a temporary (i.e., virtual) table. The Join operation uses the PROFESSOR table and concatenates a row from PROFESSOR (table alias A) with a row from the temporary table created by the subquery (table alias B) when (a) the department number of the row from A matches the department number of a row from B, and (b) the salary of the professor in the row from A exceeds the average salary of the professors in his or her department.

Result:

Pr_name	Pr_dpt_dcode	Pr_salary	Department Average
Mike Faraday	1	92000	67666.6667
Chelsea Bush	3	77000	68000
Tony Hopkins	3	77000	68000
Alan Brodie	3	76000	68000
Marie Curie	4	99000	88333.3333
John Nicholson	4	99000	88333.3333
Ram Raj	6	44000	44000
Prester John	6	44000	44000
Sunil Shetty	7	64000	58000
Katie Shef	7	65000	58000
Mike Crick	9	69000	57000

Subqueries in the Column List The column list of a query can also include a subquery expression. A subquery in the column list must return a single value.

Example 3.1.5.1. Display all professors with a salary that is equal to or exceeds the average salary of all professors in their department along with the amount by which the average is exceeded.

SQL SELECT Statement:

```
1.  SELECT A.PR_NAME, A.PR_SALARY, A.PR_DPT_DCODE,
2.  ROUND((SELECT AVG(B.PR_SALARY)
3.  FROM PROFESSOR B
4.  WHERE A.PR_DPT_DCODE = B.PR_DPT_DCODE),0) AS "Avg Dept Salary",
5.  ROUND((A.PR_SALARY - (SELECT AVG(B.PR_SALARY) FROM PROFESSOR B
6.  WHERE A.PR_DPT_DCODE = B.PR_DPT_DCODE)),0) AS "Deviation"
7.  FROM PROFESSOR A
8.  WHERE A.PR_SALARY IS NOT NULL
9.  AND ROUND((A.PR_SALARY - (SELECT AVG(B.PR_SALARY) FROM PROFESSOR B
10. WHERE A.PR_DPT_DCODE = B.PR_DPT_DCODE)),0) > 0
11. ORDER BY "Deviation" DESC;
```

In order to explain this query, each line has been numbered. Two SELECT statements, each of which is the same, appear in the column list. The first (shown in *italics* on lines 2–4) determines the average salary for the professors in the department for a given professor, while the second (shown highlighted on lines 5 and 6) recalculates this average salary and uses it to determine the amount of the deviation between the salary of the professor and the average salary for the professors in the department. The WHERE clause in the main query (a) excludes from consideration those professors with a null salary (see line 8), and (b) includes only those professors whose salary exceeds that of their average salary in their department (note that the average salary of all professors in the professors' department is calculated a third time). The ORDER BY clause on line 11 allows the result to be displayed in descending order by the amount of the deviation between the salary of the professor and the average salary of the professors in their department.

Result:

Pr_name	Pr_salary	Pr_dpt_dcode	Avg Dept Salary	Deviation
Mike Faraday	92000	1	67667	24333
Mike Crick	69000	9	57000	12000
Marie Curie	99000	4	88333	10667
John Nicholson	99000	4	88333	10667
Chelsea Bush	77000	3	68000	9000
Tony Hopkins	77000	3	68000	9000
Alan Brodie	76000	3	68000	8000
Katie Shef	65000	7	58000	7000
Sunil Shetty	64000	7	58000	6000

Subqueries in the HAVING Clause In addition to appearing in the WHERE clause, the FROM clause, and in the column list, a subquery can also be used in the HAVING clause. As an example, consider the following:

Example 3.1.6.1. Display the name and average professor salary (for all departments) whose average salary exceeds the average salary paid to all professors at Madeira College.

SQL SELECT Statement:

```
SELECT DEPARTMENT.DPT_NAME, AVG(PROFESSOR.PR_SALARY)
FROM DEPARTMENT JOIN PROFESSOR
ON DEPARTMENT.DPT_DCODE = PROFESSOR.PR_DPT_DCODE
GROUP BY DEPARTMENT.DPT_NAME
HAVING AVG(PROFESSOR.PR_SALARY) >
    (SELECT AVG(PROFESSOR.PR_SALARY) FROM PROFESSOR);
```

Result:

Dpt_name	AVG(PROFESSOR.Pr_salary)
Economics	78000
QA/QM	68000

The SELECT statement in the HAVING clause acts as a filter that insures the selection of only those departments (i.e., groups) with an average salary greater than the average salary of the entire college.

11.2.3.2 Multiple-Row Correlated Subqueries

A correlated subquery can be used if it is necessary to check if a nested subquery returns no rows. Correlated subqueries make use of the EXISTS operator, which returns the value of true if a set is non-empty.

Example 3.2.1. Display the names of professors who have offered at least one section.

```
SELECT PROFESSOR.PR_NAME
FROM PROFESSOR
WHERE EXISTS
        (SELECT *
        FROM SECTION
        WHERE PROFESSOR.PR_EMPID = SECTION.SE_PR_PROFID);
```

Result:

```
Pr_name
---------------
Ram Raj
Tony Hopkins
Katie Shef
Cathy Cobal
```

A correlated nested subquery is processed differently from an uncorrelated nested subquery. Instead of the execution of the subquery serving as input to its parent query (i.e., the outer query), in a correlated subquery the subquery is executed once for each row in the outer query. In addition, execution of the subquery stops and the EXISTS condition of the main query is declared true for a given row should the condition in the subquery be true. For example, using the data in the PROFESSOR and SECTION tables, for Ram Raj the execution of the subquery stops when the fourth row of the PROFESSOR table is evaluated against the seventh row of the SECTION table because **PROFESSOR.Pr_empid = SECTION.Se_pr_empid** at this point. Thus, in essence each value of **PROFESSOR.Pr_name** is treated as a constant during the evaluation. If the NOT EXISTS operator were used instead of the EXISTS operator, the names of professors who have not offered a section would be displayed.

The NOT EXISTS operator can be used as a way to express the DIVIDE operator in SQL. Recall the example in Section 11.1.2.4 which poses the question: What are the course numbers and course names of those courses offered in all quarters during which sections are offered? An SQL statement that produces an answer to this query is given below. The individual lines have been numbered to facilitate discussion of the execution of the query.

SQL SELECT Statement:

```
1. SELECT COURSE.CO_COURSE#, COURSE.CO_NAME
2. FROM COURSE
3. WHERE NOT EXISTS
4.        (SELECT DISTINCT A.SE_QTR
5.        FROM SECTION A
6.        WHERE NOT EXISTS
7.              (SELECT *
8.              FROM SECTION B
9.              WHERE COURSE.CO_COURSE# = B.SE_CO_COURSE#
10.             AND A.SE_QTR = B.SE_QTR));
```

Result:

```
Co_course# Co_name
---------- ----------------------
22QA375    Operations Research
```

Both the subquery that begins on line 4 and the subquery that begins on line 7 refer to the SECTION table. To prevent ambiguity, these two uses of the SECTION table have been assigned the aliases of A and B, respectively. Since the two uses of the NOT EXISTS operator may make this query difficult to understand, let's begin by assuming the first row retrieved from the COURSE table as part of the execution of lines 1-3 defines **COURSE.CO_COURSE#** as 15ECON112. Replacing **COURSE.CO_COURSE#** in line 9 with '15ECON112' causes the execution of lines 4-10 to generate the result shown below.

```
4.   (SELECT DISTINCT A.SE_QTR
5.    FROM SECTION A
6.    WHERE NOT EXISTS
7.    (SELECT *
8.    FROM SECTION B
9.    WHERE '15ECON112' = B.SE_CO_COURSE#
10.   AND A.SE_QTR = B.SE_QTR));
```

Result:

```
Se_qtr
------
A
S
U
W
```

Since Course# 15ECON112 does not appear at all in the SECTION table, the NOT EXISTS condition is true for each of the four quarters. On the other hand, when Course# 22QA375 replaces Course# 15ECON112 line 9, the NOT EXISTS condition is false for all four quarters (note that a section of Course# 22QA375 is offered during each quarter in the SECTION table) and thus "no rows selected" is the result when lines 4-10 are executed.

```
4.   (SELECT DISTINCT A.SE_QTR
5.    FROM SECTION A
6.    WHERE NOT EXISTS
7.    (SELECT *
8.    FROM SECTION B
9.    WHERE '22QA375' = B.SE_CO_COURSE#
10.   AND A.SE_QTR = B.SE_QTR));
```

Result:

```
No rows selected.
```

In other words, the NOT EXISTS condition in line 3 is true for Course# 22QA375. This SQL formulation corresponds to the following informal statement: "Display the course numbers of those courses such that there does not exist a quarter during which the course is not offered." It is left as an exercise for the reader to determine the result when lines 4-10 are executed for other courses (e.g., Course# 22IS330).

11.2.3.3 Aggregate Functions and Grouping

In SQL, an aggregate function takes as input a set of values, one from each row in a group of rows, and returns one value as the result. As illustrated in Section 11.2.1.4, the COUNT function, one of the most commonly used aggregate functions, counts the non-NULL values in a column. Other aggregate functions include retrieving the sum, average, maximum, and minimum of a series of numeric values. Another type of request involves the grouping of rows in a table or tables by the value of some of their attributes and then applying an aggregate function independently to each group.

Example 3.3.1. Using the data in the STUDENT and TAKES tables, count the number of sections taken by each student. Be sure to include those students who have never taken a class. The individual lines have been numbered to facilitate the discussion of the execution of the query.

SQL SELECT Statement:

```
1.  SELECT TAKES.TK_ST_SID, STUDENT.ST_NAME, COUNT(*) AS "Sections Taken"
2.  FROM STUDENT JOIN TAKES
3.  ON STUDENT.ST_SID = TAKES.TK_ST_SID
4.  GROUP BY TAKES.TK_ST_SID, STUDENT.ST_NAME
5.  UNION
6.  SELECT STUDENT.ST_SID, STUDENT.ST_NAME, 0
7.  FROM STUDENT
8.  WHERE STUDENT.ST_SID NOT IN
9.  (SELECT TAKES.TK_ST_SID FROM TAKES)
10. ORDER BY "Sections Taken" DESC;
```

In addition to illustrating the use of a grouping operation and an aggregate function, this query also contains a join, a nested subquery, and a union. Execution of the query begins with lines 1-4, which count the number of sections taken by those students who have taken at least one class. Note that a value of COUNT(*) is obtained for each combination of a **TAKES.Tk_st_sid**, and **STUDENT.St_name**. On the other hand lines 6-10, when executed, begin with the execution of line 9 and identify those students who have not taken a section. For each qualifying student, the numeric literal zero (0)[22] is displayed along with the value of **STUDENT.St_sid**, **STUDENT.St_name**. Since the content of columns 1 and 2 on lines 1 and 6 share the same domain and COUNT(*) and 0 also share the same domain (i.e., both are numeric values), the UNION operation on line 5 can take place. When a UNION operation takes place in a query, the sorting operation (i.e., represented by the ORDER BY clause on line 10) applies to the collective results from all SELECT statements involved in the UNION.

Result:

Tk_st_sid	St_name	Sections Taken
KS39874	Sweety Kramer	3
BE76598	Elijah Baley	2
BG66765	Gladis Bale	2
KP78924	Poppy Kramer	2
GS76775	Shweta Gupta	1
KJ56656	Joumana Kidd	1

---

[22] In addition to listing column names, expressions (e.g., see Section 11.2.1.2 and 11.2.1.3), functions (e.g., COUNT(*)), and SELECT statements (e.g., see Section 11.2.3.1), the SELECT list of a SELECT statement may also contain either a numeric constant (e.g., the numeric literal 0) or a string constant.

```
AJ76998    Jenny Aniston                0
DT87656    Tim Duncan                   0
FR45545    Rick Fox                     0
FV67733    Vanessa Fox                  0
HJ45633    Jenna Hopp                   0
HT67657    Troy Hudson                  0
JD35477    Diana Jackson                0
OD76578    Daniel Olive                 0
SD23556    David Sane                   0
SW56547    Wanda Seldon                 0
```

The same result could be obtained using the Left Outer Join shown below, since the value of COUNT(**TAKES.Tk_st_sid**) is zero when a row for a student not enrolled in a section is concatenated with the row of null values in TAKES.

SQL SELECT Statement:

```
SELECT STUDENT.ST_SID, STUDENT.ST_NAME, COUNT(TAKES.TK_ST_SID)
AS "Sections Taken"
FROM STUDENT LEFT OUTER JOIN TAKES
ON STUDENT.ST_SID = TAKES.TK_ST_SID
GROUP BY STUDENT.ST_SID, STUDENT.ST_NAME
ORDER BY "Sections Taken" DESC;
```

Example 3.3.2. Display the maximum, minimum, total, and average salary for the professors affiliated with each department. In addition, count the number of professors in each department as well as the number of professors in each department with a not null salary.

SQL SELECT Statement:

```
SELECT DEPARTMENT.DPT_NAME AS "Dept Name", DEPARTMENT.DPT_DCODE
AS "Dept Code",
MAX(PROFESSOR.PR_SALARY) AS "Max Salary", MIN(PROFESSOR.PR_SALARY)
AS "Min Salary",
SUM (PROFESSOR.PR_SALARY) AS "Total Salary",
ROUND(AVG(PROFESSOR.PR_SALARY),0) AS "Avg Salary", COUNT(*) AS "Size",
COUNT(PROFESSOR.PR_SALARY) AS "# Sals"
FROM DEPARTMENT JOIN PROFESSOR
ON DEPARTMENT.DPT_DCODE = PROFESSOR.PR_DPT_DCODE
GROUP BY DEPARTMENT.DPT_NAME, DEPARTMENT.DPT_DCODE
ORDER BY "# Sals" DESC;
```

Result:

Dept Name	Dept Code	Max Salary	Min Salary	Total Salary	Avg Salary	Size	# Sals
Economics	4	99000	67000	265000	88333	3	3
QA/QM	3	77000	43000	340000	68000	5	5
Economics	1	92000	45000	203000	67667	3	3
IS	7	65000	45000	174000	58000	3	3
Philosophy	9	69000	45000	114000	57000	3	2
Mathematics	6	44000	44000	88000	44000	3	2

Since department names are not unique but required as part of the output, grouping must be done on both **DEPARTMENT.Dpt_name** and **DEPARTMENT.Dpt_dcode**. In addition, instead of grouping by **DEPARTMENT.Dpt_dcode**, grouping could have been by **DEPARTMENT.Dpt_college** had the name of the college housing the department had been required as opposed to the department code.

Chapter Summary

The tuples of a relation can be considered elements of a set and thus can be involved in operations. In the same way that algebra is a system of operations on numbers, relational algebra is a system of operations on relations. Expressed in terms of the relations R and S, the basic operations of relational algebra are Union (R $\cup$ S), Difference (R $-$ S), Selection (σ _{\<selection condition>}(R)), Projection (π _{\<attribute list>} (R)), and Cartesian Product (R(A_1, A_2, ..., A_n) $\mathbf{X}$ S(B_1, B_2, ..., B_m)), where A_i and B_j are attributes of R and S respectively. Certain combinations of these five operations can be used to define three other basic operations. When the Union of R and S is formed, an Intersection operation identifies those tuples common to both R and S. A Join operation consists of a Cartesian Product followed by a Selection. A Division operation can be expressed as a sequence of Projection, Cartesian Product, and Difference operations. A summary of the basic relational algebra operations discussed in this chapter appears in Figure 11.2.

The SQL SELECT statement is used to express a query and is the most important statement in the language. Every SELECT statement, when executed, produces as its result a table that consists of one or more columns and zero or more rows. Six clauses make up a SELECT statement. Two of these clauses, the SELECT clause and the FROM clause, are required. The SELECT clause identifies the columns, calculated values, and literals to appear in the result table. All column names that appear in the SELECT clause must have their corresponding tables or views listed in the FROM clause.

The other four clauses, WHERE, GROUP BY, HAVING, and ORDER BY, are optional. The WHERE clause of the SELECT statement includes a search condition that consists of an expression involving constant values, column names, and comparison operators. The ORDER BY clause allows the result table to be sorted on the values that appear in the SELECT clause. If specified, the ORDER BY clause must be the final clause in the SELECT statement. The GROUP BY clause is used to form groups of rows of the result table based on column values. When grouping of rows occurs, all aggregate functions (e.g., COUNT, SUM, AVG) are computed on the individual groups and not the entire table. If used, the HAVING clause follows the GROUP BY clause. The HAVING clause functions as a WHERE clause for groups, keeping some groups and eliminating other groups from further consideration.

A data field without a value in it is said to be a null value. A null value can occur in a data field where a value is unknown or where a value is not meaningful. In an SQL SELECT statement, the only comparison operators that can be used with null values are IS NULL and IS NOT NULL. Any other operator (e.g., =, >, <, etc.) used with a null value will always produce an unknown (i.e., false) result.

SQL queries are based on one or more tables or views and often take the form of subqueries and joins. A subquery is an SQL SELECT statement embedded within another query or even another subquery. Subqueries may appear in the FROM clause, the column list, the WHERE clause, and the HAVING clause. The SQL SELECT statement is used to implement each of the relational algebra operations including both inner and outer joins. Use of the SQL SELECT statement in joining tables is required for all queries where the result comes from more than one table.

A complete list of the SQL SELECT statement features appears in Appendix D. After studying this chapter, it is hoped that readers will be able to use the features discussed here and, where necessary, adapt them to their specific database platform with a minimum of difficulty.

Exercises

1. What constitutes union compatibility?

2. What is the purpose of the Union, Intersection, and Difference operations?

3. What is a Cartesian Product operation?

4. What is the difference between the result obtained from a Selection operation versus a Projection operation?

5. Why does the relation created as result of a Projection operation on a relation that includes the primary key not contain fewer tuples than those in the source relation?

6. What is a Join operation? What is meant when it is said that two relations are join compatible? What is the difference between an Inner Join operation and an Outer Join operation?

7. What is meant by the term division compatibility?

8. Describe the six clauses that can be used in the syntax of the SQL SELECT statement. Which two clauses must be part of each SELECT statement?

9. What is the difference between a SELECT statement used in conjunction with the relational algebra Selection operation and a SELECT statement used in conjunction with the relational algebra Projection operation?

10. Of what value is the use of parentheses when making use of the rules of operator precedence?

11. What is the difference between a character field that contains a null value and a character field that contains a single blank space?

12. Which comparison operators can be used when searching for null values? Which comparison operators cannot be used when searching for null values? What is the result when these unacceptable comparison operators are used when searching for null values?

13. What is the result of an attempt to add, subtract, multiply, or divide two number fields, one of which contains a null value?

14. How are null values treated when one or more appears during the execution of a group function?

15. When must a GROUP BY clause be used in a query?

16. What SQL operator (i.e., keyword) is used in conjunction with pattern matching?

17. What is the difference between a SELECT statement that uses COUNT (*) and a SELECT statement that uses COUNT (column name)? How does COUNT (column name) differ from COUNT (DISTINCT column name)?

18. Why is it important to be aware of the distinction between the CHAR and VARCHAR data types?

19. What is the difference between a Cross Join, an Inner Join, and an Outer Join?

20. What is the difference between the JOIN ... USING and the JOIN ... ON approaches for joining tables? Which approach must be used if the requirement that attributes have unique names over the entire relational schema is enforced?

21. What is a subquery and where can subqueries appear within an SQL SELECT statement?

22. What do the ALL and ANY operators do when used in a subquery?

23. You must have completed Exercise 11 in Chapter 10 for beginning this exercise, and thus have used the SQL Data Definition Language to create tables for the three relations DRIVER, TICKET_TYPE, and TICKET.

   a. Use the SQL INSERT statement to populate these tables with the following data.

Dr_license_no	Dr_name	Dr_city	Dr_state
MVX 322	E. Mills	Waller	TX
RVX 287	R. Brooks	Bellaire	TX
TGY 832	L. Silva	Sugarland	TX
KEC 654	R. Lence	Houston	TX
MQA 823	E. Blair	Houston	TX
GRE 720	H. Newman	Pearland	TX

Ttp_offense	Ttp_fine
Parking	15
Red Light	50
Speeding	65
Failure To Stop	30

Tic_Ticket_no	Tic_ticket_date	Tic_dr_license_no	Tic_ttp_offense
1023	2007-12-20	MVX 322	Parking
1025	2007-12-21	RVX 287	Red Light
1397	2007-12-03	MVX 322	Parking
1027	2007-12-22	TGY 832	Parking
1225	2007-12-22	KEC 654	Speeding
1212	2007-12-06	MVX 322	Speeding
1024	2007-12-21	RVX 287	Speeding
1037	2007-12-23	MVX 322	Red Light

Note: When entering the date of the ticket, in the INSERT statement enclose the entire date in single quotes.

b. Once the three tables have been populated, write SQL Select statements to satisfy the following information requests:

1. Display the names of all drivers.
2. Display the license numbers of all drivers who have been issued a ticket.
3. Display the names of all drivers who have been issued a ticket.
4. Display the license numbers of all drivers who have never been issued a ticket.
5. Display the names of all drivers who have never been issued a ticket.
6. Count the number of tickets issued for each offense. Include as part of what you display any offense for which a ticket has not been issued.
7. For each ticket issued, display the name of the driver, the ticket number, and the nature of the offense. Order the results in ascending order by the name of the driver and within each driver order the results by ticket number.

24. This exercise is based on the data sets associated with Figure 2.25 in Chapter 2.

a. Use the SQL Data Definition Language to create a relational schema that consists of the following three relations:

COMPANY (Co_name, Co_size, Co_headquarters)
STUDENT (St_name, St_major, St_status)
INTERNSHIP (In_co_name, In_st_name, In_year, In_qtr, In_location, In_stipend)

When you create a table for each relation, in addition to defining its primary key, define all the appropriate referential integrity constraints. Assume that Co_name is a character data type of size 5, Co_size is an integer data type of size 4, Co_headquarters is a character data type of size 10, St_name is a Varchar data type of size 10, St_major is a character data type of size 20, St_status is a character data type of size 2, In_co_name is a character data type of size 5, In_st_name is a Varchar data type of size 10, In_year is an integer data type of size 4, In_qtr is a character data type of size 10, In_location is a character data type of size 15, and In_stipend is an integer data type of size 4. In_stipend represents the monthly stipend associated with the internship.

b. Use the SQL Insert statement to populate the three tables with the following data:

Co_name	Co_size	Co_headquarters
A	1000	Boston
B	500	Chicago
C	1000	Boston
D	400	Houston

St_name	St_major	St_status
Michelle	Communications	SR
Chris	Chemistry	JR
Andy	Finance	SO

Data Manipulation: Relational Algebra and SQL

St_name	St_major	St_status
Anna	Communications	SR
Amy	Communications	FR

In_co_name	In_st_name	In_year	In_qtr	In_location	In_stipend
A	Chris	2006	Fall	Concord	1000
A	Anna	2006	Fall	Concord	1000
B	Chris	2006	Fall	Concord	600
C	Amy	2006	Fall	South Bend	900
D	Andy	2006	Spring	South Bend	1000
A	Chris	2005	Spring	Concord	1200
D	Anna	2006	Spring	Houston	

c. After populating the three tables, write SQL Select statements to satisfy the following information requests:

1. Display the total monthly stipend received by each student.

2. For each internship, display the year and quarter offered, headquarters of the company offering the internship, and location of the internship. The output should be displayed in ascending order by the headquarters of the company offering the internship.

3. Display the names of all students who have not participated in an internship.

4. Display the number of internships offered for each year, quarter, and internship location.

5. Use pattern matching to display the names of those students whose name begins with an upper-case A and ends with some letter other than a lower-case a.

6. Use a natural join to display the name, major, and status of those students with an internship in the same city where the company is headquartered.

7. Use a subquery to display the name, major, and status of those students with an internship in the same city where the company is headquartered.

8. Display the difference between the average stipend offered by Company A and the average stipend offered by all other companies excluding Company A.

9. Use a left outer join to display the total monthly stipend received by each student including those students who have not participated in an internship. What, if anything, makes you uncomfortable about the result obtained?

10. Use a union to display the total monthly stipend received by each student including those students who have not participated in an internship.

SQL Projects

Three comprehensive SQL projects that incorporate topics covered in this chapter and Chapter 12 are included at the end of Chapter 12.

Selected Bibliography

Connolly, T. M. and Begg, C. E. (2002) *Database Systems: A Practical Approach to Design, Implementation, and Management*, Third Edition, Addison-Wesley.

Date, C. J. (1995) *An Introduction to Database Systems*, Sixth Edition, Addison-Wesley.

Date, C. J. and Darwen, H. (1997) *A Guide to the SQL Standard*, Fourth Edition, Addison-Wesley.

Elmasri, R., and Navathe, S. B. (2003) *Fundamentals of Database Systems*, Fourth Edition, Addison-Wesley.

Gennick, J. (2004) *SQL Pocket Guide*, O'Reilly Media, Inc.

Groff, J. R. and Weinberg, P. N. (2002) *SQL: The Complete Reference*, McGraw-Hill/Osborne, Second Edition.

Gulutzan, P. and Pelzer, T. (1999) *SQL-99 Complete, Really*, R&D Books.

Johnson, J. L. (1997) *Database Models, Languages, Design,* Oxford University Press, Inc.

Kifer, M., Bernstein, A. and Lewis, P. M. (2005) *Databases and Transactions Processing: An Application-Oriented Approach*, Second Edition, Addison-Wesley.

Kroenke, D. M. (2004) *Database Processing: Fundamentals, Design, and Implementation*, Ninth Edition, Pearson Prentice Hall.

Lorie, R. A. and Daudenarde, J. J. (1991) *SQL & Its Applications*, Prentice-Hall, Inc.

Morris-Murphy, L. L. (2003) *Oracle9i: SQL with an Introduction to PL/SQL,* Course Technology.

Rob, P. and Coronel, C. (2004) *Database Systems: Design, Implementation, and Management*, Course Technology, Sixth Edition.

Sunderraman, R. (2003) *Oracle9i Programming: A Primer*, Addison-Wesley.

CHAPTER **12**

Advanced Data Manipulation Using SQL

SQL for data manipulation is covered extensively in the previous chapter. In this chapter, we move on to study some of the advanced features of SQL. In Chapter 6, it is pointed out that while it is possible to specify all the integrity constraints as a part of the conceptual modeling process, some cannot be expressed explicitly or implicitly in the schema of the data model. Chapter 10 presented implementation of a majority of the schema-based or declarative constraints (such as domain constraints, key constraints, entity integrity constraints, and referential integrity constraints). It is also pointed out in Chapter 6 that some business rules pertaining to a valid state of a database or legal transitions of a database state may require procedural intervention. While application programs can be used to handle such procedural constraints, most DBMS products offer general-purpose procedural language support capable of implementing these constraints within the database. Assertions and triggers are SQL-92 facilities for capturing sophisticated declarative constraints (table-level constraints) and procedural constraints, respectively. In a database environment, views play a number of roles. As discussed in Chapter 6, views are "virtual tables" created using SQL/DDL. Since the script used in the creation of each of these database objects, i.e., ASSERTION, TRIGGER, VIEW, also requires an understanding of the SQL data manipulation statements, these topics are covered in this chapter.

The SQL-92 standard includes a number of built-in functions that can be used in an SQL statement anywhere that a constant of the same data type can be used. Section 12.2 covers a number of these functions while Section 12.3 focuses more on functions that facilitate the manipulation of dates and times. Discussion of SQL for data manipulation and retrieval concludes in Section 12.4 with a series of examples that apply a number of the SQL features introduced in this chapter and in Chapter 11.

12.1 Assertions, Triggers, and Views

This section revisits SQL/DDL because the DDL constructs ASSERTION and VIEW presented here embed the data retrieval construct of SQL/DML, namely, the SELECT statement discussed in Chapter 11.

12.1.1 Specifying an Assertion in SQL

There are times when a user-specified business rule may entail definition of a constraint not covered by the declarative constraints illustrated in Chapter 10 (for example, a constraint to be satisfied by a table as opposed to individual rows of a table or a constraint that spans multiple tables). The SQL-92 standard offers a means to specify such a "general" constraint via what is called a **declarative assertion** using a CREATE ASSERTION construct. Let us review an example to understand the utility of this construct using the PATIENT, MEDI-CATION, and ORDERS tables from Chapter 10. For convenience of the reader, Figure 10.1 is reproduced here as Figure 12.1. In addition, the SQL/DDL script from Box 3 in Section 10.1.1.1 is reproduced here as Figure 12.2.

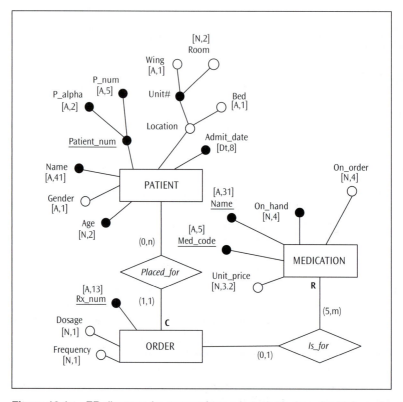

Figure 12.1a ER diagram: An excerpt from a hypothetical medical information system

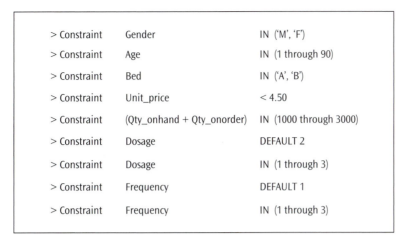

> Constraint	Gender	IN ('M', 'F')
> Constraint	Age	IN (1 through 90)
> Constraint	Bed	IN ('A', 'B')
> Constraint	Unit_price	< 4.50
> Constraint	(Qty_onhand + Qty_onorder)	IN (1000 through 3000)
> Constraint	Dosage	DEFAULT 2
> Constraint	Dosage	IN (1 through 3)
> Constraint	Frequency	DEFAULT 1
> Constraint	Frequency	IN (1 through 3)

Figure 12.1b Semantic integrity contstraints for the Fine-granular Design-Specific ER model

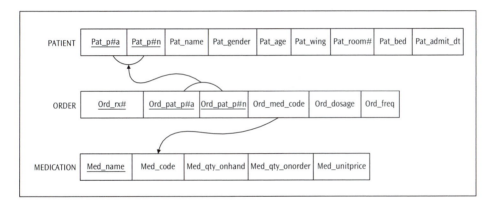

Figure 12.1c Relational schema for the ERD in Figure 12.1a

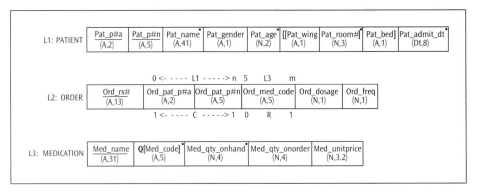

Figure 12.1d An information-preserving logical schema for the ERD in Figure 12.1a

```
CREATE TABLE patient
(Pat_p#a        char (2),
Pat_p#n         char (5),
Pat_name        varchar (41) constraint nn_Patnm not null,
Pat_gender      char (1),
Pat_age         smallint constraint nn_Patage not null,
Pat_admit_dt    date constraint nn_Patadmdt not null,
Pat_wing        char (1),
Pat_room#       integer,
Pat_bed         char (1),
CONSTRAINT pk_pat PRIMARY KEY (Pat_p#a, Pat_p#n),
CONSTRAINT chk_gender CHECK (Pat_gender IN ('M', 'F')),
CONSTRAINT chk_age CHECK (Pat_age IN (1 through 90)),
CONSTRAINT chk_bed CHECK (Pat_bed IN ('A', 'B'))
);


CREATE TABLE medication
(Med_code       char (5) CONSTRAINT nn_medcd not null CONSTRAINT unq_med UNIQUE,
Med_name        varchar (31) CONSTRAINT pk_med  PRIMARY KEY,
Med_unitprice   decimal (3,2) CONSTRAINT chk_unitprice CHECK (Med_unitprice < 4.50),
Med_qty_onhand  integer CONSTRAINT nn_medqty not null,
Med_qty_onorder integer,
CONSTRAINT chk_qty CHECK ((Med_qty_onhand + Med_qty_onorder) BETWEEN 1000 AND 3000)
);


CREATE TABLE orders
(Ord_rx#        char (13) CONSTRAINT pk_ord  PRIMARY KEY,
Ord_pat_p#a     char (2) CONSTRAINT nn_ord_pat_p#a not null,
Ord_pat_p#n     char (5) CONSTRAINT nn_ord_pat_p#n not null,
Ord_med_code    char (5) CONSTRAINT fk_med REFERENCES medication (Med_code)
ON DELETE RESTRICT ON UPDATE RESTRICT,
Ord_dosage      smallint DEFAULT 2 CONSTRAINT chk_dosage CHECK (Ord_dosage BETWEEN 1 AND 3),
Ord_freq        smallint DEFAULT 1 CONSTRAINT chk_freq CHECK (Ord_freq IN (1, 2, 3)),
CONSTRAINT fk_pat FOREIGN KEY (Ord_pat_p#a, Ord_pat_p#n)
REFERENCES patient (Pat_p#a, Pat_p#n) ON DELETE CASCADE ON UPDATE CASCADE
);
```

Figure 12.2 Chapter 10, Box 3 reproduced

Suppose a business rule says that at any given time there must be at least 100 orders present in the system; otherwise, the in-house pharmacy may not be cost-justifiable. A syntactically correct CREATE TABLE script specifying this rule is:

```
CREATE TABLE orders
(Ord_rx#              char (13) CONSTRAINT pk_ord  PRIMARY KEY,
Ord_pat_p#a           char (2)  CONSTRAINT nn_ord_pat_p#a not null,
Ord_pat_p#n           char (5)  CONSTRAINT nn_ord_pat_p#n not null,
Ord_med_code          char (5)  CONSTRAINT fk_med REFERENCES medication (Med_code)
ON DELETE RESTRICT ON UPDATE RESTRICT,
Ord_dosage            smallint CONSTRAINT chk_dosage CHECK (Ord_dosage
BETWEEN 1 AND 3),
Ord_freq              smallint CONSTRAINT chk_freq CHECK (Ord_freq IN (1, 2, 3)),
CONSTRAINT Chk_orders CHECK (SELECT COUNT (*) FROM orders >= 100),
CONSTRAINT fk_pat FOREIGN KEY (Ord_pat_p#a, Ord_pat_p#n)
REFERENCES patient (Pat_p#a, Pat_p#n) ON DELETE CASCADE ON UPDATE CASCADE
);
```

The highlighted CHECK constraint *Chk_orders* is intended to impose the business rule stated. Unfortunately, *Chk_orders* does not achieve the intended goal because a CHECK constraint in a table is expected to be satisfied by every row in the table where the constraint resides—not by the table itself. More importantly, if the ORDERS table is empty, then *Chk_orders* will *not* fail because an empty table, by definition, always satisfies all CHECK constraints in that table as there are no rows to check. The same CHECK constraint incorporated within an ASSERTION becomes a component of the database schema similar to a TABLE or a DOMAIN. Accordingly, the following statement imposes the constraint as desired since the constraint now applies to the ORDERS table instead of the individual rows of the ORDERS table.

Assertion Example 1

```
CREATE ASSERTION Chk_orders
CHECK (SELECT COUNT (*) FROM orders >= 100);
```

The general syntax for the CREATE ASSERTION construct is:

```
CREATE ASSERTION assertion_name CHECK (conditional-expression);
```

where:

- *assertion_name* is a user-supplied name for the declarative assertion
- *conditional-expression* is an expression of arbitrary complexity referring to one or more base tables in the database

Recall from Chapter 10 that enforcement of *total participation* of a parent in a relationship (i.e., min > 0) requires a general constraint specification mechanism. A *general constraint* may be described as one that applies to arbitrary combinations of columns in arbitrary combinations of base tables (Date and Darwen, 1997). In that case, we ought to be able to enforce this constraint using a declarative assertion. For example, the ER diagram and the information-preserving logical schema shown in Figure 12.1 indicates that every medication should be included in at least 5 orders. This cannot be implemented through a table creation DDL. A declarative assertion that defines this constraint is of the following form.

Assertion Example 2

```
CREATE ASSERTION Chk_ordr_per_med
CHECK (NOT EXISTS
  (SELECT *
  FROM medication
  WHERE medication.Med_code NOT IN
        (SELECT Ord_med_code
        FROM orders
        GROUP BY Ord_med_code
        HAVING COUNT (*) >= 5))
  );
```

This example seeks to generate a list of medications that violate the condition specified and checks for the return of an empty set as the result. Thus, the assertion is violated if the result of the conditional expression (i.e., the query) is not an empty set. A critical question here is what happens if at the time of specifying the assertion *Chk_ordr_per_med* (a) the ORDERS table is empty, or (b) it already has data that violates the condition specified in the assertion. According to the SQL-92 standard, if a new constraint is defined and the existing database state does not satisfy the constraint, the constraint fails (i.e., is not created). At this point, it is the responsibility of the database designer to discover the cause of the violation and either amend the constraint or rectify the data in the database.

The participation of PATIENT in the *Placed_for* relationship with ORDERS is optional (see Figure 12.1a or 12.1d). This constraint is in force by default. Suppose, for some strange reason, a business rule dictates that there *must* always be some patients who have not placed any orders for medication. First of all, this rule cannot be shown in the ERD diagram and should therefore be included in the list of semantic integrity constraints in Figure 12.1b. That said, how is such a constraint enforced? Let us evaluate the declarative assertion in the next example.

Assertion Example 3

```
CREATE ASSERTION Chk_no_ordr_pats
CHECK (NOT EXISTS
    (SELECT *
    FROM patient
    WHERE (Pat_p#a, Pat_p#n) NOT IN
        (SELECT Ord_pat_p#a, Ord_pat_p#n
        FROM orders))
  );
```

This assertion seeks to generate a list of patients that violate the condition specified and checks for the return of an empty set as the result. Thus, the assertion is violated if the result of the conditional expression (i.e., the query) is not an empty set.

As another illustration of the utility of a declarative assertion, observe that **Unit#** of the PATIENT entity type in the ER diagram (see Figure 12.1a) is a composite attribute. This information is preserved in the logical schema (Figure 12.1d) by enclosing the atomic attributes **Pat_wing** and **Pat_room#** in brackets [ ]. While it is impossible to map this information to the table definition, is it possible to capture the business rule that **Unit#** and therefore **[Pat_wing, Pat_room#]** is a *mandatory* composite attribute via a declarative assertion?

The answer is a qualified "Yes"—i.e., it is possible to specify that **Pat_wing** and **Pat_room#** together cannot have a null value, while individually each of these two columns may have a null value in any row of the PATIENT table. It is not possible, however, to specify that **[Pat_wing, Pat_room#]** represents a composite attribute since, by definition, there is no such thing as composite attribute/column in a relational database.

The declarative assertion that prohibits the occurrence of null values in both **Pat_wing** and **Pat_room#** at the same time in the same row of the PATIENT table is:

Assertion Example 4

```
CREATE ASSERTION Chk_unit#
CHECK (NOT EXISTS
    (SELECT *
    FROM patient
    WHERE Pat_wing IS NULL AND Pat_room# IS NULL)
);
```

Finally, let us examine this declarative assertion:

Assertion Example 5

```
CREATE ASSERTION Chk_costcontrol
CHECK ((Med_unitprice * Med_qty_onhand) < 5000
);
```

The assertion can be viewed as making sure that locked-up capital in stocking a medication is controlled at below $5,000. Do we need a declarative assertion to enforce this business rule? Not at all. This CHECK constraint is applicable at the row level of the MEDICATION table and hence can be defined as a table-level constraint.

Just as a declarative assertion can be created, it can also be deleted when no longer needed using the syntax:

```
DROP ASSERTION Assertion_name;
```

For example,

```
DROP ASSERTION Chk_costcontrol;
```

Note that there are no behavior options like CASCADE or RESTRICT associated with the DROP statement for an ASSERTION. Also, SQL-92 does not provide an ALTER construct for ASSERTIONs.

In general, implementing declarative assertions by a DBMS is not as efficient as implementing a CHECK constraint on a column, row, or domain. Therefore, it is advisable to use declarative assertions only when other declarative means will not work. Another fundamental problem with a declarative assertion is that like the other declarative constraints in a database schema, it offers only one option—that is, aborting the operation that violates the assertion. It may be useful if the DBMS offers other options whereby an action can be executed automatically (e.g., sending a message to the user) when the assertion is violated.

12.1.2 Triggers in SQL

A **trigger** allows specification of certain types of active rules. The concept of triggers is rooted in active databases. In simple terms, an active database must react to external events when the events occur. The model used for specifying active database rules is called the **Event-Condition-Action (ECA) model**. A trigger is an element in the database schema that implements the ECA model where:

    a.   an event is an external request for an operation on the database (e.g., insert a row into a table),

    b.   a condition is a Boolean expression that must evaluate TRUE in order for the action to be executed, and

    c.   an action is a procedure that stipulates what needs to be done when the trigger is fired (e.g., insert a row, display a message).

In general, the possible events are INSERT and DELETE rows in a base table or UPDATE specific columns. The action can be executed BEFORE or AFTER the event based on whether the condition is to be tested in the database state that exists before or after the event. The trigger granularity defines what constitutes an event. For instance, a **row-level granularity** implies that a change to a single row is an event, while a **statement-level granularity** indicates that the statement (e.g., INSERT, DELETE, UPDATE) is the event regardless of the number of rows affected by the execution of the trigger.

Interestingly triggers existed in early versions of SQL, but are not part of the SQL-92 standard. However, most DBMS products support some level of trigger processing. Triggers are part of the SQL-99 standard. This section contains examples of simple triggers created and executed using an Oracle9i database platform. None of these examples illustrates the many nuances that complicate the application of triggers. For more information, the reader is directed to the reference material in the selected bibliography at the end of this chapter (Kifer, et. al, 2005; Gulutzan and Pelzer, 1999).

12.1.2.1 Trigger Fundamentals

The syntax for creating a trigger is as follows:

```
CREATE [OR REPLACE] TRIGGER trigger_name
[BEFORE|AFTER]
[DELETE|INSERT|UPDATE [OF column]]
ON table_name
[FOR EACH  {ROW|STATEMENT}]
[WHEN condition]
statement-list
```

There are 12 basic types of triggers. A trigger's type is defined by the type of triggering transaction, and by the level at which the trigger is executed. The 12 basic types of triggers are derived from the following three categories: row-level triggers, statement-level triggers, and BEFORE and AFTER triggers. Before and after triggers execute immediately before or after inserts, updates, and deletions. Since a trigger can be (a) either a row-level or statement-level trigger, and (b) can execute immediately before or after inserts, updates, and deletes, the following 12 possible configurations arise.

BEFORE INSERT row	AFTER UPDATE row
BEFORE INSERT statement	AFTER UPDATE statement
AFTER INSERT row	BEFORE DELETE row
AFTER INSERT statement	BEFORE DELETE statement
BEFORE UPDATE row	AFTER DELETE row
BEFORE UPDATE statement	AFTER DELETE statement

586

Row-level triggers execute once for each row in a transaction, whereas statement-level triggers execute once for each transaction. For example, if a single transaction updates 100 rows in a table, a row-level trigger on the table would be executed 100 times (once either before or after the update of a row) while a statement-level trigger would be executed only one time (either before or after the execution of the updates to the 100 rows).

12.1.2.2 The AFTER DELETE Trigger

A common use of a trigger is to create a historical record of a transaction. The following example illustrates the creation of a database trigger that serves as an audit trail by inserting a row into the PATIENT_AUDIT table whenever a patient is deleted from the clinic (i.e., discharged).

AFTER DELETE Trigger Example.

```
CREATE TRIGGER PATIENT_AFT_DEL_ROW
AFTER DELETE ON PATIENT
FOR EACH ROW
BEGIN
INSERT INTO PATIENT_AUDIT VALUES (:OLD.PAT_P#A, :OLD.PAT_P#N, CURRENT_
DATE);
END;
```

The following comments discuss specific aspects involved in the creation of this trigger.

1. The PATIENT_AUDIT table contains three columns: **Pau_p#a**, **Pau_p#n**, and **Pau_datetime**. The CREATE TABLE statement shown below was used in the creation of the table.

    ```
    CREATE TABLE patient_audit
    (Pau_p#a char(2),
    Pau_p#n char(5),
    Pau_datetime date);
    ```

2. Observe that the name of the trigger (PATIENT_AFT_DEL_ROW) reflects its nature. The component, PATIENT, suggests its association with the PATIENT table, _AFT indicates that the trigger is an after trigger, _DEL indicates that the trigger involves a delete operation, and _ROW indicates that the trigger is a row-level trigger. It is important that the content of the remainder of the trigger be consistent with the name of the trigger if the name of the trigger is to accurately reflect the nature of the trigger.
3. As indicated in the syntax, in the creation of a trigger, the name of the trigger must be followed by the BEFORE or AFTER keyword, indicating whether the trigger should be fired before or after the operation that causes it to be fired. Immediately following the BEFORE or AFTER keyword is the action (or actions) with which the trigger is associated. This can be INSERT, UPDATE, or DELETE, or any combination of these separated by OR. This trigger is to be executed after the deletion of each row from the PATIENT table.
4. The FOR EACH ROW keyword defines the behavior of the trigger when it is fired by a statement affecting multiple rows. The default behavior is to fire the trigger only once, regardless of the number of rows affected (i.e., a statement-level trigger is assumed unless the FOR EACH ROW keyword is used).

Every trigger must contain a series of one or more statements[1] bracketed by the words BEGIN and END.[2] In this example there is only one statement, an INSERT statement that inserts a row into the PATIENT_AUDIT table. The :OLD keyword is used in the INSERT statement as a prefix to refer to the old value of the **PATIENT.Pau_p#a** and **PATIENT.Pau_p#n** columns. Although not used in this example since a delete operation cannot result in a new value for a column, the :NEW keyword is used as a prefix in both insert and update operations to refer to the new value of a column. CURRENT_DATE in the INSERT statement is an SQL-92 function that records the current date. Section 12. 3 contains additional information about the CURRENT_DATE and other SQL-92 functions.

To illustrate the execution of the trigger, observe the content of the PATIENT and PATIENT_AUDIT tables before and after the deletion of the row for the patient Bill Davis.

Content of Tables Prior to Deletion

```
SELECT * FROM PATIENT;

Pat_p#a  Pat_p#n  Pat_name          Pat_gender    Pat_age Pat_admit_dt Pat_wing  Pat_room# Pat_bed
-------- -------- ---------------- ----------- ----------- ------------ -------- ---------- -------

DB       77642    Davis, Bill      M                  27 2007-07-07  B              108 B
GD       72222    Grimes, David                       44 2007-07-12

SELECT * FROM PATIENT_AUDIT;

no rows selected
```

Deletion of Patient Bill Davis

```
DELETE FROM PATIENT WHERE PATIENT.PAT_NAME LIKE '%Davis, Bill%';
1 row deleted.
```

Contents of Tables After Deletion

```
SELECT * FROM PATIENT;

Pat_p#a  Pat_p#n  Pat_name          Pat_gender    Pat_age Pat_admit_dt Pat_wing  Pat_room# Pat_bed
-------- -------- ---------------- ---------- ---------- ------------ -------- ---------- -------

GD       72222    Grimes, David                       44 2007-07-12

SELECT * FROM PATIENT_AUDIT;

Pau_p#a  Pau_p#n  Pau_datetime
-------- -------- ------------
DB       77642    2006-07-22
```

---

[1] In Oracle9i, these statements can be a combination of SQL and PL/SQL statements. This section contains only enough information about PL/SQL to explain the examples. For a comprehensive discussion of PL/SQL, refer to Sunderraman (1999) or books published by Oracle Press.

[2] The words BEGIN and END enclose the executable portion of what is known as a PL/SQL block.

12.1.2.3 The BEFORE INSERT Trigger

The following is an example of a trigger that disallows the insertion of an order for a patient who has already received three orders for the same medication (a business rule that cannot be specified by a declarative constraint when defining the ORDERS table). The individual lines have been numbered to facilitate the discussion but otherwise are not part of the trigger.

BEFORE INSERT Trigger Example

```
1.   CREATE OR REPLACE TRIGGER ORDERS_BEF_INS_ROW
2.   BEFORE INSERT ON ORDERS FOR EACH ROW
3.   DECLARE
4.       NO_ORDERS NUMBER;
5.       TEMP_MED_NAME VARCHAR(20);
6.   BEGIN
7.       SELECT COUNT(*) INTO NO_ORDERS
8.       FROM ORDERS
9.       WHERE ORDERS.ORD_PAT_P#A = :NEW.ORD_PAT_P#A
10.      AND ORDERS.ORD_PAT_P#N = :NEW.ORD_PAT_P#N
11.      AND ORDERS.ORD_MED_CODE = :NEW.ORD_MED_CODE;
12.          IF NO_ORDERS >= 3 THEN
13.              SELECT MED_NAME INTO TEMP_MED_NAME FROM MEDICATION
14.              WHERE MED_CODE = :NEW.ORD_MED_CODE;
15.              RAISE_APPLICATION_ERROR (-20000, 'Patient '||:NEW.ORD_PAT_P#A||:NEW.ORD_PAT_P#N||
16.              ' has had too much '||TEMP_MED_NAME);
17.          END IF;
18.  END;
```

1. As indicated in the trigger syntax at the beginning of Section 12.1.2.1, the OR REPLACE clause on line 1 is optional. It is included here because triggers, like many application programs, can be created and thus become part of the database schema while still containing logic errors necessitating their re-creation. Since an ALTER TRIGGER statement does not exist, the CREATE OR REPLACE clause allows an existing trigger in need of modification to be dropped (i.e., deleted from the database schema) and recreated from scratch. Had CREATE TRIGGER instead of CREATE OR REPLACE TRIGGER been used, a DROP TRIGGER statement would have to have been used first to drop the erroneous trigger before a CREATE TRIGGER statement could have been used to recreate it.

2. This trigger is divided into two sections. The DECLARE section on lines 3–5 defines two variables[3] (NO_ORDERS and TEMP_MED_NAME) that will subsequently be referenced by the executable statements contained on lines 6–18.

3. The first SELECT statement on lines 7–11 searches the ORDERS table and counts the number of orders that involve the medication code for the patient for whom an attempt is being made to insert a new order into the ORDERS table. Since the SELECT statement here is embedded in a host language (in this case PL/SQL), the INTO clause must be used to allow the value returned by the COUNT(*) function to be referenced in statements written in the host language. Observe that the host language variable, NO_ORDERS, that appears in the INTO clause is subsequently referenced by the IF statement on line 12.

---

[3] These variables are technically referred to as host variables.

4. :NEW.ORD_PAT_P_P#A, :NEW.ORD_PAT_P#N, and :NEW.ORD_MED_CODE referenced on lines 9–11 are the values assigned to the **Ord_pat_p#a**, **Ord_pat_p#n** and **Ord_med_code** columns (columns 2, 3, and 4 of the ORDERS table) in the INSERT statement that is the subject of this trigger. The prefix :NEW is used to refer to the new value associated with the row that caused the execution of the trigger.

5. Lines 12–17 contain a host language IF statement that checks to see if the patient has already received three orders for the same medication (i.e., it checks to see if the host language variable NO_ORDERS is greater than or equal to 3). If the condition is true, then the SELECT statement on lines 13 and 14 retrieves the name of the medication from the MEDICATION table. RAISE_ APPLICATION_ERROR is an Oracle procedure which allows custom error-messages to be issued. Its syntax consists of (a) an error number that is a negative integer in the range –20000 to –20999, and (b) an error message. Note how the error message makes use of the patient number associated with the INSERT statement that is the subject of the trigger plus the name of the medication retrieved by the SELECT statements on lines 13 and 14.

To illustrate the execution of the trigger, observe the content of the PATIENT, MEDICATION, and ORDERS tables before and after attempts to insert new rows into the ORDERS table.

Content of Tables Prior to Insertion Attempts

```
SELECT * FROM PATIENT;


Pat_p#a  Pat_p#n  Pat_name          Pat_gender   Pat_age Pat_admit_dt Pat_wing  Pat_room# Pat_bed
-------- -------- ----------------  ----------   ------- ------------ --------  --------- -------
DB       77642    Davis, Bill       M                 27 2007-07-07   B               108 B
GD       72222    Grimes, David                       44 2007-07-12


SELECT * FROM MEDICATION;


Med_code Med_name             Med_qty_onhand Med_qty_onorder Med_unitprice
-------- -------------------- -------------- --------------- -------------
KEF      Keflin                          400             700             3
VAL      Valium                          500             500          3.33
ASP      Aspirin                        1200             100            .1


SELECT * FROM ORDERS;


Ord_rx#      Ord_pat_p#a  Ord_pat_p#n  Ord_med_code Ord_dosage  Ord_freq
------------ ------------ ------------ ------------ ----------  ----------
104          DB           77642        ASP                   3           1
105          DB           77642        ASP                   2           1
106          DB           77642        ASP                   2           1
108          GD           72222        KEF                   2           1
```

Insertion Attempts

```
INSERT INTO ORDERS VALUES ('109', 'GD', '72222', 'KEF', 2, 3);

1 row created.


INSERT INTO ORDERS VALUES ('110', 'DB', '77642', 'ASP', 3, 1);
*
ERROR-20000: Patient DB77642 has had too much Aspirin


INSERT INTO ORDERS VALUES ('111', 'DB', '77642', 'KEF', 3, 1);

1 row created.
```

Content of Tables After Insertion Attempts

```
SELECT * FROM PATIENT;
```

Pat_p#a	Pat_p#n	Pat_name	Pat_gender	Pat_age	Pat_admit_dt	Pat_wing	Pat_room#	Pat_bed
DB	77642	Davis, Bill	M	27	2007-07-07	B	108	B
GD	72222	Grimes, David		44	2007-07-12			

```
SELECT * FROM MEDICATION;
```

Med_code	Med_name	Med_qty_onhand	Med_qty_onorder	Med_unitprice
KEF	Keflin	400	700	3
VAL	Valium	500	500	3.33
ASP	Aspirin	1200	100	.1

```
SELECT * FROM ORDERS;
```

Ord_rx#	Ord_pat_p#a	Ord_pat_p#n	Ord_med_code	Ord_dosage	Ord_freq
104	DB	77642	ASP	3	1
105	DB	77642	ASP	2	1
106	DB	77642	ASP	2	1
108	GD	72222	KEF	2	1
109	GD	72222	KEF	2	3
111	DB	77642	KEF	3	1

12.1.2.4 Combining Trigger Types and Customizing Error Conditions

Triggers for insert, update, and delete statements on a table can be combined into a single trigger provided they are all at the same level (row-level or statement-level). The following example shows a statement-level trigger that is executed whenever an insertion, deletion, or update is attempted on the PATIENT table. Its purpose is to check two system conditions: that the day of the week is neither Saturday nor Sunday (i.e., the PATIENT table is available for only querying on these days), and that the username of the employee performing the insertion, update, or deletion begins with the letters "ACCTG". Once again, the individual lines have been numbered to facilitate the discussion of those statements or features not covered in the two previous examples.

BEFORE INSERT, UPDATE, OR DELETE Trigger Example

```
1.    CREATE OR REPLACE TRIGGER PATIENT_BEF_INS_UPD_DEL
2.    BEFORE INSERT OR UPDATE OR DELETE ON PATIENT
3.    DECLARE
4.      WEEKEND_ERROR EXCEPTION;
5.      NOT_ACCTG_USER EXCEPTION;
6.    BEGIN
7.      IF TO_CHAR(CURRENT_DATE, 'DY') = 'SAT' OR
8.        TO_CHAR(CURRENT_DATE, 'DY') = 'SUN' THEN
9.      RAISE WEEKEND_ERROR;
10.     END IF;
11.     IF SUBSTR(USER,1,5) <> UPPER('ACCTG') THEN
12.       RAISE NOT_ACCTG_USER;
13.     END IF;
14.   EXCEPTION
15.     WHEN WEEKEND_ERROR THEN
16.       RAISE_APPLICATION_ERROR(-20001,'PATIENT table not available on weekends');
17.     WHEN NOT_ACCTG_USER THEN
18.       RAISE_APPLICATION_ERROR(-20002, 'Not authorized to access PATIENT table');
19.   END;
```

1. Line 2 illustrates the use of the OR keyword to demonstrate the use of this trigger to fire before each possible action on the PATIENT table. As a statement-level trigger, this trigger is fired before the execution of an action on one row or to many rows in the PATIENT table.

2. The DECLARE section on lines 3–5 is used to define two host language (PL/SQL) variables with an exception data type. The exception data type allows the variables WEEKEND_ERROR and NOT_ACCTG_USER to function as user-defined exception handlers that will subsequently be referenced by the executable statements on lines 6–19.

3. The TO_CHAR (d [,fmt]) function is used in Oracle to extract different parts of a date/time and convert them to a character string using the format specified by the format element *fmt*. The purpose of the TO_CHAR (CURRENT_DATE, 'DY') function on line 7 is to extract from the current date the day of the week and compare its value with 'SAT'. Section 12.3 contains additional information about the TO_CHAR function.

4. The RAISE statement on line 9 stops normal execution of the PL/SQL block and transfers control to the exception handler WEEKEND_ERROR on lines 15 and 16.

5. The keyword, USER, on line 11 refers to the name of the current Oracle user. Its use with the SUBSTR function makes it possible for the userid to be checked. The UPPER function allows for a character string or column to be converted to upper case. The similar function LOWER converts a character string or column to lower case. Functions such as the UPPER and SUBSTR function are also discussed in Section 12.2.

6. A PL/SQL block can consist of up to three sections:

   - an optional declaration section that begins with the DECLARE header,

   - an executable section which starts with the word BEGIN, and

   - an optional exception section which consists of series of statements to respond to exceptions.

The AFTER DELETE trigger example in Section 12.1.2.2 consists of only one section—the required executable section and the BEFORE INSERT trigger example in Section 12.1.2.3 contains both the required executable section and the optional declare section. Lines 14–18 of this trigger illustrate an example of a trigger with all three sections. The exception section is made up of one or more exception handlers which consist of a WHEN clause specifying an exception name and a sequence of statements to be executed to handle the error when the exception is raised. The execution of this sequence of statements completes the execution of the block (and therefore the trigger). The RAISE_ APPLICATION_ERROR procedure is used to display customized error messages should one of the two system conditions be violated.

The following series of INSERT statements test various conditions associated with this trigger.

Content of PATIENT Table Prior to Tests

```
SELECT * FROM PATIENT;


Pat_p#a  Pat_p#n  Pat_name          Pat_gender   Pat_age Pat_admit_dt Pat_wing  Pat_room Pat_bed
-------- -------- ----------------- ---------- ---------- ------------ -------- ---------- ------
DB       77642    Davis, Bill       M                  27 2007-07-07  B              108 B
GD       72222    Grimes, David                        44 2007-07-12
```

Test 1. Assume Date of Saturday, August 25, 2007 and Accounting User

```
INSERT INTO PATIENT VALUES ('SJ', '12345', 'Su, John', 'M', 30, '2007-08-25', NULL, NULL, NULL);
INSERT INTO PATIENT VALUES ('SJ', '12345', 'Su, John', 'M', 30, '2007-08-25', NULL, NULL, NULL)
*
ERROR at line 1:
ERROR-20001: PATIENT table not available on weekends
```

Test 2. Assume Date of Friday, August 24, 2007 and Non-Accounting User

```
INSERT INTO PATIENT VALUES ('LB', '23451', 'Li, Belle', 'F', 27, '2007-08-25', NULL, NULL, NULL);
INSERT INTO PATIENT VALUES ('LB', '23451', 'Li, Belle', 'F', 27, '2007-08-25', NULL, NULL, NULL)

ERROR at line 1:
ERROR-20002: Not authorized to access PATIENT table
```

Test 3. Assume Date of Friday, August 24, 2007 and Accounting User

```
INSERT INTO PATIENT VALUES ('LS', '30232', 'Li, Sue', 'F', 23, '2007-08-25', NULL, NULL, NULL);

1 row created.

SELECT * FROM PATIENT;


Pat_p#a  Pat_p#n  Pat_name          Pat_gender   Pat_age Pat_admit_dt Pat_wing  Pat_room Pat_bed
-------- -------- ----------------- ---------- ---------- ------------ -------- ---------- ------
DB       77642    Davis, Bill       M                  27 2007-07-07  B              108 B
LS       30232    Li, Sue           F                  23 2007-08-25
GD       72222    Grimes, David                        44 2007-07-12
```

12.1.2.5 Some Cautionary Comments About Triggers

While individual triggers are usually relatively easy to understand, when combined with declarative integrity checking, trigger execution can be complex because the actions of one SQL statement may cause several triggers to fire. Mannino (2007, pp. 414–415) provides a helpful description of Oracle's simplified trigger execution procedure.

1. Execute the applicable BEFORE statement-level triggers
2. For each row affected by the SQL data manipulation statement:
   2.1 Execute the applicable BEFORE row-level triggers
   2.2 Perform the data manipulation operation on the row
   2.3 Perform integrity constraint checking
   2.4 Execute the application AFTER row-level triggers.
3. Perform deferred integrity constraint checking.[4]
4. Execute the applicable AFTER statement-level triggers.

The remainder of this section constitutes a simple illustration of how declarative constraints and triggers work together to enforce data integrity. Let's begin by assuming that PATIENT, MEDICATION, and ORDERS tables exist as defined by the three earlier CREATE TABLE statements and that the three tables contain the following data.

```
SELECT * FROM PATIENT;
```

Pat_p#a	Pat_p#n	Pat_name	Pat_gender	Pat_age	Pat_admit_dt	Pat_wing	Pat_room	Pat_bed
DB	77642	Davis, Bill	M	27	2007-07-07	B	108	B
LS	12345	Li, Sue	F	23	2007-08-25			
ZZ	06912	Zhang, Zhaoping	F	35	2007-08-11			
GD	72222	Grimes, David		44	2007-07-12			

```
SELECT * FROM MEDICATION;
```

Med_code	Med_name	Med_qty_onhand	Med_qty_onorder	Med_unitprice
KEF	Keflin	400	700	3
VAL	Valium	501	500	3.33
ASP	Aspirin	1200	100	.1

```
SELECT * FROM ORDERS;
```

Ord_rx#	Ord_pat_p#a	Ord_pat_p#n	Ord_med_code	Ord_dosage	Ord_freq
104	DB	77642	ASP	3	1
105	DB	77642	ASP	2	1
106	DB	77642	ASP	2	1
108	GD	72222	KEF	2	1
109	GD	72222	KEF	2	3
111	DB	77642	KEF	3	1

---

[4] Deferred integrity constraint checking involves enforcing integrity constraints at the end of a transaction rather than after each data manipulation statement. Recall from Section 6.7.1.2.3. that using mutual-referencing to enforce a 1:1 relationship with total participation of each entity type necessitates deferring the execution of at least one of the two referential constraints until run time.

Next, let's create the following after insert trigger on the ORDERS table reducing the quantity on hand of the medication associated with a new order by the number of doses ordered. Once again, the individual lines have been numbered to facilitate the discussion of the trigger and the introduction of new PL/SQL statements.

```
1.  CREATE OR REPLACE TRIGGER ORDERS_AFT_INS_ROW
2.  AFTER INSERT ON ORDERS FOR EACH ROW
3.  DECLARE
4.  TOTAL_AVAILABLE NUMBER;
5.  NAME_OF_MEDICATION VARCHAR(20);
6.  BEGIN
7.      UPDATE MEDICATION
8.      SET MEDICATION.MED_QTY_ONHAND = MEDICATION.MED_QTY_ONHAND - :NEW.ORD_DOSAGE
9.      WHERE MEDICATION.MED_CODE = :NEW.ORD_MED_CODE;
10.
11.     SELECT MEDICATION.MED_QTY_ONHAND, MED_NAME INTO TOTAL_AVAILABLE, NAME_OF_MEDICATION
12.     FROM MEDICATION
13.     WHERE MEDICATION.MED_CODE = :NEW.ORD_MED_CODE;
14.     DBMS_OUTPUT.PUT_LINE ('Quantity on hand for '||NAME_OF_MEDICATION||' is now '||
15.         TOTAL_AVAILABLE || ' doses.');
16.
17. END;
```

1. The DECLARE section on lines 3–5 defines two host language (PL/SQL) variables (TOTAL_AVAILABLE and NAME_OF_MEDICATION) that will subsequently be referenced on lines 11, 14, and 15.

2. The UPDATE statement on lines 7–9 updates the quantity on hand for the medication referenced in the INSERT statement that causes the trigger to be fired. The :NEW.ORD_DOSAGE and :NEW.ORD_MED_CODE represent the code for the medication code and dosage amount in the INSERT statement. Upon execution of this UPDATE statement, the quantity on hand for the medication code is reduced by the value stored in :NEW.ORD_DOSAGE.

3. After updating the MEDICATION table, the SELECT statement on lines 11–13 references the row just updated, and the quantity on hand and name of the medication are stored in the PL/SQL variables TOTAL_AVAILABLE and NAME_OF_MEDICATION. Lines 14 and 15 contain a DBMS_OUTPUT.PUT_LINE statement. The purpose of the DBMS_OUTPUT.PUT_LINE statement is to display the content of PL/SQL variables and constants and is used in this trigger to display the updated quantity on hand for the medication ordered.

A second trigger was also created and is shown below. The purpose of this statement-level after insert or update or delete trigger on the MEDICATION table is simply to display a message that records a successful insertion, update, or deletion to the MEDICATION table. The words INSERTING, UPDATING (and also DELETING) are reserved words in PL/SQL that represent valid transaction types.

```
CREATE OR REPLACE TRIGGER MEDICATION_AFT_INS_UPD_DEL
AFTER INSERT OR UPDATE OR DELETE ON MEDICATION
BEGIN
  IF INSERTING THEN
  DBMS_OUTPUT.PUT_LINE ('SUCCESSFUL INSERTION INTO MEDICATION TABLE');
  ELSE
    IF UPDATING THEN
        DBMS_OUTPUT.PUT_LINE ('SUCCESSFUL UPDATE OF MEDICATION TABLE');
    ELSE
        DBMS_OUTPUT.PUT_LINE ('SUCCESSFUL DELETION TO MEDICATION TABLE');
    END IF;
    END IF;
END;
```

Thus, at this point, a total of five triggers exist in the database schema:

- A row-level after delete trigger on the PATIENT table (see Section 12.1.2.2)
- A row-level before insert trigger on the ORDERS table (see Section 12.1.2.3)
- A statement-level before, insert, or update trigger on the PATIENT table (see Section 12.1.2.4)
- A row-level after insert trigger on the ORDERS table (defined towards the beginning of this section)
- A statement-level before, insert, or update trigger on the MEDICATION table (defined immediately above)

Test 1. Patient Sue Li places an order for two doses of the medication Valium.

```
INSERT INTO ORDERS VALUES ('112', 'LS', '12345', 'VAL', 2, 1);
INSERT INTO ORDERS VALUES ('112', 'LS', '12345', 'VAL', 2, 1)
*
ERROR at line 1:
ORA-02290: check constraint (CHAPTER11.CHK_QTY) violated
ORA-06512: at "CHAPTER11.ORDERS_AFT_INS_ROW", line 5
ORA-04088: error during execution of trigger 'CHAPTER11.ORDERS_AFT_INS_ROW'
```

A statement-level before, update, and delete trigger exists on the PATIENT table but not on the ORDERS table. Had a similar trigger existed on the ORDERS table, a check would have been made to make sure that the day of the week was neither Saturday nor Sunday and that the user name begins with the letters 'ACCTG'. Thus in this case, no applicable BEFORE statement-level trigger exists. However, a row-level before insert trigger on the ORDERS table does exist and evaluates whether or not Sue Li has already received three orders for the medication Valium. Since she has not, the insert operation is allowed to begin and a new row is inserted in the ORDERS table. At this point the row-level AFTER INSERT Trigger on the ORDERS table is executed and an UPDATE statement is issued to update the MEDICATION.Med_qty_onhand for MEDICATION.Med_code VAL. This update statement causes the declarative CHK_QTY constraint that requires MEDICATION. Med_qty_onhand + MEDICATION.Med_qty_onorder to be between 1000 and 3000 to be evaluated and the transaction fails as a result of the violation of this constraint. At the end of the failed transaction, the content of the MEDICATION and ORDERS tables is as follows:

```
SELECT * FROM MEDICATION;

Med_code Med_name              Med_qty_onhand Med_qty_onorder Med_unitprice
-------- --------------------  -------------- --------------- -------------
KEF      Keflin                          400             700             3
VAL      Valium                          501             500          3.33
ASP      Aspirin                        1200             100            .1


SELECT * FROM ORDERS;

Ord_rx#        Ord_pat_p#a   Ord_pat_p#n  Ord_med_code Ord_dosage   Ord_freq
-------------  ------------  -----------  ------------ ----------   ----------
104            DB            77642        ASP                   3           1
105            DB            77642        ASP                   2           1
106            DB            77642        ASP                   2           1
108            GD            72222        KEF                   2           1
109            GD            72222        KEF                   2           3
111            DB            77642        KEF                   3           1
```

Test 2. Patient Sue Li places an order for one dose of the medication Valium.

```
INSERT INTO ORDER_PATIENT VALUES ('113', 'LS', '12345', 'VAL', 1, 1);
SUCCESSFUL UPDATE OF MEDICATION TABLE
Quantity on hand for Valium is now 500 doses.

1 row created.
```

Once again, no applicable BEFORE statement-level trigger exists and the BEFORE row-level trigger on the ORDERS table verifies that Sue Li has not received three orders for the medication Valium. This allows the insert operation to begin and a new row is inserted in the ORDERS table. At this point, the row-level AFTER INSERT Trigger on the ORDERS table is executed and an UPDATE statement is issued to update the MEDICATION.Med_qty_onhand for MEDICATION.Med_code VAL. This update statement does not cause the declarative CHK_QTY constraint to be violated and thus the transaction is successful, and DBMS_OUTPUT.PUT_LINE statement indicates that the quantity on hand for Valium is now 500 doses. Since no deferred constraint checking is required, the applicable AFTER statement-level trigger on the MEDICATION table is executed. At the end of this transaction, the content of the MEDICATION and ORDERS tables is as follows:

```
SELECT * FROM MEDICATION;

Med_code Med_name              Med_qty_onhand Med_qty_onorder Med_unitprice
-------- --------------------  -------------- --------------- -------------
KEF      Keflin                          400             700             3
VAL      Valium                          500             500          3.33
ASP      Aspirin                        1200             100            .1


SELECT * FROM ORDERS;

Ord_rx#        Ord_pat_p#a   Ord_pat_p#n  Ord_med_code Ord_dosage   Ord_freq
-------------  ------------  -----------  ------------ ----------   ----------
104            DB            77642        ASP                   3           1
```

105	DB	77642	ASP	2	1
106	DB	77642	ASP	2	1
108	GD	72222	KEF	2	1
109	GD	72222	KEF	2	3
111	DB	77642	KEF	3	1
113	LS	12345	VAL	1	1

Test 3. Increase the price of each medication by 10 percent.

```
UPDATE MEDICATION SET MED_UNITPRICE = MED_UNITPRICE*1.10;
SUCCESSFUL UPDATE OF MEDICATION TABLE
```

No constraints prohibit updating the prices of these medications since the Chk_unitprice constraint only requires MEDICATION.Med_unitprice to be less than $4.50. Note, however, the applicable AFTER statement-level trigger on the MEDICATION table is executed once again. At the end of this transaction, the content of the MEDICATION table is as follows.

```
SELECT * FROM MEDICATION;
```

Med_code	Med_name	Med_qty_onhand	Med_qty_onorder	Med_unitprice
KEF	Keflin	400	700	3.3
VAL	Valium	500	500	3.66
ASP	Aspirin	1200	100	.11

Had an attempt been made to increase the price of each medication by 50 percent, the presence of the Chk_unitprice constraint would have caused the transaction to fail.

```
UPDATE MEDICATION SET MED_UNITPRICE = MED_UNITPRICE * 1.50;
UPDATE MEDICATION SET MED_UNITPRICE = MED_UNITPRICE * 1.50
*
ERROR at line 1:
check constraint (CHAPTER11.CHK_UNITPRICE) violated
```

```
SELECT * FROM MEDICATION;
```

Med_code	Med_name	Med_qty_onhand	Med_qty_onorder	Med_unitprice
KEF	Keflin	400	700	3.3
VAL	Valium	500	500	3.66
ASP	Aspirin	1200	100	.11

As stated above, this illustration is simplified and does not consider the possibility of overlapping triggers (two or more triggers with the same timing, granularity, and table). In addition, data manipulation statements in a trigger itself also complicate the description given above and may cause other triggers to fire. Kifer, et al. (2005) and Mannino (2007) each contain a good discussion of these and other issues associated with triggers.

12.1.3 Specifying Views in SQL/DDL

Views were introduced in Chapter 6 (see Section 6.4). This section outlines SQL support for defining and implementing views. An SQL view is a *single* table that is derived based on a relational expression involving one or more other base tables and/or views. In contrast to a base table that has physical existence as a file, a view does not have a physical form. It is an "empty shell" in a physical sense and so is referred to as a virtual table. A view is essentially a lens to look at data stored in base tables. Therefore, there are specific limitations in updating data in a database through views, while there are no such limitations on retrieving data (i.e., querying) through views. The tables that serve as the input to the relational expression defining a view are called **defining tables**. A view in a database environment is a convenience. A user can issue queries on a view as if it is a single table retrieval task instead of approaching the query as a multiple table retrieval operation involving a complex relational expression. Executing the retrieval operation itself entails converting the view definition (i.e., the relational expression defining the view) into equivalent operations on the base tables which are done by the DBMS. To that extent, there is a performance overhead associated with view processing. Therefore, indiscriminate use of views has database efficiency and performance implications.

A view is defined using the SQL/DDL statement, CREATE VIEW, with syntax as follows:

```
CREATE VIEW view_name [ (comma delimited list of columns) ]
AS relational expression [ WITH [CASCADED | LOCAL ] CHECK OPTION ];
```

where:

- *view_name* is a user supplied name for the view
- *comma delimited list of columns* is the unqualified names of the columns of the view
- *relational expression* defines the scope of the view in terms of an SQL query
- CHECK OPTION is an integrity constraint applicable only to SQL-updatable views

Several examples of defining a view in SQL follow.

View Example 1. *A simple view that results from a combination of the relational algebra operations of SELECTION and PROJECTION looks like:*

```
CREATE VIEW senior_citizen AS
  SELECT patient.Pat_name, patient.Pat_age, patient.Pat_gender
  FROM patient
  WHERE patient.Pat_age > 64;
```

The view definition above constructs a view structure that contains the three columns **Pat_name**, **Pat_age**, and **Pat_gender** from the PATIENT table and yields a result that contains only the rows of patients that are 65 or older from the PATIENT table. The name of the view (virtual table) is senior_citizen and the ordered column names in the view are, by default, the same as the unqualified name of the columns from source table—i.e., **Pat_name**, **Pat_age**, **Pat_gender**. The option to explicitly specify column names for the view is shown clearly in the syntax for view definition. The only time it is mandatory to specify the column name of a view is when the column results from applying a function or arithmetic operations, as in the next example.

View Example 2. *A view that shows the gender (male or female) and the number of patients in each gender that are 65 or older.*

```
CREATE VIEW senior_stat (V_gender, V_#ofpats) AS
  SELECT patient.Pat_gender, count (*)
  FROM patient
  WHERE patient.Pat_age > 64
  GROUP BY patient.Pat_gender;
```

View examples 1 and 2 use a query on a single table as the relational expression to derive a view. The third example involves more than one table in the relational expression, but the data retrieved for the view is from a single table.

View Example 3. *A view that exhibits a list of medications that are not ordered by any patient—the medication name and the quantity on hand are included in the list.*

```
CREATE VIEW unused_med AS
  SELECT medication.Med_name, medication.Med_code,
  medication.Med_qty_onhand
  FROM medication
  WHERE medication.Med_code NOT IN
    (SELECT orders.Ord_med_code FROM orders)
WITH CHECK OPTION;
```

The CHECK OPTION in a view definition applies exclusively to SQL-updatable tables and cannot be specified in a view definition if the view is not SQL-updatable. When included in the view definition, the CHECK OPTION makes sure that any row insertion or update of a column value in the view does not violate the view-defining condition. Thus, a CHECK OPTION should always be specified for an SQL-updatable view; yet, SQL does not enforce such a rule (Date and Darwen, 1997). The CASCADED | LOCAL choice is relevant only when a view definition is specified using another view instead of from base table(s). Note that any update on a view derived directly from a base table(s) is most definitely checked against the integrity constraints of the base table(s) irrespective of the specification of the CHECK OPTION in the view definition.

The next example entails retrieval of data for a view from more than one table.

View Example 4. *A view that lists patient number, names of all the medications used by the patient along with dosage and frequency.*

```
CREATE VIEW used_med AS
  SELECT orders.Ord_pat_p#a, orders.Ord_pat_p#n,
  medication.Med_name, orders.Ord_dosage, orders.Ord_frequency
  FROM medication, orders
  WHERE medication.Med_code = orders.Ord_med_code;
```

As noted earlier, views are mainly intended as a convenient mechanism for information retrieval. SQL offers very limited support for updating base tables via views. In general, updating a database through a view is usually (but not always) feasible when the view is defined on a *single table without any aggregate functions*. Since an update of a view that involves a join of multiple base tables can be mapped to update operations on the underlying tables in multiple ways, it introduces a lot of complications and ambiguities. SQL standards for view update, in general, are quite restrictive and complicated. Therefore, we do not treat this topic in any depth in this book. The reader is directed to one of the references (e.g., Date, 2004) in the selected bibliography of this chapter for further discussion of this topic.

12.1.4 The Division Operation

Recall that the Division operation is useful when there is a need to identify tuples in one relation that match all tuples in another relation, and is described in Chapter 11 as capable of being expressed as a sequence of Projection, Cartesian product, and Difference operations. Let's consider this description in the context of the following question which originally appeared in Chapter 11, referring to the Madeira College tables: *What are the course numbers and course names of those courses offered in all quarters during which sections are offered?* The following section shows how SQL views can be used to simulate the Division operation. For convenience of the reader, the content of the COURSE and SECTION tables follows.

COURSE Table

Co_name	Co_course#	Co_credit	Co_college	Co_hrs	Co_dpt_dcode
Intro to Economics	15ECON112	U	Arts and Sciences	3	1
Operations Research	22QA375	U	Business	2	3
Intro to Economics	18ECON123	U	Education	4	4
Supply Chain Analysis	22QA411	U	Business	3	3
Principles of IS	22IS270	G	Business	3	7
Programming in C++	20ECES212	G	Engineering	3	6
Optimization	22QA888	G	Business	3	3
Financial Accounting	18ACCT801	G	Education	3	4
Database Concepts	22IS330	U	Business	4	7
Database Principles	22IS832	G	Business	3	7
Systems Analysis	22IS430	G	Business	3	7

SECTION Table

Se_section#	Se_qtr	Se_year	Se_time	Se_maxst	Se_room	Se_co_course#	Se_pr_profid
101	A	2003	T1015	25		22QA375	HT54347
901	A	2002	W1800	35	Rhodes 611	22IS270	SK85977
902	A	2002	H1700	25	Lindner 108	22IS270	SK85977
101	S	2002	T1045	29	Lindner 110	22IS330	SK85977
102	S	2002	H1045	29	Lindner 110	22IS330	CC49234
701	W	2003	M1000	33	Braunstien 211	22IS832	CC49234
101	A	2003	W1800		Baldwin 437	20ECES212	RR79345
101	U	2003	T1015	33		22QA375	HT54347
101	A	2003	H1700	29	Lindner 108	22IS330	SK85977
101	S	2003	T1015	30		22QA375	HT54347
101	W	2003	T1015	20		22QA375	HT54347

12.1.4.1 Use of Views to Simulate the Division Operation

The Division operation that appears in Chapter 11 can be simulated by creating two views, T1 and T2, in SQL. View T1 creates a virtual table that contains the quarters during which sections of courses are offered, while view T2 begins by forming the product of the SECTION table and the T1 view (shown below with duplicate tuples crossed out) and then subtracts the projection of the SECTION table, which records the sections offered during various quarters.

```
CREATE VIEW T1 AS SELECT DISTINCT SECTION.SE_QTR FROM SECTION;

CREATE VIEW T2 AS
```

```
SELECT SECTION.SE_CO_COURSE#, T1.SE_QTR
FROM SECTION CROSS JOIN T1
MINUS
SELECT SECTION.SE_CO_COURSE#, SECTION.SE_QTR
FROM SECTION;
```

Se_co_course# Se_qtr ← **Projection of product of SECTION table and T1 view.**
------------- ------   Duplicates resulting from courses offered more than once
22QA375   A   in a given quarter are crossed out.   Note that 20
22IS270   A   unique Se_co_course# and Se_qtr combinations exist.
~~22IS270~~   ~~A~~
22IS330   A
~~22IS330~~   ~~A~~
22IS832   A
20ECES212   A
~~22QA375~~   ~~A~~
~~22IS330~~   ~~A~~
~~22QA375~~   ~~A~~
~~22QA375~~   ~~A~~
22QA375   S
22IS270   S
~~22IS270~~   ~~S~~
22IS330   S
~~22IS330~~   ~~S~~
22IS832   S
20ECES212   S
~~22QA375~~   ~~S~~
~~22IS330~~   ~~S~~
~~22QA375~~   ~~S~~
~~22QA375~~   ~~S~~
22QA375   U
22IS270   U
~~22IS270~~   ~~U~~
22IS330   U
~~22IS330~~   ~~U~~
22IS832   U
20ECES212   U
~~22QA375~~   ~~U~~
~~22IS330~~   ~~U~~
~~22QA375~~   ~~U~~
~~22QA375~~   ~~U~~
22QA375   W
22IS270   W
~~22IS270~~   ~~W~~
22IS330   W
~~22IS330~~   ~~W~~
22IS832   W
20ECES212   W
~~22QA375~~   ~~W~~
~~22IS330~~   ~~W~~
~~22QA375~~   ~~W~~
~~22QA375~~   ~~W~~

Se_co_course# Se_qtr ← **Projection on SECTION table over Se_co_course# and Se_qtr.**
------------- ------
22QA375   A
22IS270   A
22IS270   A
```
```

```
22IS330        S
22IS330        S
22IS832        W
20ECES212      A
22QA375        U
22IS330        A
22QA375        S
22QA375        W
```

```
Se_co_course# Se_qtr ← Content of T2 view recording quarters during which each
------------- ------        course is not offered.
20ECES212      S
20ECES212      U
20ECES212      W
22IS270        S
22IS270        U
22IS270        W
22IS330        U
22IS832        A
22IS832        S
22IS832        U
```

Using T2, the course(s) offered during all quarters can be determined using the following nested subquery. The individual lines have been numbered to facilitate the discussion of the execution of the query.

```
1.  SELECT COURSE.CO_COURSE#, COURSE.CO_NAME
2.  FROM COURSE
3.  WHERE COURSE.CO_COURSE# IN
4.      (SELECT SECTION.SE_CO_COURSE# FROM SECTION
5.      WHERE SECTION.SE_CO_COURSE# NOT IN
6.          (SELECT T2.SE_CO_COURSE#
7.             FROM T2));
```

```
Co_course# Co_name
---------- --------------------
22QA375    Operations Research
```

A nested subquery[5] is a complete query nested in the SELECT, FROM, HAVING, or WHERE clause of another query. Nested subqueries of the type shown here are executed from the bottom up.[6] In other words, the result of the query on lines 6 and 7 is used in the WHERE clause in the query on lines 4 and 5 and the result of the query on lines 4 and 5 is used in the WHERE clause in the query on lines 1 through 3. It is required that **T2.Se_co_course#** on line 6 share the same domain as **SECTION.Se_co_course#** referenced in line 5, and likewise that **SECTION.Se_co_course#** on line 4 share the same domain as **COURSE.Co_course#** referenced on line 3.

Execution of this query begins by retrieving the values of course numbers associated with view T2 (20ECES212, 22IS270, 22IS330, 22IS832). Next, the query on lines 4 and 5 identifies values of the course numbers in the SECTION table that are not associated with

---

[5] Subqueries are discussed in greater detail in Section 11.2.3.

[6] The subqueries in this example take the form of an uncorrelated subquery. The execution of a correlated subquery is discussed in Section 11.2.3.2.

T2. Finally, the query on lines 1 through 3 verifies that the value(s) of the course numbers retrieved by the query on lines 4 and 5 actually exist in the COURSE table.

12.1.4.2 A Second Query to Simulate the Division Operation

The following SQL SELECT statement can also be used to display the course(s) offered during all quarters. Three steps are required to execute this statement. First, the number of distinct **SECTION.Se_qtr** values in the SECTION table is determined by the SELECT statement:

```
(SELECT DISTINCT(COUNT(DISTINCT(SECTION.SE_QTR))) FROM SECTION)
```

Second, the COURSE and SECTION tables are joined on their course number attributes that share the same domain. This join yields a total of 11 rows. Next, the rows associated with the result of the join are logically grouped by the combination of **COURSE.Co_course#** and **COURSE.Co_name** with the HAVING used to identify the subset of groups we want to consider. Finally, since one course (course number 22QA375) is associated with all four quarters, only one course number is displayed, that for course number 22QA375.

SQL SELECT Statement:

```
SELECT DISTINCT COURSE.CO_COURSE#, COURSE.CO_NAME
FROM COURSE JOIN SECTION
ON COURSE.CO_COURSE# = SECTION.SE_CO_COURSE#
HAVING COUNT(*) =
      (SELECT DISTINCT(COUNT(DISTINCT(SECTION.SE_QTR)))
      FROM SECTION)
GROUP BY COURSE.CO_COURSE#, COURSE.CO_NAME;

Co_course# Co_name
---------- ----------------------
22QA375    Operations Research
```

Section 11.2.3.2 in Chapter 11 contains a third approach for representing the Division operation in an SQL SELECT statement.

12.2 SQL-92 Built-In Functions

A built-in function can be used in an SQL expression anywhere that a constant of the same data type can be used. A large number of built-in functions are supported by popular SQL implementations. Four of these functions (i.e., CURRENT_DATE(), TO_CHAR(), SUBSTR(), UPPER()) were used in the triggers discussed in Sections 12.1.2.2 and 12.1.2.4. As Groff and Weinberg (2002) report, the SQL-92 standard incorporates what have been judged to be the most useful of these built-in functions, in many cases with slightly different syntax. Column 1 of Table 12.1 contains the names of the most widely-used of the 20 built-in functions that comprise the SQL-92 standard.[7]

---

[7] For a complete list of SQL-92 functions, see Groff and Weinberg (2002).

Table 12.1 Selected SQL-92 built-in functions and their Oracle and MySQL equivalents

SQL-92 Function	Oracle Implementation	MySQL Implementation	Sections Containing Useful Examples in Chapter 12
CASE	DECODE	Function not available	12.4
CHAR_LENGTH	LENGTH	LENGTH**	12.2.2
Concatenation	CONCAT	CONCAT**	12.2.1
CURRENT_DATE	CURRENT_DATE	CURRENT_DATE or CURDATE()	12.3
CURRENT_TIME	CURRENT_DATE (With Mask)	CURRENT_TIME or CURTIME()	12.3
CURRENT_USER	USER	CURRENT_USER	12.1.2.4
Date/Time Conversions	TO_CHAR, TO_DATE	See below*	12.3
EXTRACT	EXTRACT	EXTRACT	12.3
LOWER and UPPER	LOWER and UPPER	LOWER and UPPER	12.1.2.4
POSITION	INSTR	INSTR** or LOCATE**	12.2.5
SUBSTRING	SUBSTR	SUBSTRING**	12.2.1
TRANSLATE	TRANSLATE	Function not available	12.2.4
TRIM	LTRIM and RTRIM	LTRIM and RTRIM**	12.2.3

*MySQL supports a series of functions to extract date and time elements.
**Both the Oracle and MySQL implementation of these functions are covered in this text to illustrate the similarities and differences in how they are implemented across products. For additional information on these as well as other SQL-92 functions not covered in this book, the reader is encouraged to refer to the considerable product-specific documentation available online and in books.

This section covers the use of many of these functions using a variety of short examples in the context of the SQL SELECT statement. In order to illustrate some of the differences in syntax and functionality between products, the examples employ the syntax of the Oracle and MySQL implementations of SQL. Note that Oracle's SQL requires use of the FROM keyword in every SQL SELECT statement. Thus, many of the Oracle SQL examples in this section make use of the DUAL table. The DUAL table has one column, DUMMY CHAR(1), and one row with a value of 'X'. As we will see, the DUAL table is useful when a SELECT statement is issued to display data that does not exist in a table. It is particularly useful when you want to display a numeric or character literal in a SELECT statement. On the other hand, MySQL's SQL does not require use of the FROM keyword.

Columns 2 and 3 of Table 12.1 contain names of the SQL-92 standard built-in functions used by Oracle and MySQL while column 4 contains the section numbers in Chapter 12 where examples of the use of these functions can be found.

The purpose of illustrating both the Oracle and MySQL implementation of some of the SQL-92 functions in this section is to sensitize the reader to the differences in syntax and sometimes functionality across database platforms. In short, the material in this chapter, along with the material in Chapters 10 and 11, is intended to be a highly useful but not necessarily standalone reference to SQL. As such, the reader may need to supplement the material in this textbook with product-specific documentation.

12.2.1 The SUBSTRING Function

The purpose of the **SUBSTRING function** is to extract a substring from a given string. The format of the SUBSTRING function in the SQL-92 standard is:

```
SUBSTRING (source FROM n FOR len)
```

where:

- *n* indicates the character position where the search begins
- *len* represents the length of the search.

Oracle implements the SUBSTRING function with the **SUBSTR (char, m [,n])** function which returns a portion of *char*, beginning at character *m*, *n* characters long (if *n* is omitted, to the end of *char*). The first position of *char* is 1. Floating point numbers passed as arguments to SUBSTR are automatically converted to integers. MySQL implements the substring function as **SUBSTRING (string FROM start FOR length)**.

The SELECT statement in SUBSTRING Example 1 goes into the character string 'ABCDEFG' beginning at character position 3 and returns the next four characters thus displaying 'CDEF.'

SUBSTRING Example 1

```
SELECT SUBSTR('ABCDEFG',3,4) "Substring" FROM DUAL;
SELECT SUBSTRING('ABCDEFG' FROM 3 FOR 4);
```

RESULT:[8]   CDEF

As shown in Examples 2 and 3, if the position where the search begins is not an integer, the function's value is truncated using Oracle SQL and rounded using MySQL.

SUBSTRING Example 2

```
SELECT SUBSTR('ABCDEFG',3.1,4) "Substring" FROM DUAL;
SELECT SUBSTRING('ABCDEFG' FROM 3.1 FOR 4);
```

RESULT: CDEF

SUBSTRING Example 3

```
SELECT SUBSTR('ABCDEFG',3.7,4) "Substring" FROM DUAL;
SELECT SUBSTRING('ABCDEFG' FROM 3.7 FOR 4);
```

---

[8] The output of each function is accompanied by a column heading. Since the format used to display column headings varies by function and by product, only the value returned by the function is displayed in this section. In addition, unless the value returned by the function differs between Oracle and MySQL, only one result is shown.

Oracle RESULT: CDEF
MySQL RESULT: DEFG

The position where the search begins can also be a negative number. In this case, characters beginning with the rightmost characters in the string are stripped off. As expected, the SELECT statement in Example 4 illustrates that a value of –5 for the starting position of the search produces the same result as when the starting position of the search has a value of 3.

SUBSTRING Example 4

```
SELECT SUBSTR('ABCDEFG',-5,4) "Substring" FROM DUAL;
SELECT SUBSTRING('ABCDEFG' FROM -5 FOR 4);
```

RESULT: CDEF

As shown in Example 5, if the number of characters to be searched is omitted, the number of characters returned extends to the end of the string.

SUBSTRING Example 5

```
SELECT SUBSTR('ABCDEFG',-1) FROM DUAL;
SELECT SUBSTRING('ABCDEFG' FROM -1);
```

RESULT: G

Observe that if the position where the search begins is a negative value and the number of characters to be stripped off is greater than the absolute value of *value where the search begins,* the number of characters returned extends only to the end of the string (see Example 6).

SUBSTRING Example 6

```
SELECT SUBSTR('ABCDEFG',-1, 3) FROM DUAL;
SELECT SUBSTRING('ABCDEFG' FROM -1 FOR 3);
```

RESULT: G

However, as indicated in Examples 7 and 8, if the number of characters to be stripped off is zero or negative, a null value is displayed for the result.

SUBSTRING Example 7

```
SELECT SUBSTR ('ABCDEFG',-1, 0) FROM DUAL;
SELECT SUBSTRING('ABCDEFG' FROM -1 FOR 0);
```

RESULT: null value

SUBSTRING Example 8

```
SELECT SUBSTR ('ABCDEFG',-1, -5) FROM DUAL;
SELECT SUBSTRING('ABCDEFG' FROM -1 FOR -5);
```

RESULT: null value

In Oracle, the SUBSTR function is often used in conjunction with the concatenation operator (||). Example 9a illustrates their use together in displaying the name and phone number of all professors with phone numbers that end with two digits ranging between 45 and 65.

SUBSTRING Example 9a

```
SELECT PROFESSOR.PR_NAME,
'('||SUBSTR(PROFESSOR.PR_PHONE,1,3)||')'||
SUBSTR(PROFESSOR.PR_PHONE,4,3)||'-'||
SUBSTR(PROFESSOR.PR_PHONE,7,4) "Phone" FROM PROFESSOR
WHERE SUBSTR(PROFESSOR.PR_PHONE, 9,2) BETWEEN '45' and '65';
```

RESULT:

Pr_name	Phone
John Smith	(523)556-7645
John B Smith	(523)556-7556
Sunil Shetty	(523)556-6764
Katie Shef	(523)556-8765
Cathy Cobal	(523)556-5345
Jeanine Troy	(523)556-5545
Tiger Woods	(523)556-5563

MySQL, on the other hand, does not support the concatenation operator (||) concatenating strings. Instead, concatenation requires use of the concatenation function.[9] The CONCAT function takes the form:

```
CONCAT(string [,string ...])
```

and thus allows for any number of string arguments. Example 9b contains a MySQL query with the same functionality as that shown in Example 9a.

SUBSTRING Example 9b

```
SELECT PROFESSOR.PR_NAME,
CONCAT('(',SUBSTRING(PROFESSOR.PR_PHONE FROM 1 FOR 3),')',
SUBSTRING(PROFESSOR.PR_PHONE FROM 4 FOR 3),'-',
SUBSTRING(PROFESSOR.PR_PHONE FROM 7 FOR 4)) AS "Phone"
FROM PROFESSOR
WHERE SUBSTRING(PROFESSOR.PR_PHONE FROM 9 FOR 2) BETWEEN '45' AND '65';
```

12.2.2 The CHAR_LENGTH (char) Function[10]

The SQL-92 **CHAR_LENGTH (*char*)**[11] function returns the length of the character string *char*. Both Oracle and MySQL use the **LENGTH (*char*)** syntax to implement the CHAR_LENGTH function in the SQL-92 standard. The LENGTH function returns a numeric value.

---

[9] Oracle also supports the CONCAT function. However, unlike MySQL's CONCAT function, the Oracle CONCAT function can only combine two string arguments. Hence, use of the concatenation operator (||) is preferred. However, nesting of one CONCAT function within another CONCAT function can be used to combine two string arguments.

[10] The LENGTH (*char*) Function is called the CHAR_LENGTH (*string*) function in the SQL-92 standard. It represents the length of a character string.

[11] The LENGTH (*char*) Function is the LEN (*char*) Function in SQL Server.

LENGTH Example 1

```
SELECT LENGTH ('ABCDEFG') FROM DUAL;
SELECT LENGTH ('ABCDEFG');
```

RESULT: 7

Using the TEXTBOOK table, the SELECT statements in Examples 2 and 3 illustrate the difference when the LENGTH function is applied to a column defined as a VARCHAR data type (**TEXTBOOK.Tx_title**) versus one defined as a CHAR data type (**TEXTBOOK.Tx_publisher**).

LENGTH Example 2

```
SELECT TEXTBOOK.TX_TITLE, TEXTBOOK.TX_PUBLISHER, LENGTH(TEXTBOOK.TX_TITLE)
FROM TEXTBOOK;
```

RESULT:

Tx_title	Tx_publisher	LENGTH(TEXTBOOK.Tx_title)
Database Management	Thomson	19
Linear Programming	Prentice-Hall	18
Simulation Modeling	Springer	19
Systems Analysis	Thomson	16
Principles of IS	Prentice-Hall	16
Economics For Managers		22
Programming in C++	Thomson	18
Fundamentals of SQL		19
Data Modeling		13

LENGTH Example 3

```
SELECT TEXTBOOK.TX_TITLE, TEXTBOOK.TX_PUBLISHER,
LENGTH(TEXTBOOK.TX_PUBLISHER)
FROM TEXTBOOK;
```

Oracle RESULT

Tx_title	Tx_publisher	LENGTH(TEXTBOOK.Tx_publisher)
Database Management	Thomson	13
Linear Programming	Prentice-Hall	13
Simulation Modeling	Springer	13
Systems Analysis	Thomson	13
Principles of IS	Prentice-Hall	13
Economics For Managers		
Programming in C++	Thomson	13
Fundamentals of SQL		
Data Modeling		13

MySQL RESULT

Tx_title	Tx_publisher	LENGTH(TEXTBOOK.Tx_publisher)
Database Management	Thomson	7
Linear Programming	Prentice-Hall	13
Simulation Modeling	Springer	8

Advanced Data Manipulation Using SQL

```
Systems Analysis        Thomson                                        7
Principles of IS        Prentice-Hall                                 13
Economics For Managers                                              NULL
Programming in C++      Thomson                                        7
Fundamentals of SQL                                                 NULL
Data Modeling                                                          0
```

Observe how the MySQL LENGTH function ignores trailing blanks in calculating the length of a CHAR data type, and returns a null value for the textbooks whose publisher consists of a null value, and a value of 0 for the textbooks whose publisher consists of a single space (as a result of ignoring trailing blanks).

12.2.3 The TRIM Function

The SQL-92 standard includes three basic **TRIM functions** for trimming characters from a string. **TRIM (LEADING unwanted FROM string)** trims off any leading occurrences of *unwanted*; **TRIM (TRAILING unwanted FROM string)** trims off any trailing occurrences of *unwanted*; and **TRIM (BOTH unwanted FROM string)** trims both leading and trailing occurrences of *unwanted*. TRIM is particularly useful with a CHAR data type in cases where it is desirable to remove unwanted blank spaces from the end of the string.

Both Oracle and MySQL support LTRIM and RTRIM functions. **LTRIM (string [,unwanted])** removes unwanted characters from the beginning of a *string* while **RTRIM (string [,unwanted])** removes *unwanted* characters from the end of a *string*. As illustrated in Trim Example 1, the *unwanted* argument is a string that contains the characters to be trimmed, and defaults to a single space. In MySQL you can trim only spaces[12] whereas in Oracle you can remove many characters at once by listing them all in the *unwanted* string. However, MySQL does offer TRIM (LEADING...), TRIM (TRAILING...), and TRIM (BOTH ...) functions that are part of the SQL-92 standard.

TRIM Example 1

```
SELECT LTRIM('     LAST WORD') FROM DUAL;
SELECT LTRIM('     LAST WORD');
```

RESULT: LAST WORD

The SELECT statement in TRIM Example 2 trims the leftmost 'x's from the character string 'xxxXxxLASTWORD'. Note that MySQL requires the use of the TRIM (LEADING ...) function to trim the leading 'x's.

TRIM Example 2

```
SELECT LTRIM ('xxxXxxLAST WORD', 'x') FROM DUAL;
SELECT TRIM(LEADING 'x' FROM 'xxxXxxLAST WORD');
```

RESULT: XxxLAST WORD

All forms of the TRIM function are case sensitive. Thus the SELECT statement in the TRIM Example 3 does not result in trimming any characters from the string 'xxxXxx-LAST WORD'.

---

[12] SQL Server and DB2 also permit only spaces to be trimmed.

TRIM Example 3

```
SELECT LTRIM ('xxxXxxLAST WORD', 'X') FROM DUAL;
SELECT TRIM(LEADING 'X' FROM 'xxxXxxLAST WORD');
```

RESULT: xxxXxxLAST WORD

TRIM Example 4 illustrates the difference between Oracle's implementation of the TRIM function and the MySQL implementation. Note how the Oracle LTRIM function not only trims the word "Systems" from the beginning of the textbook title "Systems Analysis," it also trims the leading "S" from all titles that begin with the letter "S" (i.e., Simulation Modeling). This is because in Oracle's LTRIM function it is important to note that any character string that begins with any of the characters included in *unwanted* will be trimmed.

TRIM Example 4—Oracle Version

```
SELECT TEXTBOOK.TX_TITLE, LTRIM(TEXTBOOK.TX_TITLE,
'Systems') "Trimmed Title"
FROM TEXTBOOK;
```

Oracle Result:

```
Tx_title               Trimmed Title
---------------------- ----------------------
Database Management    Database Management
Linear Programming     Linear Programming
Simulation Modeling    Simulation Modeling
Systems Analysis        Analysis
Principles of IS       Principles of IS
Economics For Managers Economics For Managers
Programming in C++     Programming in C++
Fundamentals of SQL    Fundamentals of SQL
Data Modeling          Data Modeling
```

TRIM Example 4—MySQL Version

```
SELECT TEXTBOOK.TX_TITLE,
TRIM(LEADING 'Systems'FROM TEXTBOOK.TX_TITLE) "Trimmed Title"
FROM TEXTBOOK;
```

MySQL Result:

```
Tx_title               Trimmed Title
---------------------- ----------------------
Database Management    Database Management
Linear Programming     Linear Programming
Simulation Modeling    Simulation Modeling
Systems Analysis        Analysis
Principles of IS       Principles of IS
Economics For Managers Economics For Managers
Programming in C++     Programming in C++
Fundamentals of SQL    Fundamentals of SQL
Data Modeling          Data Modeling
```

Suppose the trim function is to be applied to the character string "Systemsim". How would the results differ from that shown in TRIM Example 4 for Oracle's LTRIM function and the MySQL TRIM (LEADING ...) function?

The SELECT statements in TRIM Examples 5 through 7 provide additional details on how the Oracle LTRIM function works. No MySQL equivalents are shown here because in each case, use of the MySQL TRIM (LEADING ...) function results in no characters being trimmed.

Since there are no characters prior to the character string "LAST WORD," other than those in the set "xX", all of the leading x's and X's are trimmed in TRIM Example 5.

TRIM Example 5

```
SELECT LTRIM ('xxxXxxLAST WORD', 'xX') FROM DUAL;
```

RESULT: LAST WORD

Since the leading x's in char are in the set 'yx', they are trimmed and thus not displayed in the result shown in TRIM Example 6.

TRIM Example 6

```
SELECT LTRIM ('xxxXxxLAST WORD', 'yx') FROM DUAL;
```

RESULT: XxxLAST WORD

On the other hand, as illustrated by the SELECT statement in TRIM Example 7, since the leftmost character(s) in *unwanted* do not include a 'y' or 'X', no characters are trimmed in the result.

TRIM Example 7

```
SELECT LTRIM ('xxxXxxLAST WORD', 'yX') FROM DUAL;
```

RESULT: xxxXxxLAST WORD

The RTRIM function operates on the rightmost characters in a string in the same way that LTRIM operates on the leftmost characters in a string. It is important to be careful when using the RTRIM function in conjunction with a CHAR data type. Trim Examples 8 and 9 illustrate the difference between the application of this function to a VARCHAR versus CHAR data type column.

Since a VARCHAR data type stores no trailing blanks, the rightmost 'g' in the titles "Linear Programming" and "Simulation Modeling" are trimmed by the Oracle RTRIM and MySQL TRIM (TRAILING ...) functions in Trim Example 8.

TRIM Example 8

```
SELECT RTRIM(TEXTBOOK.TX_TITLE, 'g') "Trimmed Result" FROM TEXTBOOK;
SELECT TRIM(TRAILING 'g' FROM TEXTBOOK.TX_TITLE) "Trimmed Result" FROM TEXTBOOK;
```

```
RESULT:
Trimmed Result
---------------------
Database Management
Linear Programmin
Simulation Modelin
Systems Analysis
```

Trim Example 9 illustrates how the RTRIM function can be used by Oracle to trim (i.e., remove) unwanted blank spaces at the end of a CHAR data type. Column 3 of the Oracle result below confirms that the **TEXTBOOK.Tx_publisher** column in the TEXTBOOK table is defined as a CHAR(13) data type while column 5 verifies that the RTRIM function used in column 4 removed all trailing blank spaces from the end of all not-null publishers. Despite the definition of the **TEXTBOOK.Tx_publisher** column as a CHAR(13) data type, the LENGTH function in MySQL ignores the trailing blanks when determining the length of the character string. Thus the MySQL version of TRIM Example 9 does not require use of the TRIM function. Once again, observe how the MySQL LENGTH function returns a value of 0 for the length of the "single-space" publisher of the Data Modeling title.

TRIM Example 9—Oracle Version

```
SELECT TEXTBOOK.TX_TITLE, TEXTBOOK.TX_PUBLISHER,
LENGTH(TEXTBOOK.TX_PUBLISHER) "Length Pub",
RTRIM(TEXTBOOK.TX_PUBLISHER) "Trimmed Pub",
LENGTH(RTRIM(TEXTBOOK.TX_PUBLISHER)) "Trimmed Length"
FROM TEXTBOOK;
```

Oracle Result

Tx_title	Tx_publisher	Length Pub	Trimmed Pub	Trimmed Length
Database Management	Thomson	13	Thomson	7
Linear Programming	Prentice-Hall	13	Prentice-Hall	13
Simulation Modeling	Springer	13	Springer	8
Systems Analysis	Thomson	13	Thomson	7
Principles of IS	Prentice-Hall	13	Prentice-Hall	13
Economics For Managers				
Programming in C++	Thomson	13	Thomson	7
Fundamentals of SQL				
Data Modeling		13		

TRIM Example 9—MySQL Version

```
SELECT TEXTBOOK.TX_TITLE, TEXTBOOK.TX_PUBLISHER,
LENGTH(TEXTBOOK.TX_PUBLISHER) "Length Pub",
TX_PUBLISHER "Trimmed Pub",
LENGTH(TEXTBOOK.TX_PUBLISHER) "Trimmed Length"
FROM TEXTBOOK;
```

MySQL Result

Tx_title	Tx_publisher	Length Pub	Trimmed Pub	Trimmed Length
Database Management	Thomson	7	Thomson	7
Linear Programming	Prentice-Hall	13	Prentice-Hall	13
Simulation Modeling	Springer	8	Springer	8
Systems Analysis	Thomson	7	Thomson	7
Principles of IS	Prentice-Hall	13	Prentice-Hall	13
Economics For Managers				

Programming in C++	Thomson	7 Thomson	7
Fundamentals of SQL			
Data Modeling		0	0

12.2.4 The TRANSLATE Function

The SQL-92 **TRANSLATE function** is used to translate characters into a string. Oracle's implementation of the TRANSLATE function has the format:

```
TRANSLATE (char, from_string, to_string)
```

The function searches *char*, replacing each occurrence of a character found in *from_string* with the corresponding character from *to_string*. Characters that are in *char* but not in *from_string* are left untouched whereas characters in *from_string* but not in *to_string* are deleted. For example, the following TRANSLATE function could be used to extract the identifier of the department from each course number.

```
SELECT CO_COURSE#, TRANSLATE (CO_COURSE#,
'ABCDEFGHIJKLMNOPQRSTUVWXYZ0123456789',
'ABCDEFGHIJKLMNOPQRSTUVWXYZ') "Department"
FROM COURSE;
```

RESULT:

```
Co_course# Department
---------- ----------
15ECON112  ECON
22QA375    QA
18ECON123  ECON
22QA411    QA
22IS270    IS
20ECES212  ECES
22QA888    QA
18ACCT801  ACCT
22IS330    IS
22IS832    IS
22IS430    IS
```

None of the characters in **COURSE.Co_course#** are left untouched because each character is part of *from_string*. However since the characters '0123456789' in *from_string* do not appear in *to_string* 'ABCDEFGHIJKLMNOPQRSTUVWXYZ', the digits in **COURSE.Co_course** are not returned by the TRANSLATE function.

As indicated in Table 12.1, the TRANSLATE function is not available in MySQL.

12.2.5 The POSITION Function

The SQL-92 **POSITION (*target* IN *source*)** function returns the position where the *target* string appears within the *source* string. Both the target string and the source string are character strings that have the same character set. The POSITION (*target* IN *source*) function returns a numeric value as follows:

- If the *target* string is of length zero (i.e., it is a null value), the result returned is one.

- Otherwise, if the *target* string occurs as a substring within the *source* string, the result returned is one greater than the number of characters in the *source* string that precede the first such occurrence.
- Otherwise, the result is zero.

Both Oracle and MySQL implement the POSITION function with an INSTR function. The Oracle INSTR function has the following format:

```
INSTR (source, target [, position [, occurrence]])
```

The *position* argument is used to specify the starting position for the search in *source* and *occurrence* makes it possible for a specific occurrence to be found. If *position* is negative, the search begins from the end of the string.

The MySQL INSTR function meets, but does not go beyond the functionality of the SQL-92 standard. It takes the form: INSTR (*source*, *target*) and locates the first occurrence of the *target* in the *source*. MySQL also supports a LOCATE function of the form:

```
LOCATE (target, source [, position])
```

As is the case in Oracle's INSTR function, the position argument is used to specify a starting character position other than 1. However, like the MySQL INSTR function, the LOCATE function locates only the first occurrence of the *target* in the *source*.

The SELECT statement in INSTR Example 1 locates the character position of the second occurrence of the character 'S' in the character string 'MISSISSIPPI' beginning at character position 5. When the value of *occurrence* is changed to a 1 (see INSTR Example 2), observe that a different 'S' is located. Note that the MySQL LOCATE function can be used in INSTR Example 2 but not in INSTR Example 1. When the value of *position* is changed to 4 (see INSTR Example 3), both the Oracle INSTR and MySQL LOCATE function can be used to obtain the result.

INSTR Example 1

```
SELECT INSTR ('MISSISSIPPI','S',5,2) FROM DUAL;
```

RESULT: 7

INSTR Example 2

```
SELECT INSTR ('MISSISSIPPI','S',5,1) FROM DUAL;
SELECT LOCATE('S','MISSISSIPPI',5);
```

RESULT: 6

INSTR Example 3

```
SELECT INSTR ('MISSISSIPPI','S',4,1) FROM DUAL;
SELECT LOCATE('S','MISSISSIPPI',4);
```

RESULT: 4

The SELECT statement in INSTR Example 4 locates all textbooks that contain the character string 'ing' somewhere in the title. Since both *position* and *occurrence* are omitted, their values are assumed to be equal to 1.

INSTR Example 4

```
SELECT TEXTBOOK.TX_TITLE, INSTR(TEXTBOOK.TX_TITLE, 'ing') "Position of ing"
FROM TEXTBOOK
WHERE INSTR(TEXTBOOK.TX_TITLE, 'ing') > 0;
```

RESULT:

```
Tx_title              Position of ing
--------------------- ---------------
Linear Programming                 16
Simulation Modeling                17
Programming in C++                  9
Data Modeling                      11
```

Observe that the value of the INSTR function returned for all other titles is zero. Since the first occurrence of the character string 'ing' in the title was the requirement, both the Oracle and MySQL INSTR function could be used to satisfy the information request.

12.2.6 Combining the INSTR and SUBSTR Functions

The INSTR and SUBSTR functions are often combined. The purpose of the query in the following example is to begin with second word of each textbook title and display all titles with the character string 'ing' in the title. The individual lines have been numbered to facilitate discussion of the execution of the query.

```
1.   SELECT TEXTBOOK.TX_TITLE,
2.   INSTR(SUBSTR(TEXTBOOK.TX_TITLE, INSTR(TEXTBOOK.TX_TITLE, ' ')+1),'ing')
3.                     "ing string in word 2",
4.   INSTR(SUBSTR(TEXTBOOK.TX_TITLE,1),'ing')
5.                     "ing string in overall title"
6.   FROM TEXTBOOK
7.   WHERE INSTR(SUBSTR(TEXTBOOK.TX_TITLE, INSTR(TEXTBOOK.TX_TITLE, ' ')+1),'ing') > 0;
```

RESULT:

Tx_title	ing string in word 2	ing string in overall title
Linear Programming	9	16
Simulation Modeling	6	17
Data Modeling	6	11

The WHERE clause shown on line 7 governs the titles displayed when the query is executed. The **INSTR(TEXTBOOK.TX_TITLE, ' ')** function is evaluated first and returns the character position of the first blank space character in the title of each textbook (i.e., it is responsible for skipping over the first word of the title). By adding 1 to the value returned by the INSTR function, the character position of the first character in the second word of the title is obtained. The **SUBSTR(TEXTBOOK.TX_TITLE, INSTR(TEXTBOOK.TX_TITLE, ' ')+1)** function is evaluated next. Since a value of n is not provided, all remaining characters beginning with the second word of the title are selected. Finally, the outer INSTR function looks for the first occurrence of the character string 'ing' in the second or remaining words in the title. The values displayed in columns 2 and 3 indicate the values returned by the INSTR function when asked to find the character position of the character string 'ing' starting with the second word of the title (column 2) and when asked to find the character position of the character string 'ing' starting with the first word of the title (column 3).

Since the first occurrence of the 'ing' character string is the subject of the search, the MySQL INSTR and SUBSTRING functions can also be used to produce the results.

MySQL Equivalent

```
1.  SELECT TEXTBOOK.TX_TITLE,
2.  INSTR(SUBSTRING(TEXTBOOK.TX_TITLE FROM INSTR(TEXTBOOK.TX_TITLE, ' ')+1),'ing')
3.                   "ing string in word 2",
4.  INSTR(SUBSTRING(TEXTBOOK.TX_TITLE FROM 1),'ing')
5.                   "ing string in overall title"
6.  FROM TEXTBOOK
7.  WHERE INSTR(SUBSTRING(TEXTBOOK.TX_TITLE FROM INSTR(TEXTBOOK.TX_TITLE, ' ')+1),'ing') > 0;
```

12.3 Some Brief Comments on Handling Dates and Times

Examples 1.2.7–1.2.9 in Section 11.2.1.2 of Chapter 11 illustrate how mathematical operations can be included in the column-list of a query. In addition to operations involving columns and constants defined as numeric, as illustrated in Example 1.3.5 in Section 11.2.1.3, mathematical operations can also be performed on dates.

All DBMS vendors offer SQL functions to handle dates and times. Unfortunately, the implementation of date/time data types is far from standardized across vendors. This problem occurs because the ANSI SQL-92 standard defines the support of date data types, but does not say how those data types should be stored (Rob and Coronel, 2004). Given these differences, the material in this section is based on Oracle and contains examples that work with Oracle dates and times. Please refer to the SQL reference material for your database platform for the syntax associated with features comparable to those discussed here.

As indicated in Table 10.1, a date data type in SQL-92 is ten characters long in the format yyyy-mm-dd. While MySQL makes use of this format, Oracle uses as its default date format dd-mon-yy where "dd" represents a two-digit day, "mon" a three-letter month abbreviation, and "yy" a two-digit year (e.g., the date July 29, 2007 would be represented as 29-jul-07). The Oracle default format can be changed to that of the SQL-92 default by the following SQL statement.

```
ALTER SESSION SET NLS_DATE_FORMAT = 'yyyy-mm-dd';[13]
```

The remainder of the examples in this section make use of the SQL-92 format for representing a date.

Although referenced as a non-numeric field, a date is actually stored internally in a numeric format that includes the century, year, month, day, hour, minute, and second. Although dates appear as non-numeric fields when displayed, calculations can be performed with dates because they are stored internally as numeric data in accordance with the Julian calendar. A Julian date represents the number of days that have elapsed between a specified date and January 1, 4712 B.C. As a result, the query in Example 1.3.5 in Section 11.2.1.3 calculates the age (in years) of each professor in department 3 when hired.

SQL SELECT Statement:

```
SELECT PROFESSOR.PR_NAME,
TRUNC((PROFESSOR.PR_DATEHIRED - PROFESSOR.PR_BIRTHDATE)/365.25,0)
```

---

[13] NLS is an acronym for National Language Support.

```
"Age When Hired"
FROM PROFESSOR
WHERE PROFESSOR.PR_DPT_DCODE = 3;
```

RESULT:
```
Pr_name          Age When Hired
---------------  --------------
Chelsea Bush                 46
Tony Hopkins                 47
Alan Brodie                  56
Jessica Simpson              40
Laura Jackson                26
```

The CURRENT_DATE function is part of the SQL-92 standard and is used to record the current date and time. For example, the following SELECT Statement calculates the age (in days) and the age (in years) for each professor in department 3.

SQL SELECT Statement:

```
SELECT PROFESSOR.PR_NAME,
CURRENT_DATE - PROFESSOR.PR_BIRTHDATE "Age in Days",
TRUNC((CURRENT_DATE - PROFESSOR.PR_BIRTHDATE)/365.25,0) "Age in Years"
FROM PROFESSOR
WHERE PROFESSOR.PR_DPT_DCODE = 3
```

RESULT:
```
Pr_name          Age in Days Age in Years
---------------  ----------- ------------
Chelsea Bush     21876.8535            59
Tony Hopkins     20698.8535            56
Alan Brodie      22839.8535            62
Jessica Simpson  18598.8535            50
Laura Jackson    11971.8535            32
```

The reason why the age in days contains a fractional component is due to fact that when referenced in a query, the CURRENT_DATE function retrieves not only the current date but also the time of day. Thus the previous query was executed when 85.35 percent of a 24 hour day had elapsed (i.e., the query was executed at approximately 8:29 pm).

Oracle uses the TO_CHAR and TO_DATE functions with dates and times. The TO_CHAR function is used to extract the different parts of a date/time and convert them to a character string[14] while the TO_DATE function is used to convert character strings to a valid date format. Both functions make use of a format element (also known as a format mask). Table 12.2 contains a sample of format elements commonly used in conjunction with the TO_CHAR and TO_DATE functions. Table 12.3 shows the number format elements used with the TO_CHAR function.

---

[14] The TO_CHAR function can also be used to convert a number to a formatted character string. Format masks used in conjunction with numbers appear in Table 12.3.

Table 12.2 Selected date and time format elements used with the TO_CHAR and TO_DATE functions

Element	Description	Example
MONTH, Month, or month	Name of the month spelled out—padded with blank spaces to a total width of nine spaces; case follows format.	JULY, July, or july (5 spaces follows each representation of July)
MON, Mon, or mon	Three-letter abbreviation of the name of the month; case follows format.	JUL, Jul, or jul
MM	Two-digit numeric value of the month.	7
D	Numeric value of the day of the week.	Monday = 2
DD	Numeric value of the day of the month.	23
DAY, Day, or day	Name of the day of the week spelled out—padded with blank spaces to a length of nine characters.	MONDAY, Monday, or monday (3 spaces follows each representation of Monday)
fm	"Fill mode." When this element appears, subsequent elements (such as MONTH) suppress blank padding leaving a variable-length result.	fmMonth, yyyy produces a date such as March, 2007
DY	Three-letter abbreviation of the day of the week.	MON, Mon, or mon
YYYY	The four-digit year.	2007
YY	The last two digits of the year.	07
YEAR, Year, or year	Spells out the year; case follows year.	TWO THOUSAND SEVEN
BC or AD	Indicates B.C. or A.D.	2007 A.D.
AM or PM	Meridian indicator	10:00 AM
J	Julian date. January 1, 4712 B.C. is day 1.	July 27, 2007 is Julian date 2454309
SS	Seconds (value between 0 and 59)	21
MI	Minutes (value between 0 and 59)	32
HH	Hours (value between 1 and 12)	9
HH24	Hours (value between 0 and 23)	13

Table 12.3 Selected number format elements used with the TO_CHAR function

Element	Description	Example
9	Series of 9s indicates width of display (with insignificant leading zeros not displayed).	99999
0	Displays insignificant leading zeros.	0009999
$	Displays a floating dollar sign to prefix value.	$99999
.	Indicates number of decimals to display.	999.99
,	Displays a comma in the position indicated.	9,999

The following query illustrates the use of the TO_CHAR function to display the date of birth and salary of each professor in department 3 using the default format and an alternative format.

```
SELECT PROFESSOR.PR_NAME,
PROFESSOR.PR_BIRTHDATE "SQL-92 Date Format",
TO_CHAR(PROFESSOR.PR_BIRTHDATE,'DD-MON-YYYY') "Alternate Format",
PROFESSOR.PR_SALARY "Default Format",
TO_CHAR(PROFESSOR.PR_SALARY, '$99,999.00') "Alternate Format"
FROM PROFESSOR
WHERE PROFESSOR.PR_DPT_DCODE = 3
```

```
RESULT:
Pr_name          SQL-92 Date Format  Alternate Format   Default Format  Alternate Format
---------------  ------------------  -----------------  --------------  -----------------
Chelsea Bush     1946-09-03          03-SEP-1946                 77000  $77,000.00
Tony Hopkins     1949-11-24          24-NOV-1949                 77000  $77,000.00
Alan Brodie      1944-01-14          14-JAN-1944                 76000  $76,000.00
Jessica Simpson  1955-08-25          25-AUG-1955                 67000  $67,000.00
Laura Jackson    1973-10-16          16-OCT-1973                 43000  $43,000.00
```

When inserting a date in a table, Oracle assumes a default time of 12:00 AM (midnight). Should it be necessary to associate a time other than 12:00 AM with a date, the TO_DATE function can be used. For example, suppose we wish to insert the date and time of admission for each new patient into the PATIENT table. The INSERT statement that appears below illustrates how this could be done. Prior to and following the INSERT statement are two SELECT statements. The first two SELECT statements display the name and date of admission of each patient prior to the insertion of the new patient. Note that the first SELECT statement displays the date of admission using the default date format while the second displays the date of admission using a format mask that includes the time portion of the date of admission. The use of the TO_DATE function in the INSERT statement for patient Zhaoping Zhang allows both the date and time of her admission to be recorded.

Name and Date of Admission of Patients in PATIENT Table Prior to Insertion

```
SELECT PATIENT.PAT_NAME, PATIENT.PAT_ADMIT_DT "Date of Admission"
FROM PATIENT;


Pat_name              Date of Admission
--------------------  -----------------
Davis, Bill           2007-07-07
Li, Sue               2007-08-25
Grimes, David         2007-07-12



SELECT PATIENT.PAT_NAME, TO_CHAR(PAT_ADMIT_DT, 'fmMonth dd, yyyy HH:
MI AM') "Date of Admission"
FROM PATIENT;


Pat_name              Date of Admission
--------------------  -------------------------
Davis, Bill           July 7, 2007 12:00 AM
Li, Sue               August 25, 2007 12:00 AM
Grimes, David         July 12, 2007 12:00 AM
```

Insertion of New Patient

```
INSERT INTO PATIENT VALUES ('ZZ', '06912', 'Zhang, Zhaoping', 'F', 35,
TO_DATE('2007-08-11 10:15 AM', 'YYYY-MM-DD HH:MI AM'), NULL, NULL, NULL);

1 row created.
```

Name and Date of Admission of Patients in PATIENT Table After Insertion

```
SELECT PATIENT.PAT_NAME, PATIENT.PAT_ADMIT_DT "Date of Admission"
FROM PATIENT;


Pat_name              Date of Admission
--------------------  -----------------
Davis, Bill           2007-07-07
Li, Sue               2007-08-25
Zhang, Zhaoping       2007-08-11
Grimes, David         2007-07-12



SELECT PATIENT.PAT_NAME, TO_CHAR(PAT_ADMIT_DT, 'fmMonth dd, yyyy HH:
MI AM') "Date of Admission"
FROM PATIENT;

Pat_name              Date of Admission
--------------------  -------------------------
Davis, Bill           July 7, 2007 12:00 AM
Li, Sue               August 25, 2007 12:00 AM
Zhang, Zhaoping       August 11, 2007 10:15 AM
Grimes, David         July 12, 2007 12:00 AM
```

Suppose patient Zhaoping Zhang is discharged on August 12 at 3:35 pm (i.e. the value of the CURRENT_DATE function is 3:35 pm). The following SELECT statement records her length of stay.

```
SELECT PATIENT.PAT_NAME, CURRENT_DATE - PATIENT.PAT_ADMIT_DT "Length of Stay"
FROM PATIENT
WHERE PATIENT.PAT_P#A = 'ZZ' AND PATIENT.PAT_P#N = '06912'

Pat_name              Length of Stay
-------------------   --------------
Zhang, Zhaoping             1.2222338
```

Note that the 5 hours and 20 minutes (the difference between the time of day when she was discharged on August 12 and the time of day when she was admitted on August 11) is 22.22 percent of one day.

12.4 A Potpourri of Other SQL Queries

Date (1995) has described SQL as an *extremely redundant* language in the sense that there are almost always a variety of ways to formulate the same query in SQL. The advantage of this flexibility is that users have the option of choosing the approach with which they are most comfortable. For example, a join can be used as an alternative way of expressing many subqueries. In addition, as we have seen, subqueries and join conditions may appear in the FROM clause or even in the column list of the SELECT clause.

The declarative, nonprocedural nature of SQL, which allows the user to specify what the intended results of the query are, rather than specifying the details of how the result should be obtained, can also be a disadvantage. Ideally, the user should be concerned only with specifying the query correctly. The DBMS, on the other hand, should then take the query and execute it efficiently. While DBMS products all make use of a variety of techniques to process, optimize, and execute queries written in SQL, in practice it helps if the user is aware of which types of constructs in a query are more expensive to process in terms of performance than others. See Chapter 15 of Elmasri and Navathe (2004) for an interesting discussion of query processing and optimization.

The remainder of this section contains six examples of SQL queries that combine a number of the features introduced in Chapter 11 as well as in this chapter. Their purpose is not so much to specify the most efficient way of expressing the query as to illustrate the application of several SQL features and functions. Since each example makes use of SQL running on an Oracle9i platform, readers working with some other database platform may wish to refer to their SQL reference material to resolve any syntactical differences created by syntax unique to Oracle9i, such as differences in function names.

12.4.1 Concluding Example 1

Suppose Madeira College is interested in finding out how many professors have been hired each month. A review of the data in the PROFESSOR table identifies two interesting problems. First there have been months during which no professors have been hired (i.e., February, March, April, and July). Second, there are two professors (John B. Smith and Jeanine Troy) whose hire date is unavailable (i.e., unknown). The following SQL SELECT statement represents one way to count the number of professors hired during each month and also includes (a) one row for each month during which no professor has been hired, and (b) a row that records the number of professors for which a date hired is unavailable. The individual lines have been numbered to facilitate the discussion.

SQL SELECT Statement:

```
1.   SELECT MONTHS.MNUM,
2.   NVL(TO_CHAR(PROFESSOR.PR_DATEHIRED,'Month'),MONTHS.MNAME) as "Month",
3.   COUNT(PROFESSOR.PR_DATEHIRED) as "Number Hired"
4.   FROM MONTHS LEFT OUTER JOIN PROFESSOR
5.   ON MONTHS.MNUM = TO_CHAR(PROFESSOR.PR_DATEHIRED,'mm')
6.   GROUP BY MONTHS.MNUM,
7.   NVL(TO_CHAR(PROFESSOR.PR_DATEHIRED,'Month'),MONTHS.MNAME)
8.   UNION
9.   SELECT '13', 'Unknown', COUNT(*)
10.  FROM PROFESSOR
11.  WHERE PROFESSOR.PR_DATEHIRED IS NULL
12.  GROUP BY '13', 'Unknown';
```

1. Since the months February, March, April, and July do not appear in the PRO-
    FESSOR table, the MONTHS Table was created using the following CREATE
    TABLE statement:

    `CREATE TABLE MONTHS (MNUM VARCHAR(2), MNAME VARCHAR(10));`

    and a row containing a two-character month number (e.g., 01 for January, ...,
    12 for December) and the full name of the month inserted for each of the 12
    months of the year.
2. Lines 6 and 7 indicate that grouping is first done by month number and then
    by month name. Grouping by just month name alone would cause the months
    to be displayed in alphabetical (i.e., April, August, December, etc.) as opposed
    to chronological (i.e., January, February, March, etc.) order.
3. Lines 2 and 7 make use of the null value function NVL (expression1, expres-
    sion2) as a way to replace a null value with a string in the results of a query.
    If expression1 is a null value, the NVL function returns expression2. If expres-
    sion1 is not null, the NVL function returns expression1.
4. Line 4 indicates a left outer join involving the MONTHS and PROFESSOR
    tables. This allows an unmatched month number in the MONTHS table to be
    combined with the row of null values in the PROFESSOR table as a result of
    the left outer join. Use of the NVL function in lines 2 and 7 is required so that
    the name of the month in the MONTHS table can replace the null value asso-
    ciated with the **PROFESSOR.Pr_datehired** on the row of null values in the
    PROFESSOR table.
5. COUNT(PROFESSOR.Pr_datehired) is used on line 3 so that the null value in
    the **PROFESSOR.Pr_datehired** column on rows associated with months during
    which no professor was hired will allow the initial value of the accumulator to
    remain at zero. If COUNT(PROFESSOR.Pr_datehired) is replaced by
    COUNT(*), the number of professors hired during each of the months of Feb-
    ruary, March, April, and July will be 1.

623

6. Execution of the query on lines 1–7 will not include a row that records the number of professors for which a date hired is unavailable. However a query that contains the union of the SELECT statement on lines 1–7 with the SELECT statement on lines 9–13 will include a final row that shows the number of professors for which a date hired is unavailable (i.e., unknown).

7. Execution of the query that appears above generates the result:

RESULT:

```
Mnum Month        Number Hired
---- ---------- ------------
01   January              2
02   February             0
03   March                0
04   April                0
05   May                  5
06   June                 5
07   July                 0
08   August               2
09   September            1
10   October              1
11   November             1
12   December             1
13   Unknown              2
```

Different database products have a number of supporting commands that control the display attributes for a single column or all columns. For example, in Oracle9i's SQL*Plus environment, adding the command:

COLUMN Mnum NOPRINT

before line 1 would suppress the display of the column headed Mnum and leave only the Month and Number Hired columns displayed.

12.4.2 Concluding Example 2

Based on just the data in the TAKES table, display the number of grade points, classes taken, and grade point average for the classes taken for each graduate student. Include not only those graduate students who have taken classes but also those graduate students who have not taken any classes. In addition, display the name and major of each graduate student. The following SQL SELECT statement represents one way to express this information requirement. The individual lines have been numbered to facilitate the discussion.

SQL SELECT Statement:

```
1.   SELECT STUDENT.ST_NAME, GRAD_STUDENT.GS_UGMAJOR,
2.   SUM (DECODE (RTRIM(TAKES.TK_
     GRADE), 'A', 4, 'B', 3, 'C', 2, 'D', 1, 'F', 0)) "Grade Points",
3.   COUNT (TAKES.TK_ST_SID) "Classes Taken",
4.   SUM (DECODE (RTRIM(TAKES.TK_
     GRADE), 'A', 4, 'B', 3, 'C', 2, 'D', 1, 'F', 0))/COUNT (TAKES.TK_ST_
     SID) "GPA"
5.   FROM (TAKES JOIN STUDENT ON STUDENT.ST_SID = TAKES.TK_ST_SID)
6.   JOIN GRAD_STUDENT ON STUDENT.ST_SID = GRAD_STUDENT.GS_ST_SID
7.   GROUP BY STUDENT.ST_NAME, GRAD_STUDENT.GS_UGMAJOR
8.   UNION
9.   SELECT STUDENT.ST_NAME, GRAD_STUDENT.GS_UGMAJOR, 0, 0, 0
```

```
10.    FROM STUDENT JOIN GRAD_STUDENT
11.    ON STUDENT.ST_SID = GRAD_STUDENT.GS_ST_SID
12.    AND  STUDENT.ST_SID NOT IN
13.       (SELECT TAKES.TK_ST_SID FROM TAKES)
14.    ORDER BY 5 DESC, 3 DESC, 2 ASC
```

1. Line 2 introduces the DECODE function as the Oracle implementation of the SQL-92 CASE function (see Table 12.1). The DECODE function[15] facilitates character-by-character substitution. For every value it sees in a field DECODE checks for a match in a series of *if/then* tests. The format of the DECODE function is:

```
DECODE (expr, search1, result1, [search2, result2,] ... [default])
```

The purpose is to compare *expr* to each search value and to return the result if *expr* equals the *search* value. If no match is found, the DECODE function returns the *default* value. If a default is not supplied, the default will return null. In line 2, if **TAKES.Tk_grade** is equal to 'A', the value returned to the DECODE function is 4; if **TAKES.Tk_grade** is equal to 'B', the value returned to the DECODE function is 3, etc. Note that before the DECODE function is applied to **TAKES.Tk_grade**, the RTRIM function is applied. This is necessary because **TAKES.Tk_grade** is defined as a CHAR(5) data type, and thus removal of the trailing blanks is necessary before **TAKES.Tk_grade** is equal to any of the five possible letter grades. Had **TAKES.Tk_grade** been defined as a VARCHAR data type, use of the RTRIM function would have been unnecessary.

2. Use of the STUDENT table is required because the GRAD_STUDENT table does not contain the name of student.

3. Grouping by **STUDENT.St_name** and **GRAD_STUDENT.Gs_ugmajor** on line 7 is necessary to allow (a) both columns to be displayed as part of the result, and (b) number of grade points, number of classes taken, and grade point average of each student to be calculated.

4. Execution of the SELECT statement on lines 1–7 only permits the results for those graduate students who have taken a course to be displayed. The SELECT statement on lines 9–13 allows the results for those graduate students who have never taken a course to be displayed.

5. Note that the ORDER BY clause applies to the collective results from all SELECT statements involved in the query. The ORDER BY clause on line 14 causes the results to be displayed in descending order by the student's grade point average and classes taken, and in ascending order by undergraduate major.

6. Execution of the query shown above produces the following result.

---

[15] The DECODE function is similar to the CASE or IF...THEN...ELSE structures found in many programming languages.

```
St_name          Gs_ugmajor  Grade Points Classes Taken      GPA
--------------   -----------  ------------ -------------   ----------
Gladis Bale      Archeology            6             2            3
Elijah Baley     Marketing             6             2            3
Joumana Kidd     History               3             1            3
Shweta Gupta     Archeology            2             1            2
Jenny Aniston    Child Care            0             0            0
Wanda Seldon     Finance               0             0            0
Jenna Hopp       History               0             0            0
Diana Jackson    Mathematics           0             0            0
Tim Duncan       Physics               0             0            0
```

12.4.3 Concluding Example 3

Display all professors whose salary exceeds that of their department head. For each qualifying professor, display his or her name and salary along with the name and salary of the department head. The following query represents one way to satisfy this information request. The individual lines have been numbered to facilitate the discussion.

SQL SELECT Statement:

```
1.  SELECT A.PR_NAME AS "Prof Name", A.PR_SALARY AS "Prof Salary", DEPARTMENT.DPT_NAME,
2.  B.PR_NAME AS "Dept Head", B.PR_SALARY AS "Dept Head Salary", DEPARTMENT.DPT_NAME
3.  FROM (PROFESSOR A JOIN DEPARTMENT
4.  ON A.PR_DPT_DCODE = DEPARTMENT.DPT_DCODE)
5.  JOIN PROFESSOR B ON  DEPARTMENT.DPT_HODID = B.PR_EMPID
6.  AND A.PR_SALARY > B.PR_SALARY
```

1. Since both the salary of a professor and the salary of the professor's department head are recorded in the PROFESSOR table, this query references the PROFESSOR table twice (once using table alias A and once using table alias B) in order to compare the salary of a professor with that of his or her department head.

2. The ON clauses that appear on lines 4 and 5 work together to identify the professor in copy B of PROFESSOR serving as department head of the professor under investigation in copy A of PROFESSOR. The condition on line 6 ensures that the salary of the professor exceeds that of his or her department head.

3. Execution of the query yields the result shown below.

RESULT:

```
Prof Name       Prof Salary Dpt_name      Dept Head       Dept Head Salary Dpt_name
--------------  ----------- -------------- --------------- ---------------- --------------
Chelsea Bush         77000 QA/QM          Alan Brodie                76000 QA/QM
Tony Hopkins         77000 QA/QM          Alan Brodie                76000 QA/QM
Sunil Shetty         64000 IS             Cathy Cobal                45000 IS
Katie Shef           65000 IS             Cathy Cobal                45000 IS
```

12.4.4 Concluding Example 4

Display the names of all professors who have the same first name as at least one other professor.

SQL SELECT Statement:

```
SELECT PROFESSOR.PR_NAME FROM PROFESSOR
WHERE SUBSTR(PROFESSOR.PR_NAME, 1, INSTR(PROFESSOR.PR_NAME,' ')-1) IN
     (SELECT SUBSTR(PROFESSOR.PR_NAME, 1, INSTR(PROFESSOR.PR_NAME,' ')-1)
      FROM PROFESSOR
      GROUP BY SUBSTR(PROFESSOR.PR_NAME, 1, INSTR(PROFESSOR.PR_NAME,' ')-1)
      HAVING COUNT(*) > 1);
```

Since the PROFESSOR table does not have separate columns for first name, last name, and middle initial, the subquery makes use of the SUBSTR and INSTR functions along with the GROUP BY and HAVING clauses to retrieve the first names that appear more than one time in the PROFESSOR table. The main query, when executed, follows by displaying the names of all professors whose first name appears in the set of first names retrieved by the subquery. Execution of this query produces the following result. It is left as an exercise for the reader to order the result in ascending order by last name.

RESULT:
```
Pr_name
---------------
John Smith
John B Smith
John Nicholson
Mike Faraday
Mike Crick
```

12.4.5 Concluding Example 5

Display the names of those department chairs who do not teach a section of a course.

SQL SELECT Statement:

```
SELECT PROFESSOR.PR_NAME
 FROM PROFESSOR
 WHERE PROFESSOR.PR_EMPID  IN
 (SELECT DEPARTMENT.DPT_HODID FROM DEPARTMENT
 WHERE NOT EXISTS
 (SELECT * FROM SECTION
 WHERE DEPARTMENT.DPT_HODID = SECTION.SE_PR_PROFID))
```

Although this query involves three tables (PROFESSOR, DEPARTMENT, and SECTION), it is rather straightforward. The two subqueries of the main query identify the department heads who do not teach a section and provide input to the main query that displays their names.

RESULT:
```
Pr_name
---------------
Mike Faraday
Alan Brodie
Marie Curie
Mike Crick
```

12.4.6 Concluding Example 6

Display the number of professors born during each decade of the 1900s.

SQL SELECT Statement:

```
SELECT 19||SUBSTR(PROFESSOR.PR_BIRTHDATE,8,1)
||0||'''s' "Decade of",  COUNT (*) "Number Born"
FROM PROFESSOR
WHERE PROFESSOR.PR_BIRTHDATE IS NOT NULL
GROUP BY SUBSTR(PROFESSOR.PR_BIRTHDATE,8,1)
```

This query illustrates the use of the concatenation operator (||) and how to embed a single quote within a string. In order to display the column of values that contains 's (1940's, 1950's, etc.), it is necessary to follow the single quote that would normally precede the literal s with two single quotes. SQL will allow the two additional single quotes to be treated as one single quote within the literal. Note that the WHERE clause excludes those professors whose date of birth is not available.

RESULT:

```
Decade of     Number Born
-----------   -----------
1940's                  4
1950's                  2
1960's                  6
1970's                  6
```

It is left as an exercise for the reader to revise the query to include the fact that there were no professors born during the decades of the 1900s, 1910s, 1920s, ..., 1980s, etc.

Chapter Summary

The SQL-92 standard goes beyond the three types of declarative constraints for enforcing data integrity discussed in Chapter 10—column constraints, table constraints, and domain constraints, and offers means to create a constraint that specifies a relationship among data values that cross multiple tables within a database. Such a constraint is known as an assertion and is defined by a CREATE ASSERTION statement. While not part of the SQL-92 standard, the creation of stored procedures, known as triggers, is an especially useful part of database definition because triggers are activated when modifications (insertions, deletions, and updates) are attempted on tables in a database. Triggers essentially embody business rules that govern the data in the database and modifications to that data. These rules are embedded in the database in trigger definitions created by the CREATE TRIGGER statement. Oracle provides perhaps the most complex trigger facility and the use of its PL/SQL (Programming Language/SQL) language to create triggers is introduced in Chapter 12. Views allow stored data to be looked at in other ways by defining alternative views of the data. A View is a named SQL query that is permanently stored in the database. Among important uses of views are their use in (a) tailoring the appearance of the database so that different users see it from different perspectives, (b) enforcing data integrity by restricting access to data so that different users are permitted to see only certain rows or columns of a table, and (c) simplifying access to data by allowing the structure of the stored data to be presented in a way that is most natural to the user. The CREATE VIEW statement is used to create a view.

In addition to the aggregate functions introduced in Chapter 11, the SQL-92 standard contains a number of built-in functions for working with strings, dates, and times. Although the functionality associated with these functions is available in practically all database products, different products contain slightly different syntax from that specified in the standard (e.g., Oracle's SUBSTR function versus the SUBSTRING function in the SQL-92 standard). While such variations in syntax limit the portability of SQL SELECT statements, after studying Chapters 10, 11, and 12, it is hoped that the reader will be able to use the features discussed here and, where necessary, adapt them to their specific database platform with a minimum of difficulty.

Exercises

1. What is the difference between a declarative constraint that can be defined as part of a CREATE TABLE or ALTER TABLE statement and a declarative assertion?

2. Example 3 in Section 12.4 indicates that it is possible for the salary of a professor to exceed that of their department head. Create an assertion that would prohibit this from occurring.

3. What is the difference between a declarative assertion and a trigger?

4. Using the tables associated with Figure 10.1 (reproduced in this chapter as Figure 12.1) and shown in Box 3 of Chapter 10 (also reproduced in this chapter), create a BEFORE INSERT trigger that prohibits the assignment of two patients to the same room where the difference in their ages is greater than 10 years. In addition, the trigger should also prohibit two patients with allowable ages from being assigned to the same bed.

5. Using the tables associated with Figure 10.1 (reproduced in this chapter as Figure 12.1) and shown in Box 3 of Chapter 10 (also reproduced in this chapter), create a BEFORE INSERT

trigger that permits the insertion of a patient into the PATIENT table between only 8:00 am and 8:00 pm.

6. What is a view and how does it differ from a table? Under what conditions is a view updatable?

7. Which statement is used to copy data from one table and insert it into an existing table?

SQL Project 1[16]

This project is based on seven tables (RENTAL, VIEWING, RENTPROPERTY, OWNER, CLIENT, STAFF, and BRANCH) that contain data about the Parks, Cooper, and Adams (PCA) Company. The script required to create and populate these seven tables can be downloaded from *www.course.com* (search on the ISBN of this book), or obtained from your instructor. After creating and inserting rows into the seven tables, it may be useful to print a copy of the structure and content of each table to serve as a reference when trying to validate the result of a particular query.

The following comments provide some background about the Parks, Cooper, and Adams (PCA) Company. PCA manages properties on behalf of property owners. The company offers a complete service to owners who wish to rent out their furnished property. Services provided by PCA include interviewing prospective renters (known as clients), organizing visits to the property by prospective renters, and negotiating the lease agreement. Once rented, PCA assumes responsibility for the property. Listed below is a description of the data recorded, maintained, and accessed at each branch office to support the day-to-day operation and management of PCA.

Branch Offices

PCA has several branch offices located in Houston and surrounding communities. Each branch office is identified by a unique branch number and also has a name, an address (street, city, state, and zip code), telephone number, fax number, email address, and a base management fee that it charges for managing a piece of rental property. Each branch office has staff members assigned to it.

Staff

Each PCA branch office has a manager responsible for overseeing the operations of the office. Likewise, some but not all PCA branch offices have one or more staff members with the job title of supervisor. Supervisors are responsible for the day-to-day activities of a group of staff responsible for the management of property for rent. Each member of the staff is given a staff number (which is unique across all branch offices). Information on each member of the staff includes the name (first, last, and middle initial or name), job title (position), gender, date of birth, date hired, commission percentage for listing a piece of rental property, annual salary, and the branch office to which the staff member is assigned.

[16] This project is an adaptation of an example database that appears in Connolly and Begg (2002).

630

Property for Rent

Each PCA branch office has properties for rent that are identified by a property number which is unique across all branch offices. The details of property for rent include the full address (street, city, state, zip code), type of property (e.g., apartment, house, townhouse), number of rooms, monthly rent, and the date the property was listed with PCA. Each property for rent is assigned to a specific member of the staff who is responsible for the management of that property and also includes reference to the owner of property and the branch office controlling the property.

Property Owners

PCA manages property for private or business owners. Each private owner or business owner is uniquely identified by an owner number, which is unique across all branch offices. Additional information on private owners includes the owner's name (first, last, and middle initial or name), address and telephone number.

Clients/Renters

Data is collected upon a client's initial contact with any one of the PCA branch offices. This includes the client's name (first, last, and middle initial or name), telephone number, preferred type of accommodation (e.g., apartment, house, townhouse), and the maximum rent the client is prepared to pay. As a prospective renter, each client is given a unique number called the client number, which is unique across all branch offices.

Property Viewings

In most cases, a prospective renter will request to view one or more properties before renting. The details of each viewing are recorded and include the date of the viewing and any comments by the prospective renter regarding the suitability or otherwise of the property.

Rental

When a piece of property is rented, a row is added to the rental table. Data for each piece of property rented includes a unique rental number, the property number of the piece of property rented, the client number of the renter, the move-in date, the length of the lease in months, the monthly rental rate agreed upon by the client and staff member handling the lease, and the staff number of the staff member responsible for the piece of property.

An ER diagram that describes the relationship among these tables follows.

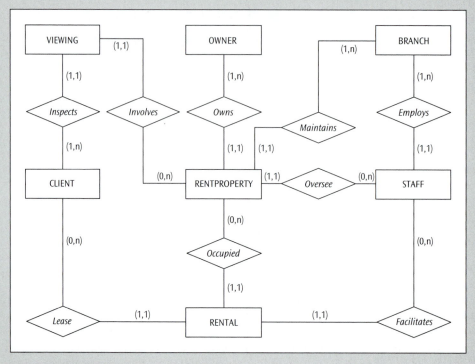

ERD for SQL Project 1

The 98 queries constituting this project will collectively provide a good introduction to SQL. Please note that in some cases it is quite clear which columns (fields) need to be displayed. In other cases the columns to be displayed are left to individual judgment. Use DISTINCT to minimize the repetition of duplicate rows, and use column aliases (especially when a column heading would otherwise be the content of a numeric or character function); they can make the output of a query more readable. In all cases, please refrain from the temptation that the results are correct the first time output from a query is obtained. In other words, please try to display enough columns (fields) and study the output in order to verify the accuracy of the result.

In order to challenge the student, the 98 queries are for the most part randomly ordered. In other words queries 1–10 are not necessarily the easiest to write nor is query 98 necessarily the most difficult to write. We suggest that students begin by working on those queries that seem to involve either (a) one table, or (b) two or three tables where some kind of join or subquery might be of use. We suggest that queries 2, 3, 6, 8, 9, 10, 13, 15, 16, 17, 18, 19, 21, 24, 25, 26, 27, 31, 38, 39, 40, 63, 64, 65, 68, and 69 be among the first queries attempted. These queries involve the use of the LIKE operator, numeric functions (e.g., COUNT, MIN, MAX, AVG, etc.), grouping, and joins.

1. List the name, position, salary, and commission percentage of each staff member eligible to receive a commission. The commission percentage should be displayed as a percentage (as well as with a % sign) rather than as a decimal value. For example, Manjari Mehta's commission percentage should be displayed as .25%. Use a column alias for the commission percentage.

2. List the salary details of all staff, showing only the staff number, the name, and the annual and monthly salary. Use appropriate column aliases when displaying the monthly and annual salary figures. Please display two digits to the right of the decimal point for each monthly salary.

3. List the property number and address of all properties that have been viewed in ascending order by city and within city by zipcode. If a property has been viewed more than one time, it should be displayed only once.

4. List the salary details of all staff, showing only each staff member's staff number, name, gender, annual salary, monthly salary, and age. Use appropriate column aliases when displaying the monthly and annual salary figures. Please display two digits to the right of the decimal point for each monthly salary and truncate each age to an integer. The results should be displayed in ascending order by gender and within gender in descending order by date of birth.

5. List the staff member with the position other than manager with the highest salary.

6. List the names and addresses of all branch offices located in Brenham or Houston.

7. List the names and salaries of all supervisors with an annual salary less than the average annual salary of supervisors.

8. List the names of all managers and supervisors without a birthdate.

9. List all owners with the string 'Houston' in their address.

10. List the details of all viewings on a property where a comment has not been supplied.

11. List the salaries for all staff who have worked for the company for more than five years in descending order by salary.

12. List all properties in order by property type and within property type in descending order by the number of days listed. Display the street address, city, state, zip code, property type, and number of days listed. Display the number of days listed as an integer value.

13. How many properties cost more than $2000.00 per month to rent?

14. How many different (i.e., distinct) properties were viewed in July, 2006?

15. Find the number of managers and the sum of their salaries.

16. Find the minimum, maximum, average, the difference between the minimum and maximum salary, the difference between the maximum and average salary and the difference between the minimum and average staff salary. Use column aliases to display the column headings.

17. Find the number of staff working in each branch and the sum of their salaries. The branch name of each branch (not the branch number) should be displayed in ascending order by branch name.

18. For each branch office with more than one staff member, find the number of staff working in each branch, the sum of their salaries, and their average salary. Once again, the branch name of each branch (not the branch number) should be displayed in ascending order by branch name.

19. List the names and positions of staff members who work in the Brenham branch. Assume that the branch number of the Brenham branch is not known (i.e., more than one table needs to be referenced in this query).

Advanced Data Manipulation Using SQL

20. List all staff whose salary is greater than the average salary, and list by how much their salary is greater than the average. *Suggestion*: Consider the possibility of using a subquery in the column list.

21. Display the property number, complete address, and property type of properties that are handled by staff who work in the West branch. Assume that the branch number of the West branch is not known.

22. Find staff members whose salary is larger than the salary of at least one member of the staff at The Woodlands branch. Display at least the names and salaries of the qualifying staff members.

23. Find staff members whose salary is larger than the salary of every salary of every member of the staff at the Memorial branch. Display at least the names and salaries of the qualifying staff members.

24. Find the names of all clients who have viewed a rental property along with any comments supplied. Display the results in ascending order by the last name of the client and within client in ascending order by property number.

25. For each branch office, list the names of staff who manage properties and the property numbers of the properties they manage. Order the result by branch office number.

26. For each branch office list the staff who manage properties, including the name of the staff member, the address of the property along with its city and state, and city in which the branch is located. Display the results in ascending order by rent property number.

27. Find the number of properties handled (i.e., managed) by each staff member. Display the names of the staff member and order the results by staff number. Do not include those staff members not managing any property.

28. List the branch offices and the properties that are in the same city along with any unmatched branches (i.e., branches with no properties in their city). Display the name of the branch, the city in which the branch is located, the property number, street address and city of the property. The requirement to list branches with no properties in their city suggests the possible use of some type of outer join.

29. List the branch offices and properties in the same city and any unmatched properties (i.e., properties located in cities without a branch office). Display the name of the branch, the city in which the branch is located, the property number, street address, and city of the property.

30. List the branch offices and properties in the same city and any unmatched branches or properties. Display the name of the branch, the city in which the branch is located, the property number, street address, and city of the property.

31. Find all staff members who work in a Houston branch. Display the name and position of the staff member and the name of the branch where he or she works.

32. Construct a list of all cities where there is either a branch office or a rental property. Think about which type of relational algebra operation is suggested by this question.

33. Construct a list of all cities where there is both a branch office and a rental property.

34. Construct a list of all cities where there is a branch office but no rental property.

35. Display the name and age of the oldest staff member. Exclude from consideration all members with the position of assistant.

36. Display the last name, first name, and middle initial of those owners whose rent property has never been viewed.

37. Display the number of staff members born in each year. Display 'Year Unavailable' to record the number of staff members whose birth date is not available. The NVL and TO_CHAR functions may be useful here.

38. Display the property number and address of those rental properties which when viewed solicited a comment that they were either too close or too far from something. Be careful not to exclude any qualifying property numbers.

39. Display the first name, last name, and middle initial of all owners whose last name begins with the letter "C" and who have a Houston address.

40. Display the names of all owners whose address contains a non-Houston zip code. All Houston zip codes begin with 770.

41. Find all staff members whose salary is less than the average salary of all staff members with their gender. Display the result in ascending order by gender and within gender in ascending order by salary.

42. Display the number of times each property has been viewed. The result should include properties that have never been viewed. For each property display its property number, street, city, state, and zip code. Order the results in ascending order by city.

43. Display the names and hiredates of all staff members hired between January 1, 1995 and December 14, 2002. Order the result in ascending order by hiredate.

44. Display the names of those owners whose rental property was viewed during July, 2006.

45. Display the names of those staff members who have never listed a rental property viewed by a client.

46. Display the names of all staff members hired prior to when the manager of their branch was hired. The result should include both the date that the staff member was hired and the date the staff member's manager was hired. *Suggestion*: Consider use of a subquery in a column list.

47. Display the names of owners of more that one piece of rental property that is a house. Include in the result the number of houses owned.

48. Display the pieces of rental property that have been on the market for six months or more without having been rented. Include the property number, address, date listed, and number of months on the market. Order the output in descending order by time on the market.

49. Display the name of the staff member with the lowest salary among those staff members hired prior to January 1, 2000.

50. Display the names of those staff members who identify themselves through the use of their middle name (e.g., someone with the name W. Lamar Hicks is identified by the name, Lamar. Assume that if there was a staff member with the name Billy Bob Thornton that he would be listed as B. Bob Thornton or Billy B. Thornton).

51. Display the names of those staff members without a middle initial. Anyone whose first name is an initial would be considered as someone without a middle initial.

52. Display the total of the bonus received by each staff member for listing pieces of property. Include in the result all staff members who have listed one or more pieces of property except for those who serve as branch managers. The bonus received by a staff member for listing a piece of property is the annual salary multiplied by their commission percentage. Staff without a commission percentage should be treated as if their commission percentage is zero.

53. Display the names of those owners whose rental property was viewed during July, 2006. Include in your result the name of the client viewing the property.

54. Display the property number and address of all townhouses that have never been viewed.

55. Display all branches where the telephone number differs from the fax number by plus or minus 1. Include in the result the branch name, telephone number, and fax number.

56. Display the number of characters that precede the @ in each branch email address.

57. Count the number of owners located in each zip code.

58. Display the rental property located in the same zip code as that associated with the address of the owner. Include in the result the name of the owner, the owner's address, and the address of the rental property.

59. Display the number of times a property of each owner has been viewed. Include each owner in the result, even those who have not had any property viewed because at the moment they either own no property or none of the property they own has been viewed, and order the result in descending order in accordance to the number of times a property of the owner has been viewed. The output should include the name of the owner plus the number of times one of his or her properties has been viewed.

60. Display just the first name, last name, city, state, and zip code of each owner.

61. Display the name and phone number for each owner. Use the format (999) 999-9999 (where 9 represents a digit) to display each phone number. For those owners without a phone number, display "No Phone Available." The results should be in ascending order by the last name of the owner.

62. Display the name of each client with only one space between the client's first name and middle initial and only one space between the client's middle initial and last name. If the client does not have a middle initial, only one space should be displayed between the first and last name of the client. In addition, all initials (either a middle initial or the first name) displayed should be followed by a period. *Suggestion*: Consider use of a union and some form of the CHAR_LENGTH function for this query.

63. Display the name of the branch responsible for each piece of property viewed. Display the property number, the address of the property, and name of the branch. Order the result in ascending order by branch name and within branch name in ascending order by property number.

64. Display the names and viewing dates of those clients who have viewed a piece of rental property located outside of Houston. Include in the display the street, city, state, and zip code of the piece of property viewed.

65. Count the number of pieces of each type of rental property located in each zip code.

66. Display the number of properties shown by each staff member affiliated with each branch who have shown at least one property. *Note*: Two viewings of the same piece of property count as having shown two properties.

67. During the time that a property has not been rented, the owner is assessed the management fee of the branch. This fee is a percentage of the monthly rent. Calculate the monthly assessment for each piece of rental property in the RENTPROPERTY table that has not been rented (i.e., does not appear in the RENTAL table). For each piece of property, display the name of the owner, the address of the property, and the amount of the monthly assessment. Display the results in ascending order by owner name.

68. Display the address of all rental properties recorded as having been viewed prior to even being listed (i.e., in this case we are checking for the presence of an error in the data).

69. Display the rental properties that have been leased to a client at a monthly rate that exceeds posted monthly rate in the RENTPROPERTY table. Display the address of the property plus the two rental amounts.

70. Oftentimes the staff member who leases a piece of property is not the person who listed the property. Display the rental properties both listed and leased by the same staff member. As a way to check the result, include in the output the name of the staff member that listed the property and the name of the staff member who leased the property.

71. Display the rental property rented without having ever been viewed. Include in the result the first and last name of the owner of the property, the address of the property, and the name of the client who rented the property.

72. For each property rented, display the number of days between the move in date and the date at which the property was last viewed. Since a piece of property may have been viewed several times, the student may find the MIN function of use here. Include in the result property rented without being viewed.

73. For each property rented, the first rental payment is due on the first Monday after the move in date. Display this date for each piece of property that has been rented. The NEXT_DAY function may be useful on this query. Go to Google and enter NEXT_DAY to learn about the NEXT_DAY function.

74. During the first month a property is rented, the client is charged only from the move in date to the end of the month. Display the rent owed by each client during the first month. Assume each month has 30 days. Include in the result the client name, property address, monthly rental rate, and rent owed during the first month. The SUBSTR function may be helpful here when stripping out the day of the month on which the client moved in.

75. Count the number of properties rented where the zip code of the property is the same as the zip code of the owner.

76. Display the names of those staff members who have listed at least one of each type of property.

77. Display the names of the branches with more than one supervisor.

78. Display the number of employees in each position at each branch. A combination of the DECODE and SUM functions may be useful here. The output should have four columns. The first column should contain the branch name, the second column should contain the number of managers in each branch, the third column should contain the number of supervisors in each branch, and the fourth column should contain the number of assistants in each branch. *Suggestion*: Review Concluding Example 2 in Chapter 12.

79. Display the names of those branches that have at least one staff member in each position. The result of this query should be consistent with the result from query 78.

80. Display the names of those branches that do not have at least one staff member in each position.

81. Display the names of the owners who identify themselves by their middle initial/middle name instead of by their first name. The result should be displayed in the following order: (a) a period follows their first name (i.e., initial), (b) next, preceded by only one space should come their middle name, and (c) next, preceded by only one space should come their last name. The string concatenation operator || or the CONCAT function should be of use in this query.

82. Display the management fee earned by each branch for managing rent properties during 2005. The management fee for a piece of property is calculated each month and is the product of the rental fee times the percentage management fee charged by the branch. A management fee is charged beginning with the month during which the property is first listed.

83. Display the names of the branches in ascending order by street number. *Note*: 14210 is the largest street number and thus the Memorial branch should be the final branch listed. Use Google to find out about SQL functions that can be used to pad character strings.

84. Display the commission earned by each staff member when a piece of rental property is rented. A commission is earned when a piece of rental property listed by a staff member is rented as long as the staff member both listed and was the staff member associated with the rental of the property. For an individual piece of property, the commission is based on the number of days between when the property was listed and when it was finally rented. The commission percentage is commission for listing a piece of property that appears in the STAFF table.

85. Display the names of all clients with a middle initial that consists of more than one letter and whose first name also consists of more than one letter. Each portion of the person's name should be separated from the next portion by a single space. The LENGTH function and the string concatenation operator || might be of use in this query.

86. For each client, display only the first initial of the person's name, followed by a period and a space, followed by the first initial of the middle name, followed by a period and a space, followed by the last name. A union operation may be useful here since some clients have a middle initial while others do not.

87. For each staff member who is not a branch manager, display the name of the staff member along with the name of the branch manager. Confine attention to those staff members who work in a branch with a branch manager.

88. Display the names of the branches that do not have a branch manager.

89. Display the names of those staff members who work in a branch without a branch manager.

90. List the salary details of all staff, showing only the staff number, the first and last names, and the annual and monthly salary of all staff members without an initial in their name. Use appropriate column aliases when displaying the monthly and annual salary figures. *Note*: This is the same as query 2 except that those staff members with an initial anywhere in their name should not be displayed. Admittedly, this is not very practical. A more practical query would be one that suppresses any initial that appears in a person's name.

91. List all properties located in the same zip code as the branch office managing the property. Display the property number, street address, city, state, and zip code of each piece of qualifying property along with the zip code of the branch office.

92. List all managers who were hired after all of the employees that they manage.

93. List the highest (i.e., maximum) rent in each category (i.e., apartment, house, and townhouse).

94. List all properties where the move in date is within 30 days of being listed. Display the address of the piece of property along with both the move in date and date listed.

95. List the number of properties listed by each staff member including those staff members who have not listed any properties. Order the results in descending order by the number of properties listed.

96. List the branch office(s) serving each client. The reason for the office(s) is that it is possible that a client may have viewed several pieces of properties and thus may have been served by more than one branch office.

97. On which days during the year 2006 was there more than one viewing.

98. Construct a list of all cities where there is a piece of rental property but no branch office.

SQL Project 2

This project is based on seven tables (CAB, SHIFT, DRIVER, MAINTAIN, INCIDENT, QUALIFICATION, and FUELUSE) that contain data about a cab company. The script required to create and populate these seven tables can be downloaded from *www.course.com* (search on the ISBN of this book), or obtained from your instructor.

Four comments about the data in the tables. The **Sh_wkflag** column of the SHIFT table indicates whether a shift scheduled to be driven was actually driven. A value of T indicates that the shift was actually driven while a value of F indicates that the shift was not driven. Unless specified as part of the query, please do not consider the content of the **Sh_wkflag** column. The QUALIFICATION table records which drivers are qualified to drive which cabs. The **Ma_lastdrvnum** column of the MAINTAIN table indicates the driver number of the driver who brought in the cab for maintenance. The **Dr_salary** column of the DRIVER table represents the annual salary of the driver. Thus, the monthly salary of each driver can be obtained by dividing the annual salary by 12.

An ER diagram that describes the relationship among these tables follows.

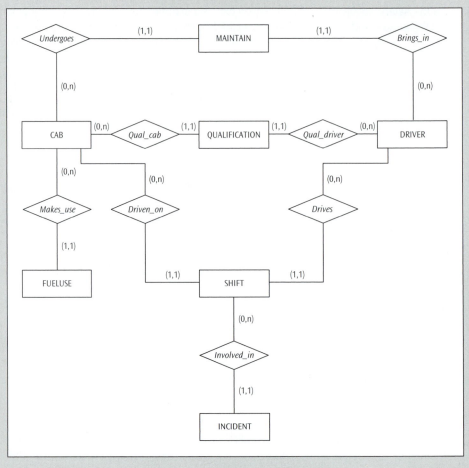

ERD for SQL Project 2

The 75 queries that constitute this project should collectively provide a good introduction to SQL. Some are quite easy while others are a bit more challenging. Please note that in some cases it is quite clear which columns (fields) need to be displayed. In other cases the columns to be displayed are left to individual judgment. Use DISTINCT to minimize the repetition of duplicate rows and don't be afraid to use column aliases (especially when a column heading would otherwise be the content of a numeric or character function); they can make the output of a query more readable. In addition, please try to avoid wraparound as much as possible. In all cases, please refrain from the temptation that the results are correct the first time output from a query is obtained. In other words, please try to display enough columns (fields) and study the output in order to verify the accuracy of the result.

In order to challenge the student, the 75 queries are for the most part randomly ordered. In other words, queries 1–10 are not necessarily the easiest to write nor is query 75 necessarily the most difficult to write. It may be a good idea to begin working with queries that seem to involve either one or two tables, such as 2, 3, 11, 15, 20, 22, 25, 30, 31, 34, 44, 46, 60, and 61. These queries involve the use of the LIKE operator, numeric functions (e.g., COUNT, MIN, MAX, AVG, etc.), grouping, nested subqueries, and joins.

1. Display the name and age of the oldest driver

2. Display all drivers who earn less than $56,000 a year. The resulting list should be ordered on driver status in ascending order and within each driver status in descending order by salary.

3. Display the name and hiredate of all drivers who were hired between January 1, 1984 and December 12, 1990. Display the qualifying drivers in ascending order by name.

4. Display the cab number, make, and model of all cabs that are not reserved for either November 15, 2007 or November 16, 2007. In order to be displayed, a cab cannot be reserved on either of these two days. In other words, if a cab is reserved on November 15 but not on November 16, it would not be displayed. Likewise, if the cab were reserved on November 16 but not on November 15, it would not be displayed. Thus the cab would only be displayed if it were not reserved on November 15 and also not reserved on November 16.

5. Display the name of the driver along with the make and model of the cab for all shifts that were scheduled but not driven on November 15, 2007. The scope of this query is limited to just the shifts on November 15, 2007.

6. Display the names of the drivers who were involved in an incident on November 15, 2007, ordered by driver number. In addition, please indicate the cab driven and the type of incident.

7. Display the cab number, make, model, and year of all cabs that have had maintenance where something was either not "replaced" or not "new".

8. Display the maximum, the minimum, and the average salary for all drivers in each driver status.

9. Display the names of those drivers who have never been assigned to drive a shift.

10. Display the names of all drivers under the age of 55 who have driven one of the Checker cabs at least one time.

11. Display the names of all drivers who have a jobcode of 52 and who have a driver status other than "Do Not Reserve". Please display the first name followed by the middle initial and the last name of each driver.

12. Display the phone numbers of all drivers with a **jobcode** of 53 or 55. This listing should be ordered by **jobcode** and within **jobcode** by driver name and driver phone. In addition, the area code of the phone number should be separated from the first three digits by a hyphen, and the first three digits of the phone number should be separated from the last four digits by another hyphen.

13. Display the number of drivers born in each year.

14. Display the names of all drivers in order by driver name who are scheduled to drive the day shift between November 14, 2007 and November 18, 2007 inclusive.

15. Display a list of all cabs not reserved for the day shift on November 15, 2007. Include in the list the number, make, model, and year of the cab. The scope of this query is limited to just the shifts on November 15, 2007.

16. Display the total number of miles driven during each day and shift combination. Exclude all shifts that have not been driven (i.e., where **Sh_wkflag** = 'F').

Advanced Data Manipulation Using SQL

17. Display the average mileage driven during each day and shift for only the day and evening shifts.

18. Display all cabs with a cab number that begins with the digit zero (0). The results should be displayed in order by the make of the cab and then by the date of purchase.

19. Display the name, age, and tenure with the company for all drivers scheduled to drive the night shift on November 15, 2007.

20. Display the make, model, and year for all cabs that have had some type of "replacement."

21. Display the license numbers of all Checker cabs made after 1985. Separate the state from the license number by a space.

22. Display all drivers whose name starts with the letter "C" and who live in Houston.

23. Display all drivers who live in Houston who earn less than the average salary of all drivers. The list should be displayed in ascending order by the driver name.

24. Issue a query that will list drivers who meet the following conditions:
    - the driver must have been hired after August 1, 1984, but before the end of 1988,
    - the jobcode must not be 52 or 55, and
    - the salary must be less than $57,500.

    For the drivers meeting these conditions, display their name, salary, jobcode, age, and length of service.

25. Display the names of those drivers not qualified to drive any cabs.

26. Display the total accumulated maintenance cost associated with each cab. For purposes of this query, the cost of fuel should not be included.

27. Display a list of those drivers who on December 1, 2007 will have worked for the company for less than 10 years. The listing should include the driver number, driver name, the number of years the driver will have worked for the company as of December 1, 2007, and the age of the driver when hired. Please order the results in descending order by tenure with the company.

28. Display the maintenance data for all cabs with Texas license numbers. In addition, include the cab number and license number in the result. Please order the results by maintenance date.

29. Display the names of drivers who have been involved in an incident and who have brought in at least one car for maintenance.

30. Display the unique day and shift combinations for the cabs that have been reserved. Order the results in ascending order by work date and work shift.

31. Display all drivers with a "Pay After" status whose monthly salary is between $4000 and $4700.

32. Produce a list showing the total number of drivers hired in each year. The list should also contain the calculated average and total salary for each year. The resulting list should be in order by year.

33. Display the monthly salary of all drivers rounded to the nearest integer. The list should be in ascending order by monthly salary.

34. Display the names of all drivers who live in a non-Houston zip code. Incidentally, all Houston zip codes begin with 770.

35. Each driver has a driver status. Find all drivers who earn more than the average salary of all drivers with their driver status. Display the result in order by driver status.

36. Display the difference in age between the oldest and youngest drivers.

37. Display the total fares for each cab.

38. Calculate the total miles driven for each person driving a shift between November 15, 2007 and November 19, 2007. Display the name of each driver along with the miles driven.

39. Display the names of drivers who have never been involved in an incident. Order the results by driver name.

40. Display the make of cab (i.e., not the cab) with the highest accumulated maintenance cost.

41. For each shift driven (i.e., **Sh_wkflag** = 'T'), calculate the total fare. The answer should clearly identify the shift and the total fare.

42. Display the cab number, make, and model of the cabs that have never been involved in an incident.

43. Display the names of those drivers who are qualified to drive all cabs.

44. Display the names of all drivers with a status other than "Do Not Reserve" qualified to drive Checker cabs.

45. For each driver involved in two or more shifts, display the driver name and the number of shifts either driven or scheduled to be driven.

46. Display the number of maintenance activities performed on each day in ascending order by maintenance date.

47. Display the cab number, make, and model of all cabs either driven or scheduled to be driven on two or more shifts.

48. Calculate the average fuel capacity for each make of cab. Display the make and average fuel capacity.

49. What are the names of the drivers involved in at least one incident?

50. Display the number of incidents associated with each driver. If a driver has never been involved in an incident, display a zero. (*Suggestion*: Use of an outer join might be helpful here.)

51. Display the total fares for each driver.

52. Calculate the total fare for all of the day, all of the evening, and all of the night shifts driven.

53. Calculate the total fuel expenditures for each fuel code.

54. Count the number of days each cab has been in for maintenance. In addition to the number of days it has been in for maintenance, display the cab number along with its make and model.

55. Display the cab number, license number, make, and model of those cabs that have never received any maintenance?

56. Display the cab number, license number, make, and model of those cabs that have never received any maintenance, been driven on a shift, or been involved in an incident.

57. Display the names of drivers who have driven shifts on more than one day, along with the number of days on which they have driven.

58. Display the cabs actually driven by each driver (i.e., **Sh_wkflag** = 'T'). Include in the results the name of the driver along with the cab number, year, make, and model of each cab driven. Order the results by the name of the driver. If a driver has driven a cab more than once, display the information requested just once (i.e., use DISTINCT).

59. Display the total of the maintenance and fuel costs for each cab.

60. Display the cab number, make, model, and license number of all cabs with either a Texas or Louisiana license number. The result should be displayed in ascending order by license number.

61. Display all shifts assigned to drivers who have a "DO NOT RESERVE" status.

62. For each cab, including those never driven on a shift, display the total mileage driven.

63. Display the number of shifts actually driven (i.e., **Sh_wkflag** = T) by each driver. The result should include the name of each driver along with the number of shifts driven.

64. Display the number of shifts actually driven (i.e., **Sh_wkflag** = T) by each driver, even those who have never driven a shift (i.e., for whom the number of shifts driven is zero). The result should include the name of each driver along with the number of shifts driven.

65. The rows in the FUELUSE table have some inaccuracies. Display which rows in the FUELUSE table are associated with instances where the number of gallons of gas exceeds the capacity of the tank.

66. Display the names of all drivers who are qualified to drive a Checker cab. Please display the first name followed by the middle initial and the last name of each driver.

67. Display the name of the driver with the lowest salary among those drivers younger than 40 years of age.

68. Display the total fare earned by each driver. The result should be displayed in descending order (i.e., the name of the driver who has earned the most should be displayed first, the name of the driver who has earned the second most should be displayed second, etc.).

69. Display the number of incidents associated with each make of cab. The results should be in ascending order by make of cab.

70. Display the names of those drivers without a middle initial.

71. Display the name of the driver, cab number, make, and model for all shifts driven using the cab with the smallest fuel capacity.

72. Display the total cost of the maintenance done on cabs currently judged to be in Excellent condition.

73. Display the names of those drivers born in the month of December.

74. Display all Checker cabs that have never been involved in an incident.

75. Append the text "Do Not Reserve" to the names of all drivers with a 'Do Not Reserve' status.

SQL Project 3[17]

This project is based on eight tables (AIRPORT, FLIGHT, DEPARTURES, PASSENGER, RESERVATION, EQUIP_TYPE, PILOTS, and TICKET) that contain data about Belle Airlines. The script required to create and populate these eight tables can be downloaded from *www.course.com* (search on the ISBN of this book), or obtained from your instructor.

Some Background on Belle Airlines

Belle Airlines is a regional carrier that operates primarily in the southwestern United States. At the present time, Belle Airlines operates its own reservation information system. To simplify our analysis, we will assume that all reservations on Belle Airlines flights are placed through Belle Airlines employees. Flights are not booked through travel agents and Belle Airlines does not participate in industry-wide reservation services. Each flight is assigned a unique flight number and has its own set of flight characteristics (*i.e., flight number, origin, destination, departure time, arrival time, meal code, base fare, mileage between origin and destination, and number of changes in time zone between the origin and destination of the flight*). Departures of each flight are stored in the Departures table. Each departure contains four attributes (*flight number, departure date, pilot id, and equipment number*).

Belle Airlines flies out of airports located all over the country. Data on these airports is stored in the Airport table. Data on these airports includes: *a three-character airport code, location of the airport, elevation, phone number, hub airline that operates out of the airport*. Since Belle Airlines flies out of airports located all over the country, Belle Airlines pilots live all over the country. Data on these pilots is stored in the Pilots table which contains the following attributes: *pilot id, pilot name, social security number, street address, city, state, zip code, flight pay, date of birth, and date hired. The company also owns its own fleet of airplanes.* Data on these airplanes is stored in the Equip_Type table which contains the following attributes: *equipment number, equipment type, seating capacity, fuel capacity, and miles per gallon.*

Three additional tables populate the Belle Airlines database: the Passenger table (with attributes: passenger name, itinerary number, and confirmation number), the Reservation table (with attributes: confirmation number, reservation date, reservation name, reservation phone, reservation flight number, and reservation flight date), and the Ticket table (with attributes: itinerary number, flight number, flight date, and seat assignment). See the ER diagram that follows for a description of the relationship among these eight tables. In addition, once the eight tables are created and their contents printed, the following two examples can be used to illustrate the relationship between the Passenger, Reservation, and Ticket tables. The key notion here is that one passenger can make a reservation for any number of other passengers on a flight (i.e., departure).

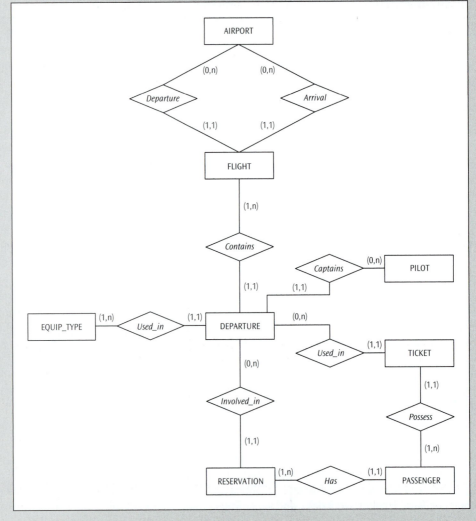

ERD for SQL Project 3

Example 1. *On April 1, Ole Olson (assigned confirmation number 1 in the Reservation table) reserved two tickets on Flight Number 15 scheduled to leave on April 1, and return via Flight Number 329 on April 1 and April 10. As a result the Passenger table shows Ole Olson as Itinerary Number 1 and his wife, Lena Olson, as Itinerary Number 2. Observe that both have Confirmation Number 1 since both reservations were booked at the same time by Ole. Observe that the Ticket Table shows that Ole (Itinerary Number 1) is sitting in Seat 10D on Flight Number 15 and in Seat 12D in Flight Number 329. On the other hand, his wife, Lena (Itinerary Number 2) is sitting in Seat 10E on Flight Number 15 and in Seat 12E in Flight Number 329.*

---

[17] This project is an adaptation of an example that appears in Lorents and Morgan (1998).

Example 2. *On April 17, Andy Anderson (assigned confirmation number 6 in the Reservation table) reserved two tickets on Flight Number 102 scheduled to leave on April 18. As a result, the Passenger table shows Andy Anderson as Itinerary Number 12 and his wife, Gloria Anderson, as Itinerary Number 13. Once again, observe that both have Confirmation Number 6 since both reservations were booked at the same time by Andy. Observe that the Ticket table shows that Andy (Itinerary Number 12) has ticket on Flight Number 102 with seat assignment 10B. The same is true for Gloria (Itinerary Number 13). She has a ticket on Flight Number 102 and seat assignment 7C.*

The 101 queries that follow should collectively provide a good introduction to SQL. Please note that in some cases it is quite clear which columns (fields) need to be displayed. In other cases the columns to be displayed are left to individual judgment. Use DISTINCT to minimize the repetition of duplicate rows and don't be afraid to use column aliases (especially when a column heading would otherwise be the content of a numeric or character function); they can make the output of a query more readable. In addition, please try to avoid wraparound as much as possible. In all cases, please refrain from the temptation that the results are correct the first time output from a query is obtained. In other words, please try to display enough columns (fields) and study the output in order to verify the accuracy of the result.

In order to challenge the student, the 101 queries are for the most part randomly ordered. In other words queries 1-10 are not necessarily the easiest to write nor is query 101 necessarily the most difficult to write. One may want to begin working with queries that seem to involve either one or two tables, such as FLIGHT and PASSENGER. These queries involve the use of the LIKE operator, numeric functions (e.g., COUNT, MIN, MAX, AVG, etc.), grouping, nested subqueries, and joins.

1. Display the origin, destination, departure time from origin, and arrival time at destination for all flights that occur in the same time zone. Your results should be displayed in order by flight number.

2. Display the code, location, and elevation of all airports without a hub airline. Your results should be in descending order by elevation.

3. Display the departures originating from Los Angeles, CA. Include in your results fights from Los Angeles for which no departures currently exist. Los Angles, CA and not LAX should be used in the WHERE clause of your query.

4. Display the flight numbers and the codes for the origin and destination of all flight reservations made by Andy Anderson.

5. Display the seating capacity, fuel capacity, and miles per gallon for all aircraft manufactured by Boeing. Information about each equipment type should be displayed only once.

6. Display the names of all pilots who live outside of the state of Texas. Order the results in alphabetical order by last name.

7. Display the flight number, flight date, fare, origin, and destination for all tickets with a flight date of July 2006. Use the fare in the FLIGHT table as the fare for the ticket. Order your results in ascending order by flight date and within flight date by flight number.

8. Display all flights that originate at an airport without a hub airline.

9. Display all flights that arrive at an airport without a hub airline.

10. Display all flights that both originate and arrive at an airport without a hub airline.

11. Display all departures that are flown by an aircraft not manufactured by Boeing. Your results should be in ascending order by departure date and within departure date by flight number.

12. Display the distance divided by the fare for each flight. For each flight, display the flight number, the origin, the destination, the distance, the fare, and the quotient. Your results should be in descending order by the quotient and rounded to two places to the right of the decimal point. Create a descriptive column alias for the quotient.

13. Display the total number of flights that originate from each point of origin.

14. Revise the previous query so that instead of displaying the code for each point of origin, the location of each point of origin from the AIRPORT table is displayed.

15. Revise the previous query to also include the display of those locations where no flights originate.

16. Display the average flight pay for pilots that live in each state.

17. Display the name and flight pay for those pilots whose flight pay exceeds the average flight pay for all pilots.

18. Display the name and flight pay for those pilots whose flight pay exceeds the average flight pay for all pilots in the state in which they reside.

19. Display the date of the most recent departure flown by each pilot. Include in what you display the name of the pilot.

20. Display not only the date of the most recent departure by each pilot, but also the number of days since the last departure date as well. Truncate the number of days (i.e., if number of days is 37.67655, display 37) to zero places to the right of the decimal point. Order the result in descending order by the number of days.

21. Display the number of departures that involve flights for each of the three time zone differences.

22. Display the number of airports located in each state.

23. Display the number of departures where the distance flown is greater than or equal to 1000 miles.

24. Display the difference in age between the oldest and youngest pilot.

25. For each type of aircraft, display the total distance that can be flown before refueling. Display your results in descending order by total distance that can be flown.

26. For each passenger listed in the PASSENGER table, display the name of the person responsible for his or her reservation.

27. For each passenger listed in the PASSENGER table, display the name of the person responsible for his or her reservation only if the passenger himself or herself was not responsible for making the reservation.

28. For each reservation in the RESERVATION table, display the name of the pilot who will be piloting the flight.

29. Display those tickets that include only one flight.

30. Display the name of the passengers whose tickets include only one flight.

31. What flights leave Phoenix for Los Angeles between 3:00 pm and midnight? Display each flight's flight number, city name of the flight's origin, city name of the flight's destination, departure time, and arrival time.

32. What are the fares from Phoenix to Los Angeles if Belle Airlines is running a 20 percent discount special off the current fares? Display each flight's flight number, city name of the flight's origin, city name of the flight's destination, and discounted fare.

33. Andy Anderson wants to know the passenger and ticket information on all passengers for which he has made reservations. For each reservation, display the passenger name, flight number, and flight date and seat assignment.

34. Display the maximum fare for flights originating at each airport if that fare is greater than $100. For each qualifying airport, display the airport code, location, and maximum fare. Display the results in descending order by fare.

35. Display the flight number of the flights that have no ticketed passengers scheduled on April 18, 2006. Display the results in ascending order by flight number.

36. What is the passenger count for each flight that has more than one ticketed passenger scheduled? For each qualifying flight, display the flight number and number of passengers.

37. Display the city name of the flight's origin and city name of the flight's destination as well as all data about the flight for all tickets held by Pete Peterson.

38. Display the names and phone numbers of persons who have reservations on flights leaving Phoenix, Arizona on May 17, 2006 for each flight booked with fewer than three passengers.

39. Display all flights where the origination time is later than at least one of the Minneapolis to Phoenix flights.

40. Display all flights that leave later than all flights going from Phoenix to Los Angeles.

41. Display the flight information (origination, destination, and times) on all passengers who are flying under a reservation made by Pete Peterson. The origin and destination should include the entire name of the city.

42. Display the number of departures associated with each pilot in ascending order by pilot name. Include all pilots, even those without any departures, in your results.

43. Display the names of those pilots who were not assigned to a departure during April 2006.

44. A passenger wants to fly from Phoenix to Los Angeles and back in a single day. He needs at least five hours in Los Angeles to get to and from the airport and conduct his business. List the flight numbers, origin times, and destination times of the flights that will accommodate his schedule.

45. What flights from Flagstaff to Phoenix have connecting flights in Phoenix going on to Los Angeles? Allow 40 minutes for a connection.

46. Display the total of the fares for all tickets related to Pete Peterson. Use the fare in the FLIGHT table in your calculation.

47. Display the total number of tickets sold for each flight across all dates.

48. Display the total of the fares collected for each flight on each date. Assume all tickets were sold at full fare and that the fares come from the FLIGHT table.

49. Display the maximum fare for flights between each origin and destination airport (e.g., Phoenix to Los Angeles, Phoenix to Flagstaff, Phoenix to San Francisco, etc.). As part of your result, display the name of the city where the airport is located—not the code.

50. Display the total miles flown by each pilot. Display the name of each pilot along with the miles flown. Display the results in descending order by miles flown. Include all pilots in your result—even those who have not yet flown a flight.

51. Display the names of all pilots who have flown a Boeing 727. Please display the first name followed by the middle initial and the last name of each pilot.

52. Display the names of the passengers with tickets on the July 23, 2006 departure of flight 104. Your results should be displayed in ascending order by seat number.

53. Display the flight number and date of all departures originating from Phoenix that serve either a snack or nothing to passengers.

54. Display all flights with either a California origination or destination.

55. Display all flights that depart on one day and arrive the next day.

56. Display the total compensation to each pilot in April 2006. The total compensation for a pilot is the product of the pilot's flight pay times the number of flights flown. Include pilots who did not fly any flights in April 2006.

57. Under the assumption that fuel costs $2.31 per gallon, display the total cost of fuel for all departures flown by each aircraft. Include the equipment number and equipment type of each aircraft.

58. Display the name and age of the youngest pilot.

59. Display the name and hiredate of all pilots who were hired during the 1990s. The qualifying pilots should be displayed in descending order by length of service.

60. Display the name and age of all pilots who were less than 40 years of age when hired and are now older than 47 years of age.

61. Display the names of those passengers whose last name begins with the letter "A". Each unique passenger name should be displayed only one time.

62. Display the confirmation number, reservation date, reservation name, phone number of the person making the reservation, and flight number for those reservations made during April 2006 for flights scheduled to depart sometime after April 2006. In addition, the area code of the phone number should be separated from the first three digits by a hyphen, and the first three digits of the phone number should be separated from the last four digits by another hyphen.

63. In a Boeing 727, seats A and E are window seats. Display the name, flight number, and departure date for all passengers who have reserved a window seat.

64. Display the names of passengers who have flown on a flight piloted by William B. Pasewark. Include in your results the flight number as well as the date of the flight. Display the result in order by passenger name.

65. Produce a list showing the total number of pilots hired in each year. The list should contain the calculated current average flight pay for all pilots hired during that year. The resulting list should be in ascending order by year.

66. Display the total number of pilots who live in each state in alphabetical order by state.

67. Display the total number of departures associated with each meal type.

68. Display the names of those pilots scheduled for a departure in an airplane manufactured by Boeing Corporation during April 2006.

69. Display the equipment number and equipment type of airplanes flown by Stuart Long.

70. Display the first and last name of all pilots who have flown more than one type of aircraft. Display the results in ascending order by last name and include in the result the equipment types each qualifying pilot has flown.

71. Display just the first and last name (first name followed by last name with no middle initial) of pilots who have not been assigned to a flight during May 2006. *Note*: This is a challenging query.

72. Display the first name and last name of those passengers who have a ticket with at least two departures.

73. For the oldest pilot, display his or her name, and aircraft flown.

74. Display the names of those passengers who at one time or another have made at least one reservation.

75. Display the name and age of the pilot with the highest pay among those pilots younger than 50 years of age.

76. Display information about all departures flown by the aircraft with the largest fuel capacity. Include in your results the flight number, departure date, origin, and destination.

77. Display the first and last name of those passengers whose reservation was made at least one month in advance of their departure. In addition, display the flight date, origin, and destination.

78. How many flights are flown within the same state? Display the flight number, origin city, destination city, and type of aircraft used.

79. Display the total number of miles flown on ticketed departures by each pilot. If a pilot has never flown a ticketed flight, display a zero.

80. Display the names of those pilots who live in a city that does not have an airport.

81. Display the types of aircraft used to fly to or from the cities with the highest or lowest elevation.

82. Display the name and phone number of the passenger booked on the most reservations.

83. Display the name, age, and hiredate of the pilots who fly either to or from the city with the highest elevation. The result should include the name of the city.

84. Display the name(s) of all persons making a reservation for Lena Olson.

85. Display information on tickets associated with a person whose last name is Peterson.

86. Display the number of tickets associated with those people who make reservations.

87. Display the number of tickets sold on each row. The results should be displayed in ascending order by row number.

88. Display the number of tickets associated with each seat in ascending order by seat.

89. Display the names of passengers who have never made a reservation themselves.

90. Display the total fare associated with each reservation. Base your calculation on the fair associated with each flight.

91. Display the total number of tickets sold during each month.

92. For each ticket, display the name of the passenger, the name of the person making the reservation, the name of the pilot, the fare, the fuel capacity of the airplane assigned to the departure, and the flight's point of origination and point of destination.

93. For each flight, display the number of departures associated with flights within the same time zone.

94. Display the location of those airports that have been neither the point of origination nor the point of destination for any Belle Airlines flights.

95. Display flights without any departures in May 2006.

96. Calculate the number of scheduled departures of flights from each airport during May 2006 that include lunch or dinner. Display the number of flights for each qualifying airport.

97. Assume that airports with a hub airline offer flights for just that airline (e.g., the only airline flying out of Phoenix is Belle Airlines, the only airline flying out of Minneapolis is Northwest, etc.). Calculate the number of departures associated with each hub airline.

98. Display the flying time associated with each flight. Remember to take into account the difference in time zones from the point of origination versus the point of destination.

99. Display the number of tickets associated with each flight that has a California destination.

100. Calculate the total fare paid by each passenger. Use the fare in the FLIGHT table.

101. Display the number of departures associated with each date in ascending order by date.

Selected Bibliography

Connolly, T. M. and Begg, C. E. (2002) *Database Systems: A Practical Approach to Design, Implementation, and Management*, Third Edition, Addison-Wesley.

Date, C. J. (1995) *An Introduction to Database Systems*, Sixth Edition, Addison-Wesley.

Date, C. J. and Darwen, H. (1997) *A Guide to the SQL Standard*, Fourth Edition, Addison-Wesley.

Elmasri, R., and Navathe, S. B. (2003) *Fundamentals of Database Systems*, Fourth Edition, Addison-Wesley.

Gennick, J. (2004) *SQL Pocket Guide*, O'Reilly Media, Inc.

Groff, J. R. and Weinberg, P. N. (2002) *SQL: The Complete Reference*, McGraw-Hill/Osborne, Second Edition.

Gulutzan, P. and Pelzer, T. (1999) *SQL-99 Complete, Really*, R&D Books.

Johnson, J. L. (1997) *Database Models, Languages, Design*, Oxford University Press, Inc.

Kifer, M., Bernstein, A. and Lewis, P. M. (2005) *Databases and Transactions Processing: An Application-Oriented Approach*, Second Edition, Addison-Wesley.

King, K. (2002) *SQL Tips & Techniques*, Premier Press.

Kroenke, D. M. (2004) *Database Processing: Fundamentals, Design, and Implementation*, Ninth Edition, Pearson Prentice-Hall.

Lorents, A. C. and Morgan, J. N. (1998) *Database Systems: Concepts, Management and Applications*, The Dryden Press.

Lorie, R. A. and Daudenarde, J. J. (1991) *SQL & Its Applications*, Prentice-Hall, Inc.

Morris-Murphy, L. L. (2003) *Oracle9i: SQL with an Introduction to PL/SQL*, Course Technology.

Rob, P. and Coronel, C. (2004) *Database Systems: Design, Implementation, and Management*, Course Technology, Sixth Edition.

Sunderraman, R. (2003) *Oracle9i Programming: A Primer*, Addison-Wesley.

Data Modeling Architectures Based on the Inverted Tree and Network Data Structures

Appendix A begins with illustrations of the inverted tree and network data structures—the two basic data structures that underlie the data models used today for designing databases. Discussion of these data structures is followed by a brief overview of how the hierarchical and CODASYL data model architectures express the inverted tree and network data structures, respectively.

A.1 Logical Data Structures

From a data modeling and database design perspective, there are just two basic logical data structures—the inverted tree structure and the network data structure—and the latter subsumes the former. Nonetheless, an exposure to both logical data structures is worthwhile because the three major logical data modeling architectures, viz., relational, hierarchical, and CODASYL, implement one or both of these structures.

A.1.1 Inverted Tree Structure

The inverted tree is a data structure in which the elements of the structure have only **1:1** and **1:n** relationships with one another. Figure A.1 contains a variation of the Presentation Layer ER diagram in Chapter 3, Figure 3.3 that takes the form of an inverted tree. An inverted tree consists of a hierarchy of nodes connected by branches. Each node in the tree, except the root node located at the top of the tree, has exactly one parent node and zero or more child nodes. Child nodes with the same parent are called twins or siblings. A root node has no parent, and a node with no child nodes is called a leaf node.

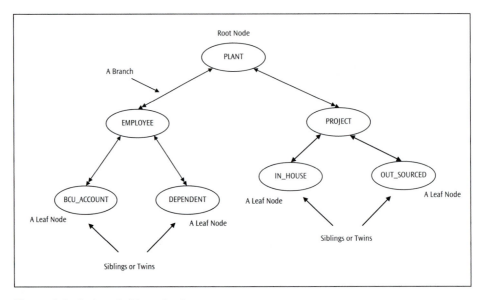

Figure A.1 An inverted tree structure

In an inverted tree, a **1:n** relationship is referred to as a **parent-child relationship type (PCR type)**.[1] Six PCRs appear in Figure A.1: a PLANT may employ many employees,[2] an EMPLOYEE may have many bank accounts, an EMPLOYEE may have many dependents, a PLANT may control many projects, a PROJECT may be done in-house, and a PROJECT may be outsourced.[3] Observe that while it is possible for nodes to have more than one child (e.g., both EMPLOYEE and PROJECT are child nodes of PLANT; BCU_ACCOUNT and DEPENDENT are child nodes of EMPLOYEE), each node has exactly one parent. If this were not true (e.g., if BCU_ACCOUNT is a child of both EMPLOYEE and DEPENDENT), an inverted tree structure would not exist. Thus, clearly, the Presentation Layer ER diagram in Figure 3.3 does not reflect an inverted tree structure. In short, in an inverted tree structure a node can participate as a parent in several PCRs; but as a child in only one PCR. That is, a parent node can have several child nodes, while a child node can have only one parent.

A.1.2 Network Data Structure

A network or plex structure is also composed of nodes and branches but unlike an inverted tree structure, a child node in a network structure can have multiple parents. Figure A.2 shows how the network data structure allows BCU_ACCOUNT to have two parents

---

[1] When two nodes exhibit a **1:1** relationship, either can be designated as the parent or the child.

[2] Observe that the line connecting EMPLOYEE to PLANT has a single arrowhead on the PLANT end indicating that an employee works for (is associated with just one PLANT); since a PLANT can have many EMPLOYEEs, the same line connecting PLANT to EMPLOYEE has two (i.e., multiple) arrowheads on the EMPLOYEE end.

[3] The semantics of this illustration allow for a project to be an in-house or an outsourced project or contain no more than one in-house and one outsourced component.

(EMPLOYEE and DEPENDENT). Likewise, ASSIGNMENT is a child node with two parents, viz., EMPLOYEE and PROJECT.[4] In essence, in a network data structure, a node can participate in more than one PCR as a parent as well as a child.

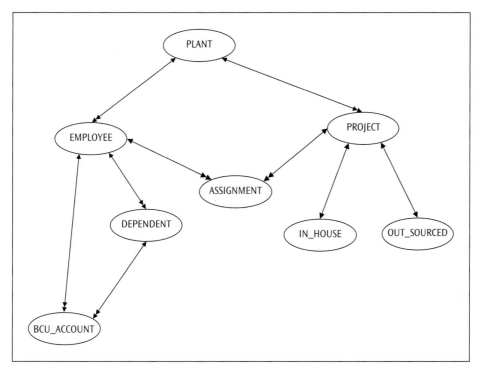

Figure A.2 A network structure

Thus, both inverted tree and network data structures permit a node to participate as a parent in multiple PCRs. That a node can participate as a child in only one PCR is an additional constraint specific to the inverted tree structure. When this constraint is relaxed and a node is permitted to also participate as a child in multiple PCRs, a network structure eventuates. From this it should be clear that an inverted tree structure is a special case of the network structure. In other words, implementation of the latter implicitly implements the former.

A.2 Logical Data Model Architectures

There are three major architectures for logical data models. The **relational data model**, discussed in Chapter 6, literally monopolizes contemporary DBMS architectures. However, the **hierarchical** and **CODASYL data models** precede the relational data model and have historical value in that many legacy systems still run on DBMS platforms that implement hierarchical or CODASYL architectures. Two more architectures are currently

[4] Observe that the network data structure is similar to the EER construct Specialization Lattice (see Section 4.1.3 in Chapter 4).

Data Modeling Architectures Based on the Inverted Tree and Network Data Structures

emerging and are often referred to as post-relational data models: the **object-oriented data model** and the **object-relational data model.** The former is considered an aggressive competitor to the relational data model, while the latter containing the features of both relational and object-oriented constructs is seen at least by some [e.g., Date (2004)] as the future of practical data modeling for database design. These models are the subject of Appendix B.

A.2.1 Hierarchical Data Model

The hierarchical data model was developed in the early 1960s as a joint effort between IBM and two of its customers, Rockwell and Caterpillar. This architecture essentially implements the inverted tree structure. Both manufacturing organizations needed a bill-of-materials processor that would print the total number of each component required to meet a given production schedule for a particular product.[5] This narrow problem required a specific solution. Solving the problem led IBM to market a DBMS known as IMS that could handle the bill-of-materials problem well but was limited in flexibility since it was not only incapable of representing anything other than an inverted tree structure, but also required regeneration of the whole data model whenever addition of node(s) became necessary.

In IBM's implementation of the hierarchical model, files are called segments or nodes and record occurrences are called segment occurrences or segment instances. Fields are called fields or data items. Relationships between segments are called parent-child relationships. In a **1:n** relationship, the segment occurring on the one side is the parent segment and the segment on the many side is called the child segment.

Mapping (or transforming) a conceptual schema to a logical schema using the hierarchical architecture requires four steps. First, each entity type must be mapped to a segment type. Second, since the hierarchical data model does not allow for the direct representation of network structures, any network structure must be reduced to a collection of inverted trees.[6] Third, after transforming the conceptual schema to one or more inverted tree structures, a decision must be made as to the hierarchical ordering of the segments.[7] The final step involves mapping each attribute to a field of a segment.

Figure A.3 shows two occurrences (one for PLANT11 and a second for PLANT12) of the inverted tree structure that appears in Figure A.1. Figure A.4 shows two occurrences of the network structure that appears in Figure A.2 implemented in a hierarchical architecture. Since in the case shown in Figure A.2 it is possible for a bank account to be associated with both an employee and a dependent, the segment occurrences for bank accounts 11a and 11b must appear redundantly as children of both EMPLOYEE11 and

---

[5] A bill of materials is sequence of 1:n relationships between the subassemblies necessary for producing an item. (Shepherd, 1990, p. 67).

[6] For example, if BCU_ACCOUNT can be a child of both EMPLOYEE and DEPENDENT, Figure A.1 would have to be revised to show a BCU_ACCOUNT node under DEPENDENT as well as a second BCU_ACCOUNT node under EMPLOYEE, thus introducing possible redundant bank account data. However, in practice this data redundancy is handled through the use of logical pointers. Figures A.3 and A.4 illustrate a situation of this nature.

[7] This is an issue of navigational efficiency through the hierarchy. Several strategies are available for ordering the segments. However, this topic is beyond the scope of this book.

DEPENDENT11A (one of the three dependents of EMPLOYEE11). Likewise, the segment occurrence for assignment 11-11 also appears redundantly under EMPLOYEE11 and PROJECT11. While this exemplifies how a network structure is accommodated in a hierarchical architecture, from a practical standpoint, the resulting redundancy often induces processing inefficiencies that are bound to reduce the database system performance.

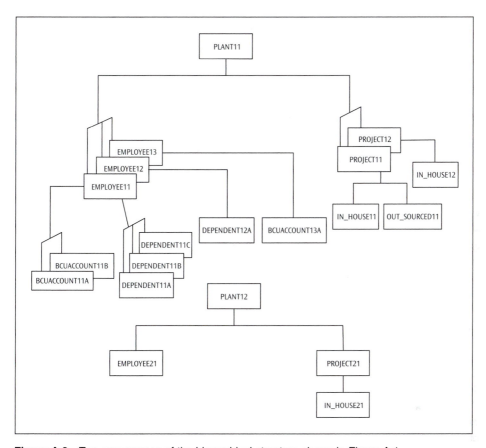

Figure A.3 Two occurrences of the hierarchical structure shown in Figure A.1

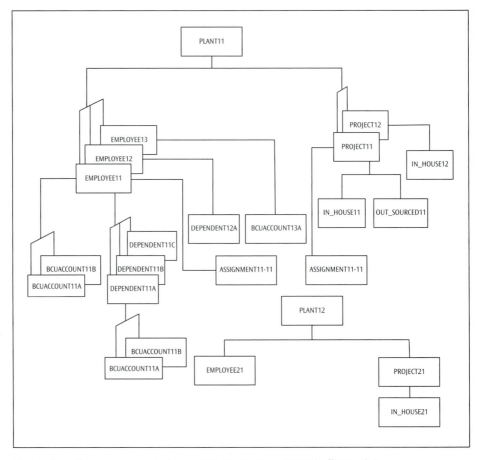

Figure A.4 Two occurrences of the network structure shown in Figure A.2

The Data Definition Language for IMS is Data Language/1 (DL/1). DL/1 makes it possible to map a conceptual schema to a logical schema as well as to a physical schema. In DL/1, a field is the smallest unit of data, a segment is a group of related fields, a database record is a hierarchically structured group of segments, and a database is a collection of database record occurrences of one or more database record types. DL/1 makes use of a database description (DBD) to define the way data is stored for use by IMS. In addition, the DBD also defines the format, length, and location of each data item to be accessed by the DL/1 data manipulation language. The following lines illustrate a DBD that reflects the hierarchical structure in Figure A.1. It should be noted that many of the details associated with a complete DBD (e.g., the length and location of each field) are not shown.

```
DBD            NAME = BEARCAT
SEGM           NAME = PLANT
                          PLANT Field Descriptions
                                  .
                                  .
                                  .

SEGM NAME = EMPLOYEE, PARENT = PLANT
                          EMPLOYEE Field Descriptions
                                  .
                                  .
                                  .

SEGM NAME = BCU_ACCOUNT, PARENT = EMPLOYEE
                          BCU_ACCOUNT Field Descriptions
                                  .
                                  .
                                  .

SEGM NAME = DEPENDENT, PARENT = EMPLOYEE
                          DEPENDENT Field Descriptions
                                  .
                                  .
                                  .

SEGM NAME = PROJECT, PARENT = PLANT
                          PROJECT Field Descriptions
                                  .
                                  .
                                  .

SEGM NAME = IN_HOUSE, PARENT = PROJECT
                          IN_HOUSE Field Descriptions
                                  .
                                  .
                                  .

SEGM NAME = OUT_SOURCED, PARENT = PROJECT
                          OUT_SOURCED Field Descriptions
                                  .
                                  .
                                  .
```

The DBD begins with a DBD macro[8] which among other things assigns a name to the hierarchical structure. Each segment description is headed by a SEGM macro that names the segment, indicates its total length in bytes (not shown here), and gives the name of its

---

[8] A macro is a short program consisting of several operations saved in a file under a certain name, which can be invoked from within another program.

parent. The first segment, or root, has no parent. Each field within a segment is represented by a FIELD macro. Applications access an IMS database through DL/1 data manipulation language commands embedded in a host language such as COBOL, PL/1, and C. These commands navigate the hierarchical structure of the database to retrieve, insert, update, or delete segments. For this reason, IMS and other implementations of the hierarchical data model are sometimes referred to as navigational systems.

The hierarchical data model allows only one relationship between any two segments and requires that each relationship be explicitly defined when the database is created. While satisfactory for applications that are inherently hierarchical in nature and for which query transactions are stable, to a great extent, databases based on the hierarchical data model are used today only in ongoing legacy systems that must be maintained. It should be noted that the hierarchical data model does not permit the direct representation of a network data structure such as the one that appears in Figure A.2. Approaches for transforming data relationships that involve network structures into a hierarchical architecture involve the use of logical pointers and are discussed in Johnson (1997).

A.2.2 CODASYL Data Model

During the 1960s efforts to develop a standard database theory were evolving along many paths. In 1965, the Conference on Data Systems Languages (CODASYL) established a Database Task Group (DBTG) to develop a database model for processing using COBOL. Such a model was developed and submitted first in 1971 to the American National Standards Institute (ANSI) for its consideration as a national standard. Although revised several times throughout the 1970s and during the first part of the 1980s, the model was never accepted as a national standard. Nevertheless, during this time several commercial database management systems based on what came to be known as the CODASYL data model (or DBTG network model) were implemented by some vendors (e.g., IDS by Honeywell, DMS-1100 by Univac, IDMS by Cullinet Corporation).

The DBTG network model adheres to the ANSI/SPARC three-schema architecture described in Section 1.5 of Chapter 1. The conceptual level (the logical view of all the data and relationships in the database) is called the **schema**. The external level (the users' views of the data needed for various applications) is called the **subschema**. The internal level (the physical details of storage) is implicit in the implementation.

In the CODASYL data model, the database is a collection of records and sets controlled by a single schema. A record type is a named collection of data items, whereas a set type is a named relationship between an owner record type and a member record type.[9] A set occurrence consists of one owner occurrence and all member occurrences of that set. Within a set occurrence, members are ordered on a member data item(s) or arranged in accordance with a time sequence.

---

[9] A record type is equivalent to a segment type in a hierarchical architecture. Likewise, a set type is the same as a PCR, an owner represents the parent and the member is the child.

Figure A.5 depicts the nine sets that relate plant, project, employee, dependent, BCU_ account, assignment, in_house, and out_sourced from Figure A.2 together. Observe that unlike the hierarchical data model, the CODASYL data model allows both an inverted tree structure and a network structure to be directly represented. This makes it possible for BCU_ACCOUNT to be shown as a member of two sets (the EMPLOYEE/BCU_ACCOUNT set and the DEPENDENT/BCU_ACCOUNT set).

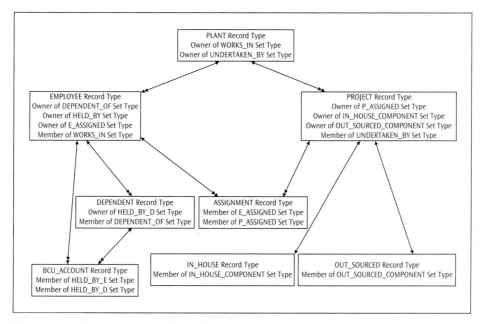

Figure A.5 The set representation of Figure A.2

Like the hierarchical data model, implementations of the CODASYL data model include a data definition language. The following lines illustrate how the eight record types and nine set types for the structure in Figure A.5 can be declared:

SCHEMA NAME IS BEARCAT
RECORD NAME IS PLANT
RECORD NAME IS EMPLOYEE
RECORD NAME IS PROJECT
RECORD NAME IS DEPENDENT
RECORD NAME IS BCU_ACCOUNT
RECORD NAME IS ASSIGNMENT
RECORD NAME IS IN_HOUSE
RECORD NAME IS OUT_SOURCED
SET NAME IS WORKS_IN
   OWNER IS PLANT
   MEMBER IS EMPLOYEE

```
SET NAME IS UNDERTAKEN_BY
        OWNER IS PLANT
        MEMBER IS PROJECT
SET NAME IS DEPENDENT_OF
        OWNER IS EMPLOYEE
        MEMBER IS DEPENDENT
SET NAME IS HELD_BY_E
        OWNER IS EMPLOYEE
        MEMBER IS BCU_ACCOUNT
SET NAME IS HELD_BY_D
        OWNER IS DEPENDENT
        MEMBER IS BCU_ACCOUNT
SET NAME IS E_ASSIGNED
        OWNER IS EMPLOYEE
        MEMBER IS ASSIGNMENT
SET NAME IS P_ASSIGNED
        OWNER IS PROJECT
        MEMBER IS ASSIGNMENT
SET NAME IS IN_HOUSE_COMPONENT
        OWNER IS PROJECT
        MEMBER IS IN_HOUSE
SET NAME IS OUT_SOURCED_COMPONENT
        OWNER IS PROJECT
        MEMBER IS OUT_SOURCED
```

As was the case for the DBD for the hierarchical data model that appears in Section A.2.1, many of the details associated with the definition of the record types and sets in this schema are not provided. For example, the definition of a record type includes several clauses that specify the scheme used to locate the record and to define the individual data items that comprise the record type. Besides identifying the owner and member of each set type, the definition of each set type includes rules for the insertion of new member records and for moving existing records from one set occurrence to another. Interested readers may wish to refer to Shepherd (1990) for an excellent introduction to the CODA-SYL as well as the hierarchical data model.

Summary

The hierarchical data model is the oldest of the data models and organizes data in the form of an inverted tree consisting of a hierarchy of parent and child segments, where a child is allowed to have only one parent. Coinciding with the development of the hierarchical data model was the CODASYL data model, which allowed more than one parent per child. Both of these models were used primarily during the mainframe era as vehicles for describing the structure of data as well as data manipulation operations but are no longer used as the basis for database systems today. While a number of legacy systems structured in accordance with these models remain in use today, many predict that they will be phased out over time as the number of qualified staff declines due to retirement and retraining.

Selected Bibliography

Date, C. J. (2004) *An Introduction to Database Systems*, Eighth Edition, Addison-Wesley.

Kroenke, D. M. (1977) *Database Processing*, Science Research Associates, Inc.

Shepherd, J. C. (1990) *Database Management: Theory and Application*, Richard D. Irwin, Inc.

Object-Oriented Data Modeling Architectures

Object-oriented concepts have drawn considerable attention among researchers and practitioners since the late 1980s and have significantly influenced efforts to incorporate in the DBMS the ability to process complex data types beyond just storage and retrieval. Appendix B briefly introduces the reader to object-oriented concepts exclusively from a database or, to be more precise, from a data modeling perspective.

B.1 The Object-Oriented Data Model

The needs of most business database applications to date are reasonably satisfied using simple data types (e.g., numbers and character strings). To this extent, the traditional database system architectures overviewed thus far (i.e., hierarchical, CODASYL, and relational) have been capable of providing adequate support to the commercial needs of modern enterprises. As the use of database systems spreads to wider domains (e.g., engineering and medicine), the need for handling complex data types has become apparent (e.g., CAD/CAM, CIM, CASE,[1] image processing, document handling) and the limitations imposed by the currently prevalent relational models have emerged as obstacles in these newer application domains. In current database environments this problem is somewhat mitigated by storing complex data outside the purview of the DBMS for concomitant access by application programs. Recent improvements in relational database systems allow for storing complex data types as large objects (LOBs)[2] in the database. However, the DBMS support involves only storage and retrieval. All other processing of these complex data types is still the responsibility of the application programs—i.e., the host language in which SQL is embedded, for storage and retrieval support only. Another related issue of concern has been the inadequacy of the database query language (i.e., SQL) in handling complex data types.

---

[1] CAD/CAM stands for Computer-aided Design/Computer-aided Manufacturing, CIM is the acronym for Computer-integrated Manufacturing and CASE is the acronym for Computer-aided Software Engineering)

[2] SQL:1999 supports the Large Object (LOB) data type with two possible variants—Binary Large Object (BLOB) and Character Large Object (CLOB). A LOB has a unique id called a *locator* which allows LOBs to be manipulated without extensive copying. LOBs are typically stored separately from the tuples in whose attributes they appear.

As a consequence, object-oriented (OO) concepts have drawn considerable attention among researchers and practitioners since the late 1980s and have significantly influenced efforts to incorporate the ability to process complex data types beyond just storage and retrieval by the DBMS itself. Object orientation in software engineering emerged originally in the programming languages arena (e.g., C++, SMALLTALK, Java). The initial motivation for OO database and Object DBMS (ODBMS) has essentially been due to a desire to seamlessly integrate DBMS functionality in the programming language environment so that the limitation of lack of persistence[3] of data in programming languages can be overcome. Several experimental ODBMS prototypes (e.g., ORION, IRIS, ODE)[4] and commercial products (e.g., GEMSTONE/OPEL of Gemstone Systems, ONTOS of Ontos Corporation, and Versant of Versant Object Technology) have been developed. Yet none have found widespread usage in the business application domain (Elmasri and Navathe, 2004). Our interest in this book is to provide a general understanding of OO concepts exclusively from a database or, to be more precise, from a data modeling perspective—certainly not from the programming perspective; in the OO world there is a difference in these perspectives.

B.1.1 Overview of OO Concepts

The core concept of the OO approach is the idea of "bundling" data and the operations pertaining to the data as integrated units called 'objects.' The internal structure of an object is not made visible to any user of the object; the users only know the attributes and the methods (i.e., operations) that the object is capable of executing on the data (i.e., attribute values). The expectation is that the objects more closely resemble their counterparts in the world at large and the user should not have to be overly concerned about the technology-oriented constructs (e.g., bits, bytes, fields, records, and even entities and relationships). For instance, the general mental image of a Flight often includes an Aircraft, Passengers, and Crew as a unified object. Join operations that connect a specific aircraft, a crew, and a group of passengers to form a flight is viewed as an implementation detail from which the users can be spared. While the simplicity of this example may not quite reflect the complexity equivalent to CAD/CAM or medical imaging applications, the example clearly demonstrates the concept. In essence, the goal is to raise the level of abstraction in data modeling and database design an additional notch. The concept is referred to as **encapsulation** and represents a paradigm shift from traditional database principles where data and methods are treated independently as complementary units.

From a data modeling perspective, object orientation espouses four notions: object structure, object class, inheritance, and object identity.

---

[3] Persistence in the OO paradigm refers to continued existence of data even after the program that created it has terminated.

[4] ORION was developed at Microelectronic and Computer Technology Corporation (MCC), Austin, TX; IRIS was developed by Hewlett-Packard; and ODE was developed at AT&T Bell Labs, now a part of Lucent Technologies.

B.1.1.1 Object Structure

So, what is an object? In the basic OO approach, "everything is an object." Some objects never change and are knows as **immutable objects** (e.g., integers like 7, 1, 43 and strings like "Cincinnati", "Star Trek"); others are **mutable**—i.e., variable (e.g., Flight, Airport). In a loose sense, an object corresponds to an entity in the ER modeling grammar except that an entity does not encapsulate methods with its data. An object constitutes a *state* (value) and a *behavior* (operations). In contrast with the OO programming languages where objects are transient—i.e., exist only for the duration of program execution, the objects in a OODB are persistent—i.e., they exist beyond the program termination and can be retrieved and shared by other programs at a later time. A set of variables analogous to attributes in the ER modeling grammar carry the data for an object, a set of pre-defined messages (with or without parameters) invoke methods for the objects, and a set of pre-defined methods (a body of code to execute operations) respond to the messages. Together, the variables, messages, and methods constitute an object. All interactions between an object and the rest of the system (essentially, other objects) are via messages.

B.1.1.2 Object Class

Every object belongs to a *type*, and this is usually referred to as an **object class**. Objects that have variables of the same name and type, respond to the same set of messages, and use the same set of methods belong to the same object class. Individual objects belonging to an object class are also called **object instances** or just **instances**. The notion of an object class is equivalent to the notion of an entity type/class in the ER modeling grammar except for the methods component of an object class. A consortium of ODBMS vendors and users called the Object Management Group (OMG) has proposed an *Object Model* which forms the basis for an *Object Definition Language* (ODL) and an *Object Query Language* (OQL). Detailed discussion of the object model is beyond the scope of this book. The reader is directed to the references in the Selected Bibliography for additional material.

B.1.1.3 Class Inheritance

Class inheritance is analogous to type inheritance in the EER modeling grammar except that in class inheritance, objects of the subclass not only inherit the public instance variables[5] of the superclass, but also the methods associated with the superclass. The former is called *structural inheritance* and the latter *behavioral inheritance*. The ability to reuse methods of a superclass with the related subclasses because of the behavioral inheritance property (*principle of substitutability*) is referred to as the property of **polymorphism**. In general, an OO database schema tends to employ an unusually large number of classes.

The inheritance principle goes hand in hand with the concept of **class hierarchies**. Once again, a class hierarchy in an OO database schema is analogous to type hierarchy (Specialization/ generalization) in the EER modeling grammar. Some object systems also support *multiple inheritance*—i.e., the ability of a subclass to inherit variables and methods from multiple superclasses. Modeled as a directed acyclic graph (DAG), this concept is

---

[5]   In a pure sense, all instance variables are hidden from the user. While logically unnecessary, in practice, objects typically expose physical representation of some instance variables usually via some special syntax and these are called **public instance variables**. Therefore, the truly hidden instance variables are labeled **private instance variables**.

similar to that of a specialization lattice in the EER modeling grammar (see Section 4.1.3 in Chapter 4).

Another related concept that is rather useful is known as *object containment*. The idea essentially conceptualizes objects as containing (i.e., being a **part of**) other objects in addition to public and private instance variables and methods. Sometimes these objects are referred to as complex objects or composite objects. Portrayal of multiple levels of containment leads to a **containment hierarchy**. Containment allows different users to view objects at different granularities. The concept essentially replicates the aggregation constructs in the EER modeling grammar in the OO context. Figure B.1 shows a containment hierarchy for a complex object called **Information Systems**. In this containment hierarchy, a business analyst can focus attention on the **Procedure** objects without any concern about **Information Technology**, **Data**, and **Personnel** objects. Likewise, a computer engineer can choose to limit the scope of his/her analysis to the hardware objects. The Information Systems manager, on the other hand, may use the containment hierarchy to monitor the whole **Information Systems**. In applications where an object is a part of several objects, the containment relationship can be portrayed as a DAG instead of a hierarchy.

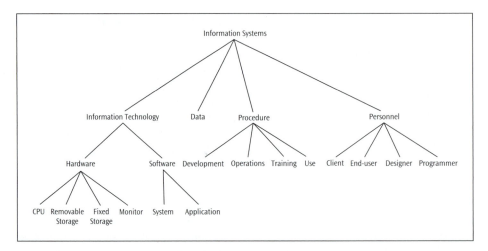

Figure B.1 Containment hierarchy in an object database

B.1.1.4 Object Identity

In an object database every mutable object that is stored (i.e., persistent) is uniquely identified by an Object ID (OID) which is typically a system-generated (conceptual) address. As a consequence, OIDs can be used elsewhere in the database as a (conceptual) pointer (e.g., in containment hierarchies). The OID is intended for independent identification of an object in the database and for managing inter-object references. The value of an OID is therefore meant for internal use by the database system and so is not visible to the user. The fundamental property of an OID is that it is immutable. Therefore, it is preferred that an OID value be retired when the associated object is removed from the database instead being reassigned to another object. For these reasons, the value of OIDs should not be a function of any variable in the database schema. Likewise, basing the value of an OID on a physical storage address is also discouraged. A commonly practiced strategy in object databases is the use of system-generated long integers as OID values and using an index (or hash table) to map the OID values to a physical storage address. Immutable objects like numbers and character strings usually do not have OIDs since they are typically stored within an object and cannot be referenced from outside the object. Note that OIDs do not eliminate the need for user-defined keys (e.g., candidate keys or primary key) because OIDs are not only prohibited for use in external interactions, but also are often not user-friendly means of external interaction. With respect to inter-object reference, the use of OID is somewhat similar to the use of a foreign key in a relational data model, except that an OID can point to an object anywhere in the OODBMS while a foreign key in an RDBMS is constrained to reference an attribute in a specific referenced relation. Lack of such a restriction in OODBMS imposes the responsibility for proper references on the application program. Use of OIDs for inter-object reference as in containment hierarchies is essentially equivalent to the low-level pointer mechanism originally defined in the CODASYL data model. Date (1998) asserts that OIDs have no place in the data model as far as the user is concerned.

B.1.2 A Note on UML

With the advent of the OO approach, Unified Modeling Language (UML) is becoming increasingly popular in the software engineering discipline. The attraction of UML may be attributed to the fact that it seeks to combine data and process (method) specifications in a single unified grammar. From a non-object paradigm data modeling and database design perspective, only the class diagram of UML is of some relevance. A class diagram is similar in many ways to the EER diagram except that the class diagram has provisions to specify methods along with attribute specifications in line with the OO approach. The Object Management Group has also endorsed UML as the standard for object modeling. Since UML combines commonly accepted concepts from several OO approaches, it is applicable to almost any application domain and is also programming language and operating system platform independent. A UML class diagram for a slightly enhanced Bearcat, Inc. ER diagram (see Figure 3.3) is shown in Figure B.2. Detailed treatment of UML grammar is outside the scope of this book[6] and can be found in software engineering textbooks that employ the OO development approach.

---

[6] Rational Rose is one of the popular CASE tools for drawing UML diagrams.

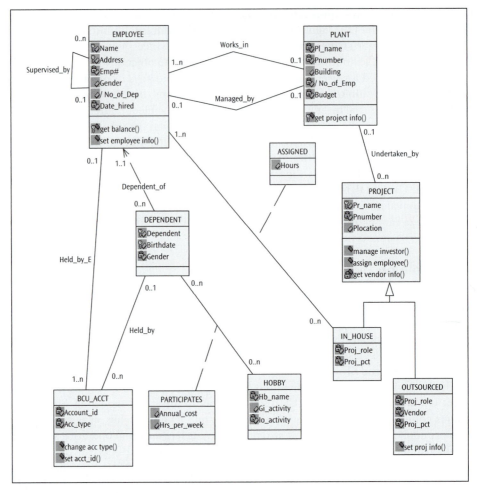

Figure B.2 UML class diagram for Bearcat, Inc.

In essence, OO databases are simply persistent extensions to OO programming languages. While persistent memory of programming languages may perhaps be viewed as databases, they are *application-specific*—i.e., they must first be tailored to an application, say, CAD/CAM; such a tailored DBMS is essentially useless for another application, say, a medical imaging system. OODBMS is not a shared, general purpose DBMS as a DBMS is expected to be. An OODBMS may be thought of as a sort of DBMS construction kit (Date, 2004, p. 846), while a second generation DBMS (e.g., RDBMS), once installed, is ready for use to build databases for various applications. There are other issues pertaining to *ad hoc* queries, treatment of derived objects, specification and enforcement of declarative integrity constraints that are application-independent, data independence and view definitions, application-independent database design, performance considerations, etc., where an object database may be inadequate. For a thoughtful treatise of these issues, see Date (2004).

B.2 The Object-Relational Data Model

In this section, we present a brief overview of the conceptual foundation for a class of DBMS called the object-relational system. OO databases discussed in the previous section are simply persistent extensions to OO programming languages like C++, SMALLTALK, and Java. In contrast, object-relational data models seek to extend the relational data model to support complex data types and other OO features. Proposing the object-relational data model, Stonebraker (1990) asserted that "second-generation systems made a major contribution in two areas, non-procedural data access and data independence, and these advances must not be compromised by third-generation systems." The underlying idea is that third-generation DBMSs should subsume rather than displace their second-generation counterparts lest the proven features of the second-generation DBMS be lost forever.

We briefly mentioned the problems associated with OODBMS (e.g., *ad hoc* queries, treatment of derived objects, specification and enforcement of declarative integrity constraints that are application-independent, data independence and view definitions, application-independent database design, performance considerations) in the last section. The principal motivation behind object-relational database systems (ORDBMS) is one of extending the RDBMS to support OO features without sacrificing the sound theoretical foundation of the RDBMS. OO features like (1) user-defined **abstract data types** (ADTs) that store complex data types and encapsulate methods for processing these data types, (2) inheritance, and (3) **structured types** where an attribute can capture more than an atomic value (e.g., sets, tuples, arrays, and sequences) when incorporated as extensions to the RDBMS **are** expected to overcome the current limitations of the RDBMS in supporting certain types of applications. Structured data types enable a *nested relational data model* by allowing domains to have non-atomic values. That is, an attribute value in a tuple can be a relation and that relation may contain other relations essentially enabling a nesting of relation schemas.

Several DBMS vendors have released object-relational DBMS products, also known as *universal servers* (e.g., IBM's universal database version of DB2, universal data option of Informix Dynamic Server, Oracle Universal Server). Sql:1999 has made significant strides towards incorporating advanced OO features while preserving the relational foundation of declarative access to data. For instance, SQL:1999, while including new data types for large objects and ARRAY type constructors, also enforces referential integrity on OIDs in the object-relational data model—i.e., all the OIDs that appear in an attribute of a relational schema in the object-relational data model must reference the same target relation schema. ORDBMS offers a convenient migration path for current users of RDBMS and affords the flexibility of using OO features selectively for appropriate applications.

Summary

The need to handle complex data types and to represent an object as consisting of both the data structure and set of operations that can be used to manipulate it (i.e., the concept of encapsulation) led to the creation of the object-oriented data model, which is intricately woven with the OO programming languages. In fact, OODBMS in principle is simply equivalent to adding persistence to OO programming language. The object-relational data model incorporating selected OO constructs as an extension to the relational data model has been proposed as an alternative to combat this problem. This has triggered a debate between proponents of the object-oriented data model and the relational data model. Both sides agree that the relational model is capable of supporting standard business applications, but lacks the capability to support special applications using complex data types. They, however, disagree as to whether extensions to the relational model can overcome this limitation. Proponents of the relational data model claim that the relational data model is a necessary part of any database management system and believe that extensions to the relational model (i.e., the object-relational data model) can address its limitations by effectively incorporating OO constructs as an extension to the theoretically sound relational data model.

Selected Bibliography

Date, C. J. (2004) *An Introduction to Database Systems*, Eighth Edition, Addison-Wesley.

Elmasri, R. and Navathe, S. B. (2003) *Fundamentals of Database Systems*, Fourth Edition, Addison-Wesley.

Ramakrishnan, R. and Gehrke, J (2000) *Database Management Systems*, McGraw Hill.

Stonebraker, M. et al. (1990) "Third-Generation Database System Manifesto," *ACM SIGMOD Record*, 19, 3.

APPENDIX **C**

Overview of SQL Reserved Words

Table C.1 Words reserved in the SQL-92, 99, and 2003 standards

SQL-92	SQL-99	SQL-2003
ABSOLUTE	ABSOLUTE	
ACTION	ACTION	
ADD	ADD	ADD
	AFTER	
ALL	ALL	ALL
ALLOCATE	ALLOCATE	ALLOCATE
ALTER	ALTER	ALTER
AND	AND	AND
ANY	ANY	ANY
ARE	ARE	ARE
	ARRAY	ARRAY
AS	AS	AS
ASC	ASC	
	ASENSITIVE	ASENSITIVE
ASSERTION	ASSERTION	
	ASYMMETRIC	ASYMMETRIC
AT	AT	AT
	ATOMIC	ATOMIC
AUTHORIZATION	AUTHORIZATION	AUTHORIZATION
AVG		
	BEFORE	
BEGIN	BEGIN	BEGIN
BETWEEN	BETWEEN	BETWEEN
		BIGINT
	BINARY	BINARY
BIT	BIT	
BIT_LENGTH		
	BLOB	BLOB
	BOOLEAN	BOOLEAN
BOTH	BOTH	BOTH
	BREADTH	
BY	BY	BY
CALL	CALL	CALL
	CALLED	CALLED
CASCADE	CASCADE	
CASCADED	CASCADED	CASCADED
CASE	CASE	CASE
CAST	CAST	CAST
CATALOG	CATALOG	
CHAR	CHAR	CHAR
CHAR_LENGTH		
CHARACTER	CHARACTER	CHARACTER
CHARACTER_LENGTH		

Table C.1 Words reserved in the SQL-92, 99, and 2003 standards (continued)

SQL-92	SQL-99	SQL-2003
CHECK	CHECK	CHECK
	CLOB	CLOB
CLOSE	CLOSE	CLOSE
COALESCE		
COLLATE	COLLATE	COLLATE
COLLATION	COLLATION	
COLUMN	COLUMN	COLUMN
COMMIT	COMMIT	COMMIT
CONDITION	CONDITION	CONDITION
CONNECT	CONNECT	CONNECT
CONNECTION	CONNECTION	
CONSTRAINT	CONSTRAINT	CONSTRAINT
CONSTRAINTS	CONSTRAINTS	
	CONSTRUCTOR	
CONTAINS		
CONTINUE	CONTINUE	CONTINUE
CONVERT		
CORRESPONDING	CORRESPONDING	CORRESPONDING
COUNT		
CREATE	CREATE	CREATE
CROSS	CROSS	CROSS
	CUBE	CUBE
CURRENT	CURRENT	CURRENT
CURRENT_DATE	CURRENT_DATE	CURRENT_DATE
	CURRENT_DEFAULT_ TRANSFORM_GROUP	CURRENT_DEFAULT_ TRANSFORM_GROUP
CURRENT_PATH	CURRENT_PATH	CURRENT_PATH
	CURRENT_ROLE	CURRENT_ROLE
CURRENT_TIME	CURRENT_TIME	CURRENT_TIME
CURRENT_TIMESTAMP	CURRENT_TIMESTAMP	CURRENT_TIMESTAMP
	CURRENT_TRANSFORM_ GROUP_FOR_TYPE	CURRENT_TRANSFORM_ GROUP_FOR_TYPE
CURRENT_USER	CURRENT_USER	CURRENT_USER
CURSOR	CURSOR	CURSOR
	CYCLE	CYCLE
	DATA	
DATE	DATE	DATE
DAY	DAY	DAY
DEALLOCATE	DEALLOCATE	DEALLOCATE
DEC	DEC	DEC
DECIMAL	DECIMAL	DECIMAL
DECLARE	DECLARE	DECLARE
DEFAULT	DEFAULT	DEFAULT

Table C.1   Words reserved in the SQL-92, 99, and 2003 standards (continued)

SQL-92	SQL-99	SQL-2003
DEFERRABLE	DEFERRABLE	
DEFERRED	DEFERRED	
DELETE	DELETE	DELETE
	DEPTH	
	DEREF	DEREF
DESC	DESC	
DESCRIBE	DESCRIBE	DESCRIBE
DESCRIPTOR	DESCRIPTOR	
DETERMINISTIC	DETERMINISTIC	DETERMINISTIC
DIAGNOSTICS	DIAGNOSTICS	
DISCONNECT	DISCONNECT	DISCONNECT
DISTINCT	DISTINCT	DISTINCT
DO	DO	DO
DOMAIN	DOMAIN	
DOUBLE	DOUBLE	DOUBLE
DROP	DROP	DROP
	DYNAMIC	DYNAMIC
	EACH	EACH
		ELEMENT
ELSE	ELSE	ELSE
ELSEIF	ELSEIF	ELSEIF
END	END	END
	EQUALS	
ESCAPE	ESCAPE	ESCAPE
EXCEPT	EXCEPT	EXCEPT
EXCEPTION	EXCEPTION	
EXEC	EXEC	EXEC
EXECUTE	EXECUTE	EXECUTE
EXISTS	EXISTS	EXISTS
EXIT	EXIT	EXIT
EXTERNAL	EXTERNAL	EXTERNAL
EXTRACT		
FALSE	FALSE	FALSE
FETCH	FETCH	FETCH
	FILTER	FILTER
FIRST	FIRST	
FLOAT	FLOAT	FLOAT
FOR	FOR	FOR
FOREIGN	FOREIGN	FOREIGN
FOUND	FOUND	
	FREE	FREE
FROM	FROM	FROM
FULL	FULL	FULL
FUNCTION	FUNCTION	FUNCTION
	GENERAL	
GET	GET	GET

SQL-92	SQL-99	SQL-2003
GLOBAL	GLOBAL	GLOBAL
GO	GO	
GOTO	GOTO	
GRANT	GRANT	GRANT
GROUP	GROUP	GROUP
	GROUPING	GROUPING
HANDLER	HANDLER	HANDLER
HAVING	HAVING	HAVING
	HOLD	HOLD
HOUR	HOUR	HOUR
IDENTITY	IDENTITY	IDENTITY
IF	IF	IF
IMMEDIATE	IMMEDIATE	IMMEDIATE
IN	IN	IN
INDICATOR	INDICATOR	INDICATOR
INITIALLY	INITIALLY	
INNER	INNER	INNER
INOUT	INOUT	INOUT
INPUT	INPUT	INPUT
INSENSITIVE	INSENSITIVE	INSENSITIVE
INSERT	INSERT	INSERT
INT	INT	INT
INTEGER	INTEGER	INTEGER
INTERSECT	INTERSECT	INTERSECT
INTERVAL	INTERVAL	INTERVAL
INTO	INTO	INTO
IS	IS	IS
ISOLATION	ISOLATION	
	ITERATE	ITERATE
JOIN	JOIN	JOIN
KEY	KEY	
LANGUAGE	LANGUAGE	LANGUAGE
	LARGE	LARGE
LAST	LAST	
	LATERAL	LATERAL
LEADING	LEADING	LEADING
LEAVE	LEAVE	LEAVE
LEFT	LEFT	LEFT
LEVEL	LEVEL	
LIKE	LIKE	LIKE
LOCAL	LOCAL	LOCAL
	LOCALTIME	LOCALTIME
	LOCALTIMESTAMP	LOCALTIMESTAMP
	LOCATOR	
LOOP	LOOP	LOOP
LOWER		

Table C.1 Words reserved in the SQL-92, 99, and 2003 standards (continued)

SQL-92	SQL-99	SQL-2003
	MAP	
MATCH	MATCH	MATCH
MAX		
		MEMBER
		MERGE
	METHOD	METHOD
MIN		
MINUTE	MINUTE	MINUTE
	MODIFIES	MODIFIES
MODULE	MODULE	MODULE
MONTH	MONTH	MONTH
		MULTISET
NAMES	NAMES	
NATIONAL	NATIONAL	NATIONAL
NATURAL	NATURAL	NATURAL
NCHAR	NCHAR	NCHAR
	NCLOB	NCLOB
	NEW	NEW
NEXT	NEXT	
NO	NO	NO
	NONE	NONE
NOT	NOT	NOT
NULL	NULL	NULL
NULLIF		
NUMERIC	NUMERIC	NUMERIC
	OBJECT	
OCTET_LENGTH		
OF	OF	OF
	OLD	OLD
ON	ON	ON
ONLY	ONLY	ONLY
OPEN	OPEN	OPEN
OPTION	OPTION	
OR	OR	OR
ORDER	ORDER	ORDER
	ORDINALITY	
OUT	OUT	OUT
OUTER	OUTER	OUTER
OUTPUT	OUTPUT	OUTPUT
	OVER	OVER
OVERLAPS	OVERLAPS	OVERLAPS
PAD	PAD	
PARAMETER	PARAMETER	PARAMETER
PARTIAL	PARTIAL	
	PARTITION	PARTITION
PATH	PATH	

SQL-92	SQL-99	SQL-2003
POSITION		
PRECISION	PRECISION	PRECISION
PREPARE	PREPARE	PREPARE
PRESERVE	PRESERVE	
PRIMARY	PRIMARY	PRIMARY
PRIOR	PRIOR	
PRIVILEGES	PRIVILEGES	
PROCEDURE	PROCEDURE	PROCEDURE
PUBLIC	PUBLIC	
	RANGE	RANGE
READ	READ	
	READS	READS
REAL	REAL	REAL
	RECURSIVE	RECURSIVE
	REF	REF
REFERENCES	REFERENCES	REFERENCES
	REFERENCING	REFERENCING
RELATIVE	RELATIVE	
	RELEASE	RELEASE
REPEAT	REPEAT	REPEAT
RESIGNAL	RESIGNAL	RESIGNAL
RESTRICT	RESTRICT	
	RESULT	RESULT
RETURN	RETURN	RETURN
RETURNS	RETURNS	RETURNS
REVOKE	REVOKE	REVOKE
RIGHT	RIGHT	RIGHT
	ROLE	
ROLLBACK	ROLLBACK	ROLLBACK
	ROLLUP	ROLLUP
ROUTINE	ROUTINE	
	ROW	ROW
ROWS	ROWS	ROWS
	SAVEPOINT	SAVEPOINT
SCHEMA	SCHEMA	
	SCOPE	SCOPE
SCROLL	SCROLL	SCROLL
	SEARCH	SEARCH
SECOND	SECOND	SECOND
SECTION	SECTION	
SELECT	SELECT	SELECT
	SENSITIVE	SENSITIVE
SESSION	SESSION	
SESSION_USER	SESSION_USER	SESSION_USER
SET	SET	SET
	SETS	

Table C.1 Words reserved in the SQL-92, 99, and 2003 standards (continued)

SQL-92	SQL-99	SQL-2003
SIGNAL	SIGNAL	SIGNAL
	SIMILAR	SIMILAR
SIZE	SIZE	
SMALLINT	SMALLINT	SMALLINT
SOME	SOME	SOME
SPACE	SPACE	
SPECIFIC	SPECIFIC	SPECIFIC
	SPECIFICTYPE	SPECIFICTYPE
SQL	SQL	SQL
SQLCODE		
SQLERROR		
SQLEXCEPTION	SQLEXCEPTION	SQLEXCEPTION
SQLSTATE	SQLSTATE	SQLSTATE
SQLWARNING	SQLWARNING	SQLWARNING
	START	START
	STATE	
	STATIC	STATIC
		SUBMULTISET
SUBSTRING		
SUM		
	SYMMETRIC	SYMMETRIC
	SYSTEM	SYSTEM
SYSTEM_USER	SYSTEM_USER	SYSTEM_USER
TABLE	TABLE	TABLE
		TABLESAMPLE
TEMPORARY	TEMPORARY	
THEN	THEN	THEN
TIME	TIME	TIME
TIMESTAMP	TIMESTAMP	TIMESTAMP
TIMEZONE_HOUR	TIMEZONE_HOUR	TIMEZONE_HOUR
TIMEZONE_MINUTE	TIMEZONE_MINUTE	TIMEZONE_MINUTE
TO	TO	TO
TRAILING	TRAILING	TRAILING
TRANSACTION	TRANSACTION	
TRANSLATE		
TRANSLATION	TRANSLATION	TRANSLATION
	TREAT	TREAT
	TRIGGER	TRIGGER
TRIM		
TRUE	TRUE	TRUE
	UNDER	
UNDO	UNDO	UNDO
UNION	UNION	UNION
UNIQUE	UNIQUE	UNIQUE
UNKNOWN	UNKNOWN	UNKNOWN
	UNNEST	UNNEST

SQL-92	SQL-99	SQL-2003
UNTIL	UNTIL	UNTIL
UPDATE	UPDATE	UPDATE
UPPER		
USAGE	USAGE	
USER	USER	USER
USING	USING	USING
VALUE	VALUE	VALUE
VALUES	VALUES	VALUES
VARCHAR	VARCHAR	VARCHAR
VARYING	VARYING	VARYING
VIEW	VIEW	
WHEN	WHEN	WHEN
WHENEVER	WHENEVER	WHENEVER
WHERE	WHERE	WHERE
WHILE	WHILE	WHILE
	WINDOW	WINDOW
WITH	WITH	WITH
	WITHIN	WITHIN
	WITHOUT	WITHOUT
WORK	WORK	
WRITE	WRITE	
YEAR	YEAR	YEAR
ZONE	ZONE	

APPENDIX D

SQL SELECT Statement Features

The tables in this appendix describe common features of SQL SELECT statements, cross-referenced to useful examples in Chapters 11 and 12. Table D.1 lists keywords and symbols, Table D.2 lists single-row functions, Table D.3 lists multiple-row (group) functions, Table D.4 lists comparison operators, Table D.5 lists logical operators, and Table D.6 lists other operators.

Table D.1 Keywords and symbols

Keyword/Symbol	Description	Useful Example(s) in Chapters 11 and 12
II concatenation)	Combines the display of contents from multiple columns into a single column.	Chapter 12, SUBSTRING Example 9a
*	Returns all data in a table when used in a SELECT clause.	Chapter 11, Example 1.1.1
%	A "wildcard" character used with the LIKE operator to perform pattern searches. It represents zero or more characters.	Chapter 11, Example 1.6.3
-	A "wildcard" character used with the LIKE operator to perform pattern searches. It represents exactly one character in the specified position.	Chapter 11, Examples 1.6.1 and 1.6.3
* multiplication / division + addition - subtraction	Arithmetic operators.	Chapter 11, Example 1.2.7
, (comma)	Separates column names in a list when retrieving multiple columns from a table.	Chapter 11, Example 1.1.1
' ' (string literal enclosed in single quotes)	Indicates the exact set of characters, including spaces, to be displayed.	Chapter 12, Concluding Example 1
" " (string literal enclosed in double quotes)	Preserves spaces, symbols, or case in a column heading alias.	Chapter 11, Example 1.2.8

Table D.1 Keywords and symbols (continued)

Keyword/Symbol	Description	Useful Example(s) in Chapters 11 and 12
AS	Indicates a column alias to change the heading of a column in output—optional.	Chapter 11, Example 1.2.8
CROSS JOIN	Matches each row in one table with each row in another table. Also known as a *Cartesian product* or *Cartesian join*. Syntax: SELECT *columnname* [,...] FROM *tablename1* CROSS JOIN *tablename2;*	Chapter 11, Example 2.1.1
DISTINCT	Eliminates duplicate lists.	Chapter 11, Example 1.3.1
JOIN...ON	The JOIN keyword is used in the FROM clause. The ON clause identifies the column to be used to join the tables. Syntax: SELECT *columnname* [,...] FROM *tablename1* JOIN *tablename2* ON *tablename1.columnname* <comparison operator> *tablename2.columnname;*	Chapter 11, Example 2.3.2.1
OUTER [RIGHT| LEFT| FULL] JOIN	This indicates that at least one of the tables does not have a matching row in the other table. Syntax: SELECT *columnname* [,...] FROM *tablename1* [RIGHT| LEFT| FULL] OUTER JOIN *tablename2* ON *tablename1.columnname* = *table name2.columnname;*	Chapter 11, Examples 2.4.1.1, 2.4.2.1, and 2.4.3.1

Table D.2 Single-row functions

Function	Description	Syntax	Useful Example(s) in Chapters 11 and 12
DECODE [Case Expression in SQL-92 Standard]	Takes a given value and compares it to values in a list. If a match is found, then the specified result is returned. If no match is found, then a default result is returned. If no default result is defined, a NULL is returned as a result.	DECODE *(v, s1, r1, s2, r2,..., d)* *v* = value sought *s1* = the first value in the list *r1* = result to be returned if *s1* and *v* match *d* = default result to return if no match is found	Chapter 12, Concluding Example 2
LENGTH [CHAR_ LENGTH in SQL-92 Standard]	Returns the number of characters in a string.	LENGTH *(char)* *char* = character string to be analyzed.	Chapter 12, LENGTH Examples 1, 2, and 3
ROUND [Not included in SQL-92 Standard]	Rounds numeric fields.	ROUND *(n, p)* *n* = numeric data, or a field, to be rounded *p* = position of the digit to which the data should be rounded	Chapter 11, Example 1.2.9
RTRIM/ LTRIM [TRIM in SQL-92 Standard]	Trims, or removes, a specific string of characters from the Right (or Left) of a set of data.	LTRIM *(char [, set])* *char* = characters to be affected *set* = string to be removed from the left/right of the data	Chapter 12, Trim Examples 1 – 9
SUBSTR [SUBSTRING in SQL-92 Standard]	Returns a substring, or portion of a string, in output.	SUBSTR *(char, m,[n])* *char* = character string *m* = position (beginning) for the extraction *n* = length of output string	Chapter 12, SUBSTRING Examples 1 – 9b
TO_CHAR	Converts dates and numbers to a formatted character string.	TO_CHAR *(n, 'f ')* *n* = number or date to be formatted *f* = format model to be used	Chapter 12, Section 12.3
TO_DATE	Converts a date in a specified format to the default-date format.	TO_DATE *(d, f)* *d* = date entered by the user *f* = format of the entered data	Chapter 12, Section 12.3
TRUNC [Not included in SQL-92 Standard]	Truncates, or cuts, numbers to a specific position.	TRUNC *(n, p)* *n* = numeric data, or a field, to be truncated *p* = position of the digit to which the data should be truncated	Chapter 11, Examples 1.2.7 and 1.3.5

Table D.3 Group (multiple-row) functions

Function	Description	Syntax	Useful Example(s) in Chapters 11 and 12
AVG	Returns the average value of the selected numeric field. Ignores NULL values.	AVG ([DISTINCT | ALL] n)	Chapter 11, Example 1.4.1
COUNT	Returns the number of rows that contain a value in the identified column. Rows containing NULL values in the column will not be included in the results. To count all rows, including those with NULL values, use a * rather than a column name.	COUNT (* | [| DISTINCT | ALL] c)	Chapter 11, Examples 1.4.1, 1.4.2, 1.4.4, 1.5.8 – 1.5.19
MAX	Returns the highest (maximum) value from the selected field. Ignores NULL values.	MAX ([DISTINCT | ALL] c)	Chapter 11, Example 1.4.1
MIN	Returns the lowest (minimum) value from the selected field. Ignores NULL values.	MIN ([DISTINCT | ALL] c)	Chapter 11, Example 1.4.1
SUM	Returns the sum, or total value, of the selected numeric field. Ignores NULL values.	SUM ([DISTINCT | ALL] n)	Chapter 11, Example 1.4.1

Table D.4 Comparison operators

Operator	Description	Useful Example(s) in Chapter 11 and 12
=	Equality operator—requires an exact match of the record data and the search value.	Chapter 11, Examples 1.1.1 – 1.1.7
>	"Greater than" operator—requires a record to be greater than the search value.	Chapter 11, Examples 1.1.1 – 1.1.7
<	"Less than" operator—requires a record to be less than the search value.	Chapter 11, Examples 1.1.1 – 1.1.7
<>, !=, or ^=	"Not equal to" operator—requires a record not to match the search value.	Chapter 11, Examples 1.1.1 – 1.1.7
<=	"Less than or equal to" operator—requires a record to be less than or an exact match with the search value.	Chapter 11, Examples 1.1.1 – 1.1.7

Table D.4 Comparison operators (continued)

Operator	Description	Useful Example(s) in Chapter 11 and 12
>=	"Greater than or equal to" operator—requires record to be greater than or an exact match with the search value.	Chapter 11, Examples 1.1.1 – 1.1.7
[NOT] BETWEEN x and y	Searches for records in a specified range of values.	Chapter 11, Examples 1.2.8 and 1.2.9
[NOT] IN (x, y,...)	Searches for records that match one of the items in the list.	Chapter 11, Examples 1.2.5 and 1.2.6
[NOT] LIKE	Searches for records that match a search pattern—used with wildcard characters.	Chapter 11, Examples 1.6.1 – 1.6.10
IS [NOT] NULL	Searches for records with a NULL value in the indicated column.	Chapter 11, Examples 1.5.2, 1.5.5 and 1.5.7
>ALL	More than the highest value returned by the subquery.	Chapter 11, Example 3.1.2.1
<ALL	Less than the lowest value returned by the subquery.	Chapter 11, Example 3.1.2.3
<ANY	Less than the highest value returned by the subquery.	Chapter 11, Example 3.1.2.5
>ANY	More than the lowest value returned by the subquery.	Chapter 11, Example 3.1.3.7
=ANY	Equal to any value returned by the subquery (same as IN).	Chapter 11, Example 3.1.3.9
[NOT] EXISTS	Row must match a value in the subquery.	Chapter 11, Example 3.2.1

Table D.5 Logical operators

Operator	Description	Useful Example(s) in Chapters 11 and 12
AND	Combines two conditions together—a record must match both conditions.	Chapter 11, Example 1.2.1
OR	Requires a record to match only one of the search conditions.	Chapter 11, Example 1.2.1

Table D.6 Other operators

Operator	Description	Useful Example(s) in Chapters 11 and 12
INTERSECT	Lists only the results returned by both queries.	Chapter 11, Example 2.2.2
MINUS [EXTRACT in SQL-92 Standard]	Lists only the results returned by the first query and not returned by the second query.	Chapter 11, Examples 2.2.3 and 2.2.3
UNION	Used to combine the distinct results returned by multiple SELECT statements.	Chapter 11, Example 2.2.1
UNION ALL	Used to combine all the results returned by multiple SELECT statements.	Chapter 11, Example 2.2.1

INDEX

S